煤矿岩层控制理论与技术进展

——37届国际采矿岩层控制会议（中国·2018）论文集

主　编　张　农　[美]彭赐灯（Syd S.Peng）

副主编　李桂臣　吴锋锋　程敬义

中国矿业大学出版社

内 容 简 介

本书收录了国内外煤矿开采岩层控制领域的学术论文31篇,集中展示了国际采矿岩层控制领域近年来的理论与技术创新成果。主要包括采场岩层控制理论与技术、坚硬顶板岩层控制理论与技术、巷道围岩控制与防治技术、深部岩石力学与工程应用、大倾角煤层开采矿压控制理论与技术、煤层群协调开采矿压控制理论与技术、煤与瓦斯共采理论与技术、充填开采理论与技术、智能岩层控制、矿山灾害防治等与采矿岩层控制相关的主题等。

本书可供从事煤矿开采方面科研、设计、工程技术及管理人员阅读参考,也可供高等院校矿业工程研究领域师生参考。

图书在版编目(CIP)数据

煤矿岩层控制理论与技术进展:37届国际采矿岩层控制会议(中国·2018)论文集/张农,(美)彭赐灯(Syd S. Peng)主编.—徐州:中国矿业大学出版社,2018.9
　ISBN 978-7-5646-4126-9

　Ⅰ.①煤… Ⅱ.①张… ②彭… Ⅲ.①煤矿开采—岩层控制—文集 Ⅳ.①TD325-53

中国版本图书馆 CIP 数据核字(2018)第 223004 号

书　　名	煤矿岩层控制理论与技术进展
	——37届国际采矿岩层控制会议(中国·2018)论文集
主　　编	张　农　[美]彭赐灯(Syd S. Peng)
责任编辑	王美柱
责任校对	仓小金
出版发行	中国矿业大学出版社有限责任公司
	(江苏省徐州市解放南路　邮编221008)
营销热线	(0516)83885307　83884995
出版服务	(0516)83885767　83884920
网　　址	http://www.cumtp.com　E-mail:cumtpvip@cumtp.com
印　　刷	江苏淮阴新华印刷厂
开　　本	787×1092　1/16　印张 27.25　字数 751千字
版次印次	2018年9月第1版　2018年9月第1次印刷
定　　价	158.00元

(图书出现印装质量问题,本社负责调换)

大会学术委员会

主　席：
Syd S. Peng　钱鸣高　彭苏萍　袁　亮　何满潮　康红普　顾大钊
刘　波　葛世荣　刘　峰　祁和刚

副主席(以姓氏笔画为序)：
于　斌　才庆祥　王家臣　卞正富　叶继红　冯夏庭　冯　涛　刘泉声
齐庆新　池秀文　许升阳　李夕兵　李树刚　杨仁树　吴爱祥　张东升
张　农　周　英　孟祥瑞　郝传波　姜德义　梁卫国　梁　冰　靖洪文
窦林名　谭云亮

委　员(以姓氏笔画为序)：
丁建丽　万志军　马占国　王卫军　冯国瑞　伍永平　华心祝　刘长友
许兴亮　纪洪广　李学华　杨永辰　杨圣奇　张吉雄　张宏伟　张国华
尚　涛　柏建彪　骆振福　郭文兵　黄炳香　曹胜根　屠世浩

大会组织委员会

主　任：张　农　靖洪文

副主任：李学华　马占国　张吉雄　万志军　杨圣奇　王晓琳　浦　海
　　　　　范　军　李　亭　李桂臣　Brijes Mishra

成　员：牟宗龙　黄炳香　王旭锋　方新秋　王襄禹　杨　真　姚强岭
　　　　　周　伟　马文顶　曹安业　黄艳利　范钢伟　徐　营

秘书长：李桂臣

秘　书：吴锋锋　程敬义　杨　真　李　剑　闫　帅　王　君　吴元周
　　　　　殷　实

前　言

  国际采矿岩层控制会议(International Conference on Ground Control in Mining,简称 ICGCM)自 1981 年起在美国举办,至今已成功举办 36 届。我国煤炭开采技术飞速发展,为了便于我国学者与国际采矿岩层控制领域的学者进行广泛交流,提高我国在采矿岩层控制领域的研究与应用水平,提升我国在国际采矿行业的国际影响力,经与 ICGCM 组委会协商,将定期在中国举办"国际采矿岩层控制会议(中国)"。

  此次国际会议是第五次在中国召开,总第 37 届。会议的目标是创建一个在煤炭开采岩层控制方面技术分享与讨论的平台。会议主席团由国内外采矿岩层控制领域权威专家组成,包括中国工程院院士和美国工程院院士。会议的交流内容不仅注重采矿岩层控制的基础理论,而且也重视煤炭开采方面的实际问题与前沿技术,将为世界采矿技术的发展起到重要的理论和实际指导意义。

  会议从 2018 年 2 月开始征集论文至 2018 年 9 月,共收到来自国内外煤矿开采岩层控制领域的中文、英文论文 41 篇,通过大会学术委员会筛选,收录 31 篇论文。

  会议组委会李桂臣、牟宗龙、王襄禹、吴锋锋、程敬义等在会议的论文征集与出版等方面做了大量工作。Syd S. Peng 院士给出了许多建设性建议和指导,为此次会议的顺利举办和提升会议的国际性作出了重要贡献。

组委会

2018 年 9 月

目 录

Detection of floor water inrush risks for deep longwall mining in a Chinese coal mine ………… HE Dongsheng, ZHANG Yang, WANG Fangtian(1)

动压影响煤柱下方沿空巷道微震特征及破坏机制………………李术才,王雷,江贝,等(21)

气体吸附诱发煤体劣化的试验研究与分析………………李清川,王汉鹏,袁亮,等(38)

大倾角工作面飞矸冲击损害及其控制………………伍永平,胡博胜,皇甫靖宇,等(55)

锚注扩散及加固规律现场试验研究与应用………………王琦,许英东,许硕,等(69)

基于大范围岩层控制技术的大倾角煤层区段煤柱失稳机理研究
……………………………………伍永平,皇甫靖宇,解盘石,等(84)

大变形巷道螺纹钢锚杆外形优化设计及应用研究 …… 张明,CAO Chen,赵象卓,等(100)

单轴压缩下含瓦斯煤破坏过程能量演化规律 ……………… 张冰,王汉鹏,袁亮,等(117)

泥质胶结岩体力学特性宏细观模拟研究 ………………孙长伦,李桂臣,何锦涛,等(132)

区段煤柱稳定性研究的新探索 ………………………………………… 闫帅,柏建彪(145)

拉伸荷载下盐岩力学及声发射特征研究 ………………曾寅,刘建锋,邓朝福,等(158)

负压条件下瓦斯渗流实验及抽采模拟研究 ………………李祥春,高佳星,李安金,等(171)

静水压力下开挖卸荷快慢对硐室围岩变形-开裂影响的连续-非连续方法模拟
……………………………………王学滨,芦伟男,白雪元,等(193)

极软弱地层双层锚固平衡拱结构形成机制研究 ………孟庆彬,韩立军,梅凤清,等(204)

Deformation and stress analysis of surface subsidence at the Jingerquan Mine
……………………………………DING Kuo, MA Fengshan, ZHAO Haijun, et al(223)

光纤光栅煤矿安全智能监测系统 ………………………方新秋,吴刚,梁敏富,等(237)

Repairing technology based on analysis of deformation or failure on main roadways in Shuanglong Coal Mine ……YUN Dongfeng, WANG Zhen, WU Yongping, et al(248)

Study on the height of fractured zone in overburden at the high-intensity longwall mining panel ……GUO Wenbing, ZHAO Gaobo, LOU Gaozhong, et al(266)

断层破碎带区域巷道围岩差异性分类及关键控制对策 …… 赵启峰,张农,李桂臣,等(277)

近距离巨厚坚硬岩层破断失稳特征及分区控制 ……………………… 赵通,刘长友(289)

深井软岩下山巷道群非对称破坏机理与控制研究………………刘帅,杨科,唐春安(301)

大倾角煤层大采高工作面倾角对煤壁片帮的影响机制 ··· 王红伟,伍永平,罗生虎,等(323)

基于VRP的采矿过程矿石质量智能优化研究 ………… 李小帅,郭连军,徐振洋,等(335)

Assessing longwall shield-strata interaction from a basic understanding of
shield characteristics and a physical modeling study
………………………………………………… SONG Gaofeng,DING Kuo,SUN Shiguo(343)

Strata movement law and support capacity determination of a upward-inclined
fully-mechanized top-coal caving panel
………………………………… KONG Dezhong,LIU Yang,ZHENG Shangshang,et al(364)

厚硬岩层直覆大倾角综采顶板失稳机理分析 ……… 魏祯,杨科,池小楼,等(375)

长壁充填开采充填步距对覆岩位移的影响分析 ………… 贾林刚,高庆丰(382)

近水平厚煤层沿空巷道位置选择及支护技术研究………… 王志强,苏越,苏泽华(393)

沿空留巷可缩性墩柱破坏形态及加固分析 ………… 郭东明,凡龙飞,王晓烨,等(406)

优化露天矿山爆堆单元划分方法与出矿品位确定 ……… 王雪松,徐振洋,李小帅,等(414)

动载作用下岩石破坏模式与强度特性分析………………………………… 王军(421)

Detection of floor water inrush risks for deep longwall mining in a Chinese coal mine

HE Dongsheng[1], ZHANG Yang[2], WANG Fangtian[2]

(1. *Chengjiao Coal Mine, Henan Zhenglong Coal Industry Co. Ltd., Yongcheng* 476600, *China*;
2. *School of Mines, State Key Laboratory of Coal Resources and Mine Safety, Key Laboratory of Deep Coal Resource Mining, Ministry of Education of China, China University of Mining and Technology, Xuzhou* 221116, *China*)

Abstract: Floor pressurized water inrush is a key problem that threatens safe and efficient production in coal mines, especially in deep longwall coal mining. Depending on the conditions of the aquifer, the hydrogeological conditions become quite complex, leading to a high water pressure environment. As the risk posed by floor pressurized water increases, prevention of water inrush risk in deep coal seams mining becomes crucial. In this study, the hydrogeological conditions of deep coal seams in Chengjiao Coal Mine were analyzed alongside the composition, microstructure, and mineral disintegration of mudstone and sandstone. The two rocks from floor strata were tested in the laboratory to gain important baseline data for a scientific evaluation of the water resistance ability. The seepage resistance strength and permeability coefficient of the rocks at different floor depths were obtained through the in-situ double-hole method detection. A combination method of both fractal dimension of fault complexity and water inrush coefficient was used to evaluate the water inrush risk. As a result, the comprehensive evaluation division of the water inrush risk at Level-Ⅱ is that the danger decreasing from the eastern area, and then the western area, and last the middle area. A prevention strategy, include advanced warning monitoring, hydrological dynamic monitoring, partial discharge testing, water drainage, pressure reduction, and grouting reinforcement, was applied to enable the safe longwall mining of the deep coal seam with high water inrush risk.

Keywords: deep buried coal seam; water inrush risk; water resisting strength; field detection; safe and efficient mining

1 Introduction

Mine water damage is the "second killer" threatening mine safety production. From 2000 to 2012, there were 1,069 mine water inrush accidents in China, leading to 4,333 deaths and economic losses of more than 35 billion RMB[1]. With the continuous increase in coal mining depth, the threat of mine water hazards has been aggravated, which seriously restricts safe and efficient mining[2]. Mine floor water inrush accidents are mainly caused by the original geological structure, with the fault structure being the main precipitating factor[3,4].

Supported by the Fundamental Research Funds for the Central Universities(2018ZDPY05), the Scholarship Program from China Scholarship Council, and the Priority Academic Program Development of Jiangsu Higher Education Institutions.
Corresponding author: wangfangtian111@163.com.

Research on coal mine water inrush mechanisms and related prevention measures has been popular in China in recent years. Wu et al. systematically classified types of mine water hazards, analyzed the characteristics of coal mine hydrogeology and water disaster problems faced by four coal mine areas, and provided the basis for the classification and prevention of mine water hazards[5, 6]. Yin focused on water inrush in mines and divided floor pressurized water inrush into three models: normal floor pressurized water inrush, water inrush from fault fissure zones, and water inrush from collapse columns. Thus, he analyzed the water inrush mechanisms in detail[7]. Sun et al. presented a water-resistant key floor strata model, prior to main roof weighting, to explore the relationship between floor water inrush and the main roof weighting[8]. Xu divided floor pressurized water inrush into the complete water inrush mode and fault water inrush mode with a water-repellent layer[9]. Li et al. utilized fractal theory to conduct a relationship analysis of fault complexity degree and water irruption rate, which was found to increase with the fractal dimension[10]. After studying the hydrogeological conditions and type of water inrush passage in the Zhengzhou mining area, Dong categorized mine water inrush into three modes, which are fault-induced mode, fold-induced mode, and mining-fracture-induced mode. He also suggested prevention and mitigation measures based on the applicable water inrush mode[11]. Xu et al. proposed three types of trapezoidal broken models considering the top-down varying pattern of the lateral and longitudinal volume expansion coefficients. These models may conduce to understand the caving mechanism of overburden strata and the formation of fracture space, and thence provide guidelines for coal mining under water bodies[12]. Wang et al. constructed a secondary fuzzy comprehensive evaluation system to assess the risk of floor water invasion in coal mines. The engineering evaluations were conducted using hydrogeological data of six mining faces[13]. Predicting the water inrush potential and taking effective measures before mining activities are essential to enable safe coal mine production[14]. However, the prediction of floor pressurized water is closely related to the hydraulic characteristics of floor rock and the distributions of the fault structure.

A deep coal seam (depth over 800 m) at Chengjiao Coal Mine was taken as the research objective. The composition, microstructure, and mineral disintegration of mudstone and sandstone were tested in the laboratory to gain important baseline data for a scientific evaluation of the water resistance ability. The seepage resistance strength and permeability coefficient of the floor rocks were obtained through the in-situ double-hole method detection. To evaluate the floor water inrush risk, a combination of fault complexity fractal dimension and water inrush coefficient was applied. Consider the water inrush risk division at Level-II, a systemic prevention strategy, include advanced warning monitoring, water drainage, pressure reduction, and grouting reinforcement, can be applied to enable the safe and efficient longwall mining of the deep coal seam. The study at Chengjiao Coal Mine attempt to provide some significances for the detection and prevention of floor water inrush risks with similar conditions.

2 Hydrogeological background

The Chengjiao Coal Mine is located in Yongcheng City, Henan Province, China. It has a capacity of 5 million t/a, and the main mining of No. 2-2 coal seam is as deep as 1,000 m, the geological columnar section of the floor strata is shown in Fig. 1. The hydrogeological conditions at Level-II (elevation -800 m) are complex: (1) Aquifer development. Coal-bearing formations contain loose rock-like pore-fractured aquifers, Permian sandstone fractures, and the Taiyuan Formation limestone insolvent aquifer; (2) The geostructure water control conditions are complex. Structural networks with unequal scales and different sequence periods not only constitute a direct water-filling channel, but also disturb the stratum structure and reduce the water resisting capacity for the floor rocks, which strengthens the hydraulic links between aquifers to some extent; (3) The complex geological environment of high water pressure, high geostress, high temperature and mining-induced disturbance. The Carboniferous Taiyuan Formation contains a regional aquifer composed of limestone in the floor, which is likely to lead to a water inrush accident.

Till 2017, two water inrush accidents occurred in the Taiyuan formation:

(1) On March 29, 2004, the water inrush occurred in working face LW2205 through a fault structure, the maximum water inflow being 80 m^3/h. This accident caused the face to withdraw to a distance of 55 m from the stoppage line, resulting a large amount of coal loss.

(2) On June 24, 2008, water inrush occurred during the 1,117 m excavation of the southern area of a rail transport roadway. The initial water volume was 80 m^3/h, while the maximum water inflow was 300 m^3/h, gradually stabilizing to 230 m^3/h. The water inrush was caused by the fault F_{N-5} along the left-hand roadway. The deep limestone water from the fault plane collapsed into the working face. The increase in mining depth caused a rise in water pressure in the limestone aquifer and lead to a disturbance in the deformation degree of the mining floor. Managing such water hazards in deep coal seam mining is quite difficult. They pose major problems with respect to mine safety and encumber high-efficiency production.

3 Laboratory experiment for the floor rocks

Borehole sampling at Level-II (Fig. 1) was conducted to analyze the composition, microstructure and mineral disintegration of mudstone and sandstone from the floor strata. The samples were experimented through the use of D8 Adavance type X-Ray Diffraction, Quanta 250 type Scanning Electron Microscope in laboratory to ascertain important baseline data for scientifically evaluating the floor water resisting ability.

3.1 Rock mineral composition

As shown in Fig. 2, a comparison of X-ray diffraction results before and after mine water immersion was conducted for mudstone and sandstone samples. The two lithological formations are mainly composed of quartz, and the other minerals are mica and chlorite.

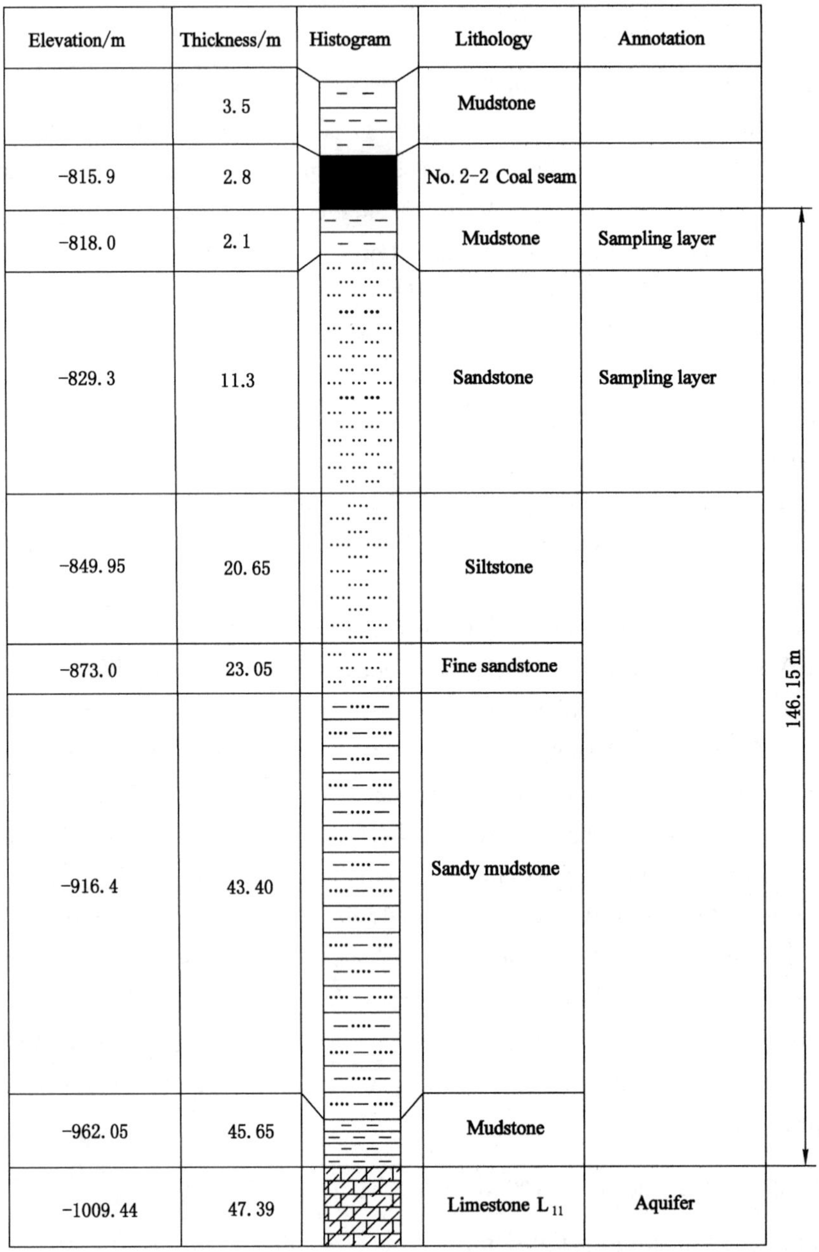

Elevation/m	Thickness/m	Histogram	Lithology	Annotation
	3.5		Mudstone	
-815.9	2.8		No. 2-2 Coal seam	
-818.0	2.1		Mudstone	Sampling layer
-829.3	11.3		Sandstone	Sampling layer
-849.95	20.65		Siltstone	
-873.0	23.05		Fine sandstone	
-916.4	43.40		Sandy mudstone	
-962.05	45.65		Mudstone	
-1009.44	47.39		Limestone L_{11}	Aquifer

(Total: 146.15 m)

Fig. 1　Geological columnar section of the floor strata

Two sets of X-ray fluorescence tests both reveal that the main components of mudstone and sandstone are silicon and aluminum. The SiO_2 contents of the mudstone and sandstone are ～57% and ～73%, respectively. Al_2O_3 being the next most prominent compound, followed by smaller amounts of iron, carbon, potassium, magnesium, sodium and calcium compounds. The high content of siliceous material is the primary reason for relatively high water resisting ability for both mudstone and sandstone. The laboratory

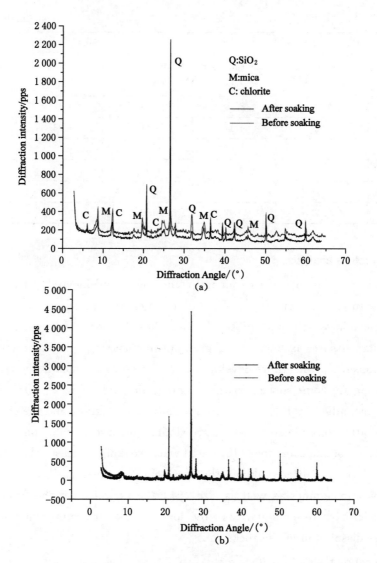

Fig. 2 Comparison of mineral compositions via X-ray diffraction before and after groundwater soaking

(a) Mudstones;(b) Sandstones

experiment results are shown in Table 1. The comparison of the molecular composition indicated that mudstone and sandstone have the same molecular formula and small change in content before and after soaking, respectively.

Table 1　X-ray fluorescence studies on the composition and content of mudstone and sandstone

Mudstone samples			Sandstone samples		
Molecular formula	Contents before soaking/%	Contents after soaking/%	Molecular formula	Contents before soaking/%	Contents after soaking/%
SiO_2	57.08	57.24	SiO_2	73.14	72.64
Al_2O_3	20.33	20.28	Al_2O_3	17.19	17.72

Continued Table 1

Mudstone samples			Sandstone samples		
Molecular formula	Contents before soaking/%	Contents after soaking/%	Molecular formula	Contents before soaking/%	Contents after soaking/%
Fe_2O_3	8.76	8.79	K_2O	3.47	3.57
CO_2	6.33	6.17	Fe_2O_3	1.38	1.42
K_2O	3.56	3.56	CaO	1.32	1.33
MgO	1.83	1.84	CO_2	1.18	0.96
Na_2O	0.66	0.65	Na_2O	1.05	1.05
CaO	0.42	0.43	MgO	0.64	0.67

3.2 Microstructure analysis

Changes in the contents of the mineral components will inevitably lead to changes in the microstructure. As per Fig. 3 (a) and (b), the mudstone is compact before groundwater soaking, only part of the bedding has developed micro-fractures, and the density of micro-fissures is low. After groundwater soaking, the microstructure of the mudstone is loose, showing more developed micro-fractures, which is the reason that easy to fracture along the joint surface in the disintegration test. Fig. 3 (c) and (d) show a micro-fissure magnified by 1,200× before groundwater soaking, and the characteristics of the sandstone after groundwater soaking are similar to those of mudstone.

The comparison indicates that (1) ground water soaking increases the micro-fissures of the rock's surface, (2) after soaking, the rock tends to break easily, (3) surface micro-fissures increase in number as well as size, (4) the rock around the fissures dissolves, (5) and the microstructure angularity of the soaked rock surface is pronounced due to the heterogeneous dissolution of the rock.

3.3 Disintegration tests of the floor rocks

Table 2 shows the disintegration test classifications according to Gamble's disintegration durability classification.

Table 2 Classification reference table for disintegration durability as per Gamble[15]

Group	Percentage left after one 10-min rotation/%	Percentage left after two 10-min rotations/%
Extremely high durability	>99	>98
High durability	98~99	95~98
Medium high durability	95~98	85~95
Moderate durability	85~95	60~85
Low durability	60~85	30~60
Extremely low durability	<60	<30

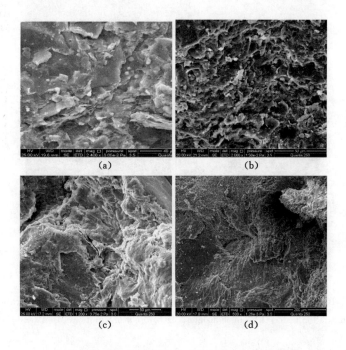

Fig. 3 Scanning electron microscopy contrast images of mudstone
and sandstone before and after ground water soaking
(a) Before soaking in mudstone;(b) After soaking in mudstone;
(c) Before soaking in sandstone (magnify 1200 times);
(d) After soaking in sandstone

The SCL-1 disintegration resistance tester (manufactured by Hangzhou Sansi Instrument Co. Ltd., China) was used for this test. The disintegration apparatus and images of sample baking are shown in Fig. 4.

(1) Mudstone: The mudstone specimens were square-shaped. The experimental phenomena related to the disintegration and the test results are shown in Table 3 and Table 4, respectively.

Table 3　　　　　　　　　　Disintegration phenomena

2nd drying	3rd drying	4th drying	5th drying	6th drying
No. 3 split into two along the cleavage plane and No. 1 cracked	No. 3 retains two pieces, and No. 1 split into two large parts along the cleavage surface and several small pieces were also formed	No changes were noted in Nos. 3 and 1	One of the two pieces of No. 3 split along the bedding plane again, and three pieces in all. No other changes were noted	A large piece of the cracked No. 1 split into two small pieces along the cleavage surface, and several small pieces were also formed

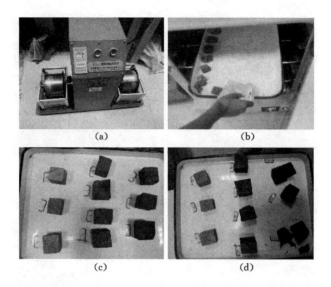

Fig. 4 Instrument and images of sample baking for the sample disintegration test
(a) Disintegration tester; (b) Sample baking; (c) Before disintegration; (d) After disintegration

Table 4　　　　　　　　　Experimental results of mudstone disintegration

No.	Sample quantity/g						Id/%
	Before the experiment A	After the 1st cycle B	After the 2nd cycle C	After the 3rd cycle D	After the 4th cycle E	After the 5th cycle F	F/A×100%
1	129.4	129.4	129.2	129.0	128.7	128.5	99.3
2	106.4	106.4	106.3	105.9	105.7	105.6	99.2
3	100.0	100.0	99.8	99.6	99.5	99.4	99.4

Note: Id is the disintegration resistance index, F is the quality after the fifth cycle, and A refers to the quality before the experiment. Id (%) = F/A×100%.

Mudstone disintegration is due to the dissolution of soluble clay minerals when interacting with water, which results in loss of adhesion and weakens the structural connection within the mudstone. Thus, the material loses strength. The mudstone presents in the deep floor, which has not been subjected to natural weathering and has undergone a relatively high degree of consolidation. Its internal structural connection strength is high and the macroscopic appearance confirms disintegration. According to Table 4, the disintegration resistance index of the collected mudstone samples ranges from 99.2% to 99.4%, indicating that the rock has extremely high durability.

(2) Sandstone: The disintegration test results for the sandstone are shown in Table 5. The disintegration resistance index lies between 99.4% and 99.9%, confirming that the rock has extremely high durability.

Table 5　　　　　　　　Experimental results of sandstone disintegration

No.	Sample quantity/g						Id (%) F/A×100%
	Before the experiment A	After the 1st cycle B	After the 2nd cycle C	After the 3rd cycle D	After the 4th cycle E	After the 5th cycle F	
A1	93.6	93.5	93.5	93.2	93.1	93.1	99.7
A2	70.2	70.0	70.0	69.9	69.8	69.8	99.4
A3	86.6	86.5	86.3	86.2	86.2	86.2	99.5
B1	112.4	112.4	112.5	112.4	112.4	112.3	99.9
B2	119.0	119.0	118.9	118.9	118.8	118.7	99.7
B3	112.9	112.9	112.8	112.8	112.8	112.6	99.7

Through the microstructure, comparison and disintegration laboratory experiment, and also the analysis of different floor rock samples before and after soaking tests, which indicate that the microstructure and composition have little change before and after ground water soaking, and the main floor rocks at Level-II have good water stability characteristics.

4　Field detection of the floor water resisting ability

Longwall mining in the normal sedimentary area is very safe at Chengjiao Coal Mine, but the situation is different in some structurally affected areas: (1) The mudstone in the floor has a good water resisting property, but its structural integrity will be damaged inordinately due to the thickness decreases and fault structure changes, therefore results in a sharp decrease of the floor water resisting strength; (2) The effects of longwall mining easily magnify at some fault structural zones and can activate fissures, thus greatly increasing the danger of underground pressured water inrush risks. Thus, the hidden dangers of longwall mining under floor pressured water mainly depend on the fault structural development. Therefore, a floor water resisting ability test is required in the field.

4.1　Main principles and equipment

This study uses the double-hole pressurization test technique. Two boreholes were arranged in the roadway, one for water injection and the other for hydraulic test (Fig. 5). By testing the water pressure conditions for low-resistance seepage in the rock formation, the permeability of the floor rock, permeation, and failure strength of a unit thickness are obtained, which provide a quantitative basis for the evaluation of floor water resisting ability. During the process of water injection, parameters such as water injection pressure, water injection flow rate, and infiltration water pressure value can be measured to obtain the water resisting strength and permeability coefficient of the rock formation at different floor depths.

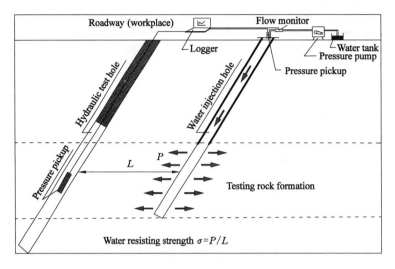

Fig. 5　Schematic diagram of the field water resisting strength test

The test equipment included borehole drilling equipment, water injection equipment, and a pressure pickup system: (1) A common underground drill rog was used as the drilling equipment; (2) The pressurizing equipment was a 2ZBQ-3/21 high-pressure pneumatic grouting system, as shown in Fig. 6(a). The adjustable flow range is 5～80 L/min, rated pump pressure is 22 MPa, and the grouting pipeline pressure no less than 15 MPa; (3) The permeability monitor system adopted for this study was an SYGJ vibrating wire type hydraulic pressure sensor [Fig. 6(b)] with high counting accuracy and a large adjustable test range. A GSJ-2A intelligent detector was used for data acquisition [Fig. 6(c)]. It can directly display the water pressure, the readings of which can be stored and viewed as required, and has the advantages of small size, light weight, high degree of integration, low power consumption, and convenience of carrying.

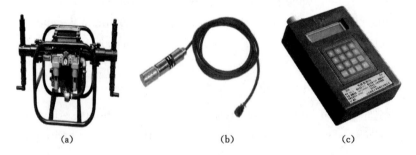

Fig. 6　Main test equipment
(a) High-pressure pneumatic pump; (b) Water pressure sensor; (c) Data acquisition instrument

4.2　Analysis of floor water resisting ability test

Considering the factors related to the roadway environment and borehole construction conditions, the test site was set to fault F_{N-6}, which located in the western belt roadway of the south area at Level-Ⅱ. Fault F_{N-6} is a normal fault with the offset of 10.3 m. The large roadway is 45 m away from limestone aquifer L_{11}. Each test hole was located at a certain distance from the fault fractured zone so as to reduce the possible impacts of aquifer L_{11} and the lower aquifer.

(1) Test hole layout. Three holes were drilled as part of the field testing, one of which was injected with water, while the other two were used for monitoring purposes. The horizontal distance between Hole C1 and monitoring point SYP23 is ~1.6 m, and that between three holes is 6.3 m (Table 6 and Fig. 7). The water resisting ability test is carried out by continuously monitoring the change in water pressure during the process of water injection.

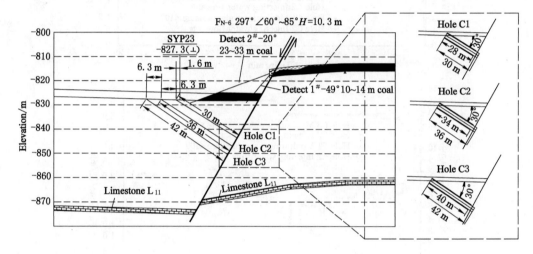

Fig. 7　The section layout of the fault test holes

Table 6　Drilling arrangement for the south lane of the double-track roadway test

Technical Parameters	Hole C1	Hole C2	Hole C3
Distance from monitoring point SYP23	1.6	7.9	14.2
Diameter of opening/depth/(mm/m)	127/28	127/34	127/40
Orifice tube diameter/length/(mm/m)	110/28	110/34	110/40
Final hole diameter/mm	89	89	89
Drilling azimuth/(°)	113	113	113
Drilling angle/(°)	−30	−30	−30
Angle with roadway/(°)	23	23	23
True inclination of the coal rock/(°)	4	4	4
Hole depth/m	30	36	42

(2) Fault water resisting ability test results analysis. Hole C2 was used for injecting the water, while Hole C1 and Hole C3 were utilized for monitoring the water pressure. The difference in the process curves for primary water pressure [Fig. 8(a)] and repeated water pressure [Fig. 8(b-e)] show that the fault zone is strongly impermeable.

Fig. 8(a) shows that between 0~1 min, the water pressure increases from 0 to 10 MPa, and the flow rate of the pressurized water increases from 0 to 3.4 L/min. However,

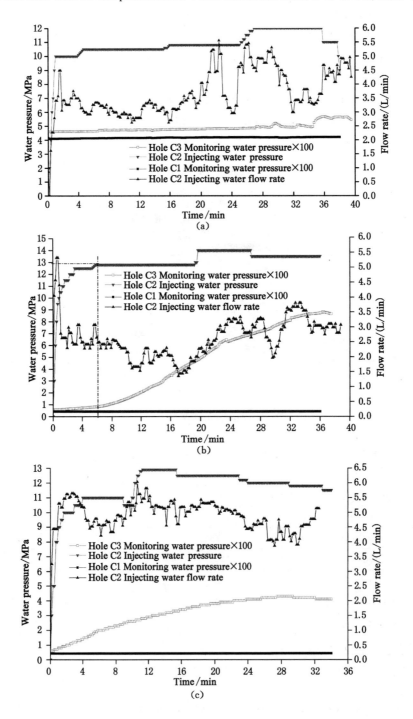

Fig. 8　Pressure-infiltration process curve for the first stage of pressure testing in the fault zone

(a) Initial pressure curve; (b) First pressure infiltration; (c) Second pressure infiltration

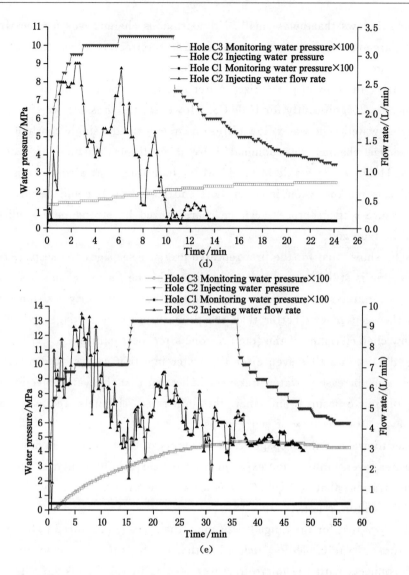

Continued Fig. 8　Pressure-infiltration process curve for the first stage of pressure testing in the fault zone

(d) Third pressure infiltration;(e) Fourth pressure infiltration

the water pressure for Holes C1 and C3 was lower than that of the injected water. The constant values for Holes C1 and C3 were 0.041 MPa and 0.046 MPa, respectively. To increase the contrast, and considering the low water pressure of Holes C1 and C3, the data for these two holes are magnified 100 times in Fig. 8(a). In the original state, the fault zone was not water-guiding and had strong resistance to seepage. With the progressive increase in water pressure from 1~34.5 min, many small cracks occurred within the rock mass, with their sizes increasing slightly at ~29 min. However, the parameters for Hole C1 remained almost unchanged. The water injection flow also fluctuated slightly in range of 3.2～3.7 L/min. showing that the scale of crack propagation and new fracture generation in the fault zone was small. No continuous channel was formed with the test hole, and a large amount of energy accumulated in the fracture zone. At ~34.5 min, the rock mass produced a relatively large cleaving, the water pressure decreased instantaneously, the flow of pressurized water increased rapidly, and the measured water pressure increased significantly for Hole C3. However, the pressure was far less than that of the injected water. The water flow diffusion circle formed a single continuous flow line, and the seepage channel was dominated by micro-fissure flow. The water pressure measured at Hole C1 was stable at 0.041 MPa, indicating that a continuous channel had not formed between the water injection hole and Hole C1. It could be inferred that the primary pressure seepage process can be characterized by strong permeability and poor seepage in the fault zone.

Fig. 8(b) shows that for the first repeated pressure seepage, the seepage resistance in the broken zone is still large, the opening degree of the fissure channel is low, seepage flow mainly constitutes micro-fissure seepage, and the permeability resistance was lower. However, the seepage was better than the primary pressure seepage, which shows the seepage flow characteristics of the fracture zone after local failure.

Fig. 8(d-e) shows that even after the water injection ceases, the monitoring holes show an obvious increase in water pressure, indicating that the water injection leads to the formation of micro-fissures and that the infiltrated channels retain certain barrier properties even after the flow of water stops.

In order to evaluate the seepage-resistant ability of the fault zone in more detail, high-pressure water reverse infiltration experiments were carried out similar to the first stage of pressure infiltration; that is, Hole C3 was used as the pressure hole, while Holes C1 and C2 were the monitoring holes.

Fig. 9 shows that for the range 0~23 min, the seepage pressure in Hole C2 gradually increased from 0.06 to 0.6 MPa, indicating that the fracture formed by reverse pressure infiltration connects with the micro-fractures formed in the first stage. The continuous through channel increases, pressure difference decreases, and permeability was relatively enhanced. The fluctuation in water injection flow ranges from 0 to 4.5 L/min, which indicates the synchronicity between water filling and hydraulic fracturing in the rock mass. A large step-up process was carried out at 23 min. The pressure of the injection water increased rapidly from 7 to 11 MPa, and the pressure pump was adjusted to keep it stable. The water pressure in Hole C2 also increased rapidly from 0.6 to 1.3 MPa, but the water injection decreased after the flow rate increased to the maximum value of 9.5 L/min. This phenomenon can be attributed to a dominant flow channel formed between Hole C2 and Hole C3. If the water injection was continued under constant pressure, the dominant flow channel becomes saturated. The formation and expansion of this independent micro-fissure network takes place to absorb the potential energy of the high-pressure water; the flow decreases during these fluctuations, and the infiltration pressure continues to rise till it

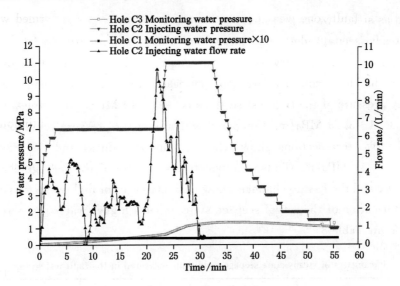

Fig. 9　Pressure-osmosis process curve of the second-stage pressure infiltration process in the fault zone

reaches a stable value.

(3) Permeability parameters in fault zones. Initial infiltration conditions: The point where the water pressure and pressure percolation volume show an obvious change is considered to be the initial permeability characteristic point, and the water injection pressure corresponding to this point is defined as the initial permeable pressure. We determined the initial permeability pressure of three test sections in fault zone F_{N-6}.

Table 7 shows that the permeability resistance of fault zone F_{N-6} was relatively strong and the initial permeability pressure was ～11 MPa. The permeability pressure gradient was 3.25 MPa/m, which indicates that the fault zone has a high initial permeability pressure. However, a large permeability pressure difference (～10.4 MPa) exists, which shows that the seepage channel is dominated by a fine fissure network. This leads to high seepage resistance and poor seepage flow characteristics.

Table 7　　　　　　　　Initial infiltration conditions of fault zone F_{N-6}

Measured location		Water pressure/MPa			Impermeability resistance	Water pressure gradient
Test venue	Test section	Steady pressure	Hole pressure	Infiltration difference		
F_{N-6} fault test section (3.2 m)	First stage	12.0	0.05	11.95	3.75	3.73
	Second stage	11.0	0.60	10.40	3.44	3.25

Steady state permeation condition: The pump pressure was continually raised to a certain level, and the water pressure and pressure-osmotic flow rate of the seepage hole showed some expansion while being relatively stable. The maximum steady water pressure was defined as steady water pressure.

The steady seepage does not mean that there is structural failure in the rock formation; rather, conductivity seepage is caused by the local fracture damage. The results show that fault zone F_{N-6} was not permeable to begin with, and that the initial permeability pressure of the two test sections is ~11~12 MPa, while the water resisting strength is 3.44~3.75 MPa/m. Compared with the water pressure conditions for steady seepage, the two test sections show relatively similar values, and the water resisting strength is ~3.44 MPa/m. Thus, the penetration degree of the seepage channel is still relatively low, and the fissure channels show characteristics of high water resisting ability. Thus, the formation of the actual seepage state under a higher water pressure gradient is necessary. As in Table 8.

Table 8　Parameters of steady-state seepage conditions measured in the fault test section

Measured location		Water pressure/MPa			Impermeability resistance	Water pressure gradient
Test venue	Test section	Steady pressure	Hole pressure	Infiltration difference		
F_{N-6} fault test section (3.2 m)	First stage	11.0	0.05	10.95	3.44	3.42
	Second stage	11.0	1.34	9.66	3.44	3.02

5　Risk analysis of floor pressurized water inrush

The water inrush coefficient refers to the maximum water pressure the aquifuge can bear and is an important indicator of the water inrush risk. According to the exploration drill data from Chengjiao Coal Mine, the most dangerous water inrush risk was likely to occur in the limestone aquifer of Taiyuan Formation. Therefore the limestone aquifer L_{11}, which has the shortest distance from the No. 2-2 coal seam and relatively stable structure, was taken as the research objective. The water resisting layer thickness between the limestone aquifer L_{11} and the No. 2-2 coal seam was determined by the exposed strata thickness of each drill borehole. The maximum water pressure of limestone aquifer L_{11} in each observation hole was selected to determine the water pressure of 4.6 MPa, and the water inrush coefficient of each drill hole was calculated. Two assessment parameters (both fractal dimension and water inrush coefficient) were then combined to comprehensively observe the water inrush risk in the study area. Thus, the possible danger posed by water inrush could be assessed.

Fig. 10 shows the structural complexity in terms of fractal contours in the deep mining area of Chengjiao Coal Mine. We observe a good positive correlation between the fractal dimension contour and water inrush coefficient contour. The inrush coefficient value of the area is obviously higher and the fault structure is complex. A large number of water inrush cases show that nearly 85% of water inrush accidents occur in normal fault structures[16].

The main reason is that, under the condition of constant water pressure, the normal fault structure results in a significant decrease of the effective water-resisting layer thickness along the fault surface[17]. The water inrush area or safe area at Level-Ⅱ is also a complex fault structure area or a simple structure area, and the two are corresponding to a certain extent. From the two water inrush risk assessment methods, it is concluded that the middle area is safe, but the floor water inrush is dangerous in the eastern and western mining areas. On the whole, the water inrush risk is decreasing from the eastern area, and then the western area, and last the middle area.

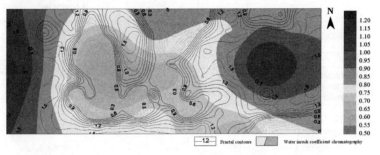

Fig. 10 Comprehensive assessment of the water inrush risk using fractal contours and water inrush factor chromatography

6 Main prevention strategies

The results of the analysis for the water inrush zone are indicative of possible prevention technologies to eliminate the water inrush hazards for the limestone aquifer at Level-Ⅱ, where undergoes high-pressure water and complex fault structures.

(1) Advanced warning monitoring: Advanced monitoring of water inrush hazards and depth of deformation testing for the floor can be performed for deep and structurally complex geological areas. The specific process is shown in Fig. 11.

(2) Hydrological dynamic monitoring: The strengthen and improvement of the Dynamic Automatic Monitoring System is needed for revealing the limestone hydrological information, so as to acquire timely information on the dynamic changes in water levels and water pressure in the limestone aquifer (which is mainly located in the Taiyuan Formation). The discovery of the abnormalities quickly would conduce to the cause analysis timely, and appropriate safety precautions can be adopted accordingly.

(3) Analysis of water quality at the outlet: Timely analysis of the water quality characteristics is necessary. Such monitoring should be conducted along the water outlets around the excavation roadway or working face. The results would help ascertain the water quality characteristics of water inrush, which is helpful to identify the water inrush source. Decisive treatment measures can then be undertaken using this information.

(4) Partial discharge tests: Aquifer discharge tests for certain sections can be conducted to quantify the corresponding hydrogeological parameters.

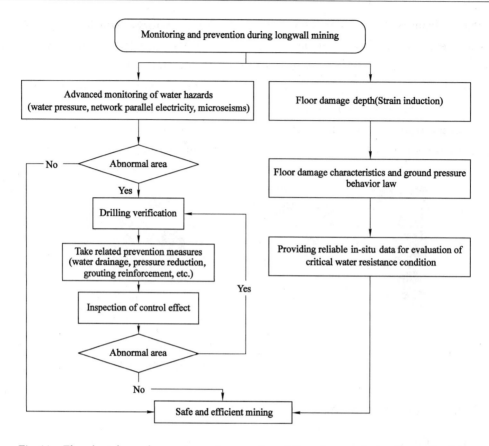

Fig. 11　Flowchart for early warning and prevention of floor damage during longwall mining

(5) Floor rock formation detection and management: There is a need to conduct detailed investigations of the hydrothermal conditions of the floor rock formation and the aquifer which bearing substantial water quantities. When the pressure head is found to be high, the water inrush risk may be eliminated through water drainage, pressure reduction, grouting reinforcement, and other measures.

(6) Improvement of water drainage system: It is crucial to establish a sound water drainage system. The working and auxiliary water pumps should be in good condition, and its double-circuit power should be readily available to meet the needs of water drainage at Level-II. All water drainage equipment and related infrastructure must be inspected regularly to ensure that the drainage system is in good condition.

7　Conclusions

(1) With the consideration of complex hydrogeological environments, include high water pressure, high geostress and longwall mining-induced disturbance, and especially the developed fault structure, the floor water inrush accident is likely occurred, which seriously restricts the safe and efficient longwall mining of deep coal seams.

(2) The composition, microstructure, and mineral disintegration of mudstone and

sandstone from floor strata were tested in the laboratory, which indicate that the microstructure and composition have little change before and after ground water soaking, and the main floor rocks at Level-Ⅱ have good water stability characteristics.

(3) Water resisting strength and permeability coefficients of the rock layers at different floor depths were obtained through the double-hole pressurization method. The permeability resistance tests in the fault zones indicated that the seepage channel is dominated by a fine fissure network, which leads to high seepage resistance and poor water flow characteristics.

(4) The comprehensive assessment of both fault complexity fractal dimension and water inrush coefficient was used to evaluate the water inrush risks. The water inrush dangerous at Level-Ⅱ is decreasing from the eastern area, and then the western area, and last the middle area. A systematic prevention strategy, include advanced warning monitoring, hydrological dynamic monitoring, water drainage, pressure reduction and grouting reinforcement, was implemented to enable the safe and efficient longwall mining at deep coal seam.

8 Conflicts of Interest

The authors declare there is no conflicts of interest regarding the publication of this paper.

References

[1] Zhang P S, Zhang W Q, Yan W. Research and application of fault waterproof coal pillars based on mining activation induced fault law[M]. Xuzhou: China University of Mining and Technology Press, 2015.

[2] Zhang W Q, Meng L L. Numerical simulation on the protective coal pillars for floor fault under the influence of subfault[J]. Mining Research & Development, 2016, 36(3):36-39.

[3] Wu Q, Wang M, Wu X. Investigation of groundwater bursting into coal mine seam floors from fault zones [J]. International Journal of Rock Mechanics & Mining Sciences, 2004, 41(4):557-571.

[4] Guo W J, Zhang S C, Sun W B, et al. Experimental and analysis research on water inrush catastrophe mode from coal seam floor in deep mining[J]. Journal of China Coal Society, 2018, 43(1):219-227.

[5] Wu Q, Cui F P, Zhao S Q, et al. Type classification and main characteristics of mine water disasters[J]. Journal of China Coal Society, 2013, 38(4):561-565.

[6] Wu Q, Zhao S Q, Sun W J, et al. Classification of the hydrogeological type of coal mine and analysis of its characteristics in China[J]. Journal of China Coal Society, 2013, 38(6):901-905.

[7] Yin S X. Modes and mechanism for water inrushes from coal seam floor[J]. Journal of

Xi'an University of Science and Technology,2009,29(6):661-665.

[8] Sun J, Hu Y, Zhao G M. Relationship between water inrush from coal seam floors and main roof weighting[J]. International Journal of Mining Science and Technology, 2017,27(5):873-881.

[9] Xu Z M. Mining-induced floor failure and the model, precursor and prevention of confined water inrush with high pressure in deep mining[D]. Xuzhou: China University of Mining and Technology, 2010.

[10] Li R, Wang Q, Wang X,et al. Relationship analysis of the degree of fault complexity and the water irruption rate, based on fractal theory[J]. Mine Water & the Environment,2015,36(1):1-6.

[11] Dong D L, Sun L K, Ma J H, et al. Water inrush mode and counter measures for Zhengzhou mining area[J]. Journal of Mining and Safety Engineering, 2010,27(3): 363-369.

[12] Xu D J, Peng S P, Xiang S Y, et al. A novel caving model of overburden strata movement induced by coal mining[J]. Energies,2017,10(476):1-13.

[13] Wang Y, Yang W F, Li M,et al. Risk assessment of floor water inrush in coal mines based on secondary fuzzy comprehensive evaluation[J]. International Journal of Rock Mechanics and Mining Sciences, 2012,52(6):50-55.

[14] Bukowski P. Water hazard assessment in active shafts in Upper Silesian coal basin mines[J]. Mine Water & the Environment,2011,30(4):302-311.

[15] Chai Z Y, Zhang Y T, Zhang X Y. Experimental investigations on correlation with slake durability and mineral composition of mudstone[J]. Journal of China Coal Society,2015,40(5):1188-1193.

[16] Zhang S J, Ding Y H, Zhu S Y,et al. Quantization division of structure complexity for deep extension No. 2 mining level[J]. Coal Technology, 2015,34(4):193-195.

[17] Zhao D, Shen H Y, Chen Z Y,et al. Study on damage depth of work face floor under safe water pressure of aquifer and water inrush risk prediction[J]. China Coal,2017, 43(11):106-110.

动压影响煤柱下方沿空巷道微震特征及破坏机制

李术才[1], 王雷[1], 江贝[1,2], 张晓[1], 王琦[1,2], 张皓杰[1], 刘博宏[1], 许硕[1]

(1. 山东大学 岩土与结构工程研究中心, 山东 济南, 250061;
2. 中国矿业大学(北京)深部岩土力学与地下工程国家重点实验室, 北京, 100083)

摘 要: 煤层采动过程中底板产生一定的变形和破坏, 底板破坏深度对煤柱下方巷道稳定性产生重要影响。以梁家煤矿 2200 工作面和煤柱下方 4106 巷道为研究背景, 采用高精度微震监测系统, 对 2200 工作面回采过程中底板微震活动进行连续性监测, 分析 2200 工作面回采过程中底板微震活动特征, 圈定底板破坏范围。结合 4106 巷道变形监测, 研究 2200 工作面回采过程中 4106 巷道变形破坏机制, 结果表明: (1) 2200 工作面采动过程中微震事件在煤层底板内的分布呈现非对称的特征, 底板破坏呈现下大上小的形态; (2) 在 4106 材料巷上方, 2200 工作面采动微震事件集中发生深度约 24 m, 4106 材料巷距 2200 工作面约 80 m, 远大于微震事件集中发生范围, 4106 材料巷受到 2200 工作面采动影响较小; (3) 4106 运输巷变形速率影响区域与 2200 工作面采动影响范围基本一致, 2200 工作面采空区对 4106 运输巷影响范围为 120~140 m, 4106 运输巷受到 2200 工作面采动影响较大; (4) 随着煤柱宽度的增加, 4106 运输巷变形呈现先增大后减小的变化, 最后趋于稳定, 4106 运输巷变形破坏表现明显的非对称特征和区域化分布。研究结果可为动压影响煤柱下方沿空巷道加固提供重要参考。

关键词: 沿空巷道; 动压影响; 微震监测; 破坏深度; 变形破坏机制

Microseismic characteristic and failure mechanism of gob side entry under the influence of dynamic pressure

LI Shucai[1], WANG Lei[1], JIANG Bei[1,2], ZHANG Xiao[1],
WANG Qi[1,2], ZHANG Haojie[1], LIU Bohong[1], XU Shuo[1]

(1. Research Center of Geotechnical and Structural Engineering,
Shandong University, Jinan 250061, China;
2. Key Laboratory for Geomechanics and Deep Underground Engineering, China
University of Mining & Technology(Beijing), Beijing 100083, China)

Abstract: In the process of coal seam mining, roadway floor has a certain deformation and damage, and the depth of floor damage has an important influence on the stability of roadway below the pillar. Taking the 2200 workface and 4106 roadway of Liangjia coal mine as background of the research, the high-precision

---

基金项目: 国家自然科学基金资助项目(51674154, 51474095, 51704125); 中国博士后科学基金资助项目(2017T100116, 2017T100491, 2016M590150, 2016M602144); 山东省重点研发计划资助项目(2017GGX30101, 2018GGX109001); 山东省自然科学基金资助项目(ZR2017QEE013)。

通信作者: 江贝(1985—), 女, 山东省济南市人, 讲师, 博士, 主要从事地下工程围岩控制机理与技术方面的研究工作。Tel:18253195766, E-mail: jiangbei519@163.com。

microseismic monitoring system is used to monitor the continuous seismic activity of roadway floor during the process of 2200 working face. Characteristic of seismic activity of roadway floor during the process of 2200 working face is analyzed. The failure depth of coal seam floor is determined. The deformation and failure mechanism of 4106 roadway during the process of 2200 working face is studied though by using 4106 roadway deformation monitoring. The results show that:(1) the distribution of microseismic events in the coal seam floor is asymmetrical in the process of the mining of 2200 working face, and the floor damage presents an asymmetric form;(2) the depth of coal seam floor is 24 m in the process of the mining of 2200 working face. The material roadway in mining face 4106 is 80 m from 2200 working face, This distance exceeded the failure depth of coal seam floor, then the material roadway in mining face 4106 is not affected by mining influence of 2200 working face;(3) influence distance of deformation rate in 4106 haulage roadway is basically the same as influence distance of 2200 working face. 4106 haulage roadway is affected by 2200 goaf, the influence distance is 120~140 m, 4106 haulage roadway is affected by mining influence of 2200 working face;(4) With the increase of the width of coal pillar, haulage roadway in mining face 4106 first increase and then decrease, and finally tend to be stable, deformation and failure of haulage roadway in mining face 4106 show obvious asymmetric characteristics and regional distribution. The results can provide some guidelines for gob side entry under the influence of dynamic pressure.

Keywords：gob side entry; dynamical pressure impact; MS monitoring; damage depth; failure mechanism

  随着经济的发展,能源的需求越来越多,煤炭资源大量的开采,造成矿井开采深度不断增加,开采条件越来越复杂,煤炭资源减少,出现工作面接续紧张,严重影响矿井的正常生产[1-3]。当上层工作面开采时,在下层工作面进行巷道布置,协调煤层群安全高效开采,能够解决回采工作面接续紧张的问题。上层工作面回采过程中,将引起围岩应力重新分布,造成采煤工作面前方和侧向应力集中,而集中应力将向煤层底板围岩传递,造成煤层底板应力先后经历升高、降低和恢复3个不同的阶段,底板变形出现压缩、膨胀和再压缩的变化过程,导致工作面底板岩体变形破坏,形成一定深度的底板破坏带[4-6]。

  长期以来,国内外许多学者对采动影响煤柱下方巷道破坏特征进行了广泛的研究,取得了很多重要的科研成果。蒋金泉等[7]针对跨采巷道的矿压特点,采用模糊聚类分析法研究了跨采巷道稳定性分类模式。李学华等[8]对高水平应力跨采巷道围岩的稳定性进行了模拟研究,得出巷道肩角处是发生破坏的弱结构部位,即影响围岩稳定性的关键部位,提出了在关键部位采取非均匀支护技术。李桂臣等[9]以多次跨采巷道为研究对象,总结了该类巷道4种典型的破坏特征,研究了多次跨采巷道的变形破坏机理,提出了该类巷道围岩强化控制技术。谢文兵等[10]分析了工作面开采引起的围岩应力演化过程及特点,讨论了工作面开采影响巷道的围岩位移特征,研究了围岩应力和跨采巷道位移之间的关系。

  采动底板变形破坏规律的研究,主要是通过现场应变(感应)法测试、数值模拟、相似模拟和震波CT探测等技术确定底板变形破坏规律。张蕊等[11]采用现场应变实测和数值模拟相结合的方法,确定了大采深厚煤层底板破坏深度,揭示了矿山压力在采动煤层底板中的传播规律。谢广祥等[12]研究了采场底板围岩三维应力场特征及演化规律,提出了工作面推进不同距离时,采场底板围岩应力壳力学特征及几何形态。张平松等[13]利用震波CT探测技术获得煤层采动过程中底板破坏的动态发育规律及特征。

  本文在上述研究成果的基础上,以典型三软煤层龙口矿区梁家煤矿上工作面2200和下工作面4106为工程背景,分析2200工作面回采过程中底板软弱围岩微震事件特征,圈定2200工作面底板破坏范围,研究2200工作面回采时4106材料巷和运输巷变形破坏机制。

1 工程地质条件

1.1 工程概况

梁家煤矿位于山东省龙口市,是我国最大的海滨煤矿,设计生产能力300万t/a,为典型三软地层矿区。矿井主要含煤地层为古近系李家崖组,主采煤$_1$层、煤$_2$层和煤$_4$层。

4106工作面位于2200工作面下方,垂直距离80 m。4106材料巷位于2200工作面内部,采用内错式布置,内错距离80 m,距2200运输巷平距150 m,4106材料巷为实体煤巷道。4106运输巷位于2200工作面外部,采用外错式布置,外错距离33~50 m,4106运输巷为沿空巷道,煤柱宽度为6~37 m,如图1所示。

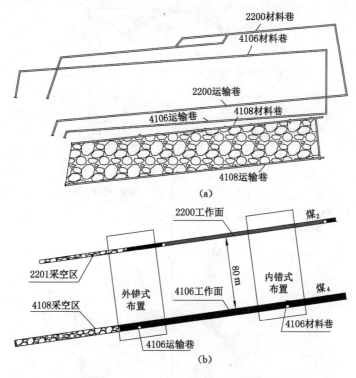

图1 4106和2200工作面巷道布置图

Fig.1 Roadway layout in mining face 4106 and 2200

(a) 平面图;(b) 剖面图

2200工作面紧邻2201采空区,2200工作面倾斜长度200 m,走向长度960 m;2200工作面材料巷为实体煤巷道,2200工作面运输巷为沿空巷道,煤柱宽度为30~60 m。

煤$_4$走向近东西向,倾向南倾,倾角3°~8°,煤层平均厚度11.78 m。煤$_2$走向近东西向,倾向南倾,倾角8°~12°,煤层平均厚度4.04 m。该区域内断层为主要构造形式,断层均为正断层,其走向整体为近东西向,本区派生小断层发育。煤$_4$与煤$_2$之间岩层依次为泥砂互层、砂岩和油$_3$,为典型三软不稳定煤层。

4106材料巷和运输巷支护方案均采用"锚网喷＋锚梁＋锚索"支护形式。锚杆采用MSGLD-335/18×2250螺纹钢锚杆,间、排距为650 mm×650 mm,底脚锚杆在巷道两帮底板以上150 mm处,底脚锚杆倾角15°,间距1 000 mm,预紧力30 kN;锚索长5 m,直径22 mm,间、排距为2 000 mm×1 300 mm;混凝土喷层采用C20混凝土,厚度120 mm;锚梁排距1 300 mm,采用U25型钢加工而成,具体支护参数详见图2。

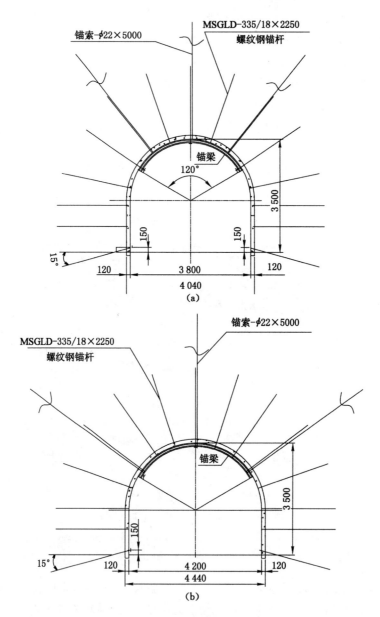

图2 4106工作面巷道支护方案设计断面图

Fig. 2 Cross-section design diagram of supporting scheme of roadway in mining face 4106

(a) 4106材料巷;(b) 4106运输巷

1.2 采煤方法

煤$_4$为平均厚度 11.78 m 的特厚煤层,煤$_2$为平均厚度 4.04 m 的厚煤层,煤$_4$与煤$_2$的平均层间距为 80 m,煤层普氏系数 $f=1.5$。梁家煤矿煤$_2$和煤$_4$开采顺序:采用下行式开采顺序,先开采上层煤$_2$的 2200 工作面,后开采下层煤$_4$的 4106 工作面,均采用综放开采。

2 微震监测系统构建

2.1 微震监测系统

项目监测设备采用加拿大 ESG 公司生产的高精度矿用微震监测系统,系统主要包括传感器、Paladin 数字信号采集系统、光缆数据通信系统、Hyperion 数字信号处理系统。传感器采用不锈钢材质,具有抗震和防水性能,能够适应井下恶劣的工作环境。Paladin 数字信号采集系统提供 24 位模数转换,采集频率 50 Hz~10 kHz。Hyperion 数字信号处理系统包括 HNAS 软件(微震信号实时采集与记录)、WaveVis 软件(波形处理及微震事件定位)、SeisVis 软件(微震事件的三维可视化),能够实现微震事件的定位和分析,煤矿微震监测系统网络拓扑如图 3 所示。传感器通过锚杆接收围岩破裂的振动信号,并将其转换为电信号,电信号通过电缆传输到 Paladin,Paladin 内的光电转换器将电信号转换为光信号,利用光纤传送到煤矿井下数据采集分站,采集分站内的光电转换器将光信号转换为电信号,通过煤矿工业环网传送到井上 Hyperion 接收系统,完成微震事件的记录[14-15]。

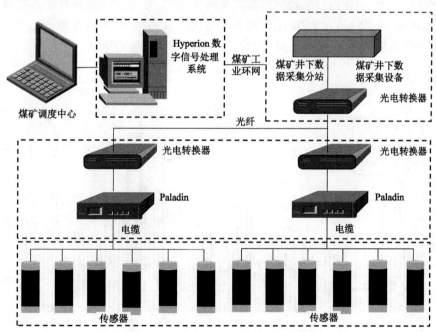

图 3 煤矿微震监测系统网络拓扑图

Fig. 3 Network topology of microseismic monitoring system

2.2 微震监测方案

根据 2200 工作面和 4106 工作面空间位置关系,在 4106 材料巷和运输巷布置 6 个传感器,随着2200工作面的回采,传感器依次向2200工作面回采的方向移动,保证2200工作面在微震监测范围内,实现2200工作面回采期间微震信号的连续监测,如图4所示。

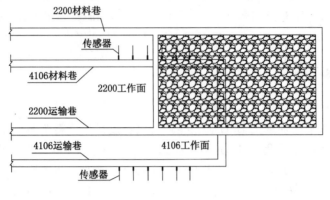

图 4 传感器布置方案

Fig. 4　Layouts of sensors

3　采动影响下围岩微震活动规律

3.1　微震事件的时间分布规律

梁家煤矿2200工作面回采期间的监测系统自2016年03月01日开始正常监测运行,截至2016年05月31日,在2200工作面回采期间经过识别、除噪最终得到微震事件1 048个。微震事件活动率和累积能量损失与时间关系分布如图5所示,2016年03月01日,

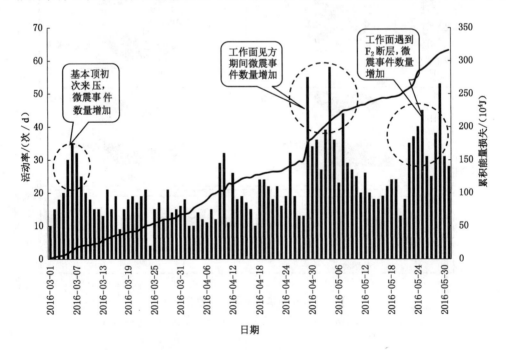

图 5　微震事件的活动率和累积能量损失与时间关系分布图

Fig. 5　Relationship chart between accumulative energy loss of microseismic events activity rate and time

2200工作面开始回采,微震事件较少,平均活动率10次/d。随着工作面推进,微震事件增多。2016年03月05日~2016年03月08日,2200工作面初次来压,微震事件增多,微震事件最大活动率为35次/d。2016年03月09日~2016年04月28日,2200工作面正常回采,微震活动率基本维持在20次/d,累积能量损失随着时间呈线性增加。2016年04月29日~2016年05月08日,微震活动频繁,微震事件出现跳跃式增长,微震事件活动率最大值为55次/d,累积损失能量出现突增现象。根据现场初步分析,发生这种现象的重要原因是2200工作面采空区初次见方,围岩应力和能量聚集增大,围岩变形破裂增多,从而产生大量微震事件。2016年05月22日~2016年05月31日微震事件活动率和累积能量损失出现异常增加,微震事件活动率最大值为53次/d,结合2200工作面地质情况分析,工作面揭露F_2断层,造成断层区域出现应力集中,且断层围岩自身强度低,断层附近围岩破裂严重,产生大量岩体微破裂,造成微震事件增多。

3.2 微震事件的空间分布规律

微震的空间分布规律主要是研究一段时间内围岩微破裂集中分布。2200工作面回采期间微震事件空间分布如图6至图8所示,图中球体代表微震事件,大小代表能量,球体越大,微震事件能量越大,不同颜色表示微震事件矩震级。从图6和图7可以看出:微震事件分为三个阶段,Ⅰ阶段:2200工作面开始回采至工作面初次来压阶段。根据微震事件分布范围,该区域微震事件相对较少,2200工作面初次来压时微震事件增加。由于2200工作面刚开始回采,工作面推进速度较慢,采空区面积较小,微震事件密度较低。Ⅱ阶段:2200工作面初次来压至2200工作面采空区初次见方阶段。2200工作面进入正常开采阶段,随着工作面的推进,采空区面积逐渐增大,底板膨胀破坏产生裂隙,而直接顶和基本顶岩层周期性断裂,上覆岩层断裂下沉后,应力传递到底板岩层中,底板受到压缩、膨胀和再次压缩后,微震事件增多,密度增大,但底板大能量微震事件较少。Ⅲ阶段:2200工作面采空区初次见方阶段。该阶段2200工作面推进距离与工作面长度在平面上的投影近似正方形,围岩积累

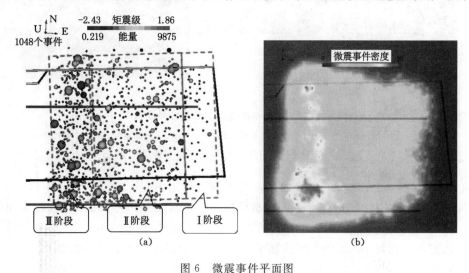

图6 微震事件平面图

Fig.6 Plane of microseismic events

(a)微震事件空间分布图;(b)微震事件密度图

的能量达到最大值[16],随工作面的推进,围岩破裂增多,积聚的能量释放,与图5中工作面采空区见方阶段累积能量损失突增一致。相比Ⅰ阶段和Ⅱ阶段,2200工作面采空区初次见方阶段微震事件数量和密度最高,能量最大,底板破坏带范围明显增大。

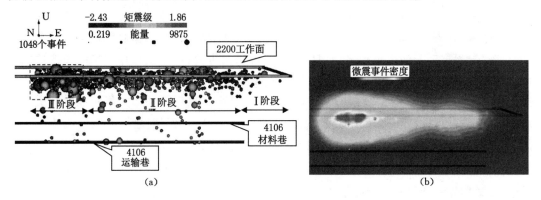

图7　沿2200工作面走向的微震事件投影图

Fig. 7　Projected of microseismic events along the advanced directing of 2200 workface

(a)微震事件空间分布图;(b)微震事件密度图

图8为沿2200工作面倾向的微震事件剖面投影图。从图8可以得到,2200工作面采动过程中微震破裂事件在煤层底板内的分布呈现非对称的特征,工作面运输巷附近区域底板内的微震破裂事件明显多于工作面材料巷附近区域底板内的微震破裂事件,工作面运输巷附近的底板相比工作面材料巷附近的底板不仅破坏深度更深,而且破坏范围更大,工作面底板破坏带呈现下大上小的非对称形态。

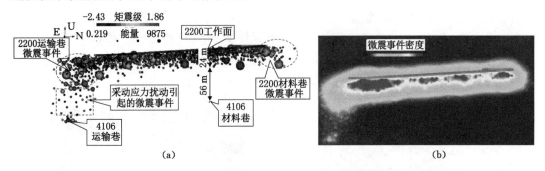

图8　沿2200工作面倾向的微震事件投影图

Fig. 8　Projected of microseismic events along the tilted directing of 2200 workface

(a)微震事件空间分布图;(b)微震事件密度图

在4106材料巷上方,2200工作面采动时微震事件集中发生深度约24 m,4106材料巷距2200工作面距离为80 m,远大于微震事件集中发生范围,未受到2200工作面采动影响。

由于4106运输巷为沿4108工作面的沿空巷道,2200工作面临近2201工作面采空区,2200工作面的采动对于4108和2201覆岩产生扰动,同时由于F_{27}、F_{2-1}断层、岩层倾斜等地质因素影响,导致在2200工作面采动过程中4106运输巷附近围岩产生较大位移,发生破裂现象,出现微震监测事件。

4 煤柱下方巷道变形破坏机理

4.1 巷道表面位移监测

巷道表面位移变化能够反映巷道稳定情况,当巷道围岩应力发生变化,围岩应力将重新调整,从而达到平衡状态,而在调整过程中,围岩将产生形变,造成巷道顶板下沉、帮部内移和底鼓[17-18]。在 4106 材料巷和运输巷各布设 24 个监测点,监测巷道两帮和顶底的收敛变化,图 9 为掘进期间和 2200 工作面回采期间 4106 材料巷各测点总变形量柱状图。由图 9 可得到,4106 材料巷顶底板变形量为 120~800 mm,平均变形量为 413 mm,两帮变形量为 18~202 mm,平均变形量为 73 mm,顶底板变形量明显大于两帮的变形量,顶底板变形以底鼓为主,顶底板变形量最大值 800 mm,两帮变形量最大值 202 mm,主要原因为 4106 材料巷底板为油页岩,易吸水膨胀,在该监测点附近进行地质勘探,巷道底板围岩遇到勘探水膨胀,变形量增大,从而造成该监测点顶底板变形量较大,2200 工作面初次见方时 4106 材料巷如图 10 所示。

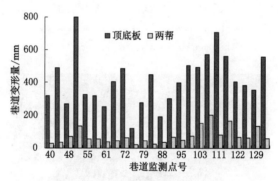

图 9 4106 材料巷变形量柱状图

Fig. 9 Histogram of displacement change of material roadway in mining face 4106

图 10 4106 材料巷变形图

Fig. 10 Deformation of material roadway in mining face 4106

图 11 为掘进期间和 2200 工作面回采期间 4106 运输巷各测点累计变形量柱状图。由图 11 可以得到,巷道顶底板变形量为 70~2 300 mm,平均变形量为 1 055 mm,两帮收敛变

形量在 22～1 467 mm,平均为 547 mm,巷道顶底板变形量大于两帮变形量,在 66 号监测点,顶底板变形量最大值 2 300 mm,主要原因为 66 号监测点位于 2200 工作面初次见方区域内,采场压力大,矿压显现剧烈,2200 工作面初次见方期间 4106 运输巷如图 12 所示。4106 工作面巷道变形量统计如表 1 所示。

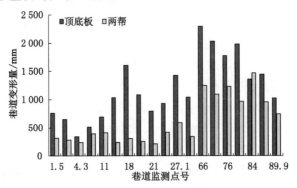

图 11　4106 运输巷变形量柱状图

Fig. 11　Histogram of displacement change of haulage roadway in mining face 4106

图 12　4106 运输巷变形破坏图

Fig. 12　Deformation and failure of haulage roadway in mining face 4106

表 1　　　　　　　　　　　4106 工作面巷道变形量统计

Table 1　Statistics of displacement change of material roadway in mining face 4106

监测部位	变形量/mm	4106 材料巷	4106 运输巷	增长率/%
顶底板	最大变形量	800	2 300	187
	平均变形量	413	1 055	155
两帮	最大变形量	202	1 467	626
	平均变形量	73	547	649

注:增长率＝[(4106 运输巷变形量－4106 材料巷变形量)/4106 材料巷变形量]×100%。

对比分析 4106 材料巷与 4106 运输巷的变形量,4106 运输巷变形量大于 4106 材料巷。

4106 材料巷和运输巷顶底板平均变形量分别为 413 mm、1 055 mm,4106 运输巷顶底板平均变形量比 4106 材料巷增加了 155%;4106 材料巷和运输巷两帮平均变形量分别为 73 mm、547 mm,4106 运输巷两帮平均变形量比 4106 材料巷增加了 649%。以上分析结果表明,4106 运输巷两帮变形量增长率比 4106 运输巷顶底板大,主要原因是 4106 运输巷为沿空巷道,煤柱处于塑性状态,煤柱破碎,围岩自身承载能力弱,巷道变形严重。

根据 2200 工作面与 4106 巷道空间位置关系,4106 材料巷位于 2200 工作面内部,采用内错式布置,4106 运输巷位于 2200 工作面外部,采用外错式布置,为研究 2200 工作面回采对两条不同布置方式巷道影响程度,在 4106 材料巷选取 125# 监测点,对应相同距离在 4106 运输巷选取 76# 监测点,分别统计掘进期间巷道变形量和 2200 回采期间的变形量,如表 2 和图 13 所示。

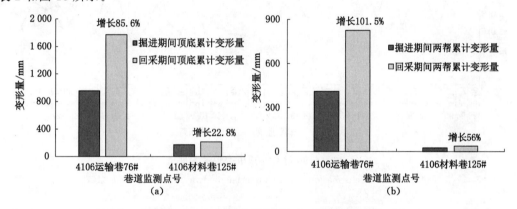

图 13　2200 工作面回采前后 4106 巷道变形量

Fig. 13　Roadway deformation in mining face 4106 before and after mining of 2200 face

(a) 巷道顶底板变形量；(b) 巷道两帮变形量

由表 2 和图 13 可知,掘进期间 4106 材料巷顶底板和两帮变形量分别为 171 mm、25 mm,2200 回采期间 4106 材料巷顶底板和两帮变形量分别为 210 mm、39 mm,分别增加了 22.8% 和 56%。2200 回采期间与掘进期间相比,4106 运输巷顶底板和两帮变形量分别由 409 mm、824 mm 增加至 955 mm、1 773 mm,分别增加了 85.6% 和 101.5%。由以上数据分析可知 2200 工作面回采对外错式布置的 4106 运输巷稳定性影响较大,对内错式布置的 4106 材料巷影响较小,因此应加强 4106 运输巷支护强度。

表 2　2200 工作面回采前后 4106 巷道变形量统计

Table 2　Statistics of roadway deformation in mining face 4106 before and after mining of 2200 face

监测部位	4106 材料巷			4106 运输巷		
	掘进期间累计变形量/mm	2200 回采后累计变形量/mm	增长率/%	掘进期间累计变形量/mm	2200 回采后累计变形量/mm	增长率/%
顶底板	171	210	22.8	955	1 773	85.6
两帮	25	39	56	409	824	101.5

注:增长率=[(2200 回采后累计变形量－掘进期间累计变形量)/掘进期间累计变形量]×100%。

4.2 巷道顶板变形速率监测

顶板变形速率能够反映矿压活动特征和规律,是评价围岩稳定性的重要指标。在4106材料巷和运输巷分别安装顶板动态监测仪,监测2200工作面采动前后巷道顶板变形速率的变化规律。

图14为4106材料巷120#监测点顶板变形速率曲线,由现场监测数据可知,顶板变形速率为0.6~1.2 mm/d,在2200工作面回采前后,4106材料巷顶板变形速率未发生变化,2200工作面回采对4106材料巷未产生影响。

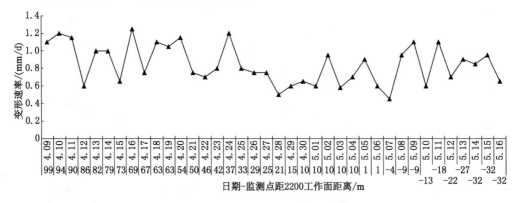

图14 4106材料巷顶板变形速率曲线

Fig. 14 Roof deformation rate of material roadway in mining face 4106

4106运输巷分别在64.5#和74#监测点安装顶板动态监测仪。图15(a)为4106运输巷64.5#监测点顶板变形速率曲线,由现场监测数据可知,当该测点距2200工作面120 m,4106运输巷顶板变形速率开始增加,由5 mm/d增加到28 mm/d,在2200工作面回采过后10 m,顶板变形速率达到最大值。2200工作面停采期间,4106运输巷变形速率显著降低,2200工作面恢复生产时,4106运输巷变形速率增加。由巷道变形速率分析,4106运输巷受2200工作面采动影响显著,且4106运输巷顶板变形速率影响范围与2200工作面超前支承压力范围一致。

图15(b)为4106运输巷74#监测点顶板变形速率曲线,由现场监测数据可知,2200工作面回采到该监测点前,4106运输巷顶板变形速率变化规律与64.5#变化规律一致,且在2200工作面回采到该监测点后,4106运输巷变形速率明显受到2200工作面停采的影响。2200工作面回采后围岩应力重新调整,顶板变形速率减小,最后趋于稳定,2200工作面采空区对4106运输巷影响范围为120~140 m。

4.3 不同煤柱宽度巷道变化规律

4106运输巷为沿空巷道,煤柱宽度为6~37 m,为研究煤柱对4106运输巷稳定性的影响,在煤柱宽度6 m、12 m、18 m、24 m和30 m的4106运输巷布置监测点,监测不同煤柱宽度下4106运输巷变形规律,如图16所示。

由图16分析可知,随着煤柱宽度的增加,4106运输巷变形呈先增大后减小的变化趋势,最后趋于稳定,当煤柱宽度为12 m时,4106运输巷变形量达到最大值。结合4106运输巷现场变形破坏情况(见图17),将4106运输巷划分为两个区域,Ⅰ区域:煤柱宽度6~

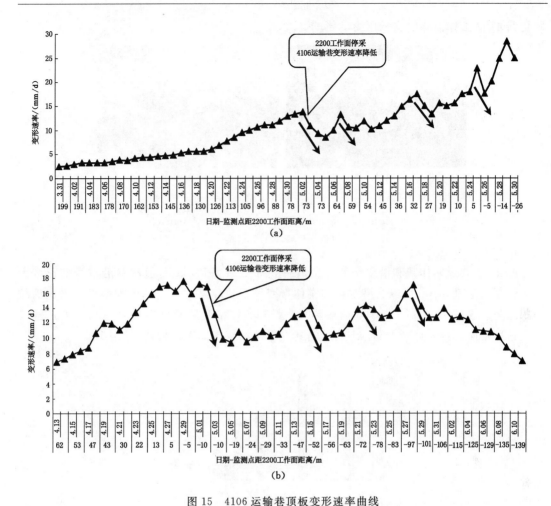

图 15 4106运输巷顶板变形速率曲线

Fig. 15 Roof deformation rate of haulage roadway in mining face 4106

(a) 64.5#监测点;(b) 74#监测点

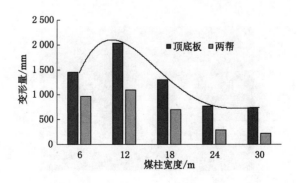

图 16 不同煤柱宽度巷道变形量柱状图

Fig. 16 Histogram of roadway displacement change of different pillar width

24 m,该区域巷道变形大,支护构件失效严重,巷道稳定性差,且明显受到4108工作面侧向支承压力的影响;Ⅱ区域:煤柱宽度24 m以上时,该区域巷道变形较小,稳定性较好,未明显受到4108工作面侧向支承压力的影响。

图17　不同区域巷道变形破坏图

Fig. 17　Deformation and failure of roadway in different areas

(a) Ⅰ区域；(b) Ⅱ区域

图18为巷道实体帮和沿空帮变形量，由图18可以得到，4106运输巷沿空帮变形量大于实体帮变形量，4106运输巷沿空帮和实体帮变形量随着煤柱宽度的增加呈现先增大后减小的趋势。煤柱为12 m时，4106运输巷沿空帮和实体帮变形量达到最大值。4106运输巷沿空帮变形破坏比实体帮严重，4106运输巷变形破坏呈现非对称特征，见图19。

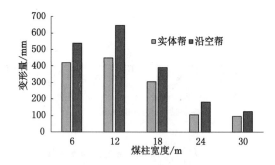

图18　巷道帮部变形量柱状图

Fig. 18　Histogram of displacement change of roadway side

图19　巷道帮部变形破坏图

Fig. 19　Deformation and failure of roadway side

4.4　动压影响煤柱下方巷道变形机制分析

根据4106材料巷与运输巷变化破坏规律和高精度微震活动特征可知，动压影响下煤柱下方沿空巷道变形机制在于：

(1) 2200 工作面回采后破坏了岩体中原岩应力平衡状态,引起围岩应力重新分布。2200 工作面侧向支承压力造成围岩应力集中,4108 和 2201 采空区覆岩产生扰动,围岩中存在断层等地质构造,4106 运输巷为外错布置受到影响,4106 材料巷为内错布置未受到影响。

(2) 4106 运输巷为沿空巷道,与 4108 工作面存在煤柱,受到 4108 工作面侧向支承压力影响,煤柱松软破碎,且围岩自身承载能力弱,4108 采空区应力对 4106 运输巷稳定性产生扰动,4106 材料巷为实体煤巷道,未受到煤柱影响。

综上所述,通过分析 4106 运输巷变形规律,4106 运输巷变形破坏表现明显的非对称特征和区域化分布,4106 运输巷受到 4108 工作面侧向支承压力的影响,而 2200 工作面回采前后,4106 运输巷变形速率产生明显变化,4106 运输巷受到 2200 工作面回采的影响。因此,2200 工作面回采时,引发 4108 上覆岩层和采空区扰动,围岩应力的重新分布,同时受 4106 运输巷煤柱宽度的影响,造成 4106 运输巷产生大变形破坏。

5 结 论

(1) 2200 工作面采动过程中微震破裂事件在煤层底板内的分布呈现非对称的特征,底板破坏呈现下大上小的形态,2200 工作面采动对 4108 和 2201 工作面覆岩产生扰动,同时由于断层、岩层倾斜等地质因素影响,导致 4106 运输巷附近围岩产生微震事件。

(2) 在 4106 材料巷上方,2200 工作面采动时微震事件集中发生深度约 24 m,4106 材料巷距 2200 工作面距离为 80 m,远大于微震事件集中发生范围,4106 材料巷附近未产生微震事件。

(3) 2200 工作面采动时 4106 运输巷变形速率增加,巷道变形速率影响区域与 2200 工作面采动影响范围基本一致,2200 工作面采空区对 4106 运输巷影响范围为 120~140 m。

(4) 随着煤柱宽度的增加,4106 运输巷变形呈现先增大后减小的变化趋势,最后趋于稳定,4106 运输巷变形破坏表现为明显的非对称特征和区域化分布。

参考文献

[1] 何满潮,景海河,孙晓明. 软岩工程力学[M]. 北京:科学出版社,2002.
HE Manchao, JING Haihe, SUN Xiaoming. Soft rock engineering mechanics[M]. Beijing:Science Press,2002.

[2] PAN Rui,WANG Qi,JIANG Bei,et al. Failure of bolt support and experimental study on the parameters of bolt-grouting for supporting the roadways in deep coal seam[J]. Engineering Failure Analysis,2017(80):218-233.

[3] WANG Qi,RAN Rui,JIANG Bei,et al. Study on failure mechanism of roadway with soft rock in deep coal mine and confined concrete support system[J]. Engineering Failure Analysis,2017(81):155-177.

[4] 李白英. 预防矿井底板突水的"下三带"理论及其发展与应用[J]. 山东矿业学院学报,1999,18(4):11-18.
LI Baiying. "Down Three Zones" in the prediction of water inrush from coalbed floor aquifer theory, development and application[J]. Journal of Shandong Institute of

Mining & Technology,1999,18(4):11-18.

[5] 雷文杰,汪国华,薛晓晓.有限元强度折减法在煤层底板破坏中的应用[J].岩土力学,2011,32(1):299-303.

LEI Wenjie, WANG Guohua, XUE Xiaoxiao. Application of finite element strength reduction method to destruction in coal seam floor[J]. Rock and Soil Mechanics,2011,32(1): 299-303.

[6] 施龙青,韩进.开采煤层底板"四带"划分理论与实践[J].中国矿业大学学报,2005,34(1):16-23.

SHI Longqing, HAN Jin. Theory and practice of dividing coal mining area floor into four-zone[J]. Journal of China university of Mining & Technology,2005,34(1): 16-23.

[7] 蒋金泉,冯增强,韩继胜.跨采巷道围岩结构稳定性分类与支护参数决策[J].岩石力学与工程学报,1999,18(1):81-85.

JIANG Jinquan, FENG Zengqiang, HAN Jisheng. Classification of structural stability of surrounding rocks of roadway affected by overhead mining and decision making of support parameters[J]. Chinese Journal of Rock Mechanics and Engineering,1999,18(1):81-85.

[8] 李学华,姚强岭,张农,等.高水平应力跨采巷道围岩稳定性模拟研究[J].采矿与安全工程学报,2008,25(4):420-425.

LI Xuehua, YAO Qiangling, ZHANG Nong, et al. Numerical simulation of stability of surrounding rock in high horizontal stress roadway under overhead mining[J]. Journal of Mining & Safety Engineering,2008,25(4):420-425.

[9] 李桂臣,马振乾,张农,等.淮北矿区多次跨采巷道破坏特征及控制对策研究[J].采矿与安全工程学报,2013,30(2):181-187.

LI Guichen, MA Zhenqian, ZHANG Nong, et al. Research on failure characteristics and control measures of roadways affected by multiple overhead mining in Huaibei mining area[J]. Journal of Mining & Safety Engineering,2013,30(2):181-187.

[10] 谢文兵,史振凡,陈晓祥,等.工作面开采对底板岩巷稳定性的影响[J].中国矿业大学学报,2004,33(1):82-85.

XIE Wenbing, SHI Zhengfan, CHEN Xiaoxiang, et al. Stability analysis of roadway surrounding rock induced by overhead mining[J]. Journal of China university of Mining & Technology,2004,33(1):82-85.

[11] 张蕊,姜振泉,岳尊彩,等.采动条件下厚煤层底板破坏规律动态监测及数值模拟研究[J].采矿与安全工程学报,2012,29(5):625-630.

ZHANG Rui, JIANG Zhengquan, YUE Zuncai, et al. In-Situ dynamic observation and numerical analysis of thick coal seam floor's failure law under the mining[J]. Journal of Mining & Safety Engineering,2012,29(5):625-630.

[12] 谢广祥,李家卓,王磊,等.采场底板围岩应力壳力学特征及时空演化[J].煤炭学报,2018,43(1):52-61.

XIE Guangxiang, LI Jiazhuo, WANG Lei, et al. Mechanical characteristics and time and space evolvement of stress shell in stope floor stratum[J]. Journal of China Coal Society, 2018, 43(1):52-61.

[13] 张平松,吴基文,刘盛东.煤层采动底板破坏规律动态观测研究[J].岩石力学与工程,2006,25(增1):3009-3013.
ZHANG Pingsong, WU Jiwen, LIU Shengdong. Study on dynamic observation of coal seam floor's failure law[J]. Chinese Journal of Rock Mechanics and Engineering, 2006, 25(S1):3009-3013.

[14] 徐奴文,李术才,戴峰,等.岩质边坡微震活动特征及其施工相应分析[J].岩石力学与工程学报,2015,34(5):968-978.
XU Nuwen, LI Shucai, DAI Feng, et al. Analysis on characteristics of microseismic activity and its response to construction at rock slopes[J]. Chinese Journal of Rock Mechanics and Engineering, 2015, 34(5):968-978.

[15] 陈炳瑞,冯夏庭,曾雄辉,等.深埋隧洞TBM掘进微震实时监测与特征分析[J].岩石力学与工程学报,2011,30(2):275-283.
CHEN Bingrui, FENG Xiating, ZENG Xionghui, et al. Real-time microseismic monitoring and its characteristic analysis during TBM tunneling in deep-buried tunnel[J]. Chinese Journal of Rock Mechanics and Engineering, 2011, 30(2):275-283.

[16] 夏永学,蓝航,毛德兵,等.基于微震监测的超前支承压力分布特征研究[J].中国矿业大学学报,2011,40(6):868-873.
XIA Yongxue, LAN Hang, MAO Debing, et al. Study of the lead abutment pressure distribution base on microseismic monitoring[J]. Journal of China university of Mining & Technology, 2011, 40(6):868-873.

[17] 刘泉声,张伟,卢兴利,等.断层破碎带大断面巷道的安全监控与稳定性分析[J].岩石力学与工程学报,2010,29(10):1954-1962.
LIU Quansheng, ZHANG Wei, LU Xingli, et al. Safety monitoring and stability analysis of large-scale roadway in fault fracture zone[J]. Chinese Journal of Rock Mechanics and Engineering, 2010, 29(10):1954-1962.

[18] 孟庆彬,韩立军,乔卫国,等.应变软化与扩容特性极弱胶结围岩弹塑性分析[J].中国矿业大学学报,2018,47(4):760-767.
MENG Qingbin, HAN Lijun, QIAO Weiguo, et al. Elastic-plastic analysis of the very weakly cemented surrounding rock considering characteristics of strain softening and expansion[J]. Journal of China university of Mining & Technology, 2018, 47(4):760-767.

气体吸附诱发煤体劣化的试验研究与分析

李清川[1]，王汉鹏[1]，袁亮[1,2]，薛俊华[3]，张冰[1]，陈本良[3]

(1. 山东大学 岩土与结构工程研究中心，山东 济南，250061；
2. 安徽理工大学 能源与安全学院，安徽 淮南，232001；
3. 淮南矿业(集团)有限责任公司 深部煤炭开采与环境保护国家重点实验室，安徽 淮南，232002)

摘　要：为定量研究气体吸附对煤体的损伤劣化作用规律，探索其劣化机制，利用可视化恒容气固耦合试验系统对相同强度的型煤标准试件充入不同吸附气体(He、N_2、CH_4、CO_2)，充分吸附后进行轴向加载试验，对比分析了气体性质和吸附压力对煤体强度、体积扩容和裂隙发育的影响规律，并采用 MATLAB 软件编程提取试验过程加载图像，基于分形理论得到了相同应力阶段煤体裂隙发育的分形维数。试验结果表明：充入无吸附性的 He 对煤体无劣化作用，充入吸附性的 CO_2 显著降低了煤体强度，其中吸附压力为 2 MPa 的煤体强度降低了 48.27%，煤体劣化率与吸附平衡压力呈对数函数关系；处于相同应力阶段的煤体裂隙分形维数 D 随着吸附量的提高而增大，幅值在 1.32~1.47 之间变化，同时煤体表面裂纹密度也不断增加，对应 2 MPa 气压时裂隙发育度 M 是不加气时的 2.28 倍并呈现"鱼鳞片"状膨胀和脱落趋势。试验结果从多角度揭示了吸附瓦斯对煤体的劣化作用特征，建立了吸附压力对煤体强度的损伤劣化机制模型，为定量研究含瓦斯煤的力学特性提供了科学依据。

关键词：含瓦斯煤；气体吸附；劣化试验；损伤扩容；分形维数

Experimental study and analysis for gas adsorption-induced coal mechanical property alterations

LI Qingchuan[1], WANG Hanpeng[1], YUAN Liang[1,2],
XUE Junhua[3], ZHANG Bing[1], CHEN Benliang[3]

(1. Research Centre of Geotechnical and Structural Engineering, Shandong
University, Jinan 250061, China; 2. Anhui University of Science and Technology,
Huainan 232001, China; 3. Huainan mining (Group) Co. Ltd., National Engineering
Research Institute of Coal Mining, Huainan 232002, China)

Abstract: In order to study the law of degradation and damage of coal by gas adsorption quantitatively and explore its mechanism, a visual constant volume gas-solid coupling test system was used to fill the standard sample of coal with the same strength with different adsorptive gases (He, N_2, CH_4 and CO_2), then the axial loading test was carried out after full adsorption, the effects of gas properties and adsorption pressure on the strength, volume expansion and fracture development of coal were compared and analyzed. MATLAB

基金项目：国家重大科研仪器设备研制项目(51427804)；山东省自然科学基金面上项目(ZR2017MEE023)。
通信作者：王汉鹏(1978—)，男，山东省济南市人，教授，博士生导师，从事岩土工程试验与灾变机理及防控方面的研究。Tel:0531-88399182，E-mail:pcwli@163.com。

numerical software was used to extract test process loading images, the fractal dimension of coal body crack development in the same stress stage was obtained based on fractal theory. The experimental results show that: The impregnation of He with no adsorption did not degrade the coal, the coal body strength was significantly reduced by adding adsorptive CO_2, The peak strength decreased by 48.27% after CO_2 adsorption at 2 MPa pressure, the deteriorate rate was logarithmically increased with adsorption equilibrium pressure. The fractal dimension D of coal at the same stress stage increased with the increase of adsorption amount, its values ranged between 1.32 and 1.47, the surface crack density of coal was also increasing, the fracture development degree M at 2 MPa pressure is 2.28 times as high as without gas, and there was a tendency of "fish scale" expansion and abscission. Experiment results from multiple perspectives revealed the degradation effect of characteristics of gas adsorption on coal, the damage deterioration mechanism model of coal body strength by adsorption pressure was established, which provided a scientific basis for quantitative evaluation of mechanical properties of coal containing gas.

Keywords: coal containing gas; gas adsorption; coal degradation; expansion and damage; fractal dimension

煤炭是我国的主体能源[1]。随着矿井开采深度不断增加,开采条件日趋复杂,高瓦斯煤层的开采无法避免[2],导致煤与瓦斯突出发生的强度及造成的伤亡比重不断增长,煤与瓦斯突出的预测与防治工作十分严峻[3,4]。含瓦斯煤作为一种复杂的混合介质,其吸附耦合状态直接影响了煤岩体的力学特性。因此,研究气体吸附对煤岩体的力学作用机制是探索煤与瓦斯突出机理的基础与前提条件。

近年来,国内外专家学者采用理论分析、室内试验、数值仿真等手段对含瓦斯煤的物理力学性质进行了深入研究,取得了丰硕成果:P. G. Ranjith、D. R. Viete[5,6]等人通过试验分析指出,气体吸附对煤岩的影响可以通过宏观强度和弹性模量来表征;J. W. Larsen[7]认为,任意可以被煤样吸附和溶解的流体,如CO_2、CH_4和N_2等,在其吸附过程均具有降低煤体自由能,释放煤颗粒膨胀诱发煤体应变的力学特性。何学秋[8]通过开展含瓦斯煤岩力学特征实验研究阐述了孔隙吸附瓦斯对煤岩的破坏作用过程,并运用表面物理化学原理解释了孔隙瓦斯对煤岩的"蚀损"机理;尹光志[9]等从内时理论出发,利用连续介质不可逆热力学的基本原理推导出了含瓦斯煤岩的内时损伤本构方程;程远平[10]等考虑了有效应力和瓦斯吸附/解吸变形等因素的、以应变为变量,研究煤体的卸荷损伤性质;刘力源[11]等基于损伤力学理论与有效应力原理研究了煤岩吸附瓦斯的特征;黄达[12]等利用PFC颗粒流程序探讨了初始单轴静态压缩的细观损伤程度对单轴动态压缩下单裂隙岩样力学性质的影响规律。

此外,李祥春[13]、梁冰[14]、聂百胜[15]、王佑安[16]等也开展了大量试验,研究验证了吸附性气体对煤岩膨胀变形的不可逆作用,但仍缺少定量描述气体吸附导致煤体损伤劣化的试验方法,本文利用自主研发的可视化恒容气固耦合试验系统[17],采用强度可调的型煤相似材料[18],开展了气体吸附对煤体的损伤劣化试验研究,通过试验过程中对试件轴向-环向变形、煤体裂隙发育与煤体峰值强度弱化规律等特征参数的可视化实时监测,得到了煤体吸附量与煤体劣化率、体积扩容、裂隙分形维数之间的影响规律,为深入探索气体吸附诱发煤体损伤劣化作用机制提供了科学手段和理论基础。

1 试验系统与测试方案

1.1 可视化恒容气固耦合试验系统

为定量研究含瓦斯煤岩体在气固耦合与应力加载过程中的失稳破坏机制与损伤劣化作用机理,配合伺服压力机研发了可视化恒容气固耦合试验系统,系统工作原理如图1所示。通过设置恒容结构,消除了加卸载过程中因压头升降导致耦合加载室内气压改变造成的试

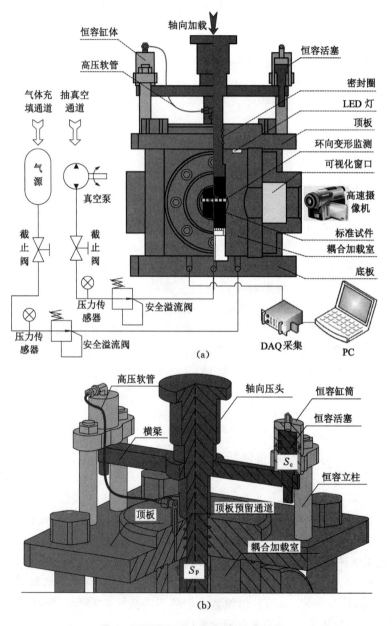

图 1 可视化恒容气固耦合试验系统

Fig.1 The gas-solid coupling test system in the visual and constant volume loading state

(a) 系统工作原理;(b) 恒容加载原理

验误差,通过环向位移测试装置和可视化窗口实现了对试件变形、煤体裂隙发育与劣化规律等特征参数的可视化实时监测。

如图1(b)所示,恒容装置工作原理为:试验过程中恒容活塞随着压头的下压同步下移,反应室内气体压力通过管路进入恒容室,压头伸入反应室部分截面积 S_p 等于两端恒容室内截面积 S_c 之和,即 $S_p=2S_c$,从而抵消了压头底部的气体压力,消除气压对压头反作用力,保证了加载过程中加载室内气体压力恒定,消除压强干扰,提高了试验准确性。

1.2 气体吸附对煤体的劣化试验方案

考虑到煤层的非均质性导致原煤性质的离散性,为统一试验变量,消除节理构造等客观因素造成的试验误差,突出主要矛盾[19,20],试验基于文献[18]选用强度可调的型煤相似材料,以淮南谢桥矿B6煤层为骨架,通过调节腐殖酸钠浓度制作了各向均质同性的相同强度标准试件,材料配比与测试参数如表1所示,型煤试件实测均值强度1.014,误差小于0.3%,精度满足试验对比分析要求,型煤试件与实测单轴应力应变曲线如图2所示。

表 1 试件参数
Table 1 Specimen parameter

预制强度/MPa	粒径分布 0～1 mm:1～3 mm	腐殖酸钠浓度/%	成型压力/MPa	重度/(N/cm³)	实测强度/MPa	弹性模量/GPa	泊松比	黏聚力/MPa	内摩擦角/(°)
1	0.76:0.24	3.25	15	12.78	1.014	0.187	0.314	0.092	29

不同吸附性气体对煤体劣化试验(以下简称试验1),采用同强度型煤试件充入高纯度 He、N_2、CH_4 和 CO_2 并与不加气单轴试验对比分析,气体压力设置1 MPa,煤体吸附时间24 h,为避免温度干扰,试验均在25 ℃的恒温试验室内开展;待充分吸附后,通过轴向加载模块的位移控制方式对试件轴向加载,加载速率1 mm/min[21],记录观察试验过程中煤体强度、体积扩容变形规律,采用高速摄像机对耦合加载过程中煤体裂隙发育特征进行实时录像,由于残余应力的存在,为方便对比,选取峰后塑性阶段中0.2倍峰值强度点为最终破坏形态,提取加载图像导入MATLAB软件进行计算,试验加载路径如图3所示。进一步,按上述试验方法,将气体固定为 CO_2,不断提高吸附压力,其他条件不变,开展了不同吸附压力下煤体劣化试验研究(以下简称试验2)。

2 试 验 结 果

2.1 不同吸附性气体中煤体劣化试验结果

通过获取环向变形和轴向变形整理得到抽真空条件下煤体全应力应变曲线和体应变变化曲线,通过高速摄像机监测得到加载过程中煤体裂隙发育特征,如图4所示。

由图4中体应变与应力应变曲线可知,在不充气常压单轴加载条件下,由于煤体内部孔隙存在,在压缩初始,轴向压缩应变大于侧向膨胀,环向变形量可视为零,煤体处于压密阶段,之后应力-应变曲线成近似直线型发展,随着外力继续增加,体应变由负值转换为正值,出现扩容,当试件进入峰后阶段,煤体内部结构遭到破坏产生宏观裂隙。

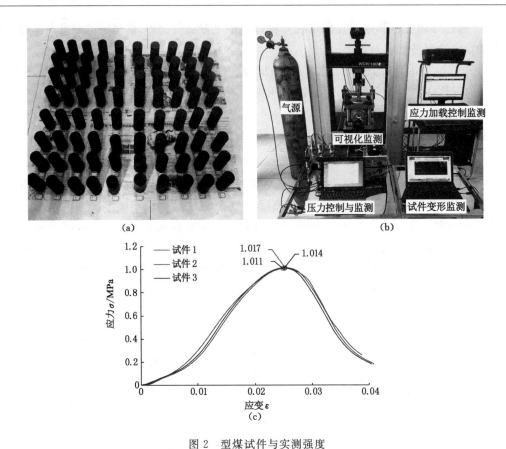

图 2 型煤试件与实测强度

Fig. 2 Coal specimen and strength

(a) 型煤标准试件；(b) 试验现场照片；(c) 实测单轴应力应变曲线

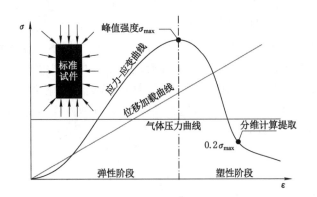

图 3 试验加载路径

Fig. 3 Test loading path

以相同试验条件,对煤体分别充入相同压力的 He、N_2、CH_4 和 CO_2,试验曲线如图 5 所示,考虑压头摩擦力迫使端部应力偏高造成的端部效应,截取试件中间高度 80 mm 范围内区域进行裂隙对比分析[22]。最终裂隙发育图像如图 6 所示。

将应力应变和体应变曲线合并处理,如图 7(a)、(b)所示；通过所得峰值强度数据及实

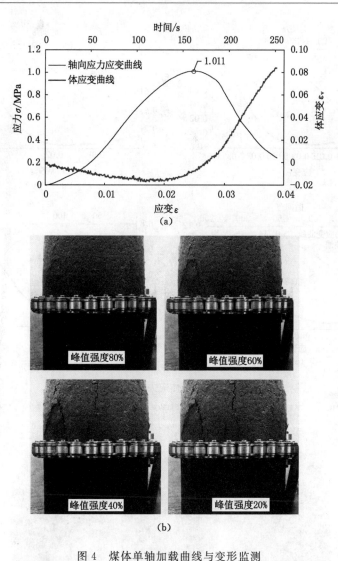

图 4 煤体单轴加载曲线与变形监测

Fig. 4 Uniaxial loading curve and deformation monitoring of coal

(a) 应力应变与体应变曲线；(b) 试件环向变形监测与裂纹发育

测煤体对 4 种气体的吸附平衡常数劣化率计算公式(1)和 Langmuir 等温吸附方程(2)得到煤体强度劣化率曲线与吸附量曲线，如图 7(c)所示。

$$f = 1 - \sigma_{max}/\sigma_i \tag{1}$$

式中，σ_{max} 为吸附后煤岩体极限承载力；σ_i 为不吸附时煤体极限承载力。

$$X = abp/(1+bp) \tag{2}$$

式中 a,b——吸附平衡常数；

p——加载时吸附压力。

由图 7 可知，在同等吸附压力下，煤体对 4 类气体的吸附量顺序为：$CO_2 > CH_4 > N_2 >$ He，随着煤体吸附量的增加，煤体强度不断降低，劣化程度增大，其中充 CO_2 煤体吸附量最大，为 23.27 cm^3/g，对应劣化率为 32.05%，充 He 吸附量为 0，可认为不吸附。

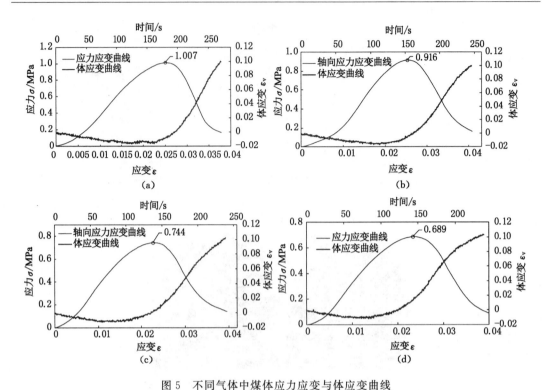

图 5 不同气体中煤体应力应变与体应变曲线

Fig. 5 Stress strain and volume strain curves of coal in different gases

(a) He; (b) N_2; (c) CH_4; (d) CO_2

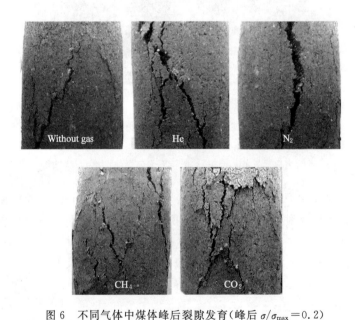

图 6 不同气体中煤体峰后裂隙发育（峰后 $\sigma/\sigma_{max}=0.2$）

Fig. 6 Fractures development after the peak in different gases

随着气体吸附性的提高，煤体更早的由压密阶段进入扩容阶段，如图 7(a)、(b)所示，煤体达到极限承载值与体积扩容点的位置不断提前，先后顺序为：$CO_2 > CH_4 > N_2 > He$。

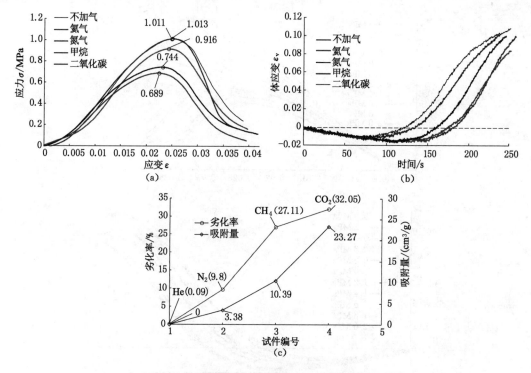

图 7 不同气体吸附后煤体劣化规律曲线

Fig. 7 Coal body deterioration law curves in different gases

(a) 应力应变对比曲线；(b) 体应变对比曲线；(c) 煤体劣化规律与吸附量关系曲线

结合图 6 煤体破坏形态特征对比图像可知，随着气体吸附性能的提高，相同应力阶段的煤体裂纹数量随之增加，裂隙发育更加丰富，在峰后强度 20％ 阶段，充入 CO_2 体表面裂纹数量最多，并呈现"鱼鳞片"状膨胀和脱落趋势。

2.2 吸附压力对煤体劣化试验结果

进一步选用吸附效果最好的 CO_2 作为试验气体，将其充入相同强度的型煤标准试件，不断提高吸附压力（0～2 MPa 范围内），得到不同吸附压力下煤体应力应变和体应变曲线，如图 8 所示。煤体强度随着 CO_2 吸附压力的提高不断降低，其中吸附压力为 2 MPa 时，煤体劣化率高达 48.27％，强度降低了近 1/2；煤体由体积压缩转化为体积扩容的节点不断提前，更容易发生失稳破坏，具体试验参数见表 2。

表 2 不同吸附压力煤体劣化率与吸附量参数

Table 2 Degradation rate and adsorption quantities in different adsorption pressure

吸附压力 p/MPa	吸附后强度 σ_{max}/MPa	降低强度 $\Delta\sigma$/MPa	裂化率 f/％	吸附量 Q/(cm^3/g)
0	1.011	0	0	0
0.2	0.929	0.082	8.11	7.19
0.4	0.841	0.170	16.82	12.66
0.6	0.781	0.230	22.75	16.96

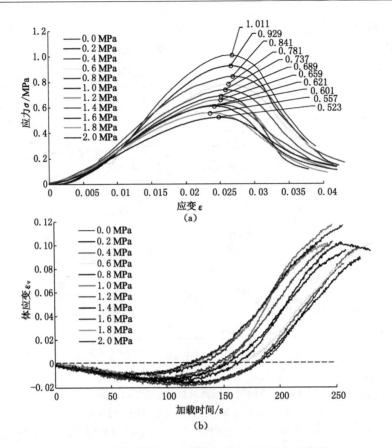

图 8 不同吸附压力下煤体应力应变与体应变曲线

Fig. 8 Stress strain and volume strain curves of coal in different adsorption pressures

(a) 应力应变曲线;(b) 体应变曲线

续表 2

吸附压力 p/MPa	吸附后强度 σ_{max}/MPa	降低强度 $\Delta\sigma$/MPa	裂化率 f/%	吸附量 Q/(cm^3/g)
0.8	0.737	0.274	27.10	20.42
1.0	0.689	0.322	31.85	23.27
1.2	0.659	0.352	34.82	25.66
1.4	0.621	0.390	38.58	27.69
1.6	0.601	0.410	40.55	29.44
1.8	0.557	0.454	44.91	30.96
2.0	0.523	0.488	48.27	32.29

3 试验结果分析

3.1 吸附量与劣化影响分析

由试验 1 结果可知,煤体的吸附程度直接影响了煤体强度。进一步,将试验 2 中所测数据代入吸附等温方程,得到煤体强度劣化率曲线与吸附量曲线,如图9所示。

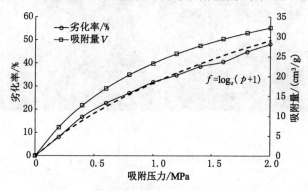

图 9 煤体劣化率与吸附量关系曲线

Fig. 9 Degradation rate and adsorption quantities curves of coal

由图9可知,煤体劣化率变化趋势与等温吸附曲线一致,即随着吸附量的增加,煤体强度逐渐降低,增长趋势逐渐变缓,通过换算得到 2 MPa 范围内煤体劣化率与吸附压力关系式:

$$f = \log_s(p+1) \tag{3}$$

式中 f——煤体劣化率;

p——加载时煤体所处气体压力;

s——劣化参数,受气体性质和煤基质性质影响,此处 $s=1.022$。

3.2 基于分形理论的裂隙扩展规律分析

众多研究表明岩石类材料的断裂变形行为具有分形特征[23],分形理论为定量认识和描述岩石复杂的断裂力学过程和物理机制提供了新途径,本文采用盒维数法,即通过不同尺寸的正方形格子($\delta \times \delta$)覆盖要测量的裂隙,得到不同尺寸下覆盖住测量物体的正方形格子数目 $N(\delta)$,最终根据格子尺寸与格子数目关系计算出分形维数,如公式(4)所示。

$$\lg N(\delta) = \lg k - D \lg \delta \tag{4}$$

式中 δ——方格尺寸;

k——常数;

D——分形维数。

运用 MATLAB 数值软件对加载过程监测图像进行处理,将裂隙逐一提取并结合公式(4)进行计算,得到不同吸附气体在相同应力状态下裂隙分形维数,其中裂纹颜色表示裂隙贯通范围,如图10所示。

由计算结果可知,拟合曲线相关性系数 R^2 数值均大于 0.98,属于显著相关,说明气体吸附作用下煤岩表面裂纹演化具有明显的分形特性。分形维数反映了煤体宏观裂隙复杂程度,裂隙密度则体现了裂隙致密程度,因此,岩石裂隙发育程度应综合考虑裂隙维数和裂隙密度[24],设裂隙发育程度 M,定义如下公式:

$$M = \rho_s D = -\frac{\sum_{i=1}^{n} S_i}{S_A} \lim_{\delta \to 0} \frac{\lg N(\delta)}{\lg \delta} \tag{5}$$

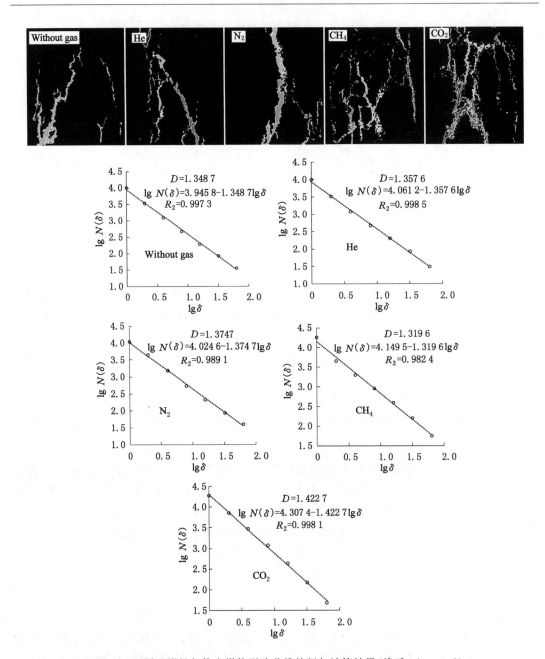

图 10　不同吸附性气体中煤体裂隙分维特征与计算结果(峰后 $\sigma/\sigma_c=0.2$)

Fig. 10　Fractal characteristics and computed results of coal in different gas

式中　S_i——单条裂纹面积；

　　　S_A——煤体基质面积。

M 值越大，表明煤体破裂程度越高，发育程度越好。

通过拟合各曲线公式得到不同气体和不同加载阶段煤体裂隙分形维数 D，按照上述方法，计算出试验 1 与试验 2 相同应力阶段的分形维数 D 和裂隙密度 ρ_s(表 3)，代入公式(5) 得到裂隙发育度曲线规律，如图 11 所示。

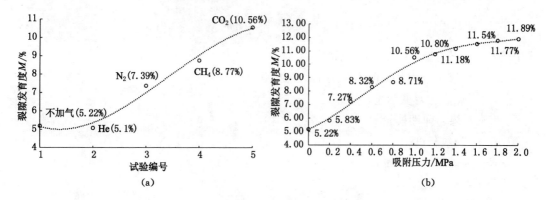

图 11 相同应力阶段煤体裂隙发育度

Fig. 11 Development degree of coal in the same loading stage

(a) 不同气体对煤体裂隙发育程度曲线；(b) 不同吸附压力对煤体裂隙发育程度曲线

表 3 不同吸附压力中煤体裂隙计算参数

Table 3 Calculation parameters of coal fracture under different adsorption pressures

吸附压力/MPa	分形维数 D	裂隙密度 ρ_s/%	裂隙发育度 M/%
0	1.325	3.942	5.22
0.2	1.368	4.261	5.83
0.4	1.379	5.260	7.25
0.6	1.382	6.023	8.32
0.8	1.402	6.214	8.71
1.0	1.423	7.420	10.56
1.2	1.428	7.563	10.80
1.4	1.442	7.752	11.18
1.6	1.453	7.940	11.54
1.8	1.460	8.066	11.77
2.0	1.462	8.128	11.89

综上可知，随着煤体吸附量的增大，裂隙发育越复杂，分形维数逐渐增大，对应幅值在 1.32~1.47 之间变化。当吸附压力小于 1.2 MPa 时，裂隙发育度与吸附压力呈正相关函数关系；当压力大于 1.2 MPa 时，裂隙发育度增长趋势减小。结合吸附压力与吸附量曲线分析表明，此时的煤样表面裂纹尺度与数量虽整体上成增长趋势，但随着吸附压力的升高，有限的煤基质逐渐逼近吸附饱和状态，导致煤体劣化程度减弱。

3.3 气体吸附造成煤体劣化的力学分析

何学秋等[8]综合岩石力学、表面物理化学和岩石断裂力学，提出吸附瓦斯对煤蚀损作用的理论模型，得出煤岩吸附瓦斯后强度性质和变形性质的劣化公式：

$$\left(\frac{\sigma_c}{\sigma_0}\right)^2 = 1 - \frac{RT}{\gamma_0 V_0 S}\int_0^p \frac{V}{p}\mathrm{d}p$$

$$\sigma_0 = \sqrt{\frac{2E\gamma_0}{\pi l}} \tag{6}$$

$$\varepsilon = \frac{\lambda RT}{SV_0} \int_0^p \frac{V}{p} dp \tag{7}$$

式中　σ_c——煤岩裂隙尖端的抗拉强度；

　　　σ_0——煤岩吸附瓦斯前裂隙尖端的抗拉强度；

　　　ε——煤岩相对变形量；

　　　E——煤岩弹性模量；

　　　l——裂纹长度；

　　　R——普适气体常量；

　　　T——绝对温度；

　　　S——比表面积；

　　　V_0——气体摩尔质量；

　　　V——吸附量；

　　　p——吸附平衡压力；

　　　$\Delta \gamma$——表面自由能变化量，$\Delta \gamma = \gamma_0 - \gamma$（$\gamma$ 是吸附后表面自由能，γ_0 是初始表面自由能）；

　　　λ——比例系数。

由式(6)、(7)可知，煤岩的强度与吸附平衡压力 p 以及吸附量 V 有关。随着 p 或 V 的增大，σ_c 是不断减小的，ε 不断增加。

文献[13]考虑气体吸附引发煤岩膨胀变形的力学特征，得到煤岩吸附瓦斯的有效应力公式：

$$\sigma'_{ij} = \sigma_{ij} - u - \sigma_p \tag{8}$$

$$\sigma_p = \frac{2a\rho RT(1-2\mu)\ln(1+bu)}{3V_0} \tag{9}$$

式中　σ'_{ij}——有效应力张量；

　　　σ_{ij}——煤岩的总应力；

　　　σ_p——吸附膨胀应力；

　　　ρ——煤岩的密度；

　　　μ——煤岩的泊松比；

　　　a——平衡吸附量（参考压力下的极限吸附量），是吸附平衡压力 p 的相关函数，p 越大，a 越大[25]。

劣化试验排除了游离气体对煤体的影响，仅考虑吸附气体的影响。事实上，当孔隙压差很小或者不存在时，渗流作用可以忽略不计，瓦斯对煤岩的劣化作用主要是吸附态瓦斯诱发的煤岩固体表面物理化学性质变化引起的。因此，式(8)可以修正为式(10)：

$$\sigma'_{ij} = \sigma_{ij} - \sigma_p \tag{10}$$

将式(7)应用到 Mohr-Coulomb 剪切破坏强度条件：

$$\tau_f = C + (\sigma_{ij} - \sigma_p)\tan \varphi \tag{11}$$

式中　τ_f——抗剪强度；

　　　C——黏聚力；

　　　φ——内摩擦角。

将式(10)应用到 Mohr 应力圆中,可以得出应力圆的新位置与大小:

$$\sigma'_m = \frac{\sigma'_{11} + \sigma'_{33}}{2} = \frac{\sigma_{11} + \sigma_{33}}{2} - \sigma_p = \sigma_m - \sigma_p$$

$$\tau'_m = 0$$

$$r'_m = \frac{\sigma'_{11} - \sigma'_{33}}{2} = \frac{\sigma_{11} - \sigma_{33}}{2} = r_m \tag{12}$$

式中 (σ'_m, τ'_m) ——应力圆圆心;

r'_m ——应力圆半径。

如图 12 所示,由式(9)可以得出 Mohr-Coulomb 强度包络线向右平移,内摩擦角 φ 不变,黏聚力 C 变小;由式(12)可以得出 Mohr 应力圆半径没有变化,而随着吸附压力的增加,应力圆向左平移;从而说明,吸附膨胀应力的存在使煤岩更容易发生失稳破坏,煤岩强度降低。

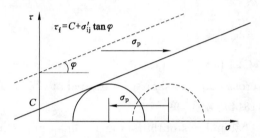

图 12 强度包络线及应力圆变化

Fig. 12 The change of strength envelope and Mohr stress circle

综合上述研究,气体吸附于煤岩裂隙或孔隙内表面,将导致煤岩裂隙尖端抗拉强度降低、煤岩黏聚力减小,使煤岩应力状态更容易进入破坏状态,从而降低型煤宏观强度。整理式(6)得:

$$\sigma_c = \sqrt{\sigma_0 - \frac{2ERT}{SV_0 \pi l} \int_0^p \frac{V}{p} \mathrm{d}p} \tag{13}$$

由式(10)和强度包络线几何关系得:

$$C = C_0 - \frac{2a\rho RT(1-2\mu)\ln(1+Bbu)}{3V}\tan\varphi \tag{14}$$

式中 C_0 ——煤岩吸附气体前的初始黏聚力。

综上分析,煤岩裂尖抗拉强度 σ_c 和煤岩黏聚力 C 是瓦斯吸附平衡压力 p 的相关函数,且 p 越大,σ_c 和 C 越小。

4 结 论

(1)采用可视化恒容气固耦合试验系统开展了不同吸附性气体和吸附压力中煤体的劣化试验研究,多角度对比分析了吸附量与吸附压力对煤体强度、体积扩容和裂隙发育的影响规律。

(2)在同等吸附压力下,充 CO_2 煤体吸附量最大,实测值为 23.27 cm³/g,对应劣化率为 32.06%,充 He 劣化率约等于 0,可视为不吸附。随着气体吸附性的提高,煤体更早地由压密阶段进入扩容阶段,煤体达到极限承载值与体积扩容点的位置不断提前,先后顺序为:$CO_2 > CH_4 > N_2 > He$。

(3) 煤体劣化率变化趋势与等温吸附曲线一致,即随着吸附量的增加,煤体强度逐渐降低,且增长趋势逐渐变缓,通过换算得到 2 MPa 范围内煤体劣化率 f 与吸附压力 p 符合 $f = \log_5(p+1)$ 函数关系。

(4) 气体吸附作用下煤岩表面裂纹演化具有明显的分形特性,煤体裂隙分形维数拟合曲线相关性系数 > 0.98,为显著相关;随着气体吸附量的提高,相同应力阶段的煤体裂纹数量随之增加,分形维数不断增加,裂隙发育更加丰富,在峰后强度 20% 阶段,充入二氧化碳煤体裂隙发育度最大,是不加气时的 2.28 倍,裂隙呈现"鱼鳞片"状膨胀和脱落趋势。

(5) 通过试验验证与力学分析得到,气体吸附于煤岩颗粒表面,降低了煤岩颗粒表面自由能,导致煤岩裂隙尖端抗拉强度降低、煤岩黏聚力减小,诱发煤体劣化,促使煤岩应力状态更容易进入失稳破坏状态。

参考文献

[1] Jiang C L, Wang C, LI X W, et al. Quick determination of gas pressure before uncovering coal in cross-cuts and shafts[J]. Journal of China University of Mining and Technology, 2008, 18(4): 494-499.

[2] Yuan L. Control of coal and gas outbursts in Huainan mines in China: A review[J]. Journal of Rock Mechanics and Geotechnical Engineering, 2016, 8(4): 559-567.

[3] 周世宁, 林柏泉. 煤层瓦斯赋存与流动理论[M]. 北京:煤炭工业出版社, 1999.
ZHOU Shining, LIN Boquan. Coal seam gas occurrence and flow theory[M]. Beijing: Coal industry press, 1999.

[4] Karacan C O. Swelling-induced volumetric strains internal to a stressed coal associated with CO_2 sorption[J]. International Journal of Coal Geology, 2007, 72(3-4): 209-220.

[5] Ranjith P G, Jasinge D, Choi S K, et al. The effect of CO_2 saturation on mechanical properties of Australian black coal using acoustic emission[J]. Fuel, 2010, 89(8): 2110-2117.

[6] Viete D R, Ranjith P G. The effect of CO_2 on the geomechanical and permeability behaviour of brown coal: implications for coal seam CO_2 sequestration[J]. International Journal of Coal Geology, 2006, 66(3): 204-216.

[7] LARSEN J W. The effects of dissolved CO_2 on coal structure and properties[J]. International Journal of Coal Geology, 2004, 57(1): 63-70.

[8] 何学秋, 王恩元, 林海燕. 孔隙气体对煤体变形及蚀损作用机理[J]. 中国矿业大学学报, 1996, 25(1): 6-11.
HE Xueqiu, WANG Enyuan, LIN Haiyan. Mechanism of action of pore gas on deformation and erosion of coal body[J]. Journal of China University of Mining and Technology, 1996, 25(1): 6-11.

[9] 尹光志, 王登科, 张东明, 等. 基于内时理论的含瓦斯煤岩损伤本构模型研究[J]. 岩土力学, 2009, 30(4): 885-889.

YIN Guangzhi, WANG Dengke, ZHANG Dongming, et al. Endchronic damage constitutive model of coal containing gas[J]. Rock and Soil Mechanics, 2009, 30(4): 885-889.

[10] 程远平, 刘洪永, 郭品坤, 等. 深部含瓦斯煤体渗透率演化及卸荷增透理论模型[J]. 煤炭学报, 2014, 39(8): 1650-1658.
CHENG Yuanping, LIU Hongyong, GUO Pinkin, et al. A theoretical model and evolution characteristic of mining-enhanced permeability in deeper gassy coal seam [J]. Journal of China Coal Society, 2014, 39(8): 1650-1658.

[11] 刘力源, 朱万成, 魏晨慧, 等. 气体吸附诱发煤体强度劣化的力学模型与数值模拟[J]. 岩土力学, 2018, 39(4): 1-9.
LIU Liyuan, ZHU Wancheng, WEI Chenhui, et al. Mechanical model and numerical analysis for gas adsorption-induced coal mechanical property alterations[J]. Rock and Soil Mechanics, 2018, 39(4): 1-9.

[12] 黄达, 岑夺丰. 单轴静-动相继压缩下单裂隙岩样力学响应及能量耗散机制颗粒流模拟[J]. 岩石力学与工程学报, 2013, 32(9): 1926-1936.
HUANG Da, CEN Duofeng. Mechanical response and energy dissipation mechanism of rock specimen with a single fissure under static and dynamic uniaxial compression using particle flow code simulations[J]. Chinese journal of Rock Mechanics and Engineering, 2013, 32(9): 1926-1936.

[13] 李祥春, 郭勇义, 吴世跃, 等. 考虑吸附膨胀应力影响的煤层瓦斯流-固耦合渗流数学模型及数值模拟[J]. 岩石力学与工程学报, 2007, 26(S1): 2743-2748.
LI Xiangchun, GUO Yongyi, WU Shiyue, et al. Mathematical model and numerical simulation of fluid-solid coupled flow of coal-bed gas considering swelling stress of adsorption[J]. Chinese journal of Rock Mechanics and Engineering, 2007, 26 (S1): 2743-2748.

[14] 梁冰, 贾立锋, 孙维吉, 等. 横观各向同性煤等温吸附变形试验研究[J]. 中国矿业大学学报, 2018, 47(1): 60-66.
LIANG Bing, JIA Lifeng, SUN Weiji, et al. Experimental study of isothermal adsorption deformation of transversely isotropic coal[J]. Journal of China University of Mining and Technology, 2018, 47(1): 60-66.

[15] 聂百胜, 卢红奇, 李祥春, 等. 煤体吸附-解吸瓦斯变形特征实验研究[J]. 煤炭学报, 2015, 40(4): 754-759.
NIE Baisheng, LU Hongqi, LI Xiangchun, et al. Experimental study on the characteristic of coal deformation during gas adsorption and desorption process[J]. Journal of China Coal Society, 2015, 40(4): 754-759.

[16] 王佑安, 陶玉梅, 王魁军, 等. 煤的吸附变形与吸附变形力[J]. 煤矿安全, 1993(6): 19-27.
WANG Youan, TAO Yumei, WANG Kuijun, et al. The adsorption deformation and adsorption deformation force of coal body[J]. Safety in Coal Mines, 1993(6): 19-27.

[17] 李清川,王汉鹏,李术才,等.可视化恒容气固耦合试验系统的研发与应用[J].中国矿业大学学报,2018,47(1):148-156.
LI Qingchuan,WANG Hanpeng,LI Shucai,et al. Development and application of a gas-solid coupling test system in the visual and constant volume loading state[J]. Journal of China University of Mining and Technology,2018,47(1):148-156.

[18] 王汉鹏,张庆贺,袁亮,等.含瓦斯煤相似材料研制及其突出试验应用[J].岩土力学,2015,36(6):1676-1682.
WANG Hanpeng,ZHANG Qinghe,YUAN Liang,et al. Development of a similar material for methane-bearing coal and its application to outburst experiment[J]. Rock and Soil Mechanics,2015,36(6):1676-1682.

[19] Harpalani S,Chen G L. Influence of gas production induced volumetric strain on permeability of coal[J]. Geotechnical and Geological Engineering,1997,15(4):303-325.

[20] 吕闰生,彭苏萍,徐延勇.含瓦斯煤体渗透率与煤体结构关系的实验[J].重庆大学学报,2012,35(7):114-118.
LV Runsheng,PENG Suping,XU Yanyong. Experiments on the Relationship Between Permeability of Gas-bearing Coal and Coal Body Structure[J]. Journal of Chongqing University,2012,35(7):114-118.

[21] 中华人民共和国国家标准.GB/T 50266—99 工程岩体试验方法标准[S].北京:中国计划出版社,1999:15-16.
The State Standard of the People's Republic of China. GB/T50266-99 Standard for test methods of engineering rock mass[S] Beijing:China planning press,1999:15-16.

[22] JAEGER J C,COOK N G W,ZIMMERMAN R W. Fundamentals of rock mechanics[M]. New York:Wiley-Blackwell,2007:90-95.

[23] 谢和平,高峰.岩石类材料损伤演化的分形特征[J].岩石力学与工程学报,1991,10(1):74-82.
XIE Heping,GAO Feng. The fractal of the damage evolution of rock materials[J]. Chinese Journal of Rock Mechanics and Engineering,1991,10(1):74-82.

[24] Hewett T A. Fractal distribution of reservoir heterogenrity and their influence on fluid transport[R]. SPE15386,1986:1-13.

[25] 赵阳升,胡耀青.孔隙瓦斯作用下煤体有效应力规律的试验研究[J].岩土工程学报,1995,17(3):26-31.
ZHAO Yangsheng,HU Yaoqing. Experimental study on effective stress of coal under pore gas[J]. Chinese Journal of Geotechnical Engineering,1995,17(3):26-31.

大倾角工作面飞矸冲击损害及其控制

伍永平[1,2],胡博胜[1],皇甫靖宇[1],汤业鹏[1],田双奇[1],刘晨光[1]

(1. 西安科技大学 能源学院,陕西 西安,710054;
2. 西部矿井开采及灾害防治教育部重点实验室,陕西 西安,710054)

摘 要:飞矸具有动态运移、危害巨大的特点,合理控制其损害是大倾角煤层安全开采的前提之一。基于卡尔曼滤波原理获取飞矸的冲击能特性,以挡矸网为控制元件,多手段综合研究了飞矸的冲击损害机制及控制元件参数,提出了一种飞矸冲击损害控制方法。研究结果表明飞矸碰撞设备时冲击能耗损为多种形式的能量,其中设备的变形能占比最大;飞矸碰撞设备前的冲击能和设备的冲击能恢复系数共同影响飞矸冲击能的耗损程度,由此飞矸损害控制可以从提高设备的冲击能恢复系数和降低飞矸的冲击能两方面入手。LS-DYNA 数值碰撞模型确定了大倾角大采高工作面控制元件参数为涤纶直径 6 mm,菱形网格大小 100 mm×100 mm。将控制元件应用于现场,大大降低了飞矸伤人毁物事故的次数,改善了防护设备使用周期短的问题。

关键词:大倾角煤层;飞矸灾害;卡尔曼滤波;LS-DYNA 碰撞模型;防护参数

Impact damage of flying gangue in steeply dipping seams and its controlling

WU Yongping[1,2], HU Bosheng[1], HUANGFU Jingyu[1],
TANG Yepeng[1], TIAN Shuangqi[1], LIU Chenguang[1]

(1. *School of Mineral Engineering, Xi'an University of Science and Technology, Xi'an 710054, China;*
2. *Key Laboratory of Western Mine Exploitation and Hazard Prevention, Ministry of Education, Xi'an 710054, China*)

Abstract: Flying gangue has the characteristics of dynamic migration and enormous detriment. Reasonably controlling its damage is one of the prerequisites for safety mining of the steeply dipping coal seam. This paper based on the Kalman filter principle, the dynamic characteristics of flying gangue were obtained, the mechanism of flying gangue hazards and parameters of control element were studied, a methodology about flying gangue damage controlling was proposed. The results show that: when the flying gangue collides with facilities, its impact energy can be dissipated into various forms of energy, of which the deformation energy of the facility accounts for the largest amount. The impact energy of flying gangue before collision and the recovery coefficient of impact energy of equipment jointly affect the degree of flying gangue energy-loss.

基金项目:国家自然科学基金重点项目(51634007);国家自然科学基金资助项目(51774230);陕西省自然基础研究计划资助项目(2016JQ5019,2016JQ5100)。

作者简介:伍永平(1962—),男,陕西汉中人,博士,教授,博士生导师。E-mail:wuyp@xust.edu.cn。

Therefore, the flying gangue damage control can be started from two aspects, including increasing the impact energy recovery coefficient of facilities and reducing the impact energy of flying gangue. Parameters of the control element for working face in steeply dipping coal seam with large mining height were determined by LA-DYNA, the diameter of the polyester was 6 mm and the size of the diamond mesh was 100 mm × 100 mm. Then the control element was applied in field, the number of casualties caused by flying gangue was reduced and the problem of short usage period was effectively improved.

Keywords: the steeply dipping coal seam; flying gangue hazard; Kalman filter; LS-DYNA collision model; protective parameters

大倾角煤层由于倾角较大，开采此类煤层时与水平和缓倾斜煤层有许多差异，例如：顶板结构、工作面布置以及衍生灾害等。众多学者对顶板结构和工作面布置开展了研究，取得的成果基本上保证了此类煤层的正常开采[1-6]。随着机械化程度和安全意识的提高，大倾角煤层开采过程中一种衍生灾害逐渐凸显——飞矸灾害[7,8]。由于大倾角工作面上下端高差大，飞矸脱离母体后在重力下加速向长壁工作面下部运移，冲击人或设备时损伤的危害巨大。众多矿井飞矸灾害毁物伤人事故频繁发生，经济损失严重，制约了矿井的效益[7-13]。

目前，飞矸灾害的治理工作主要集中在优化采煤工艺和改进防护设备，两者均存在一定的优势和弊端[8]；代表性的防护设备有：挡矸网、挡矸板、挡矸门以及具有立体防护功能的防矸液压支架[7,9-13]。但防护设备普遍存在的缺陷是：基于静态观点求解和推演防护设备参数，且大多使用类比法、经验设计法，人为主观因素占比重，很少考虑飞矸损害的动态特性对防护设备的影响。因此，目前的防护设备通常在工作面使用周期短、更换频率高，从而进一步增加了矿井的成本。

笔者之前对飞矸的形成、运移和损伤全过程进行了系统的研究，提出飞矸具有强动态性和随机性，以及冲击能对其损物伤人影响显著的观点。在此基础上，借助卡尔曼滤波算法在动态系统状态序列处理的优势，综合运用物理相似模拟实验、数值计算以及现场验证等手段，以挡矸网为控制元件，研究飞矸的冲击损害机制及控制元件参数，提供一种适用于大倾角大采高（以下简称"双大"）工作面的飞矸冲击损害控制方法。为类似地质条件和生产工艺工作面的飞矸损害控制提供借鉴。

1 飞矸冲击实验

1.1 实验布局

图1为自行研制的飞矸形成、运移和损伤全过程监测系统，具体特征见文献[8]。物理相似模拟实验平台长2 m，实际的大倾角倾斜长壁工作面长105 m，实验中飞矸的几何相似比应与工作面长度几何相似比保持一致。综合考虑各因素确定本次实验的几何相似比$C_l=50$，时间相似比$C_t=C_l^{1/2}=7.07$。平台上布置金属制液压支架，架前为煤壁。选用v411型高速摄像机记录飞矸运移过程，保存为视频文件，高速摄像机配套的PCC软件支持飞矸运移视频的回放和分帧。

1.2 实验方案

实验以"双大"工作面为背景，倾角44°，开采中煤壁易片帮，且呈现滑冒、范围广、蔓延性等特点。片帮占"双大"工作面随机分离体（煤壁片帮、冒顶、底板滑移和煤机甩煤等形式

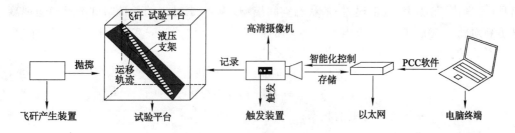

图 1 飞矸运移监测系统

Fig. 1 Monitoring system for flying gangue migration

产生的煤、岩块)总量的 60%～70%；而且随机分离体脱离母体后的运移状态和致灾过程基本相同。因此，简化工作面实际生产条件，实验着重研究煤壁片帮形式衍生的飞矸灾害的冲击损害特点。

工作面位置不同，衍生飞矸的冲击能和损害程度存在显著差异，已通过数值计算手段证实工作面上部区域衍生的飞矸危害程度最为突出[7]。为了使控制元件具有足够的冲击损害承受能力，以飞矸损伤最大化原则确定实验参数和抛矸流程如下，其中：飞矸材料是采集自工作面现场的煤样，飞矸尺寸通过实测煤壁片帮块度的加权均值换算得出(见图 2)。飞矸抛掷位置靠近工作面回风巷，距底板垂直高度 9 cm，以自由落体形式下落，冲击底板后进一步沿工作面运移。实验中飞矸与现场飞矸灾害形成、运移和损伤过程(运行状态和冲击过程)的相似性，一方面通过模拟煤层倾角、煤壁片帮、底板强度等固有属性实现；另一方面安装模型液压支架保证设备配套与尺寸和工作面实际一致。为了避免飞矸运移随机性对实验结果的影响，进行重复实验，直到找出冲击能的动态损伤特性规律为止。

图 2 飞矸的尺度及形态

Fig. 2 Lumpiness and shape of flying gangue

2 基于卡尔曼滤波的飞矸动态特性

2.1 卡尔曼滤波原理及在图像处理领域优势

卡尔曼滤波器是由 Kalman 于 1960 年提出的用于时变线性系统的递归滤波器，是一个对动态系统状态序列进行线性最小方差估计的算法。基本思想如式(1)和式(2)所示：采用信号（测量值）与噪声（测量误差）的状态空间模型，用 $t-1$ 时刻的估计值和 t 时刻的观测值来更新对状态变量的估计，求出 $t+1$ 时刻的估计值，它以"预测—实测—修正"的顺序递推，根据系统的测量值来消除随机干扰，再现系统的状态[14]。

$$X(t+1) = \boldsymbol{\Phi} X(t) + \boldsymbol{\Gamma} W(t) \tag{1}$$

$$Y(t) = \boldsymbol{H} X(t) + V(t) \tag{2}$$

式中　$X(t)$——系统在 t 时刻的状态；

$Y(t)$——状态的观测值；

$W(t)$——系统噪声，方差阵为 $\boldsymbol{Q}$；

$V(t)$——观测噪声，方差阵为 $\boldsymbol{R}$；

$\boldsymbol{\Phi}$——状态的转移矩阵；

$\boldsymbol{H}$——观测矩阵；

$\boldsymbol{\Gamma}$——系统的噪声驱动矩阵。

在图像处理领域卡尔曼滤波常被用来进行图像分割、图像复原、边缘监测以及动态目标监测等。动态目标监测是采用运动目标与背景状态空间模型，以坐标为状态变量，用 $t-1$ 时刻运动目标的坐标估计值和 t 时刻运动目标坐标的观测值来更新对状态变量（坐标）的估计。循环递推，实现动态目标的监测与跟踪。

2.2 飞矸冲击能获取方法

考虑到飞矸的动态运移特征，结合运用二值图像和动态背景差分技术[15]。编制MATLAB程序，以获得飞矸运移过程坐标点信息，实现对飞矸形成、运移和损伤全过程的监测。运动目标边缘的监测则是利用物体和背景在某种图像特性上的差异来实现的，这些特性包括灰度、颜色或纹理等，本程序中选取灰度为差异特性进行图像边缘的监测。

自编制程序的界面如图3所示，图3(a)中所示的绿色捕捉点能够检测到飞矸（黑色实心体）的边缘，捕捉点较好地反映了飞矸质心的位置。图3(b)为飞矸坐标点监测结果，监测曲线整体上光滑且连续，但有个别坐标点缺失，可能的原因是在这些坐标处由于像素、光线或重影等因素造成检测点漏测。但是缺失的坐标点对于飞矸的动态特性影响巨大，因此，运用样条插值法对缺失点的坐标进行补充，样条插值法的优点是插值后任意两个相邻的坐标点以及它的导数（速度）在缺失点处均连续。插值完成后，坐标曲线是一条光滑曲线，飞矸坐标获取后，其冲击能求法见公式(3)：

$$E = \frac{m_i \cdot C_i^5 \cdot \left(\dfrac{\sqrt{(y_{t+1} - y_t)^2 + (x_{t+1} - x_t)^2}}{(t+1) - t}\right)^2}{4 \cdot C_t^2} \tag{3}$$

式中　E——飞矸的冲击能，kJ；

x_t——t 时刻飞矸的 x 坐标，m；

x_{t+1}——$t+1$ 时刻的 x 坐标，m；

y_t——t 时刻的 y 坐标，m；

y_{t+1}——$t+1$ 时刻的 y 坐标，m；

m_i——飞矸第 i 次碰撞底板后的质量，kg。

2.3 冲击能特性分析

实验次数大于50次时，发现飞矸与支架在工作面碰撞的危害是不言而喻的。由于篇幅

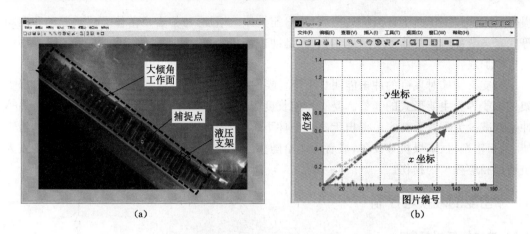

图 3 飞矸运移监测程序界面

Fig. 3 Interface of monitoring program of flying gangue migration

(a) 图像边缘检测界面；(b) 坐标监测曲线

限制，图 4 列举出了系列重复实验中第 17、22 和 28 次飞矸形成、运移并与支架碰撞的冲击能动态演化曲线。

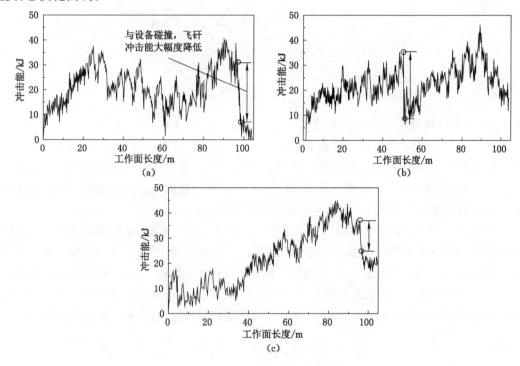

图 4 飞矸的冲击能演化特征

Fig. 4 The evolutive characteristics of flying gangue's impact energy

(a) 第 17 次；(b) 第 22 次；(c) 第 28 次

图 4 所示飞矸的冲击能演化特征曲线表明，"双大"工作面飞矸的冲击能在 0~45.9 kJ 之间。其中，飞矸在自由落体和飞溅状态未碰撞底板，曲线小幅度下降，携带的冲击能部分

损失。随着向工作面下部运移的持续,冲击能大体上增加,可知自由落体、飞溅、滑动和滚动等状态飞矸冲击能总体上不断地累积,但是冲击能极差增大,离散性增强,说明飞矸运移的随机性加强。

与支架碰撞时飞矸的冲击能明显下降(蓝色线段所示),降低部分的冲击能转化为以支架变形能等为主要形式的能量。受工作面长度限制,运移至工作面下部运输巷时逐渐停止,冲击能减少为零。重复实验得出无防护条件下,飞矸自身携带的冲击能将大比例传递给支架、刮板输送机、采煤机等设备,或者传递给随机出现在工作面的操作人员,进而对设备或人员造成相应程度的损伤。

3 冲击能的耗损与控制

3.1 冲击能耗损机制

飞矸运移过程中的极限冲击能一方面是防护设备需承受的极限值;另一方面冲击能为标量,贯穿于飞矸的整个运移过程,可以有效避免运移方向和环境(形成环境、运移环境及边界约束)对其损伤风险评估带来的不确定性和随机性。假设飞矸沿工作面倾斜向下与"三机"设备碰撞 k 次,则飞矸的冲击能累积与耗损过程可用下式进行描述:

$$E_0 + mg \cdot \sin\alpha \cdot (L+h) - \mu g \sum_{i=1}^{n} m \cdot L_i - \sum_{i=1}^{n} E^* - \sum_{j=1}^{k} E_{dj} - \sum_{j=1}^{k} E_{ej} - \sum_{j=1}^{k} E_{rj} = E + \sum_{j=1}^{k} E_{cj} \tag{4}$$

其中:
$$L = \sum_{i=1}^{n} L_i, h = \sum_{i=1}^{n} h_i$$

式中,E_0 为飞矸初始动能,kJ;m 为质量,kg;g 为重力常数,kg/m³;α 为倾角,(°);h 为弹起下落高度,m;h_i 为第 i 次碰撞后弹起下落高度,m;μ 为摩擦系数,无量纲;L_i 为第 i 次碰撞底板后滑移距离,m;n 为碰撞底板次数;E^* 为第 i 次碰撞底板时飞矸耗损的冲击能,kJ;E_{dj}、E_{ej}、E_{rj} 和 E_{cj} 分别为第 j 次碰撞设备时飞矸冲击能耗损为耗散能、声能、光能、电能、辐射能及设备变形能等的数值,kJ;E 为第 k 次碰撞设备后冲击能,kJ。

由式(4)可知飞矸冲击能碰撞设备时损失为耗散能、声能、光能、电能、辐射能以及支架的变形能,是一个伴随多种能量累积与耗损的问题。定量确定 E_{dj}、E_{ej}、E_{rj} 和 E_{cj} 的值,需要涉及众多的测量仪器,做大量的岩石力学实验[16-19],求解过程十分烦琐。因此,转换求解思路,做出下列推导,飞矸对静止设备的反复碰撞,同样可以看作是以飞矸为静态参考系,设备以不同冲击能对飞矸进行的非周期性动力加载。对于第 k 次碰撞设备的飞矸,碰撞前冲击能 E_0^k 为:

$$E_0^k = E_0 + mg(L+h) \cdot \sin\alpha - \mu g \sum_{i=1}^{n} m \cdot L_i - \sum_{i=1}^{n} E^* - \sum_{j=1}^{k-1}(E_{dj} + E_{ej} + E_{rj} + E_{cj}) \tag{5}$$

式(4)和式(5)联立得:

$$1 - \frac{E}{E_0^k} = \frac{\sum_{j=k}^{k}(E_{dj} + E_{ej} + E_{rj} + E_{cj})}{E_0^k} \tag{6}$$

注意到设备的冲击能恢复系数 φ 的物理含义为飞矸第 k 次碰撞设备后的冲击能与碰撞

前冲击能之比[8],即:

$$\psi = \frac{E}{E_0^k} \tag{7}$$

则飞矸冲击能碰撞设备时设备吸收的变形能 E_{ck} 满足:

$$E_{ck} \leqslant E_{dk} + E_{ek} + E_{rk} + E_{ck} = (1-\psi) \cdot E_0^k$$

$$E_{ck} \leqslant (1-\psi) \cdot E_0^k \tag{8}$$

从式(8)可以得出设备吸收的变形能 E_{ck} 越大,飞矸损失的能量相应越多;同时,设备吸收的变形能 E_{ck} 与其损害等级分段相关,这一结论笔者之前的研究中已进行了验证[8]。因此飞矸的冲击能耗损状态(冲击损害程度)受其冲击能和设备的冲击能恢复系数 ψ 共同影响。

3.2 冲击能控制方法

飞矸的冲击能和设备的冲击能恢复系数 ψ 共同影响碰撞时飞矸的冲击损害程度,因此,飞矸损害控制可以从提高设备的冲击能恢复系数和降低飞矸的冲击能两方面入手。

具体方法如图5所示:当某次飞矸碰撞设备时携带的冲击能很大,而且碰撞方位、接触状态均良好时,设备的冲击能恢复系数将很小,因此设备吸能的状态就很好,相应的设备损坏的风险就越高,例如图5中高风险区D。为了控制飞矸损害程度,降低设备损坏的等级,可以采取措施,提高设备的冲击能恢复系数使飞矸的损伤风险从高风险区D移至中等风险区C,或者降低飞矸的冲击能,使高风险区D移至中等风险区B。再进一步采取措施将损伤风险从中等风险区B、C移至低风险区A,从而彻底消灭飞矸的损害风险。

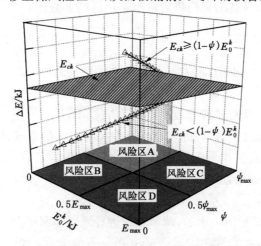

图5 飞矸损害等级区划及控制模式

Fig. 5 Regionalization for damage risk of impact objects

在飞矸冲击能动态特性物理模拟实验测定的基础上,下文以"双大"工作面飞矸灾害为例,以挡矸网为控制元件,设计合理的控制元件参数对飞矸冲击损害进行控制,控制元件参数确定流程见图6。尽管工程实际中会涉及各类地质条件和生产工艺特点的飞矸损害控制问题,本节提供的方法依然可以很好的适用。

(1) 控制元件的特点及布设

挡矸网是由一定直径的细绳扭结而成,沿大倾角工作面倾斜方向悬挂,上端悬挂于支架顶梁,下端固定于刮板输送机的溜槽内;挡矸网是阻挡飞矸从采煤作业区运移至操作作业区

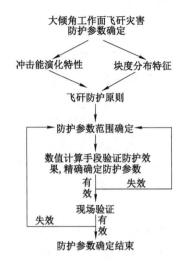

图 6 挡矸网参数确定流程

Fig. 6 The flow chart of blocking net parameters

的柔性编织物,起到隔离"两区",防止飞矸与人员或设备直接接触的作用。挡矸网为柔性体,既能强行降低飞矸的冲击能,也可以很好地隔绝飞矸与设备或人员的直接接触,挡矸网正常发挥控制效果时,可以理想地认为设备的冲击能恢复系数 ψ 为 1,此时飞矸冲击能损伤风险等级位于风险区 A,设备的损伤风险最小。挡矸网参数的合理与否对于飞矸冲击损害的控制至关重要。

(2) 控制元件参数确定

LS-DYNA 软件是分析爆炸、碰撞及地震等非线性接触问题的专业软件,用来计算飞矸碰撞防护网具有独特的优势。黄润秋基于正交设计对影响落石运动特征的 6 个因素进行现场试验,研究发现落石形状对碰撞恢复系数的影响很小,属于次要因素[20]。现场实测和模拟实验也表明有棱角(非规则形状)的飞矸经过多次与底板的碰撞,以及在煤壁与设备形成的双向约束空间运移的反复冲击后,飞矸的磨圆度和球度均较好,为了不失结论的一般性,下面用球形飞矸进行碰撞模拟;同时将物理模拟实验得到的飞矸冲击特征用于数值计算碰撞模型中。

LS-DYNA 碰撞模型如图 7 所示,球形飞矸进行六面体网格划分,涤纶质的菱形网采取

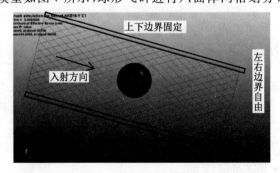

图 7 LS-DYNA 碰撞模型

Fig. 7 The collision model established by LS-DYNA

梁单元进行划分。对碰撞模型的速度、边界条件、接触参数等特征进行设置,力学参数见表1。决定挡矸网是否断裂失效的主要因素是单元体中最大拉应变,即不论是单向应力或复杂应力状态,只要单元体中的最大拉应变达到单向拉伸情况下发生断裂失效时的拉应变极限值(失效应变ε_f),材料将失效[21,22]。

表 1 　　　　　　　　　　碰撞模型的力学参数
Table 1 　　　　　　　Mechanical parameters of the collision model

材料	几何形状	密度/(tone/mm³)	弹性模量/GPa	泊松比	屈服强度/MPa	失效应变 ε_f/%
飞矸	球体	2.5e-09	2.6	0.210	/	/
涤纶质挡矸网	菱形	1.38e-09	4	0.307	200	10

(3) 控制效果分析

从图8(a)和8(b)中模拟结果可以看出,飞矸碰撞挡矸网时,压缩阶段和回弹阶段挡矸网的应变量大体上沿时间轴对称。

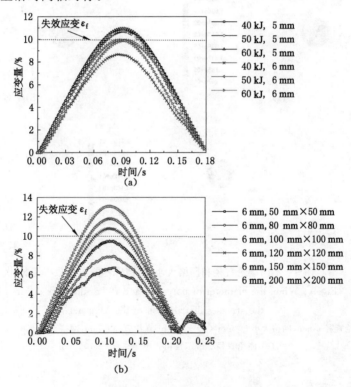

图 8　不同防护参数时挡矸网的应变量

Fig. 8　The strain of blocking net under different protective parameters

(a) 不同冲击能和涤纶直径时应变量;(b) 不同网格大小时应变量

随着冲击能的增大,应变量增大;冲击能相同时,涤纶直径越大,应变量越小。涤纶直径为 5 mm 时,挡矸网的应变量均超出其失效应变 ε_f,见图8(a)。由此确定涤纶的直径为 6 mm 时,可对飞矸的冲击能进行有效的控制。图8(b)为涤纶直径 6 mm 时,不同网格大小挡矸网的应变量曲线,随着网格的增大,应变量不断增加,网格大于 100 mm×100 mm 时,挡矸网的应变量同样超出其失效应变 ε_f。但是,需要说明的是网格同样不能太小,一方面原因是网格太小增加了挡矸网的质量,不利于推溜、移架以及检修工作的进行;另一方面网格太小,飞矸的"子弹效应"加强,反而降低了挡矸网的防护效果[23]。因此,挡矸网的最佳网格大小确定为 100 mm×100 mm,此时飞矸的冲击损害能得到有效的控制。

进一步地,对飞矸冲击时挡矸网的应力分布特征进行研究,结果如图 9 所示,极限应力出现在飞矸与挡矸网的接触区附近,且以"×"形方式向网的边缘传播,在网格的接触区域以及边界处发生应力集中,因此,在以上区域挡矸网容易发生破坏。涤纶的直径 6 mm,网格大小 100 mm×100 mm 时挡矸网的极限应力达到了 156 MPa,小于其屈服强度 200 MPa,应力值在正常范围。比较图 9(a)与图 9(b)可以看出,网格越小挡矸网能够承受的极限应力增大,对飞矸冲击损害的控制能力更强;比较图 9(b)与图 9(c)可以得出,涤纶直径越大,挡矸网能够承受的极限应力越大,控制元件达到的控制效果更好。

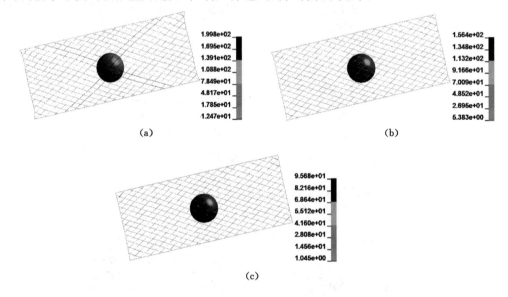

图 9　不同涤纶直径和网格大小时,挡矸网的应力分布情况

Fig. 9　When the diameter of polyester and the grid size is different, the stresses distribution of blocking net

(a) 涤纶直径 6 mm,网格大小 50 mm×50 mm;(b) 涤纶直径 6 mm,网格大小 100 mm×100 mm;
(c) 涤纶直径 5 mm,网格大小 100 mm×100 mm

4　现场应用与效果

4.1　工程背景

新疆焦煤集团艾维尔沟 2130 煤矿 5# 煤层走向 ES 12°～14°、倾向 SN 35°～45°。厚度在 4.27～7.8 m,密度 1.38 t/m³。25221 工作面为本文研究的原型,工作面倾向长 100 m,平均倾角 44°,采高为 4.5 m,采用综合机械化大采高采煤法,属于典型的"双大"工作面。煤的硬度系数 $f=0.74$;直接顶 $f=6.26$,由灰白色含砾粗砂岩,泥质胶结、风化易碎的灰白色

中砂岩组成,厚度 2.32 m;直接底以碳质泥岩为主,厚度为 2.19 m,$f=7.9$。

经调查飞矸冲击损害严重的原因如下:煤壁片帮产生大量脱离煤壁的煤块,这些煤块起初未下落,后在采煤机割煤等工序振动下由静止开始自由下落。由于煤体的安息角(30°~35°)小于煤层倾角,落体运动碰撞底板后难以停止,将沿工作面倾斜下部运动(滑动、滚动或飞溅)。重力做功下飞矸的冲击能急剧增加,防护设备的不合理或失效,致使飞矸冲击能不能被有效的控制,与人或设备直接接触(冲击)。实测发现挡矸网的破坏多集中在飞矸密集冲击的网格处以及挡矸网上下固定边界区域附近。

对 25221 工作面 20 次煤壁片帮衍生飞矸的极限冲击能 E_{max} 进行统计,具体如表 2 所示;飞矸的极限冲击能不大于 50 kJ,平均为 36.8 kJ。工作面原有的防护参数条件下控制效果如图 10 所示,挡矸网发生了失效,而且失效区域集中在飞矸密集冲击的网格处以及挡矸网上下边界固定区域。数值模拟结果与现场实测达到了高度的一致,表明用数值模拟手段确定"双大"工作面控制元件参数是可行的。

表 2 飞矸极限冲击能结果
Table 2 Test results of ultimate impact energy

编号	1	2	3	4	5	6	7
E_{max}/kJ	36.4	44.5	45.9	37.8	21.3	30.3	45.8
编号	8	9	10	11	12	13	14
E_{max}/kJ	33.7	43.2	31.5	40.9	23.7	45.6	33.1
编号	15	16	17	18	19	20	
E_{max}/kJ	36.4	43.1	44.4	34.1	30.8	44.2	

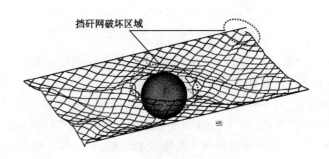

图 10 原有挡矸网失效
Fig.10 The original blocking net failure

4.2 控制效果现场验证

在以往挡矸网参数基础上,按照图 6 所示流程选定"双大"工作面挡矸网的参数为网格大小 100 mm×100 mm,涤纶直径 6 mm。此防护参数条件下挡矸网控制效果校验结果如图 11(a) 所示,挡矸网有效地控制了飞矸的冲击能,而且该参数下挡矸网的应变量小于其失效应变,有效地隔绝了飞矸与工作面设备(人员)的直接接触。

考虑到推溜、移架及检修等工序的方便,挡矸网不宜过于紧绷,在其自重作用下自然垂落,后对下边界进行固定即可。改进后的挡矸网在工作面的布置如图 11(b)所示,网格长对

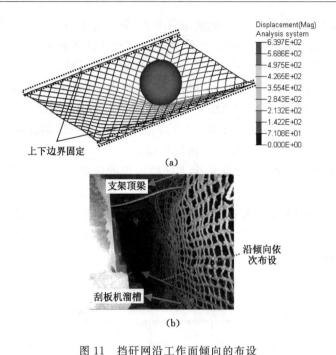

图 11 挡矸网沿工作面倾向的布设
Fig. 11 Layout of blocking nets along working face inclination
(a) 参数改进后挡矸网效果;(b) 工作面挡矸网布置

角线方向平行于底板,自回风巷倾斜向工作面下部依次设置。试验生产期间无人员伤亡事故发生;同时,防护设备的使用周期和更换频率得到大幅度的改善,节约了生产的成本。工作面的正规循环率达到 85% 以上。

5 结 论

(1) 针对目前大倾角工作面飞矸冲击损害及其控制(控制元件参数确定)人为主观因素占主导的现状,提出了一种充分考虑飞矸动态特性的损害控制方法。方法基于卡尔曼滤波原理获取飞矸的冲击能,从飞矸冲击能和设备的冲击能恢复系数两方面着手控制飞矸的损伤风险,以挡矸网为控制元件,建立 LS-DYNA 碰撞模型确立合理的控制元件参数。

(2) 飞矸碰撞设备时冲击能耗损为多种形式的能量,且设备变形能的占比最大;碰撞前飞矸的冲击能和设备的冲击能恢复系数共同影响飞矸冲击能的耗损程度。飞矸损害控制可以从提高设备的冲击能恢复系数和降低飞矸的冲击能入手,极大地简化了飞矸损伤风险评估的过程。

(3) 以"双大"工作面飞矸冲击损害及控制为例,对控制元件的效果进行碰撞模拟。综合确定控制元件参数为涤纶直径 6 mm,网格大小 100 mm×100 mm。将控制元件参数应用于现场实践当中,飞矸毁物伤人事故降低、防护设备使用周期短的问题得到了改善。

参考文献

[1] 负东风,伍永平.大倾角煤层综采工作面调伪仰斜的原理与方法[J].辽宁工程技术大学学报(自然科学版),2001(2):152-156.

[2] 周邦远,伍厚荣,聂春辉,等.绿水洞煤矿大倾角煤层综采开采实践[J].煤炭科学技术,2002,(09):21-23.

[3] 伍永平,解盘石,王红伟,等.大倾角煤层开采覆岩空间倾斜砌体结构[J].煤炭学报,2010,35(08):1252-1256.
WU Yongping,XIE Panshi,WANG Hongwei,et al. In cline masonry structure around the coal face of steeply dipping seam mining[J]. Journal of China Coal Society,2010,35(08):1252-1256.

[4] 杨科,陆伟,孙力.复杂条件大倾角大采高旋转综采矿压显现规律及其控制[J].采矿与安全工程学报,2015,32(02):199-205.
YANG Ke,LU Wei,SUN Li. Investigation into strata behaviors and ground control of high height rotary longwall mining in large inclined angle coal seam under complicated geological conditions[J]. Journal of Mining & Safety Engineering,2015,32(02):199-205.

[5] Wang Jinan,Jiao Junling. Criteria of support stability in mining of steeply inclined thick coal seam[J]. International Journal of Rock Mechanics & Mining Sciences,2016(82):22-35.

[6] Yun Dongfeng,Liu Zhu,Cheng Wendong,et al. Monitoring strata behavior due to multi-slicing top coal caving longwall mining in steeply dipping extra thick coal seam[J]. International Journal of Mining Science and Technology,2017,27(01):179-184.

[7] 伍永平,胡博胜,解盘石,等.大倾角长壁工作面飞矸灾害区域治理技术[J].煤炭科学技术,2017,45(02):1-5.
WU Yongping, HU Bosheng, XIE Panshi, et al. Flying gangue regional control technology in longwall mining steeply dipping seam[J]. Coal science and Technology,2017,45(2):1-5.

[8] 伍永平,胡博胜,王红伟,等.大倾角煤层长壁开采工作面飞矸致灾机理研究[J].煤炭学报,2017,42(09):2226-2234.
WU Yongping, HU Bosheng, WANG Hongwei, et al. Mechanism of flying gangue causing-disasters in longwall mining the steeply dipping seam[J]. Journal of China Coal Society,2017,42(09):2226-2234.

[9] 刘明申,杨峰,马怀利,等.大倾角采煤工作面防飞矸设施的研究与应用[J].山东煤炭科技,2008(5):62-63.

[10] 张建.大倾角中厚煤层综采面防飞矸液压支架的研究与应用[J].科技创新导报,2014(16):24.

[11] 曹树刚,李毅,雷才国,等.采煤工作面轻型架间挡矸装置究[J].采矿与安全工程学报,2013,30(1):51-56.
Cao Shugang, Li Yi, Lei Caiguo, et al. Research on lightweight device for blocking gangue between hydraulic supports in steeply inclined coal face[J]. Journal of Mining & Safety Engineering,2013,30(1):51-56.

[12] 吕文胜,李松强.大倾角综采工作面煤壁片帮原因分析[C].第六届天山地质矿产资源

学术讨论会论文集.乌鲁木齐:中国地质学会,2008:926-929.

[13] 张进忠,赵克俭.大倾角超长综采面飞矸事故机理分析及预防对策[J].科技创新导报,2011(22):128.

[14] (澳)古德温,孙贵生.自适应滤波、预测与控制[M].北京:科学出版社,1992.

[15] 刘鹏飞.卡尔曼滤波在运动目标跟踪问题中的研究与应用[D].哈尔滨:哈尔滨工业大学,2008.

[16] 黎立云,鞠杨,赵占文,等.静动态加载下岩石结构破坏时的能量分析[J].煤炭学报,2009,34(06):737-740.

LI Liyun, JU Yang, ZHAO Zhanwen, et al. Energy analysis of rock structure under static and dynamic loading conditions[J]. Journal of China Coal Society,2009,34(06):737-740.

[17] 赵毅鑫,龚爽,黄亚琼.冲击载荷下煤样动态拉伸劈裂能量耗散特征实验[J].煤炭学报,2015,40(10):2320-2326.

Zhao Yixin, Gong Shuang, Huang Yaqiong. Experimental study on energy dissipation characteristics of coal samples under impact loading[J]. Journal of China Coal Society,2015,40(10):2320-2326.

[18] 于水生,卢玉斌,朱万成,等.SHPB试验中花岗岩破坏程度与能量耗散关系分析[J].东北大学学报(自然科学版),2015,36(12):1733-1737.

YU Shuisheng, LU Yubin, ZHU Wancheng, et al. Analysis om relationship between degree of damage and energy dissipation of granite in SHPB tests[J]. Journal of Northeastern University(Natural Science),2015,36(12):1733-1737.

[19] 韩秀会,李成武,邢同振,等.基于变形场时空演化的煤冲击劈裂试验研究[J].煤炭学报,2016,41(11):2743-2748.

Han Xiuhui, Li Chengwu, Xing Tongzhen, et al. Experimental study on coal impact fracturing based on temporal and spatial evolution of deformation field[J]. Journal of China Coal Society, 2016, 41(11):2743-2748.

[20] 黄润秋,刘卫华.基于正交设计的滚石运动特征现场试验研究[J].岩石力学与工程学报,2009(28):882-891.

HUANG Runqiu, LIU Weihua. Insitu test of characteristics of rolling rock blocks based on orthogonal design[J]. Chinese Journal of Rock Mechanics and Engineering,2009(28):882-891.

[21] 王大刚.钢丝的微动损伤行为及其微动疲劳寿命预测研究[D].徐州:中国矿业大学,2012.

[22] Bourrier F, Lambert S, Baroth J. A reliability-based approach for the design of rockfall protection fences[J]. Rock Mech. Rock Eng.,2015(48):247-259.

[23] 周晓宇,陈艾荣,马如进.滚石柔性防护网耗能规律数值模拟[J].长安大学学报(自然科学版),2012,32(06):59-66.

ZHOU Xiaoyu, CHEN Airong, MA Rujin. Numerical simulation of energy dissipation mechanism on falling rocks protection nets[J]. Journal of Chang'an University (Natural Science Edition),2012,32(06):59-66.

锚注扩散及加固规律现场试验研究与应用

王琦[1,2],许英东[1,2],许硕[1,2],江贝[1,3],潘锐[1],刘博宏[1]

(1. 山东大学 岩土与结构工程研究中心,山东 济南,250061;
2. 中国矿业大学(北京)深部岩土力学与地下工程国家重点实验室,北京,100083;
3. 济南大学 土木建筑学院,山东 济南,250022)

摘 要:随着煤炭开采深度的增加,高地应力、复杂地质构造以及巷道围岩松散破碎现象逐渐增加,锚注作为该类巷道治理中常采用的支护手段,在该类巷道治理中取得了良好的控制效果,而锚注浆液的扩散及加固规律作为锚注研究的重要依据,相关研究目前还比较缺乏。本文以巨野矿区千米深井赵楼煤矿为工程背景,通过现场注浆前后普通锚杆拉拔力试验,定义了锚注有效加固区,得到了锚注有效加固区的陀螺形分布规律。在此基础上,开展了不同注浆锚杆布设条件下的数值模拟试验,分析了注浆锚杆长度及间排距对锚注加固效果的影响规律,提出了基于现场锚注扩散试验及数值模拟研究的锚注支护设计方法。并在赵楼煤矿5302轨道巷破碎围岩锚注支护设计中进行了应用,有效控制了围岩变形。

关键词:破碎围岩;锚注支护;浆液扩散;锚注有效加固区;影响因素

Experimental study and application of bolt-grouting slurry diffusion and support effect

WANG Qi[1,2], XU Yingdong[1,2], XU Shuo[1,2],
JIANG Bei[1,3], PAN Rui[1], LIU Bohong[1]

(1. *Geotechnical and Structural Engineering Research Center,*
Shandong University, Jinan 250061, China;
2. *Key Laboratory for Geomechanics and Deep Underground Engineering,*
China University of Mining and Technology(Beijing), Beijing 100083, China;
3. *School of Civil Engineering and Architecture, University of Jinan, Jinan 250022, China*)

Abstract: With the increase of coal mining depth, the highland stress, complex geological structure and loose breakage of the surrounding rock are increasing. As a often used support method, the bolt-grouting has achieved good control effect in the treatment of this kind of roadway. However, the law of diffusion and reinforcement of anchor grouting fluid is an important basis for bolting and grouting research. Based on the engineering background of Zhaolou coal mine in Juye mining area, the law of slurry diffusion and the influence factors of bolt-grouting reinforcement are studied. Through the test of the common bolt pulling

---

基金项目:国家自然科学基金资助项目(51674154,51704125);山东省重点研发计划资助项目(2017GGX30101,2018GGX109001);中国矿业大学煤炭资源与安全开采国家重点实验室开放基金资助项目(SKLCRSM18KF012)。

作者简介:王琦(1983—),男,山东省临沂市人,博士,副教授,博士生导师,主要从事深部巷道支护理论与技术方面的研究工作。Tel:13583120068,E-mail:chinawangqi@163.com。

force before and after the bolt-grouting, the effective reinforcement area of bolt-grouting is defined, and the distribution law of the gyroscope in the effective reinforcement area is obtained. On this basis, the numerical simulation test under the distribution condition of different grouting bolt is carried out, and the influence law of the length and spacing of the grouting bolt on the reinforcement effect is analyzed, and the design method of bolt-grouting support based on the field anchor injection diffusion test and numerical simulation is put forward. Based on the above research results, the site test of bolt-grouting support in the 5302 track Lane of Zhao Lou coal mine has been carried out. The results show that the bolt-grouting support scheme based on this design method effectively maintains the stability of the broken surrounding rock roadway.

Keywords: broken rock; bolt-grouting support; slurry flow; effective reinforcement area; influence factors

随着浅部煤炭资源的枯竭，我国中东部矿井逐步向深部发展[1]，高地应力、复杂地质构造导致的巷道围岩松散破碎现象逐渐增加。锚注作为该类巷道常见的支护形式，可有效提高围岩自承能力以及改善支护构件受力性能。目前，国内外针对锚注支护主要在锚注浆液扩散及锚注加固影响因素方面进行研究。

在锚注浆液扩散方面，王连国[2]利用连续介质假设和渗流力学理论建立了深-浅耦合锚注作用下的浆液渗流基本方程。黄耀光[3]运用多场耦合软件COMSOL建立了锚注浆液在围岩中渗透扩散的数值计算模型，系统研究了注浆时间、注浆压力以及注浆锚杆间距等锚注参数对浆液渗透扩散的影响。王琦[4]选取典型断面利用围岩钻孔电视对注浆后浆液扩散情况进行现场探测。江贝[5]采用SIR-3000地质雷达探测设备及相配套的高精度400 MHz天线对注浆前后巷道围岩的破碎范围以及注浆过程中浆液在围岩中的扩散及流动特性进行了探测。张农[6]利用钻孔窥视仪电视成像和染色剂跟踪浆液的方法对注浆后浆液扩散范围进行跟踪检测。杨坪[7]在砂卵砾石层中进行模拟注浆试验，分析了注浆因素及地层条件对浆液扩散半径的影响。葛家良[8]通过锚注加固前后围岩物理力学性能测定、声波测试和变形位移观测结果，检验了锚注加固效果。G. Lombadi[9-10]等通过在裂隙岩石中进行浆液扩散试验，推导出未考虑地下水影响及浆液黏度影响下浆液的最大扩散半径公式。H. B. 加宾[11]等在考虑裂隙倾角影响条件下，推导出浆液的注浆压力计算公式，得到注浆压力衰减规律及浆液扩散半径。

在锚注加固影响因素方面，王以功[12]通过正交设计的方法分析了影响注浆加固效果的主要围岩物理参数。孟庆彬[13-14]提出了"锚注加固体等效层"概念，并揭示了不同锚注加固体等效层物理力学参数对巷道围岩位移及塑性区的影响规律。陆士良[15]对锚注支护巷道围岩的应力和位移进行了数值计算分析，揭示了锚注支护与围岩的作用机理。Kang Zhi-qiang[16]通过对节理裂隙岩体进行了数值模拟分析，得到了节理裂隙岩体锚注加固作用机理。Wang Lian-guo[17]通过数值计算得到了锚杆支护结构的力学性能随锚注时间的变化规律。胡功笠[18]采用理论计算的方式得到了软岩巷道条件下注浆锚杆布置参数的综合确定方法。

前人所做的工作推动了锚注研究的发展，但作为锚注支护重要设计依据的浆液扩散研究方面，由于现场地质条件的复杂性，上述研究多以理论分析及数值模拟为主，即使使用钻孔电视进行现场监测，也缺乏一个定量的标准进行评价。在锚注加固影响因素研究方面，对注浆锚杆本身的布置形式尚缺乏基于现场与理论相结合的研究。

本文在前人研究成果的基础上，以千米深井赵楼煤矿5302工作面轨道巷为工程背景，通过锚注前后现场普通锚杆拉拔力试验定量确定浆液加固范围,在此基础上,以实际浆液加固范围为依据建立数值计算模型,开展不同注浆锚杆间排距及注浆锚杆长度的数值模拟试验,分析不同因素对锚注控制效果的影响,并提出相应的锚注参数设计方法,以期为类似条件下锚注支护参数设计提供借鉴。

1 工程背景

赵楼煤矿位于巨野煤田中部,矿井采用立井开拓方式,设计生产能力 3.0 Mt/a,开采水平设在 -860 m,最大水平应力的大小为 32.39~34.63 MPa,最大水平应力为垂直应力的 1.35~1.36 倍[19]。

5302 轨道巷位于五采区中部,沿煤层底板掘进,顶板以细砂岩及中砂岩为主,平均厚度为 18.66 m,底板以粉砂岩为主,平均厚度 15.70 m。煤层平均厚度为 4.60 m,工作面煤层综合柱状图如图 1 所示。

层位	岩性	柱状	厚度/m	埋深/m
顶板	细砂岩 中砂岩		$\dfrac{9.4\sim21.48}{14.82}$	848.26
	细砂岩 泥岩		$\dfrac{2.09\sim5.81}{3.84}$	852.1
煤层	煤3		$\dfrac{4.0\sim5.18}{4.6}$	856.7
	粉砂岩 泥岩		$\dfrac{0.49\sim1.05}{0.84}$	857.54
底板	细砂岩 粉砂岩		$\dfrac{13.2\sim19.28}{14.86}$	872.4

图 1 5302 轨道巷综合柱状图

Fig.1 Comprehensive histogram of 5302 track lane

1.1 原支护方案

5302 轨道巷断面形状为矩形,净宽 4 800 mm,净高 3 800 mm,原支护方案采用锚网+钢梯+锚索的全锚支护形式。其中,巷道支护所用顶板锚杆型号为 $\phi 22$ mm×2 500 mm,间、排距 850 mm×800 mm,预紧力 80 kN;两帮锚杆型号为 $\phi 20$ mm×2 500 mm,间、排距 800 mm×800 mm,预紧力 80 kN;锚索型号为 $\phi 22$ mm×6 200 mm,间、排距 1 700 mm× 1 600 mm,"2-1-2"五花形布置,预紧力 100 kN。

1.2 原支护方案下巷道破坏情况

原支护方案下,现场主要破坏区域出现在巷道两帮位置,大量出现剧烈鼓出,金属网撕裂,钢梯弯折、断裂,锚杆(索)失效等现象,如图 2 所示。

图 2 原支护方案变形破坏图

Fig.2 Deformation and failure diagram of the original support scheme

(a) 帮部鼓出严重；(b) 支护构件失效破坏

通过对原支护方案下的巷道进行现场监测，测得巷道掘进 90 d 后左帮平均变形量 441 mm，右帮平均变形量 402 mm，顶板平均变形量 223 mm。

采用钻孔窥视仪对原支护方案下巷道围岩内部破坏情况进行探测，每个检测断面设置 5 个 ϕ42 mm 探测孔。基于赵楼煤矿钻孔检测标准，对围岩完整性进行分类[20-21]。以断面 1 为例进行分析，如图 3 所示，每个钻孔中的黑色填充物代表了钻孔内壁观察到的破碎围岩的位置和程度。在每个钻孔的探测结果完成后，根据围岩破碎程度，从内到外，依次划分为严重、中等、轻微破坏区。

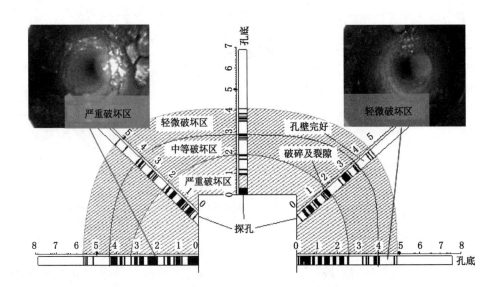

图 3 典型断面围岩破坏范围（单位：m）

Fig.3 Damage range of surrounding rock in typical section

从上述围岩钻孔电视窥探结果可知，围岩破碎范围较大，轻微破坏区的外边界距围岩平均距离为 4.88 m，中等破坏区的外边界距围岩平均距离为 3.70 m，严重破坏区的外边界距围岩平均 2.28 m。破坏区域主要出现在巷道周围，且围岩破碎程度呈现出左帮＞右帮＞顶板的特征。

通过对5302轨道巷的现场监测以及钻孔电视探测研究,分析巷道主要破坏原因如下:

(1)帮部围岩强度低,承载能力差。现场巷道所处地层为煤层,煤层较软,而顶板和底板所处地层以砂岩为主,因此出现两帮变形远大于顶底板变形的情况。

(2)现场煤层较为破碎,松动范围较大。经过前期钻孔探测研究,围岩严重破坏区最大深度可达3.30 m,超过原支护锚杆锚固长度,支护构件性能无法有效发挥,导致现场锚杆主动支护效果不足。

(3)巷道护表强度低。由于现场未对巷道喷射混凝土,导致帮部围岩护表强度低,金属网撕裂以及帮部鼓出现象严重。

为解决上述围岩控制难题,现场拟采用锚注支护方案对围岩进行支护,为合理设计锚注支护方案,现场开展锚注浆液扩散试验及不同注浆锚杆布设形式数值模拟试验,并提出基于上述试验的注浆锚杆设计方法。

2 锚注浆液加固性能现场试验

在注浆锚杆周围布置普通高强锚杆作为测试锚杆,开展注浆锚杆注浆前后测试锚杆的拉拔试验,通过评价测试锚杆注浆前后拉拔力的提高率,定量确定锚注有效加固区。

2.1 浆液加固试验方案设计

根据原设计方案及现场地质条件,选择2.5 m注浆锚杆进行现场锚注浆液扩散范围试验,注浆锚杆采用中空注浆螺纹锚杆。现场共设置3个试验段,每个试验段设置1根注浆锚杆及12根测试锚杆,分别布置于注浆锚杆上方、左方及下方0.4 m、0.8 m、1.2 m、1.6 m处,现场锚注加固试验方案布置如图4所示。

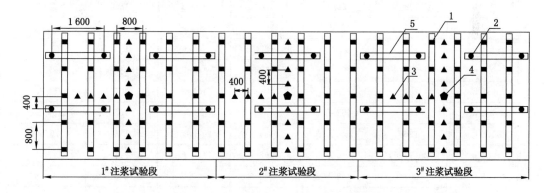

图4 浆液加固试验方案设计图

Fig.4 Design plan of slurry consolidation test

1——普通锚杆;2——普通锚索;3——测试用普通锚杆;4——注浆锚杆;5——钢带

注浆以水泥单液浆为主,单液浆水泥水灰比为0.5∶1,并加入水泥用量2%左右的高效减水剂,注浆终止压力为2 MPa。在注浆锚杆注浆28 d后对测试锚杆进行拉拔试验,现场实施如图5所示,所得试验数据列于表1。

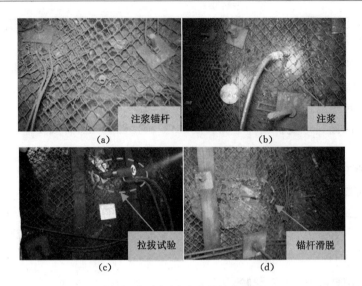

图 5　浆液加固试验现场实施图
Fig. 5　Field application of slurry reinforcement test
(a)注浆锚杆施打后；(b)注浆过程；(c)普通锚杆拉拔试验；(d)普通锚杆拉拔滑脱

表 1　　　　　　　　　　普通锚杆拉拔力试验数据
Table 1　　　　　　　　Data of tensile test of ordinary bolt

位置		1# 注浆锚杆试验段		2# 注浆锚杆试验段		3# 注浆锚杆试验段	
		拉拔力 F_t/kN	提高率 δ/%	拉拔力 F_t/kN	提高率 δ/%	拉拔力 F_t/kN	提高率 δ/%
上方	0.4 m	214(拉断)	167.5	202	152.5	215(拉断)	168.8
	0.8 m	178	122.5	174	117.5	180	125.0
	1.2 m	145	81.3	150	87.5	152	90.0
	1.6 m	117	46.3	122	52.5	126	57.5
左侧	0.4 m	210	162.5	210	162.5	218(拉断)	172.5
	0.8 m	202	152.5	201	151.3	205	156.3
	1.2 m	162	102.5	154	92.5	165	106.3
	1.6 m	125	56.3	131	63.8	136	70.0
下方	0.4 m	220(拉断)	175.0	208	160.0	216	170.0
	0.8 m	207	158.8	202	152.5	208	160.0
	1.2 m	165	106.3	163	103.8	168	110.0
	1.6 m	146	82.5	150	87.5	149	86.3

2.2　试验结果分析

根据前期现场未注浆围岩的普通锚杆平均拉拔力测试结果，确定注浆前锚杆拉拔力为 80 kN。定义测试锚杆拉拔力提高率大于 100% 的区域为锚注有效加固区，锚注有效加固区边缘距注浆锚杆距离 D_{xi}：

$$D_{xi} = \frac{(\delta_{x-l_{max}} - 100\%)}{\delta_{x-l_{max}} - \delta_{x-l_{min}}} |l_{max} - l_{min}| + l_{max}$$

式中,D_{xi} 为第 i 个试验段锚注有效加固区边缘距注浆锚杆距离;$x=u,s,p$ 分别代表锚注有效加固区上方、两侧、下方;l_{max} 代表提高率大于 100% 的测试锚杆中,离注浆锚杆最远的距离;l_{min} 代表提高率小于 100% 的测试锚杆中,离注浆锚杆最近的距离;$\delta_{x-l_{max}}$ 和 $\delta_{x-l_{min}}$ 分别代表上述位置的测试锚杆拉拔力的提高率。

$$D_x = \frac{1}{n}\sum_{i=1}^{n} D_{xi}$$

以第 1 个试验段为例计算注浆锚杆上方有效加固距离 D_{u1}:注浆锚杆上方 0.4 m、0.8 m、1.2 m 和 1.6 m 处测试锚杆的拉拔力提高率分别为 167.5%、122.5%、81.3% 和 46.3%,由此可知,$l_{max}=0.8$,相对应的 $\delta_{u-l_{max}}=122.5\%$;$l_{min}=1.2$,相对应的 $\delta_{u-l_{min}}=81.3\%$。计算得 $D_{u1}=1.02$ m,同理可得,$D_{u2}=1.03$ m,$D_{u3}=1.08$ m。从而,$D_u=\frac{1}{n}\sum_{i=1}^{n}D_{ui}=1.05$ m。

同理,可知在当前注浆条件下,$D_u=1.05$ m,$D_s=1.20$ m,$D_p=1.33$ m。锚注有效加固区呈陀螺型分布,图 6 为锚注有效加固区范围图。

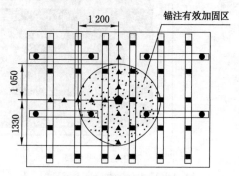

图 6 锚注有效加固区范围图

Fig. 6 Effective reinforcement area of bolt-grouting

3 锚注加固影响因素数值模拟研究

本部分拟采用数值模拟的方法,在上述注浆锚杆浆液有效加固区的基础上,研究不同注浆锚杆布设形式对围岩控制效果的影响规律,以指导现场锚注支护方案设计。

3.1 数值模拟计算模型建立

采用锚注支护的形式对原支护方案进行支护。注浆锚杆采用中空注浆螺纹锚杆,直径 28 mm,长度 2.5 m,间、排距为 800 mm×1 600 mm。注浆锚杆布置形式如图 7 所示。

按照注浆锚杆上方有效加固范围 1.05 m,横向有效加固范围 1.20 m,下方有效加固范围 1.33 m 进行计算,并以赵楼煤矿 5302 轨道巷实际开挖断面及支护设计为依据,建立模型及边界条件示意图如图 8 所示。

由图 8 可知,锚注浆液有效加固区在煤层中分布,参考破碎煤岩体注浆加固试验及前人对于锚注加固后参数的选取[22-27],可知注浆加固后,煤岩体各类物理力学参数发生一定程度的变化。其中,弹性模量、黏聚力、抗拉强度提高为原参数的 1.5~2 倍,内摩擦角提高

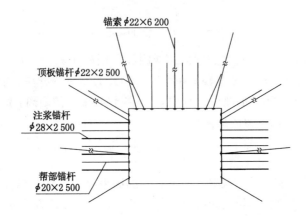

图 7 支护方案示意图

Fig. 7 Schematic diagram of support scheme

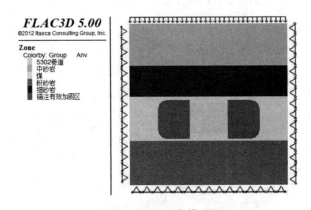

图 8 锚注方案模型图

Fig. 8 Model diagram of bolt-grouting scheme

2°～5°，泊松比降低 0.02～0.05。

本文基于赵楼煤矿 5302 轨道巷具体工程条件，保守考虑实际注浆加固效果，将锚注有效加固区内各围岩参数变化取较小值，即弹性模量、黏聚力、抗拉强度按照原参数 1.5 倍取值，内摩擦角提高 2°，泊松比降低 0.02，由此确定模型各部力学参数如表 2 所示。

表 2 5302 巷道所处地层物理力学参数表

Table 2 The table of physical and mechanical parameters of strata in 5302 laneway

围岩类型	高度 h/m	弹性模量 E/GPa	泊松比 μ	黏聚力 C/MPa	内摩擦角 φ/(°)	抗拉强度 σ_t/MPa
中砂岩	5.20	3.18	0.27	2.40	30.3	0.51
细砂岩＋泥岩	3.84	2.17	0.28	1.60	32.2	0.34
煤	4.60	0.60	0.30	0.50	34.9	0.10
锚注有效加固区		0.90	0.28	0.75	36.9	0.15
粉砂岩	6.36	2.95	0.25	2.60	34.4	0.43

3.2 锚注数值模拟方案设计

本部分设计了 2 种不同类型的数值模拟试验方案,用以研究不同因素对围岩控制效果的影响。

第一类方案 A:为分析锚注加固机理中不同注浆锚杆长度对巷道控制的影响,设计不同注浆锚杆长度进行数值模拟计算,考虑到实际工程应用中的注浆锚杆长度,设计如表 3 所示注浆锚杆长度方案。

表 3 第一类方案 A(不同注浆锚杆长度)
Table 3 The length scheme of different grouting bolts

方案编号	A1	A2	A3	A4	A5	A6
注浆锚杆长度/m	1.5	2	2.5	3	3.5	4

第二类方案 B:为分析锚注加固机理中不同注浆锚杆间排距对围岩支护效果的影响,设计不同注浆锚杆间排距数值模拟试验,为方便注浆锚杆布置,上下间距选择 400 mm 的整数倍进行计算,左右排距选择 800 mm 的整数倍进行计算。根据前述浆液有效加固区的定义,考虑到注浆锚杆的弧形扩散范围,若注浆锚杆间排距过大,则相邻注浆锚杆的浆液有效加固区无法有效重叠,如 2 000 mm×1 600 mm 方案中,每根注浆锚杆浆液有效加固区之间存在一定的空隙,如图 9 所示。因此本文对 1 600 mm×1 600 mm 以上的方案不再进行计算。根据以上分析,设计如表 4 所示 B1~B6 注浆锚杆间排距方案进行数值模拟计算。

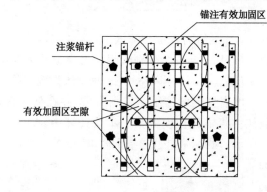

图 9 2 000 mm×1 600 mm 方案锚注有效加固区范围
Fig. 9 Area of effective reinforcement of 2 000 mm×1 600 mm scheme

表 4 第二类方案 B(不同注浆锚杆间排距)
Table 4 The plan of spacing between different grouting bolts

上下间距 \ 左右排距	800 mm	1 600 mm
800 mm	B1	B2
1 200 mm	B3	B4
1 600 mm	B5	B6

3.3 锚注数值模拟结果分析

依据上述试验方案,通过 FLAC 软件进行数值模拟,并采用均匀布点取平均值的方法得到围岩不同位置平均位移量如图 10 所示。

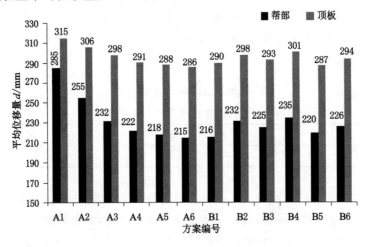

图 10 不同方案围岩位移计算结果

Fig. 10 Calculation results of surrounding rock displacement in different schemes

(1) 不同注浆锚杆长度方案中,定义围岩位移降低率为 Δ_{Aix}。

$$\Delta_{Aix} = \left| \frac{d_{Aix} - d_{A(i-1)x}}{d_{A(i-1)x}} \right| \times 100\%$$

式中,Δ_{Aix} 为 A_i 方案较 $A_{(i-1)}$ 方案的围岩位移降低率,$i = 2,3,4,5,6$;d_{Aix} 为 A_i 方案经数值计算得到的围岩不同位置变形量;$x = u, s$ 分别代表围岩顶板及帮部位置。

定义围岩位移降低率阈值 Δ 为 5%,当 $\Delta_{Aix} < \Delta$ 时,认为围岩位移变形逐渐趋于稳定,再次增加注浆锚杆长度对于围岩变形的控制效果则有限,并在低于该阈值的方案中选择最短的注浆锚杆长度即为最佳注浆锚杆长度方案。本次计算中,A4~A6 方案围岩位移变化率均低于该阈值,如图 11 所示,因此,最佳注浆锚杆长度为 A4 方案。

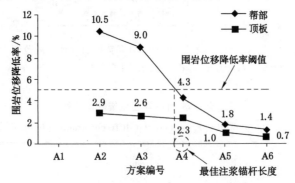

图 11 不同注浆锚杆长度方案围岩位移降低率

Fig. 11 Reduction rate of displacement of surrounding rock with different grouting bolt length schemes

（2）不同注浆锚杆间排距方案中，帮部围岩平均变形量最大 235 mm，最小 216 mm，相差 8.1%；顶板围岩平均变形量最大 301 mm，最小 287 mm，相差 4.7%。由此可见，在锚注有效加固区能够充分覆盖围岩的条件下，不同注浆锚杆间排距方案对于围岩控制效果影响有限，现场工业性应用时应尽可能选择较大的注浆锚杆间排距以节省支护成本，因此，本次注浆锚杆最佳间排距方案为 B6 方案。

4 锚注支护现场试验研究

4.1 现场锚注支护参数设计方法

基于上述试验研究，提出一种锚注参数设计方法。该方法主要流程如图 12 所示。

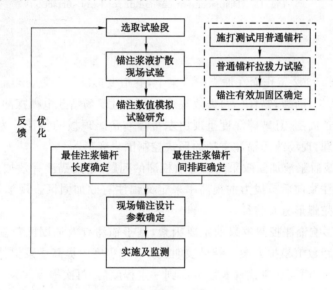

图 12 锚注参数设计流程图

Fig.12 Flow chart of bolt-grouting parameter design

4.2 5302 轨道巷现场锚注试验

为验证破碎围岩条件下锚注支护参数设计方法的可行性，在 5302 轨道巷围岩两帮进行锚注支护现场工业性试验。基于该参数设计方法，现场注浆锚杆采用长度 3 m，间排距 1 600 mm×1 600 mm 方案。

通过对煤巷收敛变形进行现场监测以验证锚注支护效果，巷道围岩表面收敛位移与时间关系曲线如图 13 所示。

由图 13 可知，经过 100 d 的监测，5302 轨道巷围岩变形趋于稳定，围岩左帮变形量 276 mm，右帮变形量 241 mm，顶板变形量 210 mm，即在两帮区域进行锚注支护有效控制了两帮破碎围岩煤巷的变形。为进一步验证锚注支护技术对破碎围岩煤巷的控制效果，在 5302 轨道巷试验段掘进六个月后对围岩变形量进行了复测，左帮位移量 310 mm，右帮位移量 260 mm，顶板变形量 226 mm。以上数据表明，基于该参数设计方法的锚注方案有效维持了破碎围岩条件下巷道的稳定。

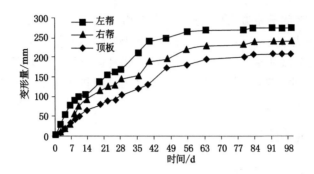

图 13 巷道表面收敛监测曲线

Fig. 13 Displacement change of roadway surface

5 结 论

(1) 赵楼煤矿 5302 工作面在原支护方案下,巷道变形破坏严重,出现帮部剧烈鼓出,金属网撕裂,钢梯弯折、断裂,锚杆(索)失效等现象。通过现场钻孔电视探测研究,表明破坏区域主要出现在巷道周围,且破碎深度呈现出左帮＞右帮＞顶板的特征。为解决破碎围岩条件下围岩控制难题,现场改用锚注支护的形式控制围岩变形。

(2) 针对注浆后浆液加固效果无法有效检测的问题,定义锚注有效加固区,通过定量分析周围普通锚杆注浆前后拉拔力的提高率来定义锚注有效加固区。现场试验结果表明,锚注有效加固区呈陀螺形分布特征。

(3) 为研究影响锚注控制效果的主要因素,基于锚注有效加固区范围进行了不同注浆锚杆长度及间排距数值模拟方案。结果表明:① 一定程度内提高注浆锚杆长度可有效提高对围岩的控制效果,而当注浆锚杆长度提高到一定程度后,再次提高注浆锚杆长度对于围岩控制效果的提高有限。② 在由单根注浆锚杆形成的锚注有效加固区能够有效覆盖围岩的条件下,不同注浆锚杆布置形式对于围岩变形量的影响较小。

(4) 提出了一种基于现场普通锚杆锚注前后拉拔测试定量确定锚注扩散范围,进而进行锚注设计的方法,并在 5302 轨道巷进行试验。现场监测表明,基于该参数设计方法的锚注方案有效维持了破碎围岩条件下巷道的稳定。

参 考 文 献

[1] 何满潮.深部软岩工程的研究进展与挑战[J].煤炭学报,2014,39(08):1409-1417.
HE Man-chao. Progress and challenges of soft rock engineering in depth[J]. Journal of China Coal Society,2014,39(08):1409-1417.

[2] 王连国,陆银龙,黄耀光,等.深部软岩巷道深-浅耦合全断面锚注支护研究[J].中国矿业大学学报,2016(01):11-18.
Wang Lian-guo, Lu Yin-long, Huang Yao-guang, et al. Deep-shallow coupled bolt-grouting support technology for soft rock roadway in deep mine[J]. Journal of China University of Mining & Technology,2016(01):11-18.

[3] 黄耀光,王连国,陆银龙.巷道围岩全断面锚注浆液渗透扩散规律研究[J].采矿与安全

工程学报,2015(02):240-246.

Huang Yao-guang, Wang Lian-guo, Lu Yin-long. Study on the law of slurry diffusion within roadway surrounding rock during the whole section bolt-grouting process[J]. Journal of Mining and Safety Engineering,2015(02):240-246.

[4] 王琦,潘锐,李术才,等. 三软煤层沿空巷道破坏机制及锚注控制[J]. 煤炭学报,2016,41(05):1111-1119.

WANG Qi, PAN Rui, LI Shu-cai, et al. Gob side entry failure mechanism and control of bolt-grouting in three soft coal seam[J]. Journal of China Coal Society,2016,41(05):1111-1119.

[5] 江贝,李术才,王琦,等. 三软煤层巷道破坏机制及锚注对比试验[J]. 煤炭学报,2015,40(10):2336-2346.

JIANG Bei, LI Shu-cai, WANG Qi, et al. Failure mechanism of three soft coal seam roadway and comparison study on bolt and grouting[J]. Journal of China Coal Society,2015,40(10):2336-2346.

[6] 张农,王保贵,郑西贵,等. 千米深井软岩巷道二次支护中的注浆加固效果分析[J]. 煤炭科学技术,2010(05):34-38,46.

Zhang Nong, Wang Bao-gui, Zheng Xi-gui, et al. Analysis on grouting reinforcement results in secondary support of soft rock roadway in kilometre deep mine[J]. Coal Science and Technology,2010(05):34-38,46.

[7] 杨坪,唐益群,彭振斌,等. 砂卵(砾)石层中注浆模拟试验研究[J]. 岩土工程学报,2006(12):2134-2138.

Yang Ping, Tang Yi-qun, Peng Zhen-bin, et al. Study on grouting simulating experiment in sandy gravels[J]. Chinese Journal of Geotechnical Engineering,2006(12):2134-2138.

[8] LOMBARDI G. Grout Line[J]. Geotechnical News,2008,30(1):3-10.

[9] LOMBARDI G. Grouting of rock masses[C]. Proceedings of the Grouting and ground treatment,F,2002,ASCE.

[10] 熊厚金,林天健,李宁. 岩土工程化学[M]. 北京:科学出版社,2001.

Xiong Hou-jin, Lin Tian-jian, Li Ning. Geotechnical Chemistry[M]. Beijing: Science Press,2001.

[11] 葛家良,陆士良. 巷道锚注加固技术及其效果的研究[J]. 化工矿山技术,1997(02):13-16.

Ge Jia-liang, Lu Shi-liang. Anchored and grouted technology in tunnel and its effect[J]. Industrial Minerals & Processing,1997(02):13-16.

[12] 王以功,林登阁,柴天星. 锚注支护机理及参数优化研究[J]. 地质与勘探,2002(03):84-86.

Wang Yi-gong, Lin Deng-ge, Chai Tian-xing. Research of bolt-grouting supporting mechanics and parameteroptimization[J]. Geology and Prospecting,2002(03):84-86.

[13] 孟庆彬. 深部高应力软岩巷道变形破坏机理及锚注支护技术研究[D]. 青岛:山东科技

大学,2011.

Meng Qing-bin. Research on deformation failure mechanism and bolt-grouting reinforcement in high-stress deep soft rock roadway[D]. Qingdao: Shandong University of Science and Technology,2011.

[14] 孟庆彬,韩立军,乔卫国,等.深部软岩巷道锚注支护机理数值模拟研究[J].采矿与安全工程学报,2016,33(01):27-34.

Meng Qing-bin, Han Li-jun, Qiao Wei-guo, et al. Numerical simulation research of bolt-grouting supporting mechanism in deep soft rock roadway[J]. Journal of Mining & Safety Engineering,2016,33(01):27-34.

[15] 陆士良,汤雷.巷道锚注支护机理的研究[J].中国矿业大学学报,1996(02):3-8.

Lu Shi-liang, Tang Lei. Study on the mechanism of bolting-grouting for roadway[J]. Journal of China University of Mining & Technology,1996(02):3-8.

[16] Kang Zhi-qiang, Han Qiang, Zhang Shu-qing. Study on numerical simulation and bolt grouting mechanism of mine fracture rock masses[C]. International Conference on Mechanics and Materials Engineering (ICMME), Xian, PEOPLES R CHINA, APR 12-13,2014.

[17] Wang Lian-guo, Li Hai-liang, Zhang Jian. Numerical simulation of creep characteristics of soft roadway with bolt-grouting support[J]. Journal of Central South University of Technology,2008,15(1):391-396.

[18] 胡功笠,田艳凤,张超,等.软岩条件下锚注加固锚杆布置参数与注浆浆液参数的确定[J].岩土力学,2003(S2):267-270.

Hu Gong-li, Tian Yan-feng, Zhang Chao, et al. The confirmation of disposal-parameters of bolting and serosity-parameter of injection-grouting of bolting-grouting reinforcement in soft-rock[J]. Rock and Soil Mechanics,2003(S2):267-270.

[19] 王琦.深部厚顶煤巷道围岩破坏控制机理及新型支护系统对比研究[D].济南:山东大学,2012.

Wang Qi. Research on Control Mechanism of Surrounding Rock Failure in Deep Roadways with Thick Top-coal and Contrast of New Support Systems[D]. Jinan: Shandong University,2012.

[20] Wang Qi, Jiang Bei, Pan Rui, et al. Failure mechanism of surrounding rock with high stress and confined concrete support system[J]. International Journal of Rock Mechanics and Mining Sciences,2018(102):89-100.

[21] 王德超,王琦,李术才,等.深井综放沿空掘巷围岩变形破坏机制及控制对策[J].采矿与安全工程学报,2014(05):665-673.

Wang De-chao, Wang Qi, Li Shu-cai, et al. Mechanism of rock deformation and failure and its control technology of roadway driving along next goaf in fully mechanized top coal caving face of deep mines[J]. Journal of Mining and Safety Engineering, 2014(05):665-673.

[22] Kang Yong-shui, Liu Quan-sheng, Gong Guang-qing, et al. Application of a combined

support system to the weak floor reinforcement in deep underground coal mine[J]. International Journal of Rock Mechanics & Mining Sciences,2014(71):143-150.

[23] Wang Fang-tian, Zhang Cun, Wei Shuai-feng, et al. Whole section anchor-grouting reinforcement technology and its application in underground roadways with loose and fractured surrounding rock[J]. Tunnelling and Underground Space Technology,2016(51):133-143.

[24] 王波,高延法,朱伟,等.龙口海域软岩巷道锚注支护试验研究[J].采矿与安全工程学报,2008(02):150-153.
Wang Bo, Gao Yan-fa, Zhu Wei, et al. Experimental study of bolt-grouting support system in soft rock roadway of longkou sea area[J]. Journal of Mining and Safety Engineering,2008(02):150-153.

[25] 孟庆彬,韩立军,乔卫国,等.大断面软弱破碎围岩煤巷演化规律与控制技术[J].煤炭学报,2016,41(08):1885-1895.
Meng Qing-bin, Han Li-jun, Qiao Wei-guo, et al. Evolution law and control technology of surrounding rock for weak and broken coal roadway with large cross section[J]. Journal of China Coal Society,2016,41(08):1885-1895.

[26] 王连国,缪协兴,董健涛,等.深部软岩巷道锚注支护数值模拟研究[J].岩土力学,2005(06):983-985.
Wang Lian-guo, Miao Xie-xing, Dong Jian-tao, et al. Numerical simulation research of bolt-grouting support in deep soft roadway[J]. Rock and Soil Mechanics,2005(06):983-985.

[27] 王连国,韩继胜,孙求知.软岩巷道锚注支护效果的数值模拟研究[J].山东科技大学学报(自然科学版),2001,20(1):53-56.
Wang Lian-guo, Han Ji-sheng, Sun Qiu-zhi. Numerical simulation research for the effect of bolt-grouting support in soft rock roadways[J]. Journal of Shandong University of Science and Technology(Natural Science),2001,20(1):53-56.

基于大范围岩层控制技术的大倾角煤层区段煤柱失稳机理研究

伍永平[1,2]，皇甫靖宇[1]，解盘石[1,2]，胡博胜[1]

(1. 西部矿井开采及灾害防治教育部重点实验室，陕西 西安，710054；
2. 西安科技大学 能源学院，陕西 西安，710054)

摘　要：为了实现大范围岩层控制技术，提高大倾角煤层长壁工作面"R-S-F"系统稳定性，本文在分析区段采空区围岩失稳机制的基础上，建立区间围岩失稳模型，研究了区段煤柱的应力分布规律和失稳破坏准则，确定了区段煤柱的合理尺寸。结果表明：大倾角煤层区段围岩连通运移可总结为"挤压-弯垮"与"压垮-倾倒"两种模式。而要实现大范围岩层控制技术，应使区段采空区围岩以"压垮-倾倒"模式进行失稳连通。覆岩结构以"梁-拱组合梁"形式存在时，区段煤柱强度决定了区段采空区的连通运移模式，以"双拱连续梁"形式存在时，区段煤柱尺寸决定了"压垮-倾倒"连通运移模式的矿压显现特征及变形破坏后的采空区三维空间形态。因而大范围岩层控制技术的关键在于区段煤柱的设计。结合区段煤柱自身强度及失稳破坏准则，确定其宽度 k 须在满足 $k>k_{ZH}$ 条件的同时尽量满足 $k<k_{SK}$。最后以甘肃东峡煤矿大倾角煤层多区段开采为例，进行理论计算判定与相似模拟实验对比分析，从而对本文提出的区段煤柱失稳机理进行了验证。

关键词：大倾角煤层；区段大范围岩层控制；区段煤柱；稳定性

Research on laws of stope inclined support pressure distribution with large dip angle and large mining height

WU Yongping[1,2], HUANGFU Jingyu[1], XIE Panshi[1,2], HU Bosheng[1]

(1. Key Laboratory of Western Mine Exploitation and Hazard Prevention, Ministry of Education, Xi'an 710054, China;
2. College of Energy Science and Engineering, Xi'an University of Science and Technology, Xi'an 710054, China)

Abstract: In order to achieve a large-scale ground control between sections technology and improve the stability of steeply dipping seam longwall face "Roof-Support-Floor" system, having based on the analysis of section of gob sur-rounding rock instability mechanism, this article established the inter section rock instability model, researched the stress distribution law and instability criterion of the section coal pillar, and determined the reasonable size of the section coal pillar. The results show that the steeply dipping seam

基金项目：国家自然科学基金重点项目(51634007,51604212)；国家自然科学基金资助项目(51774230)。
作者简介：伍永平(1962—)，男，陕西汉中人，教授，博士生导师，主要从事大倾角煤层开采研究工作。E-mail：wuyp@xust.edu.cn。
通信作者：皇甫靖宇，Tel：1599552191，E-mail：2676349112@qq.com。

section of the surrounding rock movement can be summarized as two modes of "extrusion-bend down" and "crush-dumping". In order to realize this technology,the surrounding rock of the gob will loss of stability in a "crush-dumping" mode. When the overburden structure exists in the form of "beam arch combined beam", the strength of the section coal pillar determines the movement mode of the surrounding rock in the section of the gob. And when the overburden structure is in the form of "double arch continuous beam",the size of section pillar determines the pressure behavior characteristics of the "crush-dumping" mode and the gob's three-dimensional spatial form after the deformation and failure. So the key of this technology lies in the design of the section coal pillar. According to the strength and instability criterion of the section coal pillar, it is determined that the width of k must satisfy the $k>k_{ZH}$ condition and meet the $k>k_{SK}$ condition as much as possible. Finally,taking the multi-section mining of steeply dipping seam in Gansu Dongxia coal mine as an example, the comparative analysis of the theoretical calculation and similar simulation experiment is carried out,and the mechanism of the pillar instability is verified.

Keywords：steeply dipping seam;large-scale ground control between sections;section coal pillar;stability

大倾角煤层指埋藏倾角为 35°～55°的煤层,是采矿界公认的难采煤层[1]。大倾角煤层多区段开采时,上区段回采结束后形成下部密实充填、中部部分充填、上部悬空特性的采空区。进行下区段开采时,重复开采扰动下区段煤柱可能发生破坏,上区段垮落矸石及高位岩层会向下区段采空区运移充填,将下区段工作面非均匀充填区和悬空区域上移,扩大了下区段采空区充填范围,形成了"大工作面"。基于此,有学者提出了大倾角煤层区段大范围岩层控制(large-scale ground control between sections,LGCS)概念[2],即在多区段工作面开采过程中,弱化区段煤柱的作用,使上区段经历采动的覆岩与垮落岩体参与到下区段工作面的岩层运动。LGCS 有利于提高大倾角煤层"顶板(Roof)-支架(Support)-底板(Floor)"(R-S-F)系统的完整性和稳定性[1];同时瓦斯、一氧化碳等有毒有害气体会向"大工作面"的上部悬空区转移,使下区段生产环境得到改善,利于安全回采[3]。

为了在大倾角煤层长壁开采工作面中成功实现 LGCS 技术,提高工作面"R-S-F"系统稳定性和改善工作面生产环境。必须掌握多区段开采时区段围岩失稳机制,为大倾角煤层 LGCS 技术的工程应用提供理论依据。而分隔两区段采空区的区段煤柱,不仅对区段间围岩失稳连通具有直接意义,也影响着围岩破坏运移后的三维空间形态。国内学者也对相关问题进行了研究,如高玮[4-6]研究了条带煤柱稳定性与煤层倾角及岩体层面效应的关系,魏峰远[7]总结出了煤层埋藏、工作面采高及煤层倾角等对煤柱稳定性的影响,李小军[8]探讨了倾角变化对采场区段煤柱内应力分布的影响,屠洪盛[9]分析了区段煤柱处于不同倾角、不同宽度情况下区段煤柱的破坏特征和煤柱周围应力分布规律。但这些研究均未研究区段煤柱与区段间围岩失稳机制的影响关系。因此,本文着重从大倾角煤层长壁工作面区段间围岩失稳机制入手,研究了区段煤柱的失稳机制,并基于以上研究进行了模型试验验证。为 LGCS 技术的工程应用以及大倾角煤层长壁工作面"支架-围岩"系统稳定性控制提供一种新的、科学的方法。

1 区段间围岩失稳机制

要实现大倾角煤层 LGCS 技术需要考虑两方面因素,一是相邻区段工作面采空区围岩的失稳连通方式的控制,二是采空区围岩破坏运移后的三维空间形态。前者是大范围岩层控制实施的基础,后者影响连通后上区段垮落矸石对下区段采空区的充填效果。通常区段间围岩失稳机制决定了围岩连通运移模式,围岩连通运移模式决定了围岩破坏运移后的三维空间形态。

1.1 区段间围岩区域划分及失稳模式

相邻区段采场围岩的连通运移是一个复杂的过程,围岩结构应力状态与其自身的强度决定了结构首先破坏的所在位置与之后的运动形式及其整体的变形破坏分布。相邻区段采空区围岩结构如图1所示可分为三个区域,即上区段采空区、区段煤柱区域和下区段采空区。

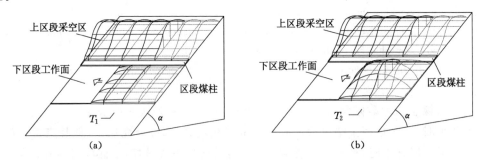

图1 大倾角煤层区段采空区围岩结构

Fig.1 The surrounding rock structure of the goaf in the section of the steeply dipping seam

(a) 下区段未周期来压时围岩空间结构;(b) 下区段周期来压后围岩空间结构

由于上区段采空区围岩结构在经历采动作用后已趋于稳定,且不受下区段工作面采动的直接影响,因此上区段采空区不会首先发生破坏,而区段煤柱区域或下区段采空区二者之一会首先破坏。根据围岩首先破坏区域发生的不同位置,综合围岩连通运移的时空逻辑,考虑首先破坏区位置对围岩失稳连通和破坏运移后三维空间形态的影响,可根据区段采空区围岩连通运移的过程,将区段间围岩连通运移总结为两种模式。

第一种是区段煤柱区域首先破坏时的围岩"挤压-弯垮"连通运移模式。即随着下区段工作面回采,围岩应力状态改变,区段煤柱区域首先破坏失稳,导致煤柱上部岩层向底板方向运动挤压下区段采空区顶板覆岩产生破坏,区段围岩整体失稳,两区段采空区连通。从岩层控制理论的角度分析,"挤压-弯垮"连通运移模式的实质是下区段采空区顶板还未达到走向方向极限状态(即未发生周期垮落),区段煤柱区域便发生失稳,如图1(a)所示。

第二种是下区段采空区首先破坏时,区段间围岩将发生"压垮-倾倒"连通运移模式。这种模式中的围岩运移过程如下:首先是回采后,下区段采空区基本顶垮落,形成特殊的覆岩空间"壳体结构",此时上区段采空区覆岩结构近似于沿走向无变化的柱壳,如图1(b)所示。然后区段煤柱区域或上下区段采空区覆岩空间结构发生失稳破坏,煤柱上部岩层沿煤层倾向向下运移,两区段采场围岩整体失稳,采空区连通。此模式实质是下区段采空区顶板首先发生周期垮落,形成新区段覆岩结构。区段煤柱区域或上下区段采空区覆岩空间结构在新区段覆岩结构状态下发生的失稳。

1.2 区段间围岩失稳模式调控

(1)"挤压-弯垮"失稳模式

"挤压-弯垮"连通运移模式中,由于区段煤柱区域首先破坏,使下区段顶板岩层受挤压被动破断,无法形成稳定的单跨梁板结构。因而下区段顶板上方岩层未经历应力集中,岩体内部损伤发育不足,无法充分破断,故形成高位残余顶板结构。同时,高位残余顶板结构的存在也导致覆岩应力无法充分向下方岩体传递,下区段采空区靠近工作面区域顶板不能充分破断,形成下部残余结构并承担部分荷载,使垮落体不能进行充分的滑移,充填状态无序化,令倾向各区域"R-S-F"系统完整性和稳定性更为混乱,充填效果不理想,如图2所示。

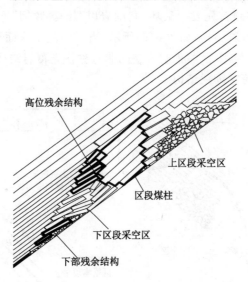

图 2 "挤压-弯垮"失稳模式空间结构简图

Fig. 2 Spatial structure diagram of "extrusion-bend down" instability mode

"挤压-弯垮"连通运移模式中这种高位与下部残余结构可能会形成无法预测时间与强度的大规模二次破断与二次充填,影响工作面支架上方顶板的力学状态,对回采工作造成危害。且这种二次破断充填的随机性,无法进行精准的预管理。故在大倾角煤层LGCS技术中应加以避免。

(2)"压垮-倾倒"失稳模式

"压垮-倾倒"围岩失稳模式中,由于围岩运移分为两个阶段,采场相应地会经历两次矿压显现,对工作面的支架"R-S-F"系统产生两次影响。但因两次矿压显现的间隔时间由第二阶段的覆岩结构状态(即下文中的"双拱连续梁"覆岩结构)下的岩层层位的力学状态及强度决定(当其强度不能够承受应力状态时,第二阶段的围岩运移会立即开始;当其强度能够承受应力状态时,覆岩须经一段时间发生流变效应后,其强度不能承受应力状态时,才发生第二阶段的围岩运移),故可通过对煤柱区域强度的控制实现对间隔时间的控制。

"压垮-倾倒"失稳模式因首先发生下区段顶板覆岩的周期来压,故下区段顶板岩层能够得到充分破断,在失稳连通后不易形成大规模的残余结构,因而不会产生残余结构的二次破断充填。所以此模式下的区段工作面充填效果优于"挤压-弯垮"失稳运动模式。

由上述分析可知,"压垮-倾倒"失稳模式更有助于改善大倾角煤层长壁工作面"R-S-F"系统完整性和稳定性。因而在大倾角煤层LGCS技术中,应使区段采空区围岩以"压垮-倾倒"连通模式发生失稳连通。

2 区段间围岩失稳力学模型

"压垮-倾倒"连通运移模式中采空区覆岩会经历两种结构状态,一是"梁-拱组合梁"覆岩结构,如图3(a)所示,下区段工作面基本顶未来压破断形成一端固支的单垮梁结构,同时上区段采空区覆岩结构在倾向方向上呈非对称拱形[10],两者组成两跨梁拱组合梁结构。随着下区段的回采,若区段煤柱区域首先发生失稳,则围岩以"挤压-弯垮"连通运移模式进行失稳连通;若下区段基本顶首先发生破坏,则围岩以"压垮-倾倒"连通运移模式进行失稳连通。二是"双拱连续梁"覆岩结构,如图6(a)所示,当下区段工作面周期来压后,下区段采空区的覆岩空间"壳体结构"与上区段采空区空间非对称柱壳覆岩结构在倾向方向上形成了由两个非对称拱组成的两跨连续梁结构。

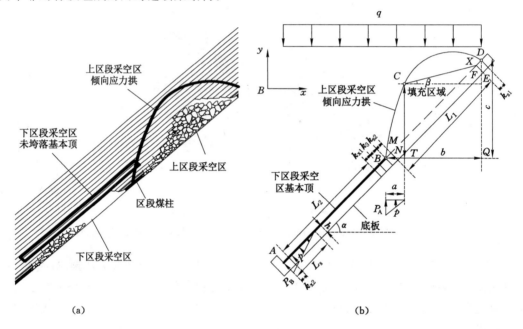

图3 "梁-拱组合梁"覆岩结构及模型

Fig. 3 The overburden structure model of "beam arch combined beam"

(a) 覆岩结构;(b) 力学模型

2.1 "梁-拱组合梁"模型

(1)模型建立

在采空区距工作面约半个来压步距处[即图1(a)中 T_1 截面,此处受前方煤壁及后方"大工作面"影响最小],沿走向取单位宽度建立"梁-拱组合梁"覆岩结构模型并进行分析,建立如图3(b)所示的双跨梁 $ABCD$ 模型。其中,下区段工作面 A 处为固定支座,B 处为单铰支座,BCD 段为倾向应力拱。煤层倾角为 α,煤层厚度为 h,上下区段工作面长度为 L_1 和 L_2,设下区段基本顶厚度为 A,弹性模量为 E,惯性矩为 I。两区段工作面上下煤柱塑性区宽度分别为 k_{s1}、k_{s2} 和 k_{x1}、k_{x2},区段煤柱弹性区宽度为 k_3,区段煤柱宽度为弹性区宽度 k_3 加两侧塑性区宽度 k_{s2}、k_{x1}。以 B 点为坐标原点,水平向右为 x 轴,垂直向上为 y 轴,建立坐标系 Bxy。

由于相邻两工作面长度在垂直方向的投影远小于煤层埋深,因此将覆岩荷载简化为均布荷载 q。上区段工作面垮落矸石充填区域为 $BCEXNB$,休止角为 β,并根据其非均匀充填特征,假设其对 BC 段的作用满足三角形载荷分布特征,即:

$$p(x) = \frac{P_A}{a}(a-x) \tag{1}$$

其中,P_A 为常量,设半拱 BC 段及 CD 段的轴线方程为:

$$y_1 = y_1(x), 0 \leqslant x \leqslant a \tag{2}$$

$$y_2 = y_2(x), a \leqslant x \leqslant b \tag{3}$$

下区段采空区基本顶下部被直接顶垮落滑移矸石填充,并沿倾斜方向向上堆积逐渐松散[11,12]。故将矸石对顶板荷载简化为沿倾向法线向上的三角形分布,设充填区域倾向长度为 L_s,矸石支撑力按照密实程度不同取上覆岩层载荷的 0.4 倍,即:

$$p_{2(x)} = 0.4 \frac{q}{L_s} \cdot [(L_1 + k_{x1} + k_{x2} + k)\cos\alpha - x] \tag{4}$$

(2)区段煤柱处约束力求解

双跨梁 $ABCD$ 为超静定结构,使用结构力学中的力法进行求解[13],取隔离体如图 4 所示。对 BCD 段建立平衡方程,求出 AB 段对 BCD 段空间壳体结的约束力 F_{Bx1}、F_{By1} 以及上区段上部拱脚 D 处的约束力 F_{Dx}、F_{Dy} 为[14]:

$$F_{Bx1} = F_{Dx} = \left[b\left(-\frac{1}{3}a^2 P_A + \frac{1}{2}a^2 q\right) - a\left(\frac{1}{6}a^2 P_A - \frac{1}{2}abP_A + \frac{1}{2}b^2 q\right)\right]/(ac-bf) \tag{5}$$

$$F_{By1} = \left[c\left(-\frac{1}{3}a^2 P_A + \frac{1}{2}a^2 q\right) - f\left(\frac{1}{6}a^2 P_A - \frac{1}{2}abP_A + \frac{1}{2}b^2 q\right)\right]/(ac-bf) \tag{6}$$

$$F_{Dy} = \left[-c\left(-\frac{1}{3}a^2 P_A + \frac{1}{2}a^2 q\right) + f\left(\frac{1}{6}a^2 P_A - \frac{1}{2}abP_A + \frac{1}{2}b^2 q\right)\right]/$$
$$(ac-bf) + \left[-\frac{1}{2}aP_A + aq + (b-a)q\right] \tag{7}$$

其中未知量 a、f 大小可由式(8)~(15)联立求得,式中,φ、c_0 为区段煤柱的内摩擦角和黏聚力。

$$y_1(x) = x\{P_A[a^2(2c+f) - bfx^2 - 3a^2(bf+cx) + ax(3bf+cx)] + 3xq[ax - a^2c + bf(b-x)]\}/[a^2(a-b)(aP_A - 3bq)] \tag{8}$$

$$y_2(x) = \{a^2 P_A(ac - bf - cx + fx) + 3xq[acx - a^2c + bf(b-x)]\}/[a^2(a-b)(aP_A - 3bq)] \tag{9}$$

$$S_{ABQ} = \frac{1}{2}bc \tag{10}$$

$$S_{MNT} = \frac{1}{2}\tan\alpha\left(a - \frac{h}{\sin\alpha}\right)^2 \tag{11}$$

$$S_{CDQT} = \frac{[f + b\tan\alpha](b-a)}{2} \tag{12}$$

$$S_{MXQT} = \frac{b-a}{2}\left(a + b + \frac{2h}{\sin\alpha}\right)\tan\alpha \tag{13}$$

$$(\lambda - 1)\int_0^a y_1 dx + \lambda\left(\int_a^b y_2 dx - S_{ABQ}\right) + S_{MNT} - S_{CDQT} + S_{MXQT} = 0 \tag{14}$$

$$F_{Ax} = F_{Ay}\tan\varphi + c_0 \tag{15}$$

下区段基本顶暨隔离体 AB 段受力情况如图 4(b)所示,F_{Bx1}、F_{By1} 与 F_{Bx2}、F_{By2} 互为反

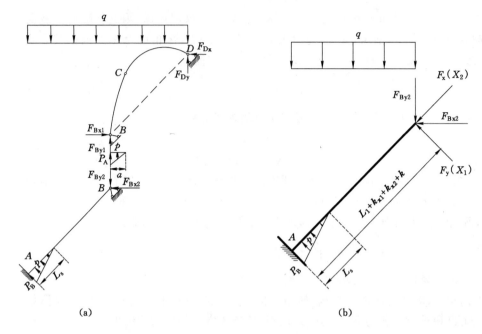

图 4 "梁-拱组合梁"结构隔离体

Fig. 4 Isolation body of "beam arch combined beam" structure

(a)"梁-拱组合梁"结构隔离体;(b) AB 段基本体系板垮落

力。将铰支座 B 去除,代以多余未知力 X_1、X_2,建立力法的典型方程:

$$\begin{cases} \delta_{11}X_1 + \delta_{12}X_2 + \Delta_{1P} = 0 \\ \delta_{21}X_1 + \delta_{22}X_2 + \Delta_{2P} = 0 \end{cases} \tag{16}$$

根据图乘法求得各项系数和自由项,可求得 X_1、X_2,即为煤柱对覆岩的约束力 F_{By3}、F_{Bx3}。

$$F_{By3} = X_1 = F_{By2}\cos\alpha - F_{Bx2}\sin\alpha + \frac{3}{8}q\cos\alpha \cdot (L_2 + k_{x1} + k_{x2} + k) +$$

$$\frac{q_B L_s^3}{6}/(L_2 + k_{x1} + k_{x2} + k)^2 - \frac{q_B L_s^4}{18}/(L_2 + k_{x1} + k_{x2} + k)^3 \tag{17}$$

$$F_{Bx3} = X_2 = -F_{By2}\sin\alpha - F_{Bx2}\cos\alpha - \frac{q}{L_2 + k_{x1} + k_{x2} + k}\sin\alpha \tag{18}$$

(3) 区段煤柱应力状态

根据大倾角煤层 LGCS 技术的要求,在此种覆岩结构下,煤柱强度应当足够承担所受荷载。而煤柱形成煤柱上侧塑性破坏区、煤柱弹性区以及煤柱下侧塑性破坏区三部分,为简化计算,现进行合理放大,即假设外部荷载均作用于煤柱弹性区。

现建立区段煤柱倾斜力学模型(如图 5 所示)对"梁-拱组合梁"覆岩结构时煤柱应力状态进行分析。设其中区段煤柱宽度 $k_3 = 2k'$、煤层厚度 $h = 2h'$。

通过对以往大倾角煤层开采的物理相似模拟实验现象进行总结[15-18],发现在常规采高的情况下,由于上区段工作面顶板在垮落后沿倾向形成了砌体梁结构,阻挡了上区段采空区垮落矸石对区段煤柱侧方的挤压。故对垮落矸石的推力忽略不计。同时对煤柱自重省略不计。

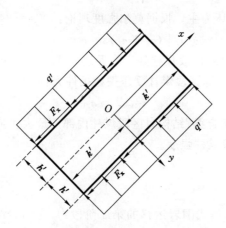

图 5 区段煤柱力学模型
Fig. 5 Mechanical model of section coal pillar

覆岩亦对区段煤柱施加作用,其等效集中荷载与覆岩受到的约束力 F_{By3}、F_{Bx3} 互为反力。且由于力的传递和反作用,底板也对煤体形成了作用,其顶底板作用效果呈轴对称。故上下两侧均布荷载 q' 及面力 F_x 为:

$$q' = \left[F_{By2}\cos\alpha - F_{Bx2}\sin\alpha + \frac{3}{8}q\cos\alpha \cdot (L_2 + k_{x1} + k_{x2} + k) + \frac{q_B L_s^3}{6} \right/$$

$$\left. (L_2 + k_{x1} + k_{x2} + k)^2 - \frac{q_B L_s^4}{18(L_2 + k_{x1} + k_{x2} + k)^3} \right]/k_3 \tag{19}$$

$$F_x = F_{By2}\sin\alpha + F_{Bx2}\cos\alpha + \frac{q}{L_2 + k_{x1} + k_{x2} + k}\sin\alpha \tag{20}$$

由于在上下边界($y = \pm h$)上,挤压应力 σ_y 主要由垂直于煤层倾角的均布荷载引起。荷载大小 q 不随 x 而变化,因而可以假设应力分量 σ_y 只是 y 的函数:

$$\sigma_y = \frac{\partial^2 \Phi}{\partial x^2} = f(y) \tag{21}$$

可得到应力函数的表达式为:

$$\Phi = \frac{x^2}{2}(Ay^3 + By^2 + Cy + D) + x(Ey^3 + Fy^2 + Gy) - \frac{A}{10}y^5 - \frac{B}{6}y^4 + Hy + Ky^2 \tag{22}$$

应力分量的表达式可由应力函数求出,并根据区段煤柱的边界条件确定常数。经计算,得应力分量的最后解答为:

$$\sigma_x = F_x\left(\frac{2y}{xh} - \frac{x}{h} + \frac{h}{6x} - \frac{k_3^2}{4xh}\right), \sigma_y = F_x(h^2/4 - y^2)/xh - q', \tau_{xy} = 2yF_x/h \tag{23}$$

(4) "梁-拱组合梁"结构区段煤柱失稳准则

将求得的应力分量式(23)代入主应力求解计算公式,即可得到区段煤柱内任意一点的主应力表达式:

$$\sigma_1, \sigma_3 = \frac{\sigma_x + \sigma_y}{2} \pm \sqrt{\left(\frac{\sigma_x - \sigma_y}{2}\right)^2 + \tau_{xy}^2} \tag{24}$$

由于大倾角煤层开采的物理相似模拟实验中,区段煤柱产生滑剪现象,可以判断出区段

煤柱的失稳模式以压剪破坏为主。根据莫尔强度理论:

$$f(\sigma_1,\sigma_3) = (\sigma_1-\sigma_3)/[(\sigma_1+\sigma_3)\sin\varphi + 2c_0\cot\varphi](\max(\sigma_1,\sigma_2)) \quad (25)$$

当函数 $f(\sigma_1,\sigma_3)=1$ 时,区段煤柱处于临界破坏状态;当 $f(\sigma_1,\sigma_3)>1$ 时,在对应的位置 (x_i,y_i) 发生压剪破坏。即区段煤柱受压失稳条件为:

$$\sigma_{ZHmax} > \sigma_{ZH} \quad (26)$$

式中,σ_{ZHmax} 为覆岩"梁-拱组合梁"结构时区段煤柱内部的最大主应力;σ_{ZH} 为覆岩"梁-拱组合梁"结构应力状态时的区段煤柱强度。

2.2 "双拱连续梁"模型

(1) 模型建立

"压垮-倾倒"连通运移模式围岩运移的第二阶段中,在围岩结构的极限状态处(下区段采空区空间壳体结构走向垮落高度最高的壳肩截面处,即图1中 T_2 截面),沿走向取单位宽度建立"双拱连续梁"覆岩结构模型并进行分析。建立如图6(b)所示的双跨梁 $AC'D'BCD$ 模型,图中各部位尺寸与"梁-拱组合梁"模型相同。下区段工作面非均匀充填特征与上区段类似,设垮落矸石充填区域休止角为 β'。

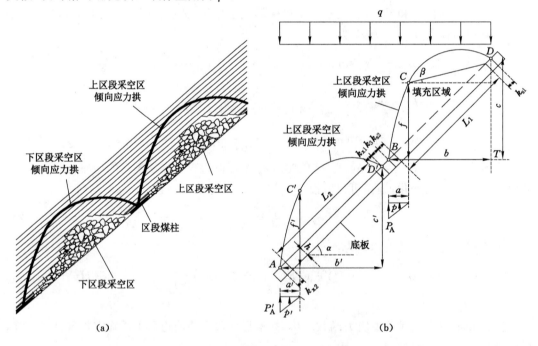

图 6 "双拱连续梁"覆岩结构及模型

Fig. 6 The overburden structure model of "double arch continuous beam"

(a) 覆岩结构;(b) 力学模型

(2) 区段煤柱处约束力求解

建立平衡方程,可求出 $AC'D'$ 段在 D' 处的约束力 $F_{D'x}$、$F_{D'y}$ 以及拱脚 A 处的约束力 F_{Ax}、F_{Ay},其表达式为:

$$\begin{cases} F_{Ax} = F_{D'x} = \left[b'\left(-\frac{1}{3}a'^2 P_A' + \frac{1}{2}a'^2 q\right) - a'\left(\frac{1}{6}a'^2 P_A' - \frac{1}{2}a'b'P_A' + \frac{1}{2}b'^2 q\right) \right] / (a'c' - b'f') \\ F_{Ay} = \left[c'\left(-\frac{1}{3}a'^2 P_A' + \frac{1}{2}a'^2 q\right) - f'\left(\frac{1}{6}a'^2 P_A' - \frac{1}{2}a'b'P_A' + \frac{1}{2}b'^2 q\right) \right] / (a'c' - b'f') \\ F_{D'y} = \left[-c'\left(-\frac{1}{3}a'^2 P_A' + \frac{1}{2}a'^2 q\right) - f'\left(\frac{1}{6}a'^2 P_A' - \frac{1}{2}a'b'P_A' + \frac{1}{2}b'^2 q\right) \right] / (a'c' - b'f') + \\ \qquad \left[-\frac{1}{2}a'P_A' + a'q + (b' - a')q \right] \end{cases}$$

(27)

区段煤柱对覆岩 BD' 处的约束力 F_{x1}、F_{y1}：

$$\begin{cases} F_{x1} = F_{By}\sin\alpha + F_{Bx}\cos\alpha - F_{D'x}\cos\alpha - F_{D'y}\sin\alpha \\ F_{y1} = F_{By}\cos\alpha + F_{D'x}\sin\alpha - F_{D'x}\cos\alpha - F_{By}\sin\alpha \end{cases}$$

(28)

（3）区段煤柱处约束力求解

同样的建立区段煤柱倾斜力学模型，得区段煤柱上应力分量为：

$$\sigma_x = -F_{x1}\left(\frac{2y^2}{xh} - \frac{x}{h} + \frac{h}{6x} - \frac{k_3^2}{4xh}\right), \sigma_y = -F_{x1}(h^2/4 - y^2)/xh + F_{y1}/k_3, \tau_{xy} = -2yF_{x1}/h$$

(29)

（4）"双拱连续梁"结构区段煤柱失稳准则

同理可得区段煤柱受压失稳条件为：

$$\sigma_{SKmax} > \sigma_{SK} \tag{30}$$

式中，σ_{SKmax} 为覆岩"双拱连续梁"结构时区段煤柱内部的最大主应力；σ_{SK} 为覆岩"双拱连续梁"结构应力状态时的区段煤柱强度。

3 区段煤柱强度分析及控制

3.1 区段煤柱失稳类型分析

上节通过对两种覆岩结构的力学分析，得到了区段煤柱在各阶段内的失稳准则。现结合大倾角煤层 LGCS 技术的要求，对可能出现的三种应力情况进行分析：

（1）当 $\sigma_{ZHmax} > \sigma_{ZH}$。此情况即覆岩"梁-拱组合梁"结构状态下区段煤柱中最大应力超过其强度，区段煤柱先于下区段基本顶失稳破坏，区段采空区围岩会以"挤压-弯垮"连通运移模式进行破坏连通充填。根据上文的分析，应避免此模式的失稳连通，需提高此状态下下区段煤柱的强度。可加强区段煤柱的支护强度，也可根据上文计算过程，合理调整两区段工作面回采设计参数，如煤柱宽度、上下区段工作面长度等。

（2）当 $\sigma_{ZHmax} < \sigma_{ZH}$，而 $\sigma_{SKmax} < \sigma_{SK}$。此情况时区段采空区围岩会以"压垮-倾倒"连通运移模式进行失稳连通，但在"双拱连续梁"结构状态下，区段煤柱未立即失稳，两次垮落有时间间隔。故需在下区段顶板垮落后进行人为控制，降低区段煤柱强度。可对采空区的区段煤柱使用爆破或水压致裂等技术手段，控制区段煤柱失稳使围岩二次垮落。但由于需进行失稳作业的煤柱位于采空区中，且长度为一个周期来压步距，在技术实施上具有难度。

（3）当 $\sigma_{ZHmax} < \sigma_{ZH}$，而 $\sigma_{SKmax} > \sigma_{SK}$。此情况即为下区段基本顶先于区段煤柱发生破坏失稳后，区段煤柱随即失稳破坏，区段采空区以"压垮-倾倒"连通运移模式自主地进行失稳连通，不需人为干预，且填充效果好。

3.2 区段煤柱合理宽度确定

由上节分析可知,在大倾角煤层 LGCS 技术中区段煤柱强度必须在满足 $\sigma_{ZHmax} < \sigma_{ZH}$ 条件的同时满足 $\sigma_{SKmax} > \sigma_{SK}$,因此可以确定 σ(即 LGCS 技术全过程中区段煤柱内部最大主应力的极值)的区间,即:

$$\sigma = (\sigma_{ZH}, \sigma_{SK}) \tag{31}$$

由于煤体的强度与其应力状态的变化过程有关,故区段煤柱的 σ_{ZH}、σ_{SK} 并不相同,但可通过区段煤柱岩样的实验室应力状态模拟实验测得。同时结合两区段工作面参数,可计算出煤柱在两种覆岩结构状态下所受的垂直于煤层倾向的集中荷载 F_{ZH}、F_{SK}。可假设煤柱在垂直于煤层倾向的集中荷载的作用下内部各处主应力值均为 σ。以此计算,可求得两种覆岩结构状态下的煤柱宽度的边界值 k_{ZH}、k_{SK}(k_{ZH}、k_{SK} 分别为覆岩"梁-拱组合梁"、"双拱连续梁"结构稳定状态时的区段煤柱最小宽度),其公式为:

$$\begin{cases} k_{ZH} = F_{ZH} K/(\sigma_{ZH} \cdot 1) + k_{x1} + k_{s2} \\ k_{SK} = F_{SK} K/(\sigma_{SK} \cdot 1) + k_{x1} + k_{s2} \end{cases} \tag{32}$$

式中 K——安全系数,可根据实验及工程实际总结确定。

结合大倾角煤层 LGCS 技术要求,可确定区段煤柱宽度 k 须在满足 $k > k_{ZH}$ 条件的同时尽量满足 $k < k_{SK}$。

4 试验验证

为验证大倾角煤层 LGCS 技术中区段煤柱失稳判别公式的准确性,并揭示区段煤柱的失稳机理与采空区围岩连通运移模式之间的关系,现选取东峡煤矿大倾角特厚易燃煤层区段开采为例进行验证,上下区段工作面斜长 110 m,走向长度 1 118.5 m,倾角 38°。

首先将东峡煤矿大倾角煤层区段工作面的实际参数代入区段煤柱失稳的理论判别公式中,其中区段煤柱宽度为 40 m,经计算可求得 $\sigma_{ZHmax} > \sigma_{ZH}$。此结果说明区段采空区围岩会以"压垮-倾倒"连通运移模式进行连通充填。

同时进行区段岩层控制物理相似模拟实验。结合实际参数,选用 2 200 mm×2 000 mm×20 mm 的平面模型架进行实验,并配合使用数码摄像机、PENTAX R-400NX 型光学全站仪进行围岩破坏运移形态的记录、顶板覆岩移动量的监测,以此构建了区段岩层控制实验系统(模型及监测系统如图 7 所示)。

对模型进行模拟开挖,在开采下区段过程中,区段煤柱破坏逐渐严重。当下区段采完时,区段煤柱已完全破坏,不能起到承担载荷的作用,但仍能阻挡上方破碎煤体或岩石向下滚(滑)动,如图 8(a)所示。模型下区段采通时,由于区段煤柱完全失稳,因而煤柱上方岩层失稳并挤压下区段顶板岩层,引起顶板覆岩垮落,并在下区段覆岩较高位与下部形成残余结构,如图 8(b)所示。一段时间后,较高位岩层的残余结构失稳向下区段采空区倾倒,对下部垮落矸石及残余结构产生作用,如图 8(c)所示。这种采空区围岩的二次来压给下区段工作面带来随机性的作用,影响工作面"R-S-F"系统稳定性。围岩变形破坏空间分布规律如图 9 所示。

模型实验结果表明,在区段煤柱首先破坏失稳的情况下,采空区围岩的失稳连通过程符合区段间围岩"挤压-弯垮"连通运移模式下的围岩连通运移演化规律、矿压显现特征及变形

图 7 实验模型及监测系统
Fig. 7 Experimental model and monitoring system

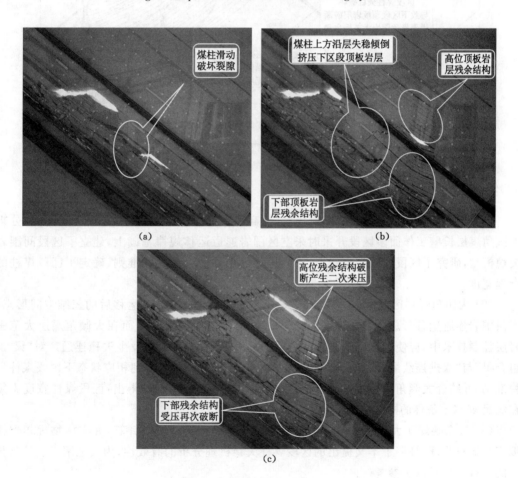

图 8 围岩破坏状态演化过程
Fig. 8 State of surrounding rock damage evolution process

破坏后的采空区三维空间形态,证明了本文提出的区段煤柱失稳机理分析对于区段采空区围岩控制的有效性。

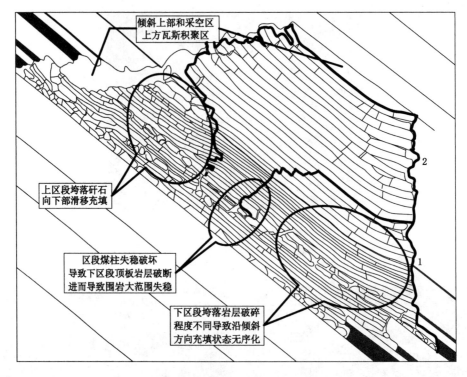

图 9　围岩变形破坏空间分布规律

Fig. 9　Spatial distribution law of deformation and failure of surrounding rock

5　结　论

（1）本文以实现大范围岩层控制技术，提高工作面"R-S-F"系统稳定性为目的，在分析大倾角煤层长壁工作面多区段开采时采空区围岩连通运移规律基础上，建立了区段间围岩失稳模型，研究了区段煤柱受力状态，得到了区段煤柱的失稳破坏准则，确定了区段煤柱的合理宽度。

（2）大倾角煤层区段采空区的围岩失稳机制决定了围岩破坏运移后的三维空间形态，区段围岩连通运移可总结为"挤压-弯垮"与"压垮-倾倒"两种模式，而在大倾角煤层大范围岩层控制技术中，应使区段间采空区围岩以"压垮-倾倒"连通模式发生失稳连通。对"梁-拱组合梁"与"双拱连续梁"两种覆岩结构状态建立力学模型，分析得到相应状态下区段煤柱失稳准则，并结合大倾角煤层大范围岩层控制技术的要求进行分析，得出：区段煤柱宽度 k 须在满足 $k>k_{ZH}$ 条件的同时尽量满足 $k<k_{SK}$。

（3）以东峡煤矿大倾角特厚易燃煤层区段开采为例进行判别计算，并实施物理相似模拟实验进行验证，证明了本文提出的区段煤柱失稳机理分析的有效性，为大范围岩层控制技术的工程应用提供了参考。

参 考 文 献

[1] 伍永平.大倾角煤层开采"R-S-F"系统动力学控制基础研究[M].西安：陕西科学出版社，2006：1-23.

WU Yongping. Study on dynamics controlling basis of system "R-S-F" in steeply dipping seam mining[M]. Xi'an:Shannxi Science and Technology Publishing House, 2006:1-23.

[2] 伍永平,刘孔智,贠东风,等. 大倾角煤层安全高效开采技术研究进展[J]. 煤炭学报, 2014,39(8):1611-1618.
WU Yongping,LIU Kongzhi,YUN Dongfeng,et al. Research progress on the safe and efficient mining technology of steeply dipping seam[J]. Journal of China Coal Society, 2014,39(8):1611-1618.

[3] 解盘石. 大倾角煤层长壁开采覆岩结构及其稳定性研究[D]. 西安:西安科技大学,2011: 26-30.
XIE Panshi. Response of overburden structure and its stability around the longwall mining face area in steeply dipping seam[D]. Xi'an:Xi'an University of Science and Technology,2011:26-30.

[4] 高玮,姜学云. 条带开采中条带煤柱塑性区宽度分析[J]. 山西矿业学院学报,1997,15 (2):142-147.
GAO Wei,JIANG Xueyun. Analysis of width of plastic zone in left strip coal pillars by strip mining[J]. Shanxi Mining Institute Learned Journal,1997,15(2):142-147.

[5] 高玮. 条带开采岩层移动机理浅析[J]. 矿山压力与顶板管理,2000,17(1):53-55.
GAO Wei. Simple study on strata movement mechanism of strip mining[J]. Ground Pressure and Strata Control,2000,17(1):53-55.

[6] 高玮. 倾斜煤柱稳定性的弹塑性分析[J]. 力学与实践,2001,23(2):23-26.
GAO Wei. The elastic-plastic analysis of stability of inclined coal pillar[J]. Mechanics and Practice,2001,23(2):23-26.

[7] 魏峰远,陈俊杰,邹友峰. 影响保护煤柱尺寸留设的因素及其变化规律[J]. 煤炭科学技术,2006,36(10):85-87.
WEI Fengyuan,CHEN Junjie,ZOU Youfeng. Influence factors and change law of protection coal pillars left[J]. Coal Science and Technology,2006,36(10):85-87.

[8] 李小军,李怀珍,袁瑞甫. 倾角变化对回采工作面区段煤柱应力分布的影响[J]. 煤炭学报,2012,37(8):1270-1274.
LI Xiaojun,LI Huaizhen,YUAN Ruifu. Stress distribution influence on segment coal pillar at different dip angles of working face[J]. Journal of China Coal Society,2012,37 (8):1270-1274.

[9] 屠洪盛,屠世浩,白庆升,等. 急倾斜煤层工作面区段煤柱失稳机理及合理尺寸[J]. 中国矿业大学学报,2013,42(1):6-12.
TU Hongsheng,TU Shihao,BAI Qingsheng,et al. Instability of a coal pillar section located at a steep mining face:Pillar size selection[J]. Journal of China University of Mining & Technology,2013,42(1):6-12.

[10] 伍永平,解盘石,任世广. 大倾角煤层开采围岩空间非对称结构特征分析[J]. 煤炭学报,2010,35(2):182-184.

WU Yongping, XIE Panshi, REN Shiguang. Analysis of asymmetric structure around coal face of steeply dip-ping seam mining[J]. Journal of China Coal Society, 2010, 35(2):182-184.

[11] 王金安,张基伟,高小明,等. 大倾角厚煤层长壁综放开采基本顶破断模式及演化过程（Ⅰ）——初次破断[J]. 煤炭学报,2015,40(6):1353-1360.
WANG Jinan, ZHANG Jiwei, GAO Xiaoming, et al. The fracture mode and evolution of main roof stratum above longwall fully mechanized top coal caving in steeply inclined thick coal seam(I)—initial fracture[J]. Journal of China Coal Society, 2015, 40(6):1353-1360.

[12] 王金安,张基伟,高小明,等. 大倾角厚煤层长壁综放开采基本顶破断模式及演化过程（Ⅱ）——周期破断[J]. 煤炭学报,2015,40(8):1737-1745.
WANG Jin'an, ZHANG Jiwei, GAO Xiaoming, et al. Fracture mode and evolution of main roof stratum above fully mechanized top coal caving longwall coalface in steeply inclined thick coal seam (Ⅱ)—Periodic fracture[J]. Journal of China Coal Society, 2015,40(8):1737-1745.

[13] 龙驭球,包世华,匡文起,等. 结构力学教程（Ⅰ）[M]. 北京:高等教育出版社,2000:315-384.
LONG Yuqiu, BAO Shihua, KUANG Wenqi, et al. Structural mechanics tuto-rial(Ⅰ)[M]. Beijing:Higher Education Press, 2000:315-384.

[14] 罗生虎,伍永平,刘孔智,等. 大倾角煤层长壁开采空间应力拱壳形态研究[J]. 煤炭学报,2016,41(12):2993-2998.
LUO Shenghu, WU Yongping, LIU Kongzhi, et al. Study on the shape of the space stress arch shell in steeply dipping coal seam mining[J]. Journal of China Coal Society,2016,41(12):2993-2998.

[15] 解盘石,伍永平,王红伟,等. 大倾角煤层长壁开采覆岩空间活动规律研究[J]. 煤炭科学技术,2012,40(9):1-5.
XIE Panshi, WU Yongping, WANG Hongwei, et al. Study on space activity law of overburden strata above longwall coal mining face in high inclined seam[J]. Coal Science and Technology, 2012,40(9):1-5.

[16] 尹光志,李小双,郭文兵. 大倾角煤层工作面采场围岩矿压分布规律光弹性模量拟模型试验及现场实测研究[J]. 岩石力学与工程学报,2010,29(S1):3336-3343.
YIN Guangzhi, LI Xiaoshuang, GUO Wenbing. Photo-elastic experi-mental and field measurement study of ground pressure of sur-rounding rock of large dip angle working coalface[J]. Chinese Journal of Rock Mechanics and Engineering, 2010, 29(S1):3336-3343.

[17] 王红伟,伍永平,解盘石,等. 大倾角煤层开采"关键域"转换与岩体结构平衡特征[J]. 辽宁工程技术大学学报(自然科学版),2016,35(10):1009-1014.
WANG Hongwei, WU Yongping, XIE Panshi, et al. Critical zone conversion and rock structure balance characteristics in mining the steeply dipping seam[J]. Journal of

Liaoning Technical University(Natural Science),2016,35(10):1009-1014.
[18] 伍永平,解盘石,王红伟,等.大倾角煤层开采覆岩空间倾斜砌体结构[J].煤炭学报,2010,35(2):1252-1256.
WU Yongping,XIE Panshi,WANG Hongwei,et al. Incline masonry structure around the coal face of steeply dipping seam mining[J]. Journal of China Coal Society,2010,35(2):1252-1256.

大变形巷道螺纹钢锚杆外形优化设计及应用研究

张明[1]，CAO Chen[2,3]，赵象卓[1]，张怀东[1]，
REN Ting[2,3]，马双文[1]，韩军[1,3]

(1. 辽宁工程技术大学 矿业学院，辽宁 阜新，123000；
2. 澳大利亚伍伦贡大学 土木、采矿与环境学院，新南威尔士 伍伦贡，2522；
3. 辽宁省煤炭资源安全开采与洁净利用工程研究中心，辽宁 阜新，123000)

摘　要：针对煤矿大变形巷道支护问题，基于锚杆支护机理，运用厚壁理论，建立了乌东煤矿地质力学条件的锚固力测试实验室模拟模型；根据锚杆锚固失效模式判定和锚固力残余强度分析，对螺纹钢锚杆肋的排列方式和肋间距等外形参数进行了优化，设计了针对大变形巷道的新型锚杆，对比乌东煤矿原常规左旋锚杆进行了实验室和现场拉拔试验。结果表明，相比原左旋锚杆，在实验室仿真条件和现场实际条件下，新型锚杆拉拔力分别提高了13%和30%，吸能能力提高了15%和47%，较原左旋锚杆更适用于乌东矿大变形巷道支护。研究提供的设计方法可用于不同地质条件下大变形巷道锚杆支护设计优化，为提高锚固力学理论水平和大变形巷道支护技术提供新的方法。

关键词：锚杆外形优化设计；锚固破坏方式；残余强度；厚壁理论；大变形巷道

Rebar bolt profile optimisation for large deformation roadway with application study

ZHANG Ming[1], CAO Chen[2,3], ZHAO Xiangzhuo[1],
ZHANG Huaidong[1], REN Ting[2,3], MA Shuangwen[1], HAN Jun[1,3]

(1. College of Mining, Liaoning Technical University, Fuxin 123000, China；
2. University of Wollongong, EIS, NSW Wollongong 2522；
3. Liaoning Province Coal Resources Safety Mining and Clean Utilization
Engineering Research Center, Fuxin 123000, China)

Abstract: Aiming at the supporting problem of the roadway with large deformational in coal mine, based on the rock bolting mechanisms, application of thick walled cylinder theory, the laboratory simulation of anchor mechanics with the geological conditions of Wudong coal mine was established. According to the analysis of rock bolting failure modes, anchorage residual strength, a new rebar bolt for large deformational roadway was designed. Laboratory and field pullout tests were conducted for the new bolts and the commonly used left spiral bolt. Results indicates that compared with the commonly used left spiral bolts,

---

基金项目：辽宁省煤炭资源安全开采与洁净利用工程研究中心开放基金资助课题(55000KY03002)；国家自然科学基金资助项目(51774174)。
作者简介：张明(1993—)，男，河南商丘人，在读研究生。
通信作者：韩军(1980—)，男，内蒙古临河人，教授。Tel:0418-3350469，E-mail: hanj_lntu@163.com。

laboratory and field drawing force of the new bolt separately increases 13％ and 30％, and the energy absorption capacity increases 15％ and 47％. The new bolt can be used for roadway support with similar geological conditions and the design principle and method provided in the study can be applicable to the optimization bolt support under different geological conditions. It provides a new method for improving the theoretical level of anchoring mechanics and supporting technology of large deformational roadway.

Keywords：bolt profile optimization；anchorage failure modes；residual strength；thick walled cylinder theory；large deformation roadway

1 引　　言

锚杆支护作为煤矿井下巷道的主要支护形式,在国内外已经得到广泛应用[1-2]。据不完全统计,我国国有煤矿每年新掘进的巷道总长达 25 000 km[3],目前很多矿区锚杆支护率达到 60％,有些矿区锚杆支护率已超过 90％,甚至达到 100％[4],锚杆支护已占我国煤矿巷道支护总量的 70％以上[5],锚杆支护研究对于我国煤炭工业的发展具有重要的意义。

目前,我国煤矿回采巷道支护一般采用螺纹钢锚杆,部分矿井在采煤帮使用玻璃钢锚杆,与相应护表材料构成巷道的基本支护,再以锚索支护、注浆、喷浆、架棚等方式作为补强支护。随着深部资源开采的进行,煤矿高地应力、易破碎围岩及软岩等大变形巷道问题逐渐增多[6]。

针对大变形巷道支护难题,我国学者采用了多种补强支护方法和特殊支护组合的岩层控制技术。何满潮[7,8]等提出了以恒阻大变形锚网索耦合支护为核心的支护技术。康红普[9]采用强力锚索、拱形大托板及钢筋网组成全断面强力锚索支护系统。郭志飚[10]等提出了以吸收变形能、控制围岩变形为核心,通过恒阻大变形锚杆耦合支护为主体的深部软岩控制对策。郑朋强[11]等采用了 U 型钢可压缩支架和泡沫混凝土填充结合预应力锚索的变形控制方法。高明仕[12]等运用了全断面、全支全让 O 型封闭控制的原理进行支护。马念杰[13]等提出了可接长锚杆支护技术。刘洪涛[14]提出了大变形巷道顶板可接长锚杆耦合支护系统。王卫军[15]提出深部高应力巷道在掘进中预留变形,采用"可接长锚杆＋刚性长螺纹钢锚杆＋锚网＋W 钢带＋喷射混凝土"并辅以可接长锚杆强化顶板支护的支护方式。大变形巷道补强支护研究现已取得了一定成果,并成功运用于工程实践中,但基础支护部分并未改变,且该类支护技术基于新型支护材料及支护组合方式,对工程技术、施工工艺要求较高,支护成本大幅增加。

锚杆外形,即肋形、肋高、肋宽和肋间距等因素对锚杆支护系统的荷载传递具有重要影响。Aydan[16-18]等研究表明,没有横肋或锚杆肋间距过小,其锚固力均明显低于普通螺纹钢锚杆,即一定地质条件下,螺纹钢锚杆存在最佳肋间距值。Fabjanczyk 和 Tarrant[19]在锚杆锚固机理研究中发现,肋高较小的锚杆锚固性能较差,且锚固力是锚杆外形、锚固剂强度、肋高等变量的函数。Aziz[20,21]、林建[22,23]对肋间距分别为 12.5 mm、25 mm、37.5 mm、50 mm 和 11.05 mm、22.10 mm、33.15 mm、44.20 mm 的锚杆试件进行了试验研究,结果表明随肋间距增大,锚杆试件拉拔力呈先增大后减小的趋势,在肋间距分别为 37.5 mm 和 33.15 mm 时拉拔力达到最大。吴涛[24,25]等对肋高 1.8 mm,肋间距 12 mm、24 mm、36 mm、48 mm 的右旋锚杆进行了锚固性能试验,结果显示,随着肋间距的增大,锚杆拉拔力、剪胀位移量和锚杆吸能均呈增大趋势。赵象卓[26]等研究了不同围岩条件下肋高为 1.0 mm,肋间距 12 mm、24mm、36 mm、48 mm 的左旋锚杆锚固性能,结果表明在肋间距 24 mm 时锚杆拉拔力达到最大。上述研究表明,锚杆外形对锚固性能具有重要影响,通过改变锚杆外形,可以改变锚杆锚固效果,以满足不同条件的支护要求。但上述研究仅限于实验室试验,并未进行现场测试,且对锚固机理分析较少,对实验室拉拔结果能多大程度客观反映现场锚固力的实际情况也尚未探究清楚。

研究针对新疆乌东煤矿巷道变形量大的问题,基于厚壁理论[27,28],建立模拟乌东煤矿地质条件的实验室试验模型;通过对锚固破坏方式和过程分析,明确锚固体破坏后影响锚固力残余强度的关键因素;对螺纹钢锚杆外形进行优化设计,控制锚固破坏方式,增加螺纹钢锚杆锚固力残余强度及系统吸能能力,从而设计开发出一种适于该地质条件及类似条件的大变形巷道锚杆,在不改变目前的锚杆支护工艺及工程成本的基础上提升巷道支护效果。最后,通过实验室及现场试验,对新型大变形巷道锚杆锚固力及吸能能力进行评估、对研究提出的设计方法进行验证,为螺纹钢锚杆外形研究提供现场试验结果。研究通过锚杆外形优化,开发适用于大变形巷道支护的螺纹钢锚杆,在不增加工程成本条件下,提高巷道的支护效果,为提升大变形巷道支护效果提供新的方法,解决大变形巷道锚杆支护研究的重大难题。

2 锚杆外形优化设计理论分析

2.1 锚杆外形与锚固破坏方式分析

康红普[29]等研究发现,影响树脂锚杆锚固力的主要因素有杆体形状与直径、锚固剂物理力学性质、围岩强度与完整性等。胡滨[30]研究了围岩强度对锚固性能的影响,发现围岩强度不同,锚固剂、围岩应力分布也不同;围岩应力随着围岩强度的降低逐渐降低,但塑性区逐渐扩大。

Kaiser[31]认为锚固剂破坏是锚固件周向受拉产生径向裂纹并拓展至围岩交界面,最终导致锚固剂劈裂,并推导锚索受拉的黏结强度模型式:

$$u_{\text{lat}} = u_0 \left(1 - \frac{p}{\sigma}\right)^{B/\sigma} \quad (1)$$

式中,u_{lat} 为径向位移;u_0 为轴向位移;B 为常数,根据实验室拉拔曲线得出;p 为交界面处压力;σ 为锚固剂抗压强度。

Hyett[32]对锚固体劈裂破坏进行了进一步研究,建立了破坏机理、锚杆轴力分布及对应拉拔曲线阶段关系(图1)。

通过理论分析,得到锚固力解析式:

$$F_a = \frac{A p_1 \tan \varphi_{\text{gs}}}{\sin \alpha} + \frac{4\pi^2 C u_a}{l^2 (u_a + L_r)} \quad (2)$$

式中,A 为锚索与锚固剂接触面面积;p_1 为交界面处压力;$\tan \varphi_{\text{gs}}$ 为锚固剂与锚索间的动摩擦因数;α 为钢绞线与轴向夹角;C 为锚索扭转刚度;u_a 为轴向位移;l 为锚索螺距;L_r 为锚索非锚固段长度。

针对拔出式破坏模式,Kilic 等[18]研究了肋为圆锥形的螺纹钢锚杆锚固效果,对肋角分别为 15°、30°和 45°的单肋、双肋和三肋锚杆进行了拉拔试验,实验结果表明,随着圆锥形肋

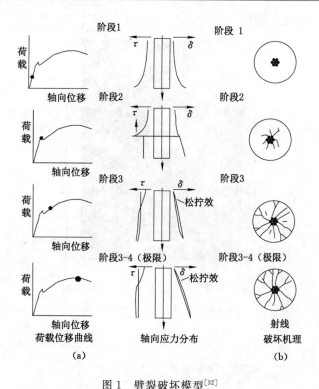

图 1 劈裂破坏模型[32]

Fig.1 Splitting failure model

(a)沿锚杆方向应力分布;(b)垂直于锚索轴向的破坏形式

的角度增大,锥形螺纹钢锚杆的承载力减小,而剪胀位移量增大。即,锚杆的外形对锚杆的承载能力和锚固体的剪胀位移量具有很大影响。Liang[33]等人建立了锥形锚杆的力学模型,通过分析锚杆的锥形功能来研究锚杆的残余强度。研究发现锚杆残余强度与锚杆外形、锚固剂强度、地应力和围岩的径向刚度密切相关。

Cao[34,35]等通过分析树脂-锚杆界面破坏来研究螺纹钢锚杆外形配置优化,通过拉拔实验确定了两种主要的破坏方式,即平行剪切破坏以及扩张滑移破坏(图 2)。

研究表明,锚固效果与锚杆外形、锚固剂的力学性能和工矿地质条件有关。通过改变螺纹钢外形,可以改变锚固段破坏方式,从而影响锚固力表现。试验结果表明,增加螺纹钢肋间距,可以使锚杆破坏方式从平行剪切转变到扩展滑移破坏,在特定地质条件下,达到增加系统吸能的目的。

针对大变形巷道锚杆支护系统,残余强度至关重要。在新锚杆外形设计中,不仅应增加锚杆支护残余强度和延伸扩张滑移长度,而且应考虑地质条件,研究运用厚壁理论对围岩应力和径向刚度,建立地质力学条件的锚固力测试实验室模拟模型。

2.2 地质力学条件的锚固力测试实验室模拟模型

加拿大学者 S. Yazici 和 P. K. Kaiser[27]首先将厚壁理论引入到锚固力学的研究领域,美国学者 A. J. Hyett[28]等进一步应用厚壁理论对锚索受拉锚固强度演化进行了理论分析与现场试验。Hyett 认为,现场锚杆表现研究[图 3(a)],可以简化为径向弹性应力的力学模型[图 3(b)],在实验室可使用钢、铝、PVC 等材料的套筒拉拔实验进行锚固力测试[图 3(c)]。

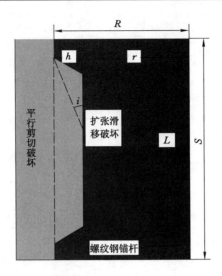

图 2 轴向载荷作用下锚杆外形与破坏模式[34,35]

Fig. 2 Failure modes and geometry of one bolt profile subjected to axial loading[36]

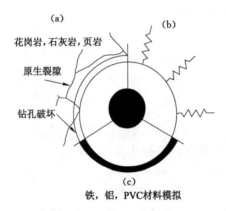

图 3 厚壁理论模型

Fig. 3 Experimental model of thick walled cylinder theory

(a) 现场模型；(b) 力学模型；(c) 实验室模型

澳大利亚学者 Cao[36] 研究将厚壁理论应用于实验室拉拔试验结果分析中，采用等效径向应力的方法对锚固在钢套筒及混凝土的锚杆进行了等效围岩强度计算。赵象卓[26] 等在锚杆肋间距优化研究中，运用厚壁理论，将壁厚为 4.5 mm 和 6.0 mm 的 20# 钢套筒对弹性模量为 53.6 GPa 和 69.7 GPa 的巷道围岩进行相似模拟，探讨了不同围岩条件下锚杆肋间距与锚固性能的关系。

计算中设定刚性套筒内径为 n，外径为 w，套筒内压 p_1，外压 p_2，如图 4 所示。

设定应力边界，得到经典 Lame 应力式：

$$\sigma_{rr} = \frac{n^2 p_1 - w^2 p_2}{w^2 - n^2} - \frac{n^2 w^2 (p_1 - p_2)}{w^2 - n^2} \cdot \frac{1}{r^2} \tag{3}$$

$$\sigma_{\theta\theta} = \frac{n^2 p_1 - w^2 p_2}{w^2 - n^2} + \frac{n^2 w^2 (p_1 - p_2)}{w^2 - n^2} \cdot \frac{1}{r^2} \tag{4}$$

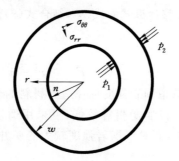

图 4 套筒截面
Fig. 4 Section of sleeve

在套筒与锚杆双重作用下，树脂受力可表示为：

$$\sigma_{\text{tension(max)}} = \sigma_{\theta\theta}(w) = \frac{2n^2 p_1 - (n^2 + w^2) p_2}{w^2 - n^2} \tag{5}$$

由于套筒两端开口，$\sigma_{zz}=0$，因此锚固单元体径向位移为：

$$u_r = \frac{r}{E(w^2 - n^2)}\left[(1-v)(p_1 n^2 - p_2 w^2) + \frac{(1+v) n^2 w^2}{r^2}(p_1 - p_2)\right] \tag{6}$$

式中　E——弹性模量，GPa；
　　　v——泊松比；
　　　r——套筒壁中某点与圆心距离。

若套筒不受外围 p_2 约束，即 $p_2 = 0$，则可得锚固树脂环外壁径向位移：

$$u_{r=n} = \frac{n\left[(1-v)n^2 + (1+v)w^2\right]}{E(w^2 - n^2)} p_1 \tag{7}$$

当 $w \to \infty$ 时，表示外径无限延伸，可认为是锚固在无限岩体或煤体内，则此时围岩径向位移为：

$$\lim_{w \to \infty} u_r = \frac{n(1+v_{围岩})}{E_{围岩}} p_1 \tag{8}$$

假设锚杆分别锚固在套筒与岩体中，锚杆受轴力产生的径向刚度相同，则可认为锚杆在套筒的锚固效果与围岩的锚固效果近似等效。即通过相同压力下径向位移相等，可计算出外壁 w 值，从而计算出不同围岩条件下的等效套筒壁厚。

$$u_{r=n} = \lim_{w \to \infty} u_r \tag{9}$$

式中，$u_{r=n}$ 为套筒 r 点处的径向位移；$\lim\limits_{w \to \infty} u_r$ 为围岩径向位移。

由式(4)~式(7)即得到平面应力条件下不同围岩等效套筒壁厚计算公式：

$$\frac{n\left[(1-v)n^2 + (1+v)w^2\right]}{E(w^2 - n^2)} = \frac{n(1+v_{围岩})}{E_{围岩}} \tag{10}$$

若从平面应变考虑，径向位移计算公式为：

$$u_r = \frac{(1+v)(1-2v)}{E} \cdot \frac{p_1 n^2 - p_2 w^2}{w^2 - n^2} \cdot r + \frac{1+v}{E} \cdot \frac{(p_1 - p_2) n^2 w^2}{w^2 - n^2} \cdot \frac{1}{r} \tag{11}$$

同理，若套筒不受外围 p_2 约束，即 $p_2 = 0$，则可得锚固树脂环外壁径向位移：

$$u_{r=n} = \frac{(1+v)(1-2v)}{E} \cdot \frac{p_1 n^3}{w^2 - n^2} + \frac{1+v}{E} \cdot \frac{p_1 n w^2}{w^2 - n^2} \tag{12}$$

当 $w\to\infty$ 时，围岩径向位移计算式同式(8)，结合式(9)得到平面应变条件下不同围岩等效套筒壁厚计算公式：

$$\frac{(1+v)(1-2v)}{E}\cdot\frac{n^3}{w^2-n^2}+\frac{1+v}{E}\cdot\frac{nw^2}{w^2-n^2}=\frac{n(1+v_{围岩})}{E_{围岩}} \quad (13)$$

对实验室拉拔试验，根据巷道围岩强度选择适当材料的套筒对围岩进行模拟。一般情况下，巷道变形较大的矿井，围岩强度相对较低，可采用适当壁厚 6061 铝套筒（$E=68.9$ GPa，$v=0.33$）对围岩进行模拟，而对于围岩强度较大的矿井，可采用适当壁厚 20# 钢套筒（$E=206$ GPa，$v=0.3$）对围岩进行模拟。

运用平面应力[式(10)]及平面应变[式(13)]计算不同围岩强度与实验室套筒壁厚的等效关系，计算结果如图 5 所示。从结果分析，弹性模量 10～30 GPa 围岩，径向刚度与壁厚 3～10 mm 铝套筒一致；弹性模量 30～70 GPa 围岩，径向刚度与壁厚 3～7 mm 钢套筒一致。平面应力与平面应变的等价计算中，二者结果差异不大，并随着围岩弹性模量增加，计算误差减小，两者误差最大为 11.4%。

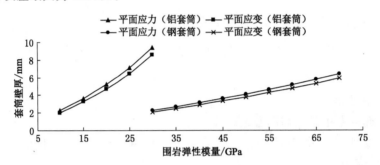

图 5 围岩弹性模量与套筒壁厚关系

Fig. 5 Relationship between the elastic modulus of the surrounding rock and the thickness of the sleeve wall

3 锚杆外形设计及实验室试验

3.1 大变形巷道锚杆外形设计

研究针对煤矿巷道变形较大的问题，为控制锚固破坏方式，增加锚固段残余强度，参考巷道变形量较大（大于 300 mm）的乌东煤矿地质条件，并结合使用的锚固剂性质，确定了锚杆肋的排列方式和肋间距，如图 6 所示，锚杆外形结构及力学参数如表 1 所示。

表 1 大变形巷道锚杆外形结构及力学参数

Table 1 Structure parameters and mechanical parameters of large deformation roadway bolt

参数	数值	参数	数值
外径 ϕ/mm	24	肋宽（顶/底）/mm	1.5/3.6
内径 h/mm	20	肋间距 L/mm	50
肋高/mm	2.0	肋坡角/(°)	60
抗拉强度/MPa	642.7	屈服强度/MPa	501.6

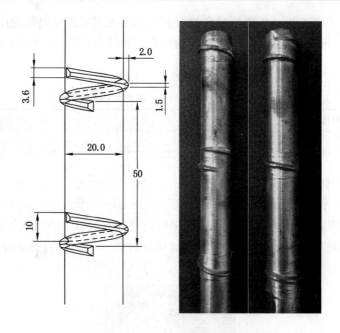

图 6 新设计大变形巷道锚杆外形

Fig. 6 New design of bolt shape in large deformation roadway

对锚杆外形优化设计,通过增加肋间距+单肋重叠的形式,达到增加残余强度和吸能能力的目的。

3.2 大变形巷道锚杆实验室试验

（1）材料准备

① 锚杆试件

试验拟对目前矿井常使用的左旋无纵筋螺纹钢锚杆和大变形巷道锚杆进行对比,矿井常使用的左旋锚杆结构及力学特性如表 2 所示。将两种锚杆都切割成长度为 280 mm 的试件,每种切割 3 个试件。

表 2 矿井常用锚杆外形结构及测试力学参数

Table 2 Structure parameters and mechanical parameters of bolt

参数	数值	参数	数值
直径 ϕ/mm	22	肋宽（顶/底）/mm	1.4/3.3
肋高/mm	1.0	肋间距 L/mm	12
横肋上升角/(°)	77.3	横肋侧面角/(°)	46.9
抗拉强度/MPa	642.7	屈服强度/MPa	501.6

② 锚固剂

试验选用乌东煤矿现使用的树脂锚固剂(MSCKa23-35),经试验测试,锚固剂抗压强度为 62.02 MPa,抗剪强度 19.5 MPa,弹性模量 13.03 GPa,泊松比 0.26。

③ 试验套筒

根据乌东煤矿巷道围岩弹性模量和锚杆钻孔实测数据,试验套筒选用 6061 铝材料车制,套筒内径 32 mm。运用厚壁理论,计算确定套筒壁厚,结果如表 3 所示。

表 3 套筒壁厚计算结果
Table 3 Calculation results of sleeve wall thickness

围岩	弹性模量/GPa	泊松比	套筒壁厚(平面应力)/mm	套筒壁厚(平面应变)/mm
顶板	23.01	0.23	5.8	5.2

采用 6061 铝材料车制适合模拟乌东煤矿围岩条件下大变形巷道锚杆锚固效果的实验室套筒装置,根据厚壁理论计算确定套筒壁厚为 5.8 mm,为保证试验过程中锚固剂-套筒壁界面不发生失效滑移,更好的研究拉拔过程中树脂锚固效果以及锚固剂的破坏方式,对套筒内壁进行了车丝处理,套筒内径、长度、内壁丝扣分别为 32 mm、100 mm 和 1.0 mm,试验套筒如图 7 所示。

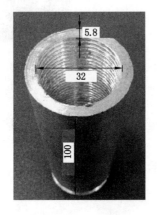

图 7 试验套筒
Fig. 7 Steel sleeve

④ 锚固试件

将切割好的锚杆制备锚固试件,锚固过程中保证锚杆平行于套筒轴线,确保锚杆在套筒中的居中度,使锚固剂环厚度一致,并将锚固试件在 22 ℃ 的保温箱中养护 2 h。

(2) 实验室拉拔试验

试验采用 WAW-600C 型微机控制电液伺服万能试验机进行拉拔试验,最大试验力为 600 kN,试样采用液压夹持,夹持范围 $\phi 13$ mm~$\phi 40$ mm,进行拉拔试验连续全程测量。为方便锚杆锚固试件的拉拔试验,选用 40Cr 钢材车制套筒拉拔工装,将制备完成的锚固试件放置于已安装在试验机夹口上的工装中,试验过程中采用位移控制进行加载,拉拔试验如图 8 所示,拉拔试验完成后的两种锚杆状态如图 9 所示。

(3) 试验结论

分别对左旋锚杆和大变形巷道锚杆进行拉拔试验,两种锚杆的拉拔力与位移关系如图 10 所示。图中 N 为新设计大变形巷道锚杆拉拔曲线,LX 为通常矿井采用的左旋锚杆拉拔曲线。

由图 10 可知,肋间距为 12 mm 的左旋锚杆拉拔力最大为 100 kN,大变形巷道锚杆拉

图 8 实验室拉拔试验

Fig. 8 Pullout test process in laboratory

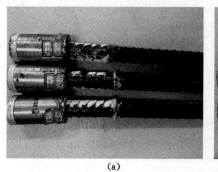 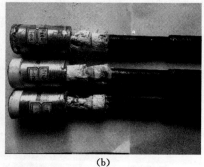

图 9 拉拔试验完成后两种锚杆状态对比

Fig. 9 Two kinds of anchor state after pullout test

(a) 左旋锚杆拉拔试验完成后；(b) 大变形巷道锚杆拉拔试验完成后

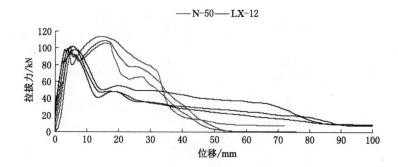

图 10 新设计锚杆与左旋锚杆拉拔力与位移关系

Fig. 10 Relationship between the pulling force and displacement of the newly designed bolt and left spiral bolt

拔力最大可达到 113 kN,拉拔力提高了 13%。锚固系统吸能能力可通过计算拉拔曲线下面积获得,从两种锚杆在拉拔过程中的吸能能力来看,肋间距为 12 mm 的左旋锚杆吸能值为 2 768 J,而大变形巷道锚杆在拉拔过程中的吸能值为 3 169 J,吸能能力较左旋锚杆提高了 15%。

4 大变形巷道锚杆现场拉拔试验

4.1 试验矿井地质条件与巷道支护现状

新设计大变形巷道锚杆在乌东煤矿进行了现场拉拔试验。乌东煤矿位于乌鲁木齐市东北部约34 km,井田基本构造形态呈一向南倾斜的单斜构造,地表被第四系堆积物大面积覆盖。煤层走向北东67°,倾向南东157°,倾角西陡东缓,煤层倾角43°~51°,主要可采煤层总厚占全区煤层总厚的74.62%。矿井主要开采煤层为42#、43#和45#煤层,煤层结构简单,含矸5~8层,单层矸石厚度多在0.3 m以下,夹矸岩性为粉砂岩、泥质粉砂岩及炭质泥岩,煤层顶板为块状粉砂岩夹有少量细砂岩,抗压强度38.06 MPa,硬度较大。煤层直接底为厚达7.0~10.0 m泥岩,并夹有46#薄煤层,遇水易变软脱落,稳定性差,底板处煤层层理、裂隙较发育,易破碎垮落。

巷道掘进断面形状为直墙三心拱形,现阶段巷道所用锚杆为肋间距12 mm的螺纹钢锚杆,规格$\phi 20$ mm×2 500 mm,支护方式为锚网索加钢筋梯支护,锚杆间、排距800 mm×800 mm,锚杆锚固方式为端部锚固。锚索为$\phi 18.9$ mm×10 000 mm,间、排距1 500 mm×1 600 mm,每排3根,所用的金属网为冷拔丝经纬网。巷道支护方案如图11所示。

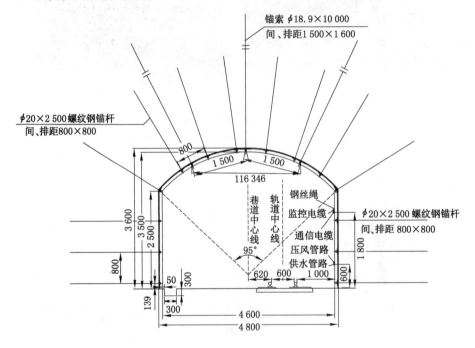

图11 乌东煤矿巷道现支护方案

Fig.11 Supporting Scheme of roadway in Wudong Coal Mine

乌东煤矿回采巷道在现支护条件下,巷道变形较大,顶板下沉、两帮收敛均超过300 mm,巷道变形状况如图12所示。

4.2 大变形巷道锚杆现场拉拔试验

拉拔试验在乌东煤矿+575水平43#煤层北巷顶板进行。为保证试验结果的可靠性,

图 12 巷道变形状况
Fig. 12 Roadway deformation condition

选取巷道顶板变形较大的典型区域,分别进行了 4 组对比试验,对目前矿井常用的左旋螺纹钢锚杆和大变形巷道锚杆进行对比研究。试验锚杆钻孔平均直径均为 32 mm,锚固长度 200 mm,测试过程中使用煤科总院 ZY 型便携式锚杆拉拔仪和百分表分别进行施压和测试锚杆位移。拉拔方式如图 13 所示,试验结果如图 14 所示。

图 13 现场拉拔试验
Fig. 13 Pullout test in field

从图 14 可知,现场试验左旋锚杆拉拔力最大为 55 kN,新设计大变形巷道锚杆最大拉拔力达到 72 kN,较左旋锚杆拉拔力提高 30%。左旋锚杆在拉拔过程中吸能值为 1 574 J,新设计锚杆的吸能值达到 2 315 J,较左旋锚杆提高 47%。从现场试验表明,新设计锚杆具有较高的承载能力和吸能能力。因此,新设计大变形巷道锚杆可以提高大变形巷道的支护效果。

新设计的大变形巷道锚杆可用于类似地质条件的巷道支护中,研究方法与试验手段可为其他不同地质条件矿井锚杆支护优化研究提供借鉴。

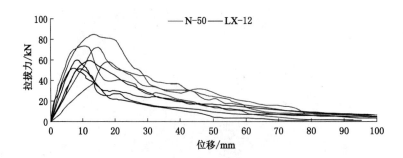

图 14 新设计锚杆和常用左旋锚杆拉拔力与位移关系

Fig. 14 Load-displacement curve for new bolt (N) and left spiral bolt (LX)

5 结 论

(1)针对乌东煤矿大变形巷道支护,通过增加肋间距+单肋重叠的锚杆外形设计,达到控制锚固破坏模式,增加螺纹钢锚杆锚固力残余强度和吸能能力的目的,在不改变目前的锚杆支护工艺及工程成本的基础上提高了巷道支护效果。

(2)对新设计大变形巷道锚杆进行了实验室和现场拉拔试验,并与国内常用的锚杆进行了比较,结果表明,新设计大变形巷道锚杆拉拔力分别提高了13%和30%,吸能能力提高了15%和47%。

(3)厚壁理论可用于锚固力学研究中,从平面应力和平面应变计算结果,给出了不同现场地质条件的实验室锚固力实验仿真条件,围岩平面应力与平面应变等价计算结果差异不大,弹性模量10～30 GPa围岩,径向刚度与壁厚3～10 mm铝套筒一致;弹性模量30～70 GPa围岩,径向刚度与壁厚3～7 mm钢套筒一致。

参考文献

[1] 林健,杨景贺,韩国强,等.不同杆体外形树脂锚杆锚固与安设性能对比试验研究[J].煤炭学报,2015,40(2):286-292.
　　LIN Jian, YANG Jing-he, HAN Guo-qiang, et al. Comparable test on anchoring and setting performance of resin bolts with different rod shapes[J]. Journal of China Coal Society, 2015, 40(2):286-292.

[2] 王洪涛,王琦,王富奇,等.不同锚固长度下巷道锚杆力学效应分析及应用[J].煤炭学报,2015,40(3):509-515.
　　WANG Hong-tao, WANG Qi, WANG Fu-qi, et al. Mechanical effect analysis of bolts in roadway under different anchoring lengths and its application[J]. Journal of China Coal Society, 2015, 40(3):509-515.

[3] KANG Hong-pu. Support technologies for deep and complex roadways in underground coal mines: a review[J]. International Journal of Coal science & Technology, 2014, 1(3):261-277.

[4] 李冲,徐金海,李明.全长锚固预应力锚杆杆体受力特征分析[J].采矿与安全工程学报,2013,30(2):188-193.

LI Chong, XU Jin-hai, LI Ming. The mechanical characteristics analysis of fully anchored pre-stressed bolts in coal mines[J]. Journal of Mining & Safety Engineering, 2013, 30(2):188-193.

[5] 康红普,王金华.煤巷锚杆支护理论与成套技术[M].北京:煤炭工业出版社,2007.
KANG Hong-pu, WANG Jin-hua. Rock bolting theory and complete technology for coal roadways[M]. Beijing: China Coal Industry Publishing Home, 2007.

[6] 康红普.我国煤矿巷道锚杆支护技术发展60年及展望[J].中国矿业大学学报,2016,45(6):1071-1081.
KANG Hong-pu. Sixty years development and prospects of rock bolting technology for underground coal mine roadways in China[J]. Journal of China University of Mining & Technology, 2016, 45(6):1071-1081.

[7] 何满潮,郭志飚.恒阻大变形锚杆力学特性及其工程应用[J].岩石力学与工程学报,2014,33(7):1297-1308.
HE Man-chao, GUO Zhi-biao. Mechanical property and engineering application of anchor bolt with constant resistance and large deformation[J]. Chinese Journal of Rock Mechanics and Engineering, 2014, 33(7):1297-1308.

[8] 何满潮,袁越,王晓雷,等.新疆中生代复合型软岩大变形控制技术及其应用[J].岩石力学与工程学报,2013,32(3):433-441.
HE Man-chao, YUAN Yue, WANG Xiao-lei, et al. Control technology for large deformation of mesozoic compound soft rock in Xinjiang and its application[J]. Chinese Journal of Rock Mechanics and Engineering, 2013, 32(3):433-441.

[9] 康红普,林健,吴拥政.全断面高预应力强力锚索支护技术及其在动压巷道中的应用[J].煤炭学报,2009(9):1153-1159.
KANG Hong-pu, LIN Jian, WU Yong-zheng. High pretensioned stress and intensive cable bolting technology set in full section and application in entry affected by dynamic pressure[J]. Journal of China Coal Society, 2009(9):1153-1159.

[10] 郭志飚,王炯,张跃林,等.清水矿深部软岩巷道破坏机理及恒阻大变形控制对策[J].采矿与安全工程学报,2014,31(6):945-949.
GUO Zhi-biao, WANG Jiong, ZHANG Yue-lin, et al. Failure mechanism and constant resistance large deformation control measures of deep soft rock in Qingshui Coal Mine[J]. Journal of Mining & Safety Engineering, 2014, 31(6):945-949.

[11] 郑朋强,陈卫忠,谭贤君,等.软岩大变形巷道底鼓破坏机制与支护技术研究[J].岩石力学与工程学报,2015(s1):3143-3150.
ZHENG Peng-qiang, CHEN Wei-zhong, TAN Xian-jun, et al. Study of failure mechanism of floor heave and supporting technology in soft rock of large deformation roadway[J]. Chinese Journal of Rock Mechanics and Engineering, 2015(s1):3143-3150.

[12] 高明仕,赵一超,李明,等.软岩巷道顶、帮、底全支全让O型控制力学模型及工程实践[J].岩土力学,2014(8):2307-2313.

GAO Ming-shi, ZHAO Yi-chao, LI Ming, et al. Roof and support and bottom yielding support with whole section and O-shape control principle for soft rock roadway and engineering practice[J]. Rock and Soil Mechanics, 2014(8):2307-2313.

[13] 马念杰,赵希栋,赵志强,等. 深部采动巷道顶板稳定性分析与控制[J]. 煤炭学报, 2015, 40(10):2287-2295.

MA Nian-jie, ZHAO Xi-dong, ZHAO Zhi-qiang, et al. Stability analysis and control technology of mine roadway roof in deep mining[J]. Journal of China Coal Society, 2015, 40(10):2287-2295.

[14] 刘洪涛,王飞,蒋力帅,等. 顶板可接长锚杆耦合支护系统性能研究[J]. 采矿与安全工程学报, 2014, 31(3):366-372.

LIU Hong-tao, WANG Fei, JIANG Li-shuai, et al. On the performance of lengthened bolt coupling support system in roadway roof[J]. Journal of Mining & Safety Engineering, 2014, 31(3):366-372.

[15] 王卫军,袁超,余伟健,等. 深部大变形巷道围岩稳定性控制方法研究[J]. 煤炭学报, 2016, 41(12):2921-2931.

WANG Wei-jun, YUAN-Chao, YU Wei-jiang, et al. Stability control method of surrounding rock in deep roadway with large deformation[J]. Journal of China Coal Society, 2016, 41(12):2921-2931.

[16] Aydan O. The stabilisation of rock engineering structures by rock bolts, Geotechnical Engineering[D]. Nagoya, Nagoya University, 1989:202.

[17] Aziz N I, Jalalifar H, Remennikov A, et al. Optimisation of the Bolt Profile Configuration for Load Transfer Enhancement[C]. Proc 8th Underground Coal Operators Conference, Wollongong, 2008: 125-131.

[18] Kilic A, Yasar E, Atis C D. Effect of bar shape on the pull-out capacity of fully-grouted rockbolts[J]. Tunnelling & Underground Space Technology Incorporating Trenchless Technology Research, 2003, 18(1):1-6.

[19] Fabjanczyk M W, Tarrant G C. Load transfer mechanisms in reinforcing tendons[C]. Proceedings of the 11th International Conference on Ground Control in Mining, The University of West Virginia, Morgantown, 1992:1-8.

[20] Aziz N. A new technique to determine the load transfer capacity of resin anchored bolts[C]. Coal Operators' Conference, 2002: 208.

[21] Aziz N, Jalalifar H, Remennikov A, et al. Optimisation of the bolt profile configuration for load transfer enhancement[C]. Coal Operators' Conference, 2008: 11.

[22] 林健,任硕,杨景贺. 树脂全长锚固锚杆外形尺寸优化实验室研究[J]. 煤炭学报, 2014, 39(6): 1009-1015.

LIN Jian, REN Shuo, YANG Jinghe. Laboratory research of resin full-length anchoring bolts dimension optimization[J]. Journal of China Coal Society, 2014, 39(6): 1009-1015.

[23] 林健,任硕. 树脂全长锚固锚杆外形尺寸优化数值模拟研究[J]. 采矿与安全工程学报,

2015, 32(2): 273-278.

LIN Jian, REN Shuo. Numerical simulation optimization research of bolt profile configuration with resin full-length anchoring [J]. Journal of Mining & Safety Engineering, 2015, 32(2): 273-278.

[24] Wu T, Chen C, Han J, et al. Effect of bolt rib spacing on load transfer mechanism [J]. International Journal of Mining Science and Technology, 2017, 27(3):431-434.

[25] 吴涛, CAO Chen, 赵象卓, 等. 不同肋间距锚杆锚固性能实验室试验研究[J]. 煤炭学报, 2017, 42(10):2545-2553.

WU Tao, CAO Chen, ZHAO Xiang-zhuo, et al. Laboratorial study of anchorage performance in different rib spacings of bolt[J]. Journal of China Coal Society, 2017, 42(10):2545-2553.

[26] 赵象卓, 张宏伟, CAO Chen, 等. 不同围岩条件下锚杆肋间距与锚固力优化研究[J]. 岩土力学, 2018, 39(4): 1-9.

ZHAO Xiang-zhuo, ZHANG Hong-wei, CAO Chen, et al. Study on optimization of bolts rib spacing and anchoring force in different conditions of surrounding rock[J]. Rock and Soil Mechanics, 2018, 39(4): 1-9.

[27] Yazici S, Kaiser P K. Bond strength of grouted cable bolts[J]. Int. J. of Rock Mechanics and Min. Sci. & Geomech. Abstrs., 1992, 29(3):279-292.

[28] Hyett A J, Bawden W F, Macsporran G R, et al, A constitutive law for bond failure of fully-grouted cable bolts using a modified Hoek cell[J]. Int. J. of Rock Mechanics and Min. Sci. & Geomech. Abstrs., 1995, 32(1): 11-36.

[29] 康红普, 崔千里, 胡滨, 等. 树脂锚杆锚固性能及影响因素分析[J]. 煤炭学报, 2014, 39(1):1-10.

KANG Hong-pu, CUI Qian-li, HU Bin, et al. Analysis on anchorage performances and affecting factors of resin bolts[J]. Journal of China Coal Society, 2014, 39(1):1-10.

[30] 胡滨. 围岩强度对树脂锚杆锚固性能影响的数值研究[J]. 煤矿支护, 2015(1):8-11.

HU Bin. Numerical study on the influence of strength of surrounding rock on anchorage performance of resin anchorage[J]. Coal Mine Support, 2015(1):8-11.

[31] Hyett A J, Bawden W F, Reichert R D. The effect of rock mass confinement on the bond strength of fully grouted cable bolts[J]. Int. J. Rock Mech. Min. Sci. & Geomech. Abst., 1992, 29(5):503-524.

[32] Kaiser P K, Yazici S, Nose J. Effect of stress change on the bond strength of fully grouted cables[J]. Int. J. of Rock Mechanics and Min. Sci. & Geomech. Abstrs., 1992, 29(3): 293-306.

[33] Liang Y, He M, Cao C, et al. A mechanical model for conebolts[J]. Computers & Geotechnics, 2017(83):142-151.

[34] Cao C, Jan N, Ren T, et al. A study of rock bolting failure modes[J]. International Journal of Mining Science and Technology, 2013, 23(1):79-88.

[35] Cao C, Nemcik J, Aziz N, et al. Analytical study of steel bolt profile and its

influence on bolt load transfer[J]. International Journal of Rock Mechanics & Mining Sciences, 2013, 60(60):188-195.

[36] Chen Cao, Ting Ren, Yi-dong Zhang, et al. Experimental investigation of the effect of grout with additive in improving ground support[J]. International Journal of Rock Mechanics & Mining Sciences, 2016, 85(2016): 52-59.

单轴压缩下含瓦斯煤破坏过程能量演化规律

张冰[1]，王汉鹏[1]，袁亮[1,2]，余国锋[3]，刘众众[1]，王伟[1]

(1. 山东大学 岩土与结构工程研究中心，山东 济南，250061；
2. 安徽理工大学 能源与安全学院，安徽 淮南，232001；
3. 淮南矿业(集团)有限责任公司 深部煤炭开采与环境保护国家重点实验室，安徽 淮南，232002)

摘 要：为了获取含瓦斯原生煤、构造煤变形破坏过程的能量演化规律及能量差异，采用原煤及性质与构造煤相近的型煤，开展了单轴加卸载试验及瓦斯释放试验，精确测定了含瓦斯原煤、型煤在不同单轴应力下的弹性能与瓦斯膨胀能。结果表明：在变形破坏过程中，型煤、原煤的弹性能最大变化量均为 4 mJ/g，原煤瓦斯膨胀能可增大 2.4 倍，型煤瓦斯膨胀能可增大 35.88 mJ/g；随着轴向应力变化，原煤、型煤的弹性能均呈现先增大后减小的趋势，瓦斯膨胀能均呈现加速增长趋势，其变化速率呈阶段性；在整个应力应变过程中，两类煤体的弹性能大小相近，而型煤瓦斯膨胀能是原煤的 10 倍左右。基于试验结果拟合了弹性能、瓦斯膨胀能与损伤变量的关系，拟合结果显示型煤、原煤具有相同的函数关系，证明基于损伤变量对应力变化过程中的含瓦斯煤能量演化进行数学描述是可行的。

关键词：原煤；型煤；弹性能；瓦斯膨胀能；损伤变量

Energy evolution of coal containing gas in failure process under uniaxial compression

ZHANG Bing[1], WANG Hanpeng[1], YUAN Liang[1,2],
YU Guofeng[3], LIU Zhongzhong[1], WANG Wei[1]

(1. *Research Centre of Geotechnical and Structural Engineering,
Shandong University, Jinan 250061, China;*
2. *School of Energy and Safety, Anhui University of Science and
Technology, Huainan 232001, China;*
3. *State Key Laboratory of Deep Coal Mining and Environmental Protection,
Huainan Mining (Group) Co. Ltd., Huainan 232002, China*)

Abstract: In order to obtain the energy evolution law and energy difference of gas-filled coal and technical coal during the deformation and failure process, raw coal and the briquette with similar properties of tectonic coal were used to conduct uniaxial loading and unloading test and gas releasing test. The elastic energy and gas expansion energy of the two coals were accurately measured under different stress conditions. The test

---

基金项目：国家重大科研仪器研制项目(51427804)；国家自然科学基金面上资助项目(41672281)；山东省自然科学基金项目(ZR2017MEE023)；淮南市科技计划项目(2016A02)。

通信作者：王汉鹏(1978—)，男，山东省济南市人，教授，博士生导师，从事岩土工程试验与灾变机理及防控方面的研究。Tel：0531-88399182，E-mail：pcwli@163.com。

results showed that: in the process of deformation and failure, the maximum changes of elastic energy of both briquette and raw coal were 4 mJ/g, the expansion energy of raw coal increased by 2.4 times, and briquette increased to 35.88 mJ/g; As the axial stress changed, the elastic energy of raw coal and briquette firstly increased and then decreased, and an accelerating growth trend with phased changing rates can be seen in the gas expansion energy of both raw coal and briquette; During the complete stress-strain process, the elastic energy of the two types of coal bodies was similar in quantity, while the expansion energy of briquette was about 10 times than that of raw coal. The relationship between elastic energy, gas expansion energy and damage variable was fitted according to the test results, which indicated the two types of coal had the same functional relationship. The fitting curves proved that it is practicable to depict the energy evolution of gas-filled coal in mathematical description during the loading process based on the damage variable.

Keywords: raw coal; briquette; elastic energy; gas expansion energy; damage variable

煤与瓦斯突出是危害巨大的煤岩瓦斯动力灾害,严重制约煤炭安全生产[1-2]。其本质是能量的意外积累与释放,可采用能量方法对其全过程进行研究与描述[3]。

在煤矿现场中,煤层应力条件复杂。由于开采扰动,采空区周围原有的应力平衡状态受到破坏,引起井巷及采场周围应力的重新分布,在掘进工作面前方将形成由卸压区、应力集中区和原始应力区组成的"三带"[4-5]。复杂的开采条件导致不同开采阶段的煤岩处于不同的应力状态下,强度较低的软弱煤体常常经历整个应力应变过程。另外,爆破、打钻等动力扰动也会打破原有的应力平衡导致煤体应力状态发生变化[6]。据研究,煤体的应力状态变化会改变能量的积聚状态,继而影响突出的准备与孕育过程,最终对突出结果造成影响[7-8]。因此,研究不同应力状态下含瓦斯煤能量的积聚、耗散及演化规律对于突出预测与防治具有重要意义。

国内外学者对上述问题开展了大量研究。其中,煤体弹性能、煤体中的瓦斯膨胀能作为煤与瓦斯突出主要的能量来源[3],具有不同的研究进展。对于煤体弹性能,文献[9]通过理论分析与试验研究得出煤体的单位体积弹性能与体应力之间呈幂函数关系;文献[10]分析了应力变化过程中含瓦斯原煤的输入能、弹性能、耗散能的演化规律;文献[11-12]通过室内试验研究了循环加卸载对煤岩弹塑性和能量积聚耗散的影响;此外,文献[13-15]研究了多种应力条件下岩石的弹性能演化规律,相关结论有助于含瓦斯煤弹性能演化规律的掌握。对于瓦斯膨胀能,文献[16-17]通过理论分析认为,煤体释放瓦斯膨胀能的大小与煤体质点受地应力破坏后的破碎程度有关;文献[18-19]通过不同破碎程度煤样的瓦斯膨胀能测试,间接研究了地应力对瓦斯膨胀能的影响规律;文献[20-22]通过含瓦斯煤解吸试验获取了不同应力状态下瓦斯解吸量,研究成果有利于含瓦斯煤体瓦斯膨胀能演化规律的掌握。

综上所述,现有研究主要围绕原生煤展开,针对突出频发的构造煤研究较少,两类煤体的能量差异尚不明确。其次,由于瓦斯膨胀能的直接测定存在一定难度,瓦斯膨胀能在各应力阶段的演化规律尚不清楚。另外,煤体变形破坏过程中两类能量均缺乏具体的数学描述。

针对现有问题,本文采用性质相近的型煤代替构造煤,通过含瓦斯煤的单轴加卸载试验及瓦斯释放试验,精确测定了原煤、型煤在不同应力条件下的弹性能与瓦斯膨胀能。基于测定结果,分析获取了含瓦斯原生煤、构造煤在破坏过程中能量演化规律及差异,并从损伤角度对含瓦斯煤能量演化进行了数学描述。

1 能量测试原理

1.1 弹性能测试原理

煤体在变形破坏过程中,会伴随弹性应变能的存储,塑性能、断裂能等能量的耗散与释放。该关系可描述如下[14]:

$$U = U_d + U_e \tag{1}$$

式中 U——总应变能,即外力对煤体做的功,J/kg;

U_d——耗散能,J/kg;

U_e——弹性能,J/kg。

对于任何受力状态的煤体,其耗散能、弹性能、总应变能可通过煤体加卸载曲线中应力与应变所围成的面积计算得到。如图1所示,总应变能可由加载应力曲线与应变坐标轴之间的面积确定,见式(2);弹性能可由卸载应力曲线与应变坐标轴之间的面积确定,见式(3);耗散能可由加载应力曲线与卸载应力曲线之间的面积确定,见式(4)。

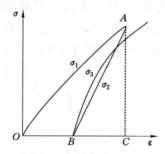

图 1 循环加卸载应力-应变曲线

Fig. 1 Cyclical loading and unloading stress-strain curve

$$U = \frac{1}{\rho} \int_O^A \sigma_1 \mathrm{d}\varepsilon \tag{2}$$

$$U_e = \frac{1}{\rho} \int_B^A \sigma_2 \mathrm{d}\varepsilon \tag{3}$$

$$U_d = \frac{1}{\rho} \int_O^A \sigma_1 \mathrm{d}\varepsilon - \frac{1}{\rho} \int_B^A \sigma_2 \mathrm{d}\varepsilon \tag{4}$$

式中 ρ——煤体视密度,kg/m^3。

1.2 瓦斯膨胀能测试原理

瓦斯膨胀能是指瓦斯在膨胀过程中所做的功。现阶段,突出耗散的瓦斯膨胀能主要基于突出过程为绝热过程或等温多变过程的假设,通过理论公式计算得到。事实上,突出过程中煤体热量传导及温度变化复杂,与以上假设有很大区别[23]。这导致了瓦斯膨胀能理论计算值存在一定误差。

针对上述问题,蒋承林教授基于气体动力学原理提出了瓦斯膨胀能的直接测定方法[18]。在该方法中,瓦斯膨胀能通过特定的测定仪器(图2)被转化为动能,其大小可通过瓦斯动能随时间变化的曲线确定。

如图2所示,测定装置主体为一煤样罐和渐缩型喷口。阀门打开后,从渐缩型喷口喷出

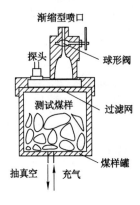

图 2　瓦斯膨胀能测定装置原理
Fig.2　Principle of measuring device for gas expansion energy

的瓦斯气流处于准定常流动状态,遵循定常流动方程。基于该方程,可得到喷出气体流量、流速与罐内气体压力、温度的定量关系:(1) 当 $p_0 > p^*$ 时,瓦斯气流处于超临界状态,其流速为声速,其流量与罐内瓦斯的压力、温度的关系如式(6)所示;(2) 当 $p_0 < p^*$ 时,瓦斯气流处于亚临界状态,喷口处瓦斯流量、流速与罐内瓦斯的压力、温度的关系如式(7)所示。其中,临界压力可由式(5)计算得到。

$$p^* = p_2\left(\frac{2}{\gamma+1}\right)^{-\frac{\gamma}{\gamma-1}} \tag{5}$$

$$p_0 > p^*,\begin{cases} m = \sqrt{\left(\dfrac{\gamma}{R}\right)\left(\dfrac{1}{\gamma+1}\right)^{\frac{\gamma+1}{\gamma-1}}}\dfrac{p_0 s}{\sqrt{T_0}} \\ v = \sqrt{\dfrac{2\gamma RT}{\gamma+1}} \end{cases} \tag{6}$$

$$p_0 < p^*,\begin{cases} m = \sqrt{\dfrac{2\gamma}{\gamma-1}\left[\left(\dfrac{p_2}{p_0}\right)^{\frac{2}{\gamma}} - \left(\dfrac{p_2}{p_0}\right)^{\frac{\gamma+1}{\gamma}}\right]}\dfrac{p_0 s}{\sqrt{RT_0}} \\ v = \sqrt{\dfrac{2\gamma}{\gamma-1}RT_0\left[1 - \left(\dfrac{p_2}{p_0}\right)^{\frac{\gamma-1}{\gamma}}\right]} \end{cases} \tag{7}$$

式中　p^*——临界瓦斯压力,Pa;

p_2——大气压力,Pa;

γ——气体绝热指数;

m——喷口处的瓦斯质量流量,kg/s;

v——瓦斯流速,m/s;

s——喷口截面积,m²;

R——气体常数,J/(kg·K);

p_0——各时刻煤样罐内瓦斯的绝对压力,Pa;

T_0——各时刻煤样罐内瓦斯的绝对温度,K。

通过测定瓦斯喷出过程中罐内的气压与温度,可以得出各时刻喷出瓦斯的流量、流速;然后由式(8)计算出各时刻由喷口流出的瓦斯膨胀能;最后通过式(9)即可获取单位质量煤体的瓦斯膨胀能。

$$W_p = \frac{1}{2}mv^2 \quad (8)$$

$$U_p = \frac{\int W_p \mathrm{d}t}{m_c} \quad (9)$$

式中 W_p——各时刻的瓦斯膨胀能，J/s；

t——瓦斯喷出持续时间，s；

m_c——煤体质量，kg；

U_p——单位质量煤体的瓦斯膨胀能，J/kg。

2 能量测试试验

2.1 试验煤样

据统计，煤与瓦斯突出经常发生在受地质构造影响的煤层中[24]。这类煤体因受到地质构造的挤压与剪切破坏作用，呈现低强度、高孔隙率、结构松散的特征，称为构造煤。与之对应的是高强度、低孔隙率、结构致密的原生煤。

探索变形破坏过程中原生煤与构造煤的能量差异是本文的目的之一，因此试验需采用 $\phi 50 \times 100$ mm 的原生煤与构造煤试件。原生煤试件可通过钻孔取芯的方法从煤层中直接获取，但构造煤试件由于其结构差异则难以取到。为此，试验采用性质相似的型煤进行构造煤性质研究。事实上，型煤已被广泛应用到揭示煤与瓦斯突出内在机制的试验中[25-26]。相关试验结果表明，型煤可以有效地模拟Ⅳ类构造煤[27]。本次试验所用型煤采用自主研制的相似材料制作[28]。型煤以粒径分布为 0～1 mm：1～3 mm＝0.76：0.24 的煤粉为骨料，以浓度为 20 % 的腐殖酸钠水溶液为胶结剂，在 15 MPa 压力下压制成型。试验所用原生煤试件及型煤制作所用煤粉均取自属于煤与瓦斯突出煤层的安徽省淮南矿区望峰岗煤矿 C_{13} 煤层。型煤与原煤的物理力学参数如表 1 所示。

表 1 煤样物理力学参数
Table 1 Physical and mechanical parameters of coal samples

煤样	单轴抗压强度/MPa	弹性模量/GPa	视密度/(kg/m³)	真密度/(kg/m³)	孔隙率/%	Langmuir体积/(cm³/g)	Langmuir压力/MPa
型煤	2.89	0.33	1 306	1 450	9.91	21.592 4	1.077
原煤	11.47	11.97	1 400	1 450	3.45	21.592 4	1.077

2.2 试验装置

现有的煤体加卸载试验主要采用不含瓦斯的煤体进行。由于没有考虑瓦斯对煤体的影响，试验测得的应力-应变曲线与含瓦斯煤体应力-应变曲线有一定区别[29]。为克服以上缺点，本次加卸载试验采用自主研发的"可视化恒容气固耦合试验仪"进行[30]。利用该仪器可进行气固耦合状态下试样的单轴、常规三轴力学试验及渗透率测定试验等多种试验。且该仪器具有试验过程可视化，加卸载过程煤样气压恒定的特点。配合伺服压力试验机，该仪器可精确高效的测定含瓦斯煤加卸载应力-应变曲线。

加卸载过程中煤样气压的恒定由仪器巧妙的结构设计实现。如图3所示,恒容室通过气体管路与试样相通,恒容活塞与轴向压头连接,同时,两个恒容活塞的总截面积等于轴向压头截面积。加载过程中,恒容活塞与轴向压头同步下移,煤样内多余气体通过管路进入恒容室,保证了煤样气压的恒定。需要说明的是,单轴与常规三轴力学试验的气体加载方式有所不同。常规三轴力学试验时,试件通过橡胶套与加载室隔离。此时,气体通道1用于试样的气体加卸载,气体通道2用于加载室的气体加卸载。单轴力学试验时,试件与加载室相通,气体通道1和气体通道2均可用于试件的气体加卸载。

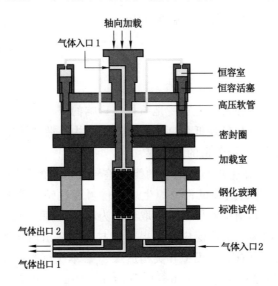

图3 可视化恒容气固耦合试验仪

Fig. 3 Gas-solid couping test system in the visual and constant volume loading state

本次瓦斯释放试验采用"瓦斯膨胀能测定仪"进行,该仪器依据蒋承林教授的思路研发而成。作为瓦斯膨胀能计算的主要参数,罐内瓦斯的压力与温度由安装在罐内的气压传感器、温度传感器获取。两套仪器的实物图如图4所示。

2.3 试验方法

试验包括含瓦斯煤体的单轴加卸载试验及瓦斯释放试验,以分别获取不同应力状态下含瓦斯煤的变形能及瓦斯膨胀能。

在单轴加卸载试验中,对达到吸附平衡的煤体试件由初始状态加载至预定轴向应力,然后卸载至0 MPa,以获取煤样在该应力状态下的完整加卸载应力-应变曲线。10个加卸载周期中,最大轴向应力设置如下:峰前的轴向力$0, 0.17R_c, 0.33R_c, 0.50R_c, 0.67R_c, 0.83R_c, 0.92R_c, R_c$(峰值);峰后的轴向力$0.75R_c, 0.4R_c$,其具体加载值如表2所示。以上应力点保证了煤体各阶段都具有充足的数据:前三个点大致处于煤体压缩阶段,第3至第6个应力点大致处于煤体弹性阶段,第6至第8个应力点大致处于煤体屈服阶段,后3个应力点处于煤体破坏阶段。

 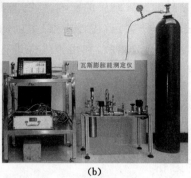

(a) (b)

图 4 试验仪器实物

Fig. 4 Photos of the experimental system

(a)可视化恒容气固耦合试验仪;(b)瓦斯膨胀能测定仪

表 2　　　　　不同应力状态下含瓦斯煤体的应变能与瓦斯膨胀能

Table 2　Strain energy and gas expansion energy of coal containing gas under different stress states

煤样	轴向应力/MPa		损伤变量	弹性能/(mJ/g)	耗散能/(mJ/g)	总应变能/(mJ/g)	瓦斯膨胀能/(mJ/g)
原煤	峰前	0.00	0.000	0.00	0.00	0.00	6.03
		1.80	0.006	0.12	0.05	0.17	6.08
		3.64	0.016	0.41	0.14	0.55	6.33
		5.71	0.031	0.94	0.27	1.21	6.55
		7.75	0.053	1.63	0.46	2.10	6.82
		9.81	0.099	2.67	0.87	3.54	7.17
		10.69	0.157	3.24	1.37	4.61	8.02
	峰值	11.26	0.250	3.99	2.18	6.17	8.96
	峰后	7.66	0.660	1.89	5.76	7.64	12.39
		4.66	0.870	0.82	7.59	8.41	14.46
型煤	峰前	0.00	0.000	0.00	0.00	0.00	67.62
		0.47	0.024	0.14	0.41	0.56	68.67
		1.00	0.058	0.60	1.00	1.61	69.79
		1.52	0.107	1.29	1.88	3.16	71.76
		2.03	0.179	2.16	3.12	5.28	74.58
		2.54	0.290	2.99	5.06	8.05	78.35
		2.78	0.394	3.48	6.88	10.36	85.37
	峰值	2.90	0.566	3.70	9.88	13.59	91.04
	峰后	2.05	0.802	2.39	14.00	16.39	99.21
		1.32	0.914	1.51	15.94	17.45	103.50

瓦斯释放试验采用加载过的煤体试件进行,其应力状态与上述单轴加卸载试验一致。

为消除煤样差异对试验结果的影响,所有的瓦斯释放试验与单轴加卸载试验均采用同一煤体试件。其操作方法为:首先,进行最大轴向应力为 $0.17R_c$ 的加卸载试验;其次,利用该煤样进行瓦斯释放试验;然后,利用该煤样进行最大轴向应力为 $0.33R_c$ 的加卸载试验及瓦斯释放试验,直至所有应力点的试验进行完毕。

甲烷是突出过程中的主要气体。以上试验均采用纯度为 99.99 % 的甲烷进行试验,以保证与实际工况条件的一致性。依据煤层气藏条件,煤样吸附平衡压力确定为 0.75 MPa(相对压力)。

3 试验结果与分析

3.1 能量测试结果

试验获取了含瓦斯原煤、型煤的循环加卸载应力-应变曲线,如图 5 所示。可以看出,由于加卸载作用,两类煤体应力-应变曲线形成了明显的滞后环,而且出现了残余变形。其中型煤的滞后环和残余变形较为明显,体现出原煤与型煤在加卸载过程中的损伤演化差异。随着应力的增加,两类煤岩加卸载时的弹性模量在峰值前均有小幅增加,过峰值后均大幅降低,试验结果与文献[31]研究成果一致。

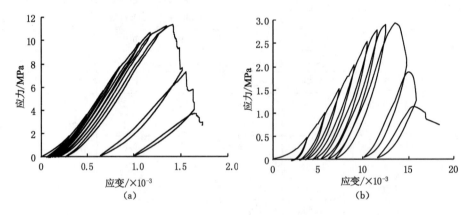

图 5 含瓦斯煤加卸载应力-应变曲线

Fig.5 Loading and unloading stress-strain curve of coal containing gas

(a) 原煤;(b) 型煤

依据试验获取的加卸载应力-应变曲线(图 5),基于式(2)~式(4),计算得到了各应力状态下含瓦斯型煤、原煤的总应变能、弹性能、耗散能;依据瓦斯释放过程的气压、温度,基于式(5)~式(9),计算得到了各应力状态下的含瓦斯型煤、原煤的瓦斯膨胀能,计算结果如表 2 所示。

3.2 弹性能演化规律

研究表明,煤体循环加卸载过程的变形特征表现出明显记忆性,加卸载过程的应力-应变外包络线与连续加载的全程应力-应变曲线相吻合[31]。为清晰展示弹性能在全应力应变过程的演化规律,将加卸载过程的应力应变外包络线及各应力条件下的弹性能一并绘入图 6 中。

试验结果显示,煤体变形破坏对弹性能有显著影响。在整个应力应变过程中,煤体的弹

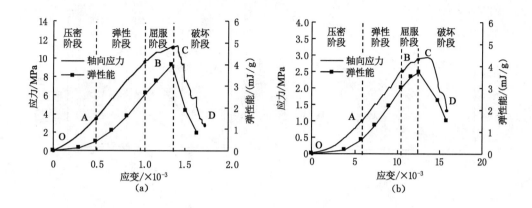

图 6 弹性能演化规律

Fig. 6 The evolution of elastic energy

(a) 原煤；(b) 型煤

性能变化量约为 4 mJ/g。煤体弹性能在突出孕育过程中有破坏煤体、激发突出的作用[32]，因此，以上因应力状态变化而导致的弹性储能增长会对突出发动产生影响。

如图 6 所示，随着煤体应力状态的改变，原煤、型煤的弹性能均呈现先增大后减小的趋势。按照变化趋势，可将两类煤体的弹性能划分为三个阶段，缓慢增长阶段、线性增长阶段、线性降低阶段，分别对应于煤体的压密阶段（OA 段）、弹性及屈服阶段（AC 段）、破坏阶段（CD 段）。

尽管构造煤、原生煤力学性质差异较大，但两类煤体在各阶段的弹性能大小几乎无差异。峰值点的弹性能为煤体弹性能极限值，反映了煤体储存变形能的能力[14]。以该点弹性能为例，型煤在该点的弹性能为 3.70 mJ/g，原煤在该点的弹性能为 3.99 mJ/g，原煤的弹性能极限值略大于型煤，但差异仅为 7%。以上试验结果说明，煤样类型对煤体弹性能影响较小。

3.3 瓦斯膨胀能演化规律

基于以上试验结果可以发现，随着煤体应力状态变化，原煤瓦斯膨胀能增大 2.4 倍，型煤瓦斯膨胀能增大 35.88 mJ/g。蒋承林教授基于试验研究认为：瓦斯膨胀能大于 42.98 mJ/g 时，煤体便会发生弱突出，大于 103.8 mJ/g 时，煤体便会发生强突出[33]。对于构造煤及吸附性较强的原生煤，以上因煤体应力状态变化导致的瓦斯膨胀能变化，会极大地影响煤层突出危险性。

由图 7 可以看出，随着煤体应力增大，原煤与型煤的瓦斯膨胀能均呈增长趋势，并具有明显的阶段性。但由于变形破坏过程中煤体孔隙、裂隙结构演化不同，两类煤体在变形破坏的四个阶段变化率有所差异。原煤的瓦斯膨胀能在四个阶段呈加速增长趋势。在压密阶段（OA 段），瓦斯膨胀能基本不变，其增量仅为 5%（0.30 mJ/g）；在弹性阶段（AB 段），瓦斯膨胀能有轻微上涨，其增量为 13%（0.85 mJ/g）；在屈服阶段（BC 段），瓦斯膨胀能明显增大，增量约为 25%（1.79 mJ/g）；在破坏阶段（CD 段），瓦斯膨胀能猛增，增量约为 61%（5.50 mJ/g）。对于型煤，在压密阶段（OA 段），瓦斯膨胀能基本不变，其增量仅为 3.2%（2.17 mJ/g）；在弹性（AB 段）、屈服（BC 段）、破坏阶段（CD 段），瓦斯膨胀能明显增长，且增量相近，分别为 12%（8.56 mJ/g）、15%（12.70 mJ/g）、11%（10.26 mJ/g）。

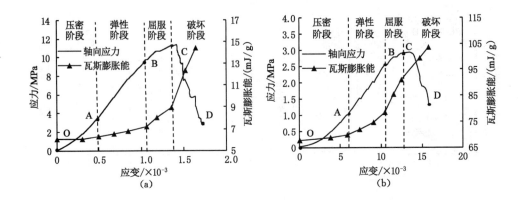

图 7 瓦斯膨胀能演化规律

Fig.7 The evolution of gas expansion energy

(a) 原煤；(b) 型煤

在整个应力应变过程中，由于煤体的孔隙结构差异，孔隙率较低的原生煤的瓦斯膨胀能远小于孔隙率较高的构造煤的瓦斯膨胀能。如表 2 所示，原煤的瓦斯膨胀能仅为 6.03～14.46 mJ/g，而型煤瓦斯膨胀能高达 67.62～103.50 mJ/g，两者相差 10 倍左右。正是瓦斯膨胀能的巨大差异，导致受构造作用影响的煤层突出发生率极高。

4 基于损伤变量的能量演化数学描述

在煤体变形破坏过程中，煤样损伤不断加剧，从微观上表现为微裂纹的萌生、扩展至贯通的过程；从宏观上表现为变形及力学性能不断变化[15]。其中，煤体孔隙、裂隙结构的演化会影响孔隙中的瓦斯解吸、裂隙中的瓦斯渗流[34]，最终对瓦斯膨胀能构成影响；煤体变形及力学性能的演化会直接导致煤体弹性能发生变化。换言之，煤体损伤程度的变化是引起煤体弹性能、瓦斯膨胀能变化的本质原因。因此，从损伤角度描述含瓦斯煤弹性能、瓦斯膨胀能演化规律具有一定可行性。

为探讨弹性能、瓦斯膨胀能与损伤关系，采用式(10)计算了原煤、型煤的损伤变量，计算结果一并列入表 2。需要说明的是，式(10)从能量角度对岩体内部裂隙发育程度及相关力学劣化效应进行量化，可以准确计算岩体各应力应变阶段的损伤程度[35]。

$$D = \frac{U_d}{U} \tag{10}$$

由损伤变量的计算结果可知，两类煤体的损伤变量在全应力应变过程差异较大，且呈现阶段性。这就解释了原煤与型煤的弹性能、瓦斯膨胀能在应力应变全过程呈现出的能量差异及分段特性。

如图 8、图 9 所示，两类煤体的弹性能、瓦斯膨胀能与其损伤变量有相同的函数关系。弹性能与损伤变量在峰值强度以前均呈二次函数递增关系，在峰值强度以后均呈线性递减关系；瓦斯膨胀能与损伤变量在峰值强度前后均呈线性递增关系。

原煤与型煤的能量测试结果与拟合函数之间均呈现出高度的相关性，证明基于损伤变量对应力变化过程中的含瓦斯煤能量演化进行数学描述是可行的。

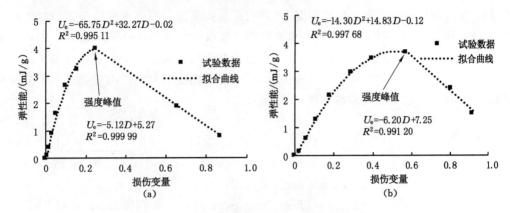

图 8 弹性能与损伤变量关系

Fig. 8 Relation between elastic energy and damage variable

(a)原煤;(b)型煤

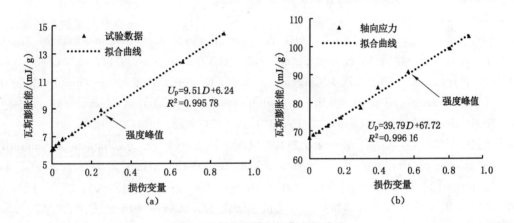

图 9 瓦斯膨胀能与损伤变量关系

Fig. 9 Relation between gas expansion energy and damage variables

(a)原煤;(b)型煤

5 结 论

(1)煤体变形破坏对型煤、原煤两类煤体的弹性能、瓦斯膨胀能均有显著影响。其中,两类煤体的弹性能变化量约为 4 mJ/g,原煤瓦斯膨胀能增大 2.4 倍,构造煤瓦斯膨胀能增大 35.88 mJ/g。以上因煤体应力状态变化导致的突出潜能变化,会极大地影响煤层突出危险性。

(2)煤体变形破坏中,型煤、原煤两类煤体的弹性能、瓦斯膨胀能变化规律一致。随着应力状态的改变,原煤、型煤的弹性能呈现先增大后减小的趋势,瓦斯膨胀能呈现加速增长趋势。另外,由于在全应力应变过程中煤体损伤演化呈现阶段性,煤体弹性能、瓦斯膨胀能在压密阶段、弹性阶段、屈服阶段、破坏阶段呈现不同的变化速率。

(3)在整个应力应变过程中,物理力学性质差异较大的两类煤体弹性能大小相近,瓦斯膨胀能却差异较大,其中型煤瓦斯膨胀能是原煤瓦斯膨胀能的 10 倍左右。正是瓦斯膨胀能的巨大差异,导致受构造作用影响的煤层突出发生率极高。

（4）通过数学方程拟合了含瓦斯煤弹性能、瓦斯膨胀能与损伤变量的关系，型煤、原煤具有相同的函数关系：弹性能与损伤变量在峰值强度以前均呈二次函数递增关系，在峰值强度以后均呈线性递减关系；瓦斯膨胀能与损伤变量在峰值强度前后均呈线性递增关系。测试结果与拟合函数之间均呈现出高度相关性，证明基于损伤变量对应力变化过程中的含瓦斯煤能量演化进行数学描述是可行的。

参 考 文 献

[1] NIE Baisheng, LI Xiangchun. Mechanism research on coal and gas outburst during vibration blasting[J]. Safety Science, 2012, 50(4)：741-744.

[2] LI Hui, FENG Zengchao, ZHAO Dongdong, et al. Simulation Experiment and Acoustic Emission Study on Coal and Gas Outburst[J]. Rock Mechanics & Rock Engineering, 2017, 50(8)：2193-2205.

[3] 谢雄刚, 冯涛, 王永, 等. 煤与瓦斯突出过程中能量动态平衡[J]. 煤炭学报, 2010, 35(7)：1120-1124.

XIE Xionggang, FENG Tao, WANG Yong, et al. The energy dynamic balance in coal and gas outburst[J]. Journal of China Coal Society, 2010, 35(7)：1120-1124.

[4] XIE Heping, XIE Jing, GAO Mingzhong, et al. Theoretical and experimental validation of mining-enhanced permeability for simultaneous exploitation of coal and gas[J]. Environmental Earth Sciences, 2015, 73(10)：5951-5962.

[5] CHEN Zuyun, XIAO Zhuxin, ZOU Ming. Research on mechanism of quantity discharge of firedamp from coal drift of headwork surface reflect coal and gas outburst[J]. International Journal of Hydrogen Energy, 2017, 42(30)：19395-19401.

[6] 孙晓元. 受载煤体振动破坏特征及致灾机理研究[D]. 北京：中国矿业大学（北京），2016：1-38.

SUN Xiaoyuan. The research on failure characteristics and disaster-causing mechanism of the loaded coal in vibration[D]. Beijing：China University of Mining & Technology (Beijing), 2016：1-38.

[7] 李成武, 付帅, 解北京, 等. 煤与瓦斯突出能量预测模型及其在平煤矿区的应用[J]. 中国矿业大学学报, 2018, 47(2)：231-239.

LI Chengwu, FU Shuai, XIE Beijing, et al. Establishment of the prediction model of coal and gas outburst energy and its application in Pingdingshan mining area[J]. Journal of China University of Mining & Technology, 2018, 47(2)：231-239.

[8] 高魁, 刘泽功, 刘健. 地应力在石门揭构造软煤诱发煤与瓦斯突出中的作用[J]. 岩石力学与工程学报, 2015, 34(2)：305-312.

GAO Kui, LIU Zegong, LIU Jian. Effect of geostress on coal and gas outburst in the uncovering tectonic soft coal by cross-cut[J]. Chinese Journal of Rock Mechanics and Engineering, 2015, 34(2)：305-312.

[9] 姜永东, 郑权, 刘浩, 等. 煤与瓦斯突出过程的能量分析[J]. 重庆大学学报, 2013, 36(7)：98-101.

JIANG Yongdong, ZHENG Quan, LIU Hao, et al. An analysis on the energy of coal and gas outburst process[J]. Journal of Chongqing University, 2013, 36(7): 98-101

[10] 滕腾,高峰,张志镇,等. 含瓦斯原煤三轴压缩变形时的能量演化分析[J]. 中国矿业大学学报, 2016, 45(4): 663-669.

TENG Teng, GAO Feng, ZHANG Zhizhen, et al. Analysis of energy evolution on gas saturated raw coal under triaxial compression[J]. Journal of China University of Mining & Technology, 2016, 45(4): 663-669.

[11] 刘江伟,黄炳香,魏民涛. 单轴循环荷载对煤弹塑性和能量积聚耗散的影响[J]. 辽宁工程技术大学学报(自然科学版), 2012, 31(1): 26-30.

LIU Jiangwei, HUANG Bingxiang, WEI Mintao. Influence of cyclic uniaxial loading on coal elastic-plastic properties and energy accumulation and dissipation[J]. Journal of Liaoning Technical University (Natural Science), 2012, 31(1): 26-30.

[12] 肖福坤,刘刚,申志亮,等. 循环载荷作用下煤样能量转化规律和声发射变化特征[J]. 岩石力学与工程学报, 2016, 35(10): 1954-1964.

XIAO Fukun, LIU Gang, SHEN Zhiliang, et al. Energy conversion and acoustic emission(AE) characteristics of coal samples under cyclic loading[J]. Chinese Journal of Rock Mechanics and Engineering, 2016, 35(10): 1954-1964.

[13] 张黎明,高速,王在泉,等. 大理岩加卸荷破坏过程的能量演化特征分析[J]. 岩石力学与工程学报, 2013, 32(8): 1572-1578.

ZHANG Liming, GAO Su, WANG Zaiquan, et al. Analysis of marble failure energy evolution under loading and unloading conditions[J]. Chinese Journal of Rock Mechanics and Engineering, 2013, 32(8): 1572-1578.

[14] 张志镇,高峰. 单轴压缩下红砂岩能量演化试验研究[J]. 岩石力学与工程学报, 2012, 31(5): 953-962.

ZHANG Zhizhen, GAO Feng. Experimental research on energy evolution of red sandstone samples under uniaxial compression[J]. Chinese Journal of Rock Mechanics and Engineering, 2012, 31(5): 953-962.

[15] 彭瑞东,鞠杨,高峰,等. 三轴循环加卸载下煤岩损伤的能量机制分析[J]. 煤炭学报, 2014, 39(2): 245-252.

PENG Ruidong, JU Yang, GAO Feng, et al. Energy analysis on damage of coal under cyclical triaxial loading and unloading conditions[J]. Journal of China Coal Society, 2014, 39(2): 245-252.

[16] 蒋承林,俞启香. 煤与瓦斯突出过程中能量耗散规律的研究[J]. 煤炭学报, 1996, 21(2): 173-178.

JIANG Chenglin, YU Qixiang. Rules of energy dissipation in coal and gas outburst[J]. Journal of China Coal Society, 1996, 21(2): 173-178.

[17] 罗甲渊. 煤与瓦斯突出的能量源及能量耗散机理研究[D]. 重庆:重庆大学, 2016: 85-105.

LUO Jiayuan. Study on energy source and energy dissipation mechanism of coal and gas outburst[D]. Chongqing:Chongqing University,2016:85-105.

[18] 李晓伟,蒋承林,季明,等.初始释放瓦斯膨胀能与煤体破碎程度的关系研究[J].煤矿安全,2008(5):1-3.
LI Xiaowei, JIANG Chenglin, JI Ming, et al. Study on relation between screening modulus of coal and expansion energy of initial released gas[J]. Safety in Coal Mines, 2008(5): 1-3.

[19] 徐乐华,蒋承林.煤样粒径对初始释放瓦斯膨胀能影响的试验研究[J].煤矿安全,2015,46(4):20-22.
XU Lehua, JIANG Chenglin. Experimental study on influence of particle diameter on initial expansion energy of released gas[J]. Safety in Coal Mines,2015,46(4): 20-22.

[20] 何满潮,王春光,李德建,等.单轴应力-温度作用下煤中吸附瓦斯解吸特征[J].岩石力学与工程学报,2010,29(5):865-872.
HE Manchao, WANG Chunguang, LI Jiande, et al. Desorption characteristics of adsorbed gas in coal samples under coupling temperature and uniaxial compression [J]. Chinese Journal of Rock Mechanics and Engineering,2010,29(5): 865-872.

[21] 唐巨鹏,潘一山,李成全,等.三维应力作用下煤层气吸附解吸特性实验[J].天然气工业,2007,27(7):35-38.
TANG Jupeng, PAN Yishan, LI Chengquan, et al. Experimental study of adsorption and desorption of coalbed methane under three-dimensional stress[J]. Natural Gas Industry,2007,27(7):35-38.

[22] 胡泊.压应力作用下煤体瓦斯解吸及渗流规律研究[D].北京:中国矿业大学(北京),2016:45-91.
HU Po. Study on the gas desorption and seepage flow law of coal under compressive stress[D]. Beijing:China University of Mining & Technology (Beijing),2016:45-91.

[23] 齐黎明,陈学习,程五一.瓦斯膨胀能与瓦斯压力和含量的关系[J].煤炭学报,2010,35(S1):105-108.
QI Liming, CHEN Xuexi, CHENG Wuyi. Relationship of expansion energy of gas with gas pressure and content[J]. Journal of China Coal Society, 2010, 35(S1): 105-108.

[24] 邵强,王恩营,王红卫,等.构造煤分布规律对煤与瓦斯突出的控制[J].煤炭学报,2010,35(2):250-254.
SHAO Qiang, WANG Enying, WANG Hongwei, et al. Control to coal and gas outburst of tectonic coal distribution[J]. Journal of China Coal Society,2010,35(2):250-254.

[25] HU Qianting, ZHANG Shutong, WEN Guangcai, et al. Coal-like material for coal and gas outburst simulation tests[J]. International Journal of Rock Mechanics and Mining Sciences,2015,74:151-156.

[26] TU Qingyi, CHENG Yuanping, GUO Pinkun, et al. Experimental study of coal and gas outbursts related to gas-enriched areas [J]. Rock Mechanics and Rock

Engineering,2016,49(9):3769-3781.

[27] CAO Yunxing,HE Dingdong,DAVID Cglick. Coal and gas outbursts in footwalls of reverse faults[J]. International Journal of Coal Geology,2001,48(1):47-63.

[28] 王汉鹏,张庆贺,袁亮,等.含瓦斯煤相似材料研制及其突出试验应用[J].岩土力学,2015,36(6):1676-1682.
WANG Hanpeng,ZHANG Qinghe,YUAN Liang,et al. Development of a similar material for methane-bearing coal and its application to outburst experiment[J]. Rock and Soil Mechanics,2015,36(6):1676-1682.

[29] 王家臣,邵太升,赵洪宝.瓦斯对突出煤力学特性影响试验研究[J].采矿与安全工程学报,2011,28(3):391-394.
WANG Jiachen,SHAO Taisheng,ZHAO Hongbao. Experimental study of effect of gas on mechanical properties of outburst coal[J]. Journal of Mining & Safety Engineering,2011,28(3):391-394.

[30] 李清川,王汉鹏,李术才,等.可视化恒容气固耦合试验系统的研发与应用[J].中国矿业大学学报,2018,47(1):104-112.
LI Qingchuan,WANG Hanpeng,LI Shucai,et al. Development and application of a gas-solid couping test system in the visual and constant volume loading state[J]. Journal of China University of Mining & Technology,2018,47(1):104-112.

[31] 苏承东,熊祖强,翟新献,等.三轴循环加卸载作用下煤样变形及强度特征分析[J].采矿与安全工程学报,2014,31(3):456-461.
SU Chengdong,XIONG Zuqiang,ZHAI Xinxian,et al. Analysis of deformation and strength characteristics of coal samples under the triaxial cyclic loading and unloading stress path[J]. Journal of Mining & Safety Engineering,2014,31(3):456-461.

[32] 郭臣业.岩石和瓦斯突出发生条件及机理研究[D].重庆:重庆大学,2010:130-142.
GUO Chenye. Research on occurrence conditions and mechanism of rock and gas outburst[D]. Chongqing:Chongqing University,2010:130-142.

[33] TIAN Shixiang,JIANG Chenglin,XU Lehua,et al. A study of the principles and methods of quick validation of the outburst-prevention effect in the process of coal uncovering[J]. Journal of Natural Gas Science and Engineering,2016(30):276-283.

[34] XU Lehua,JIANG Chenglin. Initial desorption characterization of methane and carbon dioxide in coal and its influence on coal and gas outburst risk[J]. Fuel,2017(203):700-706.

[35] 金丰年,蒋美蓉,高小玲.基于能量耗散定义损伤变量的方法[J].岩石力学与工程学报,2004,23(12):1976-1980.
JIN Fengnian,JIANG Meirong,GAO Xiaoling. Defining damage variable based on energy dissipation[J]. Chinese Journal of Rock Mechanics and Engineering,2004,23(12):1976-1980.

泥质胶结岩体力学特性宏细观模拟研究

孙长伦,李桂臣,何锦涛,孙元田,董玉玺

(中国矿业大学 深部煤炭资源开采教育部重点实验室 矿业工程学院,江苏 徐州,221116)

摘　要:对于以黏粒胶结为主要胶结方式的泥质岩体,岩体力学特征与组构密切相关。文章采用物理模拟和数值模拟手段进行泥质胶结岩体单轴压缩试验,从宏-细观尺度开展组构对力学特性影响规律研究。岩体组构对力学特性的影响规律为:水化前后泥质胶结岩体单轴抗压强度和弹性模量与黏粒含量符合指数小于0的Allometric1关系,黏粒含量越高,泥质岩体的单轴抗压强度、弹性模量越小;黏粒含量越高,泥质胶结岩体水化弱化指数越大,水化越剧烈;当黏粒体积分数增加到0.4左右时,泥质胶结岩体强度和水化弱化指数不随黏粒含量增加而变化,组分的增加不影响泥质胶结岩体的强度和水化弱化特性。岩体组构对力学性质的影响机理为:黏粒含量越高,岩体碎屑之间的强胶结数量越少,黏粒之间的弱胶结数量越多;岩体压缩破坏过程弱胶结破坏数量越多、强胶结数量越少;水化过程弱胶结进一步弱化。

关键词:泥质胶结、力学特性、宏细观、离散元

Macroscopic and microscopic simulation study on mechanical properties of argillaceous cemented rock mass

SUN Changlun,LI Guichen,HE Jintao,SUN Yuantian,DONG Yuxi

(*Key Laboratory of Deep Coal Resource Mining of the Ministry of Education*,
China University of Mining and Technology,*School of Mines*,*Xuzhou* 221116,*China*)

Abstract:For argillaceous rock masses with clay cementation as the main cementing method, the mechanical properties of rock mass are closely related to the fabric. In this paper, physics simulation and numerical simulation are used to carry out UCT of argillaceous cemented rock mass, and the influence law of fabric on mechanical properties is studied from macro-microscopic scale. The influence of rock fabric on the mechanical properties is as follows. Before and after hydration, the argillaceous cemented rock UCS, elastic modulus are in accordance with the Allometric1 relationship with the index of less than 0. The higher the clay content is, the larger the hydration weakening index of the argillaceous cemented rock mass is, and the more hydration is. When the clay volume fraction increases to about 0.4, the strength and hydration weakening index of cemented rock mass do not change with the increase of clay content. The increase of composition does not affect the strength and hydration weakening characteristics of argillaceous cemented rock mass. The mechanism of above law is as follows. The higher the clay content, the less strong cements between the cutting are and the more the weak cementation between the clays are. The higher the clay

基金项目:国家重点研发计划专项资助项目(2016YFC0600901);国家自然科学基金资助项目(51574224)。
作者简介:孙长伦(1990—),男,山东省临沂市人,硕士研究生。Tel:13115205973,E-mail:852383588@qq.com。
通信作者:李桂臣(1980—),男,河北省衡水市人,教授。Tel:15805215566,E-mai:liguichen@126.com。

content, the more the cementation failure in the compression process of the rock mass are. Weak cementation in the hydration process is further weakened.

Keywords: argillaceous cement; mechanical properties; macro-microscopic; discrete element

细观上，泥岩是指以黏土颗粒胶结为主要胶结类型的一种泥质胶结岩体，其软岩特性及遇水弱化问题是岩土工程领域研究热点。在泥岩软岩特性研究中，A. C. Guéry，F. Cormery 等[1]将泥岩视为弹塑性黏土基体和线弹性或弹塑性夹杂物组成的非均质材料；J. C. Robinet，P. Sardini 等[2]研究表明泥岩是由细粒黏土为基质，黏土颗粒之间嵌入粒径较大的非黏土矿物；G. Armand，N. Conil 等[3]研究 COx 泥岩力学行为的过程中，发现泥岩力学参数弱化程度取决于黏土矿物含量；K. Oohashi，T. Hirose 等[4]发现泥质胶结岩体蒙脱石含量从 20% 增加到 50%，摩擦系数从 0.5~0.6 逐渐下降到 0.1；M. Takahashi，K. Mizoguchi 等[5]研究发现随着蒙脱石含量的增加，断层泥的摩擦系数从 0.68 降低到 0.08；S. Tembe，D. A. Lockner 等[6]发现在伊利石和石英的混合体中，随着伊利石含量的增加，摩擦系数从 0.75 降低到 0.30。水化弱化特性研究中，周翠英、邓毅梅等[7]研究了泥岩力学特性与浸水时间的关系；A. Ghorbani[8]分析了含水率对泥岩波速的影响规律；D. Yang[9-10]研究了不同含水量对泥岩气体渗透性能、蠕变特性的影响；F. Zhang[11]研究了泥岩饱水程度与强度、弹性模量和蠕变特性等力学性质的关系；H. Tetsuka，I. Katayama 等[12]发现随着湿度增加，Na-蒙脱土的摩擦系数从 0.33（相对湿度 10%）下降到 0.06（相对湿度 93%），Ca-蒙脱土的摩擦系数从 0.22（相对湿度 11%）下降到 0.04（相对湿度 91%）。考虑黏土矿物类型[13-14]和含量[15]的研究对于泥质胶结岩体水化弱化特性具有重要意义。

关于特定组构泥质胶结岩体研究，天然的工程岩体难以满足要求，原因如下：(1) 以天然岩样为研究对象进行泥岩特性研究，存在试验结果重复性差、取样损伤等问题；(2) 在不同位置的天然岩样矿物组分差异明显[16]，相同位置取样获得的泥岩也不尽相同[17]。因此，亟待提出一种简易的泥质胶结岩体力学规律研究方法，并进行相关规律研究。

为了开展泥质胶结岩体软岩特性和水化弱化规律宏细观研究，从岩体组构特征研究泥质胶结岩体，文章首先基于离散元思想，构建泥质胶结岩体的颗粒胶结模型；然后以水泥胶结标准石英砂模拟沉积岩成岩作用胶结、膨润土为泥质弱胶结制作泥质弱胶结岩体并模拟黏粒组分含量对其软岩特性和水化弱化规律的影响；最后，采用离散元颗粒流模拟软件，从细观尺度揭示泥质胶结岩体软岩特性和水化弱化机理。

1 泥质胶结岩体胶结物理模型

1.1 泥质胶结岩体组构

根据离散元思想，泥质胶结岩体细观组构主要包括岩屑颗粒、黏粒和孔隙三种组分以及各组分之间形成的颗粒间胶结，如图 1(b)(c)(d)所示，泥质胶结岩体力学性质及水化弱化规律与其组构密切相关。在碎屑和黏粒成岩过程中，碎屑胶结（胶结Ⅰ）、黏粒胶结（胶结Ⅱ）以及碎屑、黏粒之间胶结（胶结Ⅲ）等组构特征决定这种多组分沉积岩体的软岩特征和遇水弱化规律。泥质胶结岩体以黏粒胶结（胶结Ⅱ）为主，不存在或少量存在碎屑胶结（胶结Ⅰ），其力学性质与水化弱化规律有岩体内部黏粒组分和黏粒形成的胶结（胶结Ⅱ、Ⅲ）决定。

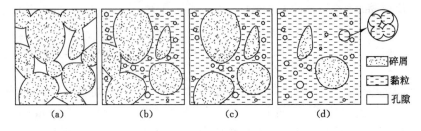

图 1　泥质胶结岩体组构

Fig.1　The fabric of argillaceous rock mass

(a) 胶结Ⅰ;(b) 胶结Ⅰ、Ⅱ、Ⅲ;(c) 胶结Ⅱ、Ⅲ;(d) 胶结Ⅱ、Ⅲ

1.2　泥质胶结岩体物理模型

泥质胶结岩体胶结强度低,遇水弱化,物理模拟需要选择合适的岩屑和胶结材料。石英是一种最为常见的造岩矿物,以标准石英砂为岩屑,岩屑之间通过普通硅酸盐水泥模拟成岩胶结作用;膨润土来源广泛,蒙脱石类黏粒含量高,可作为泥质胶结材料。所采用的物理模型材料如图 2 所示。排水法测标准砂、普通硅酸盐水泥和膨润土密度分别为 2.65、3.32、2.12 g/cm³,筛分标准砂,粒径分布如图 3 所示。

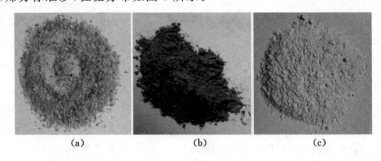

图 2　试件制样材料

Fig.2　The specimen materials

(a) 标准砂;(b) 普通硅酸盐水泥;(c) 膨润土

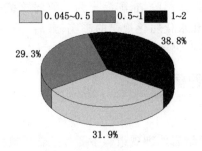

图 3　标准砂粒径分布

Fig.3　The particle diameter range of standard sand

采用 D/Max-3B X 射线衍射设备测试膨润土矿物种类,并初步量化各矿物含量,如图 4 所示。试验采用的膨润土组分为蒙脱石(Montmorillonite)、石英(Quartz)和方解石(Calcite),含量分别为 70.27%、19.24%、10.49%,泥质胶结岩体试件为蒙脱石类泥岩,其水化弱化特征由黏土矿物蒙脱石决定,具有遇水膨胀特性。

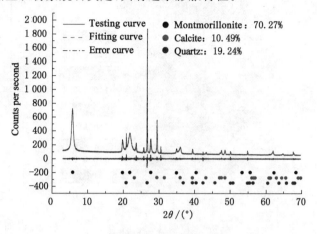

图4　膨润土X射线衍射图谱
Fig. 4　X-ray diffraction patterns of bentonite

采用电液伺服压力机配套自制成型模具压制试件,如图5所示,制样采用程序控制,2 mm/min位移控制加载,加载到设计成型压力,保载一定时间;采用该压力机进行成型试件单轴压缩试验,位移控制,加载速度2 mm/min。

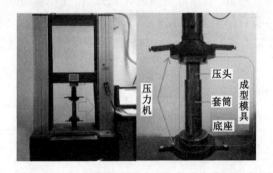

图5　试件制样设备
Fig. 5　The devices for the forming of specimen

压制不同矿物组分含量的试件,研究黏土矿物含量对岩体胶结参数的弱化规律,并进行水化弱化实验。试件压制条件为加载至5 MPa,保载5 min;试件含1份标准砂、0.5份水、0.5份水泥,膨润土与标准砂的质量比$P\_s=0.3、0.4、0.5、0.7、0.9$,成型试件5组,每组4个,如图6所示,试件成型后室温养护7 d,然后进行单轴压缩试验。根据各组试件矿物组分和试件密度计算平均三维孔隙率分别为33.6%、32.8%、31.3%、30.4%、28.7%。

2　泥质胶结岩体力学特性物理模拟

将5组20个不同黏土矿物含量试件分成浸水实验组和对照组两个部分,试验组包括5组不同黏土矿物含量试件,每组试件2个,对照组相同;实验组浸水6 h,充分饱和,对照组不浸水,研究黏土矿物含量对试件力学特性的影响规律,图7为实验组和对照组泥岩单轴压缩

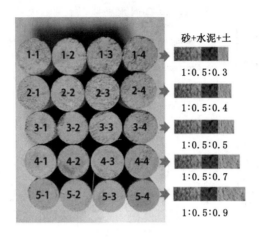

图 6　成型试件
Fig. 6　The forming specimen

应力-应变曲线。

随着黏土矿物含量的增加,试件单轴抗压强度、弹性模量降低,强度弱化;峰值应变增加,塑性破坏特性显现。与对照组相比,浸水实验组试件强度显著降低,水化弱化作用明显。图 8 为试件水化前后力学参数[单轴抗压强度(UCS)、弹性模量(E)与黏土矿物含量关系,采用 Allometric1 拟合,拟合结果如表 1 所示。试件力学参数与黏土矿物含量符合指数小于 0 的 Allometric1 函数,黏土矿物含量越高,单轴压缩强度和弹性模量越小,减小的幅度逐渐变小。

表 1　　　　　　　　　　函数拟合结果
Table 1　　　　　　　　The function fitting results

参数	组别	拟合公式	常数值		R^2
			a	b	
UCS	对照组		1.45	−0.77	0.87
UCS	实验组	$y = ax^b$	0.40	−0.94	0.99
E	对照组		116.88	−1.13	0.91
E	实验组		35.29	−1.40	0.89

为了确定试件黏土矿物含量对水化弱化程度的影响规律,采用水化弱化指数(δ)量化试件水化过程中的弱化程度,水化指数是指泥岩水化过程中力学性质降低量与初值的比值,能够确定不同黏土矿物含量试件水化弱化程度。采用拟合公式分析,对于泥岩单轴抗压强度、弹性模量的水化弱化指数(分别用 δ_{UCS} 和 δ_E 表示)与膨润土含量的关系如图 9 所示。泥岩黏土矿物含量越高,其水化弱化指数越高,说明黏土矿物遇水胶结弱化、膨胀剥离是导致泥岩弱化最主要原因。随着黏土矿物组分含量增加,水化弱化指数先快速增加,然后趋于稳定,说明当黏土矿物含量增加到一定程度($P\_s = 0.8$),试样颗粒间胶结全部为泥质胶结岩样,力学性质完全由黏粒形成的泥质胶结决定。

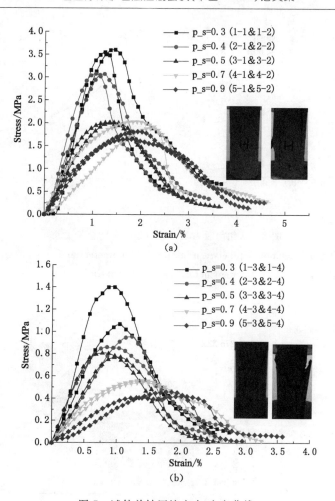

图 7 试件单轴压缩应力-应变曲线

Fig. 7 The UCT stress-strain curve of specimens

(a) 对照组；(b) 实验组

3 泥质胶结岩体力学特性数值模拟

3.1 泥质胶结岩体细观力学参数校核

泥质胶结岩体细观力学参数主要包括碎屑胶结参数Ⅰ、黏粒胶结参数Ⅱ、黏粒与碎屑之间的胶结参数Ⅲ，泥质胶结岩体细观力学参数校核基于以下假设：

（1）以球颗粒代替岩体碎屑和黏粒，不考虑颗粒形状。

（2）纳米级别的黏粒按团聚粒径处理，量级与碎屑颗粒最小粒径相同，简化计算过程。

采用线性平行接触模型(PB)进行细观力学参数校核，校核的原则是确定碎屑胶结参数Ⅰ、黏粒胶结参数Ⅱ、黏粒与碎屑之间的胶结参数Ⅲ基础上，校核随着黏粒含量增加，泥质胶结岩体力学参数弱化规律符合物理模拟结果。根据何咏睿，朱晟等[18]人研究，离散元模拟中，颗粒二维孔隙率与三维孔隙率的关系满足式(1)。

$$n_{2D} = \frac{n_{3D} - 0.08794}{1.4618} \tag{1}$$

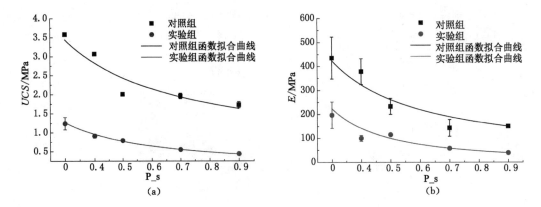

图 8 水化前后力学参数与黏土矿物含量关系

Fig. 8 Relationship between mechanical parameters and clay mineral content before and after hydration

(a) UCS; (b) E

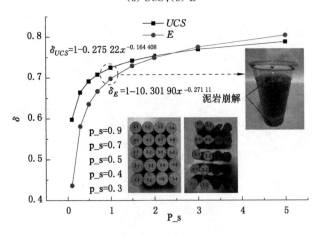

图 9 不同黏土矿物含量试件弱化指数

Fig. 9 Hydration weakening index of specimen with different clay mineral content

式中，n_{3D} 为三维孔隙率；n_{2D} 为二维孔隙率。

以上述 P_s=0.3、0.4、0.5、0.7、0.9 对应的二维孔隙率为 0.170、0.168、0.162、0.153、0.136 试件物理力学参数为校核对象，进行细观参数校核，校核结果如表 2 所示。

表 2 胶结细观力学参数

Table 2 Cementation meso-mechanics parameters

参数 \ 值	Emod /MPa	Krat	Pb_ten /MPa	Pb_coh /MPa	Pb_fa /(°)	Fric	Dp_nratio
Ⅰ	897.00	1	8.65	20.95	38	0.50	0.50
Ⅱ	14.44	1	0.54	1.13	22	0.30	0.35
Ⅲ	94.72	1	1.27	4.32	28	0.50	0.50

校核单轴压缩曲线如图10所示,模拟曲线能够在一定程度上表征岩体的应力-应变关系,根据物理模拟及数值模拟结果,随着黏土矿物含量增加,试件的强度弱化,弹性模量降低,峰值应变增加,具有向软岩演化特性。

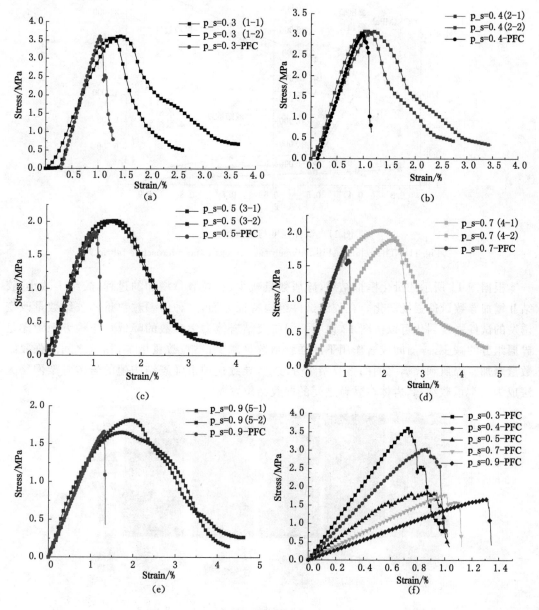

图10 数值模拟应力-应变曲线

Fig.10 The stress-strain curves of numerical simulation

3.2 泥质胶结岩体力学特性弱化机理

泥质胶结岩体压缩破坏在细观上表现为碎屑胶结Ⅰ、黏粒胶结Ⅱ、黏粒与碎屑之间的胶结Ⅲ断裂,泥质胶结岩体的力学性质随黏土矿物含量变化关系需要从其胶结类型改变讨论。通过模拟软件可以获得岩体的不同胶结数量,量化试件强度与岩体胶结类型、数量的关系,如图11所示。

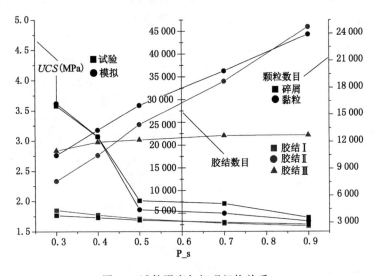

图11　试件强度与细观组构关系

Fig.11　The relationship between the strength and microscope fabric

根据图11所示的量化模拟结果,碎屑数量减少、黏粒数量增加的过程,胶结Ⅰ减小、胶结Ⅱ增加导致试件强度弱化。在胶结试件组构特征方面,这种由颗粒数量和胶结数量改变诱发的试件强度弱化可以由图12说明。泥质胶结岩体力学参数的降低从组构方面考虑是碎屑组分的减少,碎屑间成岩作用下的强胶结数量减少;黏粒数量增加,黏粒之间的强胶结数量增加。碎屑和碎屑之间的强胶结组分减少,导致这种由强胶结构建的岩体骨架网络无法成为一个承载结构,岩体在黏粒之间的弱胶结位置断裂。

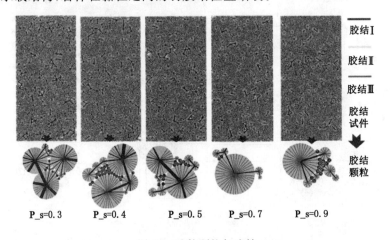

图12　试件颗粒与胶结

Fig.12　The particles and cementations of specimen

采用校核的细观胶结力学参数,在泥质胶结试件孔隙率为0.136条件下黏粒与标准砂体积比 $P\_s(v)$=0、0.1、0.2、0.3、0.4、0.5、0.6、0.7、0.8、0.9、1.0十一种方案下试件的力学参数变化规律,应力应变曲线如图13所示。根据试件应力应变曲线容易看出,对于黏粒组分增加的过程,试件应力应变曲线颗粒分成两个阶段:(1)强度、模量降低,峰值应变降低,P_s(v)=0、0.1、0.2、0.3;(2)强度不变,模量降低,峰值应变增大,P_s(v)=0.4、0.5、0.6、0.7、0.8、0.9、1.0。考虑试件组分变化,认为阶段(1)主要是由于碎屑固结的胶结数量减小,碎屑容易发生破坏,导致强度、模量、峰值应变降低;阶段(2)由黏粒之间弱胶结承载,强度随组分增加没有明显差异,但是增加的组分使试件具有软岩的大变形规律,峰值应变增大。

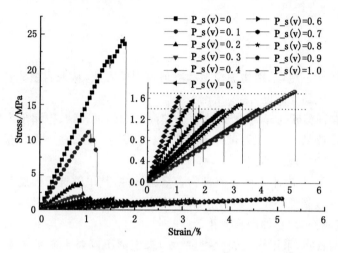

图 13　不同黏土矿物含量试件的应力-应变曲线

Fig. 13　the specimen stress-strain curve with different clay content

胶结数量的改变导致断裂位置的差异,进而影响力学测试结果。如图 14 给出试件强度与单轴压缩试验胶结断裂数量关系,容易看出,试件力学参数弱化过程是断裂强胶结(胶结

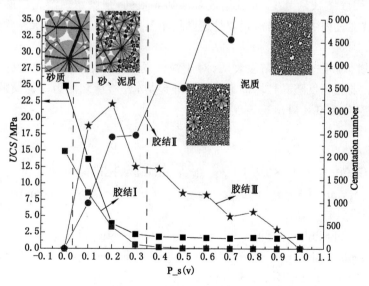

图 14　岩体强度与单轴压缩过程中胶结断裂数量的关系

Fig. 14　Relationship between specimen USC and cementation break numbers during uniaxial compression

Ⅰ)数量减少,弱胶结(胶结Ⅱ)数量增多的过程,根据岩体力学强度准则,强胶结的断裂必然需要较高的外力水平,对应的试件强度高,这是泥质胶结试件强度弱化细观机理。当岩样强度弱化到一定程度,强度值基本保持不变,胶结Ⅰ断裂数量为0,试件断裂胶结全部为黏粒形成的弱胶结。物理模拟当$P_s=0.8$,泥质胶结岩体水化弱化程度不随黏土矿物含量的增加而加剧,对应的体积分数为$P_s(v)=0.4$,数值模拟同样对应黏土矿物体积分数增加到$P_s(v)=0.4$,泥质胶结岩体强度随黏土矿物体积分数增加而不变,物理模拟与数值模拟结论一致。

4 结 论

文章基于离散元思想,采用物理模拟和数值模拟的方法,从泥岩组构特征研究泥质胶结岩体软岩特性和水化弱化问题,主要结论:

(1) 采用标准石英砂为碎屑、普通水泥为碎屑之间的强胶结、膨润土为弱胶结剂和特征黏土矿物制作泥质胶结岩体,泥岩试件不受取样扰动影响,能够进行黏土矿物组分单一且含量不同的试件力学特性规律研究。

(2) 物理模拟表明,对于不同黏土矿物含量的试件,其单轴抗压强度和弹性模量与黏土矿物含量符合幂函数指数小于0的Allometric1关系,即黏土矿物含量越高,泥岩的单轴抗压强度、弹性模量越小,减小的幅度变小。

(3) 物理模拟表明,采用水化弱化指数(δ)来表征不同黏土矿物含量泥岩水化弱化程度,黏粒含量越高,泥质胶结岩体水化弱化指数越大,水化越剧烈;当膨润土含量$P_s=0.8$左右时,泥质胶结岩体强度和水化弱化指数不随黏粒含量增加而变化,组分的增加不影响泥质胶结岩体的强度和水化弱化特性。

(4) 离散元数值模拟表明,泥质弱胶结岩体强度弱化实际是由于碎屑数量减少、黏粒数量增加对应的强胶结数量减少、弱胶结数量增加导致的,强胶结逐渐无法形成稳定承载骨架,试件压缩破坏过程中强胶结断裂数量逐渐较小至0,从细观尺度揭示泥质胶结岩体软岩特性。

(5) 物理模拟和数值模拟结果表明,对于泥质弱胶结试件,当黏粒含量增加的一定程度,岩体强度和弱化程度随黏粒增加不再有明显差异,岩体承载能力由泥质胶结决定,模拟结果显示体积分数为0.4时出现该现象。

参考文献

[1] Guéry A C, Cormery F, Shao J F, et al. A micromechanical model of elastoplastic and damage behavior of a cohesive geomaterial[J]. Physics & Chemistry of the Earth, 2008, 33(5):S416-S421.

[2] Robinet J C, Sardini P, Coelho D, et al. Effects of mineral distribution at mesoscopic scale on solute diffusion in a clay-rich rock: Example of the Callovo-Oxfordian mudstone (Bure, France)[J]. Water Resources Research, 2012, 48(5):5554.

[3] Armand G, Conil N, Talandier J, et al. Fundamental aspects of the hydromechanical behaviour of Callovo-Oxfordian claystone: From experimental studies to model calibration and validation[J]. Computers & Geotechnics, 2016.

[4] Oohashi K, Hirose T, Takahashi M, et al. Dynamic weakening of montmorillonite-bearing faults at intermediate velocities: Implications for subduction zone earthquakes [J]. Journal of Geophysical Research Solid Earth, 2015, 120(3): 1572-1586.

[5] Takahashi M, Mizoguchi K, Kitamura K, et al. Effects of clay content on the frictional strength and fluid transport property of faults [J]. Journal of Geophysical Research Solid Earth, 2007, 112(B8).

[6] Tembe S, Lockner D A, Wong T. Effect of clay content and mineralogy on frictional sliding behavior of simulated gouges: Binary and ternary mixtures of quartz, illite, and montmorillonite [J]. Journal of Geophysical Research Solid Earth, 2010, 115(B3).

[7] 周翠英,邓毅梅,谭祥韶,等. 饱水软岩力学性质软化的试验研究与应用[J]. 岩石力学与工程学报,2005,24(1):33-38.
ZHOU Cuiying, DENG Yimei, TAN Xiangshao, et al. Experimental research on the softening of mechanical properties of saturated soft rocks and application [J]. Chinese Journal of Rock Mechanics and Engineering, 2005, 24(1): 33-38.

[8] Ghorbani A, Zamora M, Cosenza P. Effects of desiccation on the elastic wave velocities of clay-rocks [J]. International Journal of Rock Mechanics & Mining Sciences, 2009, 46(8): 1267-1272.

[9] Yang D, Billiotte J, Su K. Characterization of the hydromechanical behavior of argillaceous rocks with effective gas Permeability under deviatoric stress [J]. Engineering Geology, 2010, 114(3): 116-122.

[10] Yang D, Bornert M, Chanchole S, et al. Experimental investigation of the delayed behavior of unsaturated argillaceous rocks by means of Digital Image Correlation techniques [J]. Applied Clay Science, 2011, 54(1): 53-62.

[11] Zhang F, Xie S Y, Hu D W, et al. Effect of water content and structural anisotropy on mechanical property of claystone [J]. Applied Clay Science, 2012, 69(21): 79-86.

[12] Tetsuka H, Katayama I, Sakuma H, et al. Effects of humidity and interlayer cations on the frictional strength of montmorillonite [J]. Earth Planets & Space, 2018, 70(1): 56.

[13] Matsumoto S, Shimada H, Sasaoka T. Interaction between Physical and chemical weathering of argillaceous rocks and the effects on the occurrence of acid mine drainage (AMD) [J]. Geosciences Journal, 2017, 21(3): 1-10.

[14] Li G, Jiang Z, Feng X, et al. Relation between Molecular Structure of Montmorillonite and Liquefaction of Mudstone [J]. Rsc. Advances, 2015, 5(30): 23481-23488.

[15] 何满潮,周莉,李德建,等. 深井泥岩吸水特性试验研究[J]. 岩石力学与工程学报,2008,27(6):1113-1120.
HE Manchao, ZHOU Li, LI Dejian, et al. Experimental research on hydrophilic characteristics of mudstone in deep well [J]. Chinese Journal of Rock Mechanics and Engineering, 2008, 27(6): 1113-1120.

[16] 周翠英,谭祥韶,邓毅梅,等. 特殊软岩软化的微观机制研究[J]. 岩石力学与工程学报,

2005,24(3):394-400.

ZHOU Cuiying,TAN Xiangshao,DENG Yimei,et al. Research on softening micro-mechanism of special soft rocks[J]. Chinese Journal of Rock Mechanics and Engineering,2005,24(3):394-400.

[17] Wang L L,Zhang G Q,Hallais S. Swelling of Shales:A Multiscale Experimental Investigation[J]. Energy & Fuels,2017(31):10442-10451.

[18] 何咏睿,朱晟,武利强. 粗粒料二维与三维孔隙率的对应关系研究[J]. 水力发电,2014,40(5):27-29.

HE Yongrui,ZHU Sheng,WU Liqiang. Research on the corresponding relationship between two-dimensional porosity and three-dimensional porosity of coarse materials[J]. Water Power,2014,40(5):27-29.

区段煤柱稳定性研究的新探索

闫帅[1], 柏建彪[2,3]

(1. 中国矿业大学 矿业工程学院,江苏 徐州,221116;
2. 中国矿业大学 煤炭资源与安全开采国家重点实验室,江苏 徐州,221116;
3. 新疆工程学院 矿业工程与地质学院,新疆 乌鲁木齐,830091)

摘 要: 区段煤柱作为保护井下巷道、隔离采空区的有效措施在煤矿开采中得到广泛应用。本文结合煤柱宽度、煤柱内塑性区分布范围和煤柱内应力特征三个方面开展了区段煤柱稳定性研究的新探索,基于Wilson弹性核理论,将煤柱内垂直应力特征和塑性区分布规律相结合,以煤柱稳定性和经济性为目标,建立"煤柱宽度综合指数"数学模型来量化煤柱合理宽度。煤柱宽度综合指数越大煤柱的稳定性或经济性越高,综合指数小于零,煤柱稳定性较低,易发生动力灾害。并通过工程实例对综合指数的影响因素进行了分析。

关键词: 区段煤柱;核心应力;塑性区宽度;煤柱宽度综合指数

An alternative approach for the stability of coal pillar in coal mines

YAN Shuai[1], BAI Jianbiao[2,3]

(1. School of Mines, China University of Mining & Technology, Xuzhou 221116, China;
2. State Key Laboratory of Coal Resource and Safe Mining, China University of Mining & Technology, Xuzhou 221116, China;
3. Institute of Mining Engineering and Geology, Xinjiang Institute of Engineering, Urumqi 830091, China)

Abstract: As an effective measure to protect underground roadway and isolate a gob, a coal pillar has been widely employed in underground coal mining. In this paper, the three aspects of the width of coal pillar, the distribution range of plastic zone in coal pillar and the characteristics of vertical stress of coal pillar are systematically considered, and an innovative approach for the stability of coal pillar is proposed. Based on the Wilson's elastic core theory, the synthetic factor of coal pillar is associated with vertical stress in the core of pillar and plastic zone width to evaluate the stability and efficiency of the pillar. The higher the synthetic factor is, the better the stability or efficiency of the pillar are. When the factor is less than zero, the pillar has a potential dynamic disaster. The impact factors are analyzed in detail using a case study with gob-side entry layout.

Keywords: coal pillar;core stress;plastic zone width;synthetic factor

1 引 言

井工开采面临两个突出问题:一是井下的安全生产,二是采动后的地表沉陷。这两个问题都与矿井下煤柱有着密切联系,煤柱可以保护井下巷道硐室的安全,隔离采空区瓦斯和老坑水等,从而保证井下安全生产[1,2];同时煤柱也支撑上覆岩层,影响地表沉陷规律,进而保护人工建筑和地表水等资源。

煤柱尺寸对煤柱的稳定性和巷道维护的影响主要体现在两个方面:(1)影响围岩应力重新分布和煤柱承受的载荷;(2)决定煤柱极限强度。

国内外专家学者在煤柱稳定性和煤柱宽度设计方面开展了丰富的研究,主要集中在:(1)对大量实测结果的数理统计、归纳推理得出不稳定围岩条件下护巷煤柱尺寸[3-5];(2)运用矿山压力规律留设各种煤柱的方法及经验公式对煤柱合理的尺寸进行分析[6-8];(3)用现场实测煤柱支承压力分布方法分析给出煤层回采巷道的合理煤柱宽度范围[9];(4)钻孔煤粉量变化确定煤柱宽度[10];(5)根据岩体的极限平衡理论推导出护巷煤柱保持稳定状态时的宽度计算公式[11-14];(6)从理论上推导出了三维应力状态下估算煤柱塑性区宽度的理论公式[15,16];(7)用数值模拟对煤柱护巷的围岩变形进行了计算分析确定煤柱合理尺寸[17,18,19]。

国内外在煤柱设计过程中,经验公式使用方便,但是受各矿区地质条件不同的影响,一个经验公式往往只实用于一个矿区[7]。理论分析基于严苛的条件假设,与现场实际有一定的差距[15]。物理试验受实验室条件的限制,能模拟相对简单的生产地质条件,对复杂开采条件、多地质因素影响的情况不能很好地模拟,同时受尺度效应影响比较大。数值模拟方法[19]以其试验成本低、可重复试验、多物理场耦合、模拟复杂开采条件等优势被广泛使用,目前对模拟结果的处理还不尽人意,往往是定性的分析,缺乏统一的判据。

本文依据Wilson弹性核理论,将煤柱内塑性区分布规律和应力场进行结合,提出了"煤柱宽度综合指数法"设计煤柱宽度,综合指数曲线包括全煤柱宽度,从让压煤柱、临界煤柱到承压煤柱,系统直观地反映了煤柱宽度与煤柱稳定性、经济性的综合关系。为煤柱合理宽度的确定提供了一种新方法。

2 煤柱宽度综合指数法

煤柱从作用机理上可分为承压煤柱和让压煤柱[20]。承压煤柱的强度始终大于服务年限内所承受的载荷,煤柱中央存在一定的弹性核;而让压煤柱则是在一定时间按一定的速率屈服,自身不能承受所有载荷,通过屈服让压将大部分载荷转移到相邻承载结构体上(如采空区或者实煤体),煤柱内没有弹性区。通常承压煤柱的尺寸要比让压煤柱大,尤其是在深部开采中,承压煤柱大得多。

介于承压煤柱和让压煤柱之间存在一个过渡区煤柱宽度称为临界煤柱。临界煤柱往往承载几倍原岩应力的载荷,极易发生动力灾害。

煤柱极限强度理论认为煤柱承载力大于所承受的载荷才能稳定,而大量的工程实践表明,安全系数小于1时煤柱也能稳定[21,22]。DeMarco通过对巷道矿压显现和煤柱是否稳定区分了煤柱类型和煤柱内应力状态(图1)[23]。以往研究表明:当煤柱内有弹性核时,煤柱内垂直应力呈现双峰值形态,煤柱中部垂直应力大于原岩应力;当煤柱内没有弹性核,煤柱整体进入塑性屈服状态,垂直应力曲线为单峰,煤柱中部垂直应力或大于或小于原岩应力。基于Wilson弹性核理论,将煤柱内垂直应力特征和塑性区分布规律相结合,以煤柱稳定性和经济性为目标,建立"煤柱宽度综合指数"数学模型来量化煤柱合理宽度。

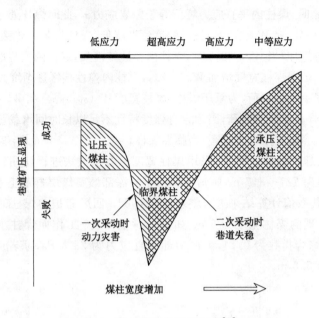

图 1 临界煤柱的作用机理[23]

Fig. 1 Mechanism of critical coal pillar [23]

以煤柱中部垂直应力 σ_{mid}、原岩应力 σ_0、煤柱单侧塑性区宽度 x_0 和煤柱宽度 W_p 为特征变量,建立煤柱宽度综合指数表达式为:

$$F_d = \frac{(\sigma_{mid} - \sigma_0)}{\sigma_0} \times \left(\frac{W_p - 2x_0}{W_p}\right) \tag{1}$$

图 2 为煤柱宽度综合指数曲线,随着煤柱宽度增加,煤柱宽度综合指数 F_d 先减小后增大再减小,曲线被横坐标轴分为 2 部分,一部分在横轴上方大于零,一部分在横轴下方小于零。将图 2 中曲线分为 3 个区:Ⅰ区为让压煤柱、Ⅱ区为临界煤柱、Ⅲ区为承压煤柱。

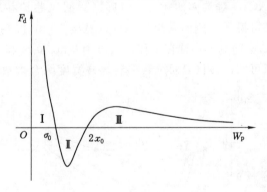

图 2 煤柱宽度综合指数曲线

Fig. 2 Curve of pillar synthetic factor

在煤柱宽度综合指数曲线中,$F_d > 0$ 表明煤柱内应力集中相对较低、煤柱自身稳定,可以选作护巷煤柱;当 $F_d < 0$ 表明煤柱内应力集中显著、稳定性差,可能发生动力型灾害,不利于维护井下巷道硐室。煤柱宽度综合指数曲线与 X 轴有两个交点($F_d = 0$),分别为 $\sigma_P(W_p) = \sigma_0$ 和

$W_p = 2x_0$,当 $\sigma_P = \sigma_0$ 时,煤柱内垂直应力峰值等于原岩应力 σ_0,此时煤柱处于塑性屈服状态,随着煤柱宽度继续减小 $W_p < W_p(\sigma_0)$,煤柱的强度逐渐降低,煤柱将载荷转移到邻近的采空区或实体煤中,成为让压煤柱;当 $W_p = 2x_0$ 时,煤柱内弹性核消失,煤柱两侧塑性区贯通,此时认为煤柱达到临界宽度,随着煤柱宽度增加 $W_p > 2x_0$,煤柱内弹性核区逐渐增大,煤柱处于弹塑性状态,煤柱强度大于承受的载荷,为承压煤柱。煤柱宽度 $W_p(\sigma_P = \sigma_0) < W_p < 2x_0$,煤柱的综合指数 $F_d < 0$,此时煤柱承受高载荷,因为自身强度特性不足以长时间承载,又不能转移到邻近承载结构上,所以最容易发生动力灾害,为临界煤柱。

结合已发表文献中的数据[24,25],采用煤柱宽度综合系数法进行煤柱稳定性分析。

文献[24]对埋深 727~817 m,倾角 17°~32°的深部倾斜煤层群沿空巷道的护巷煤柱宽度进行了研究,将其数值计算结果进行整理,得到 7113 回风巷护巷煤柱的综合指数(图 3)。文中分析确定合理的煤柱宽度为 6 m,如图 3 所示,7113 工作面煤柱小于 5 m 为让压煤柱,煤柱 6 m 时的综合指数为 0.14,煤柱中部垂直应力为 28 MPa,原岩应力为 27.1 MPa,建议煤柱宽度选为 5.5 m。

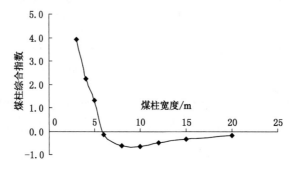

图 3 7113 回风巷护巷煤柱宽度综合指数

Fig. 3 Pillar synthetic factor of tailgate 7113

文献[25]采用 FLAC3D 对淮南谢桥煤矿 1151(3)回风巷护巷煤柱宽度进行了研究。根据文中数值模拟结果可得到一侧回采煤柱塑性区宽度 10 m,掘巷塑性区 5 m,工作面超前采动塑性区宽度 7 m,不同宽度煤柱在工作面前方 100 m、30 m 和工作面处煤柱宽度综合指数如图 4 所示。由图可知,1151(3)回风巷护巷煤柱宽度综合指数符合图 2 所示规律,由

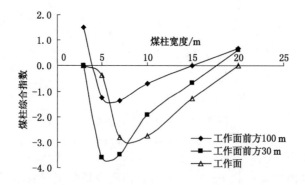

图 4 谢桥煤矿 1151(3)回风巷护巷煤柱宽度综合指数

Fig. 4 Pillar synthetic factor of tailgate 1153(3)

于采动作用不仅影响煤柱内应力特征而且也改变塑性区分布,相同宽度煤柱在距工作面不同位置处的煤柱宽度综合指数不尽相同。

3 实例分析

3.1 工程背景

某煤矿一采区平均埋深350 m,主采3#煤层厚度为2.7~3.8 m,平均3.5,煤层倾角7°~15°,平均倾角12°,赋存稳定。为结构简单-复杂层位稳定的全区可采煤层。局部有夹矸,其顶板岩性为砂质泥岩和泥岩,平均厚度为2 m,泥质胶结含植物化石,单轴抗压强度为60.3 MPa;基本顶为中粒砂岩,平均厚度为6.7 m,灰色厚层状;底板岩性为泥岩和砂质泥岩,平均厚度为2.9 m,泥质结构裂隙发育,遇水泥化;老底为细粒砂岩,平均厚度为5.3 m。

1302工作面煤层采3.5 m,工作面倾向长度为180 m,工作面可采长度1 263 m,材料巷长度1 228 m,胶运巷长度1 326 m。工作面地质类型简单,试验巷道为1303工作面材料巷,宽4.5 m,高3.2 m,试验巷道布置平面图如图5所示。

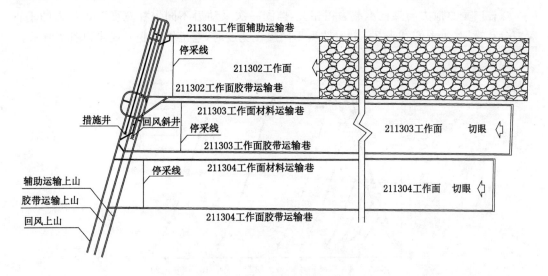

图 5 试验巷道布置平面图

Fig.5 Mining layout of test roadway

采用FLAC³ᴰ数值计算软件,根据211302工作面地质概况,取典型钻孔地质断面,分析不同宽度煤柱内应力分布规律。

模型X方向长500 m,Y方向宽160 m,Z方向高100 m。模拟巷道埋深350 m,对模型侧边限制水平移动、底部边界限制垂直移动,模型上边界施加垂直载荷模拟上覆岩层300 m自重,水平侧压系数1.2。本构模型为弹塑性模型,屈服准则选用Mohr-Coulomb准则。基于实验室岩样物理力学性能测试,结合反演1302工作面矿压显现特征,得到各地层岩石力学参数见表1。

表 1 岩石力学参数
Table 1 Physical and mechanic parameters of rock mass

岩层	岩层厚度 h/m	弹性模量 E/MPa	泊松比 ν	内摩擦角 φ/(°)	内聚力 C/MPa	抗拉强度 σ_t/MPa
粉砂岩	3.0	3 400	0.24	32	6.0	3.9
中砂岩	7.0	3 500	0.20	30	6.0	2.5
泥岩	1.5	2 500	0.20	27	3.0	1.5
13#煤	3.5	1 400	0.25	23	2.0	0.85
砂质泥岩	3.0	3 000	0.22	28	3.5	2.0
细砂岩	5.0	4 000	0.30	38	7.0	3.0

选取不同煤柱宽度 3 m、4 m、5 m、6 m、8 m、10 m、15 m、20 m 计算煤柱内应力特征和塑性区分布规律,对计算结果进行分析。

3.2 煤柱合理宽度的确定

煤柱内垂直应力与煤柱承载力正相关,监测一次采动后不同煤柱宽度下煤柱内的垂直应力分布规律,见图 6 和表 2。同时监测不同煤柱宽度时,巷道围岩变形规律如图 7 所示。

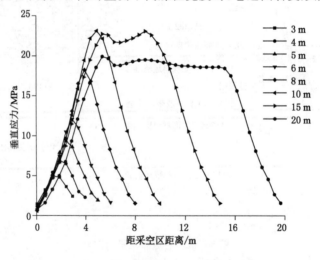

图 6 煤柱垂直应力分布

Fig.6 Vertical stresses in coal pillar

从图 6 可以看出,煤柱宽度从 3 m 变化到 20 m 时,煤柱中的垂直应力峰值分别为 5.1 MPa、7.0 MPa、9.4 MPa、12.1 MPa、18.2 MPa、23.0 MPa、22.6 MPa 和 19.5 MPa。煤柱宽度为 3～10 m 时,煤柱应力分布形状为"单峰"分布;煤柱宽度 15～20 m 时,煤柱应力分布形状为"梯形分布";前者煤柱塑性区所占比重太大,后者的弹性核区形成稳定的垂直应力分布区。煤柱宽度为 3～6 m 时,煤柱峰值应力较低,结合图 7 巷道围岩变形,煤柱宽度为 3～5 m 时,围岩移进量呈整体下降趋势,说明塑性区煤柱应力的适当增大有利于巷道围岩的稳定;煤柱宽度为 5～8 m 时,围岩移进量呈上升趋势,说明塑性区煤柱垂直应力的过度增大不利于围岩的稳定;煤柱宽度为 10～20 m 时,围岩移进量呈下降趋势,说明弹性核

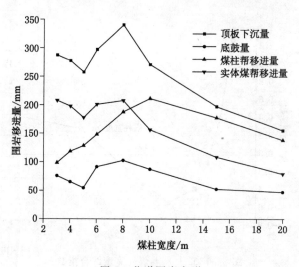

图 7 巷道围岩变形

Fig. 7 Convergence of roadway with respect to pillar width

区的高承载力有利于巷道的稳定,但煤柱宽度太大,造成资源浪费。

表 2   不同宽度煤柱内垂直应力特征

Table 2   Vertical stress in different width coal pillars

煤柱宽度/m	煤柱一侧塑性区宽度/m	煤柱中部垂直应力/MPa	峰值应力/MPa	峰值应力集中系数	综合指数
3	6	5.1	5.1	0.58	1.25
4	6	7	7	0.80	0.40
5	6	9.4	9.4	1.07	−0.10
6	6	12.1	12.1	1.38	−0.38
8	6	18.2	18.2	2.08	−0.54
10	6	23	23	2.63	−0.33
15	6	21	22.6	2.58	0.28
20	6	18	19.5	2.23	0.42

由图 8 材料巷煤柱宽度综合指数曲线可知,煤柱宽度为 6～10 m 为临界煤柱,煤柱宽小于 5 m 时为让压煤柱,煤柱宽度超过 15 m 时煤柱内有弹性核为承压煤柱。通过对围岩应力场、塑性区分布和巷道变形规律的分析,认为:15 m 以上为稳定的护巷煤柱宽度;对 3～5 m 宽度的煤柱,通过一定的加固措施,也可以保持稳定;煤柱宽度 6～10 m 时稳定性较差。综合考虑,确定 1304 工作面材料巷护巷煤柱宽度为 4～5 m 或者 15～20 m。

3.3 应用效果分析

现场沿 1302 采空区留 4 m 煤柱掘进 1303 材料巷,对巷道变形和煤柱内应力进行矿压监测。

(1) 巷道围岩变形

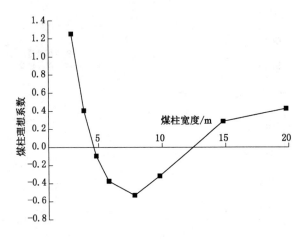

图 8　1303 材料巷煤柱宽度综合指数
Fig. 8　Pillar synthetic factor of tailgate 1303

巷道掘进过程中对围岩变形进行了连续性监测,其中顶板测点位于距实体煤帮 1.3 m 处,帮部测点位于距顶板 1.7 m 处,试验段内巷道围岩位移变化曲线如图 9 所示。由图可知,工作面前方 120 m 巷道开始变形,距工作面 40 m 时,1303 材料巷变形急剧增加,煤柱侧顶板最大变形量为 150 mm,实体煤最大变形量为 110 mm,煤柱帮最大变形量为 85 mm。距工作面 10 m 时,顶板最大下沉量约 300 mm,煤柱帮变形 150 mm,实煤体采煤帮变形约 200 mm,说明巷道围岩控制效果明显。

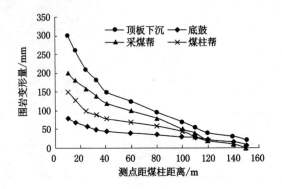

图 9　1303 材料巷变形情况
Fig. 9　Convergence of tailgate 1303

(2) 煤柱应力监测

在 4 m 煤柱内布置钻孔应力计测站,每个测站布设 7 个钻孔应力计,安装深度依次为 0.5、1、1.5、2、2.5、3、3.5 m,钻孔间距为 3 m,回采期间煤柱应力分布见图 10。由图可知,煤柱内支承压力始终呈先增大后减小的趋势,峰值位置位于距煤柱帮 2 m 测点处;在距工作面前方约 10.5 m 处,煤柱内出现了应力峰值,应力值为 6.85 MPa。可见,煤柱虽然发生较大塑性破坏,但煤体仍具有一定程度承载力。

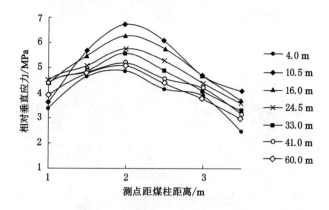

图 10 回采过程中煤柱内垂直应力

Fig. 10 Variation of vertical stress in small pillar

4 煤柱宽度综合指数影响因素分析

煤柱内核心应力特征在回采巷道合理布置的研究中,是一个很重要的参数,它对上区段工作面采空区侧煤体中的垂直应力分布规律、煤柱沿空巷道的合理位置确定及巷道外部力学环境的演化起着很大的影响作用。影响基本顶岩层断裂位置的因素很多,主要有煤层力学参数、煤层厚度、埋深、底板强度及煤层采空状况等[26]。

采用正交试验法研究四个因素对煤柱宽度综合指数的影响规律,确定所研究的四个因素的水平均为三水平,如表 3 所示。

表 3 试验因素及水平

Table 3 Factors and levels of orthogonal experiment

水平 \ 因素	煤层性质 /MPa	煤层厚度 /m	埋深 /m	直接顶岩性 /MPa
1	软煤	3	300	软煤直接顶
2	中硬煤	6	500	中硬煤直接顶
3	硬煤	9	1 000	硬煤直接顶

研究认为,以上四个因素对煤柱宽度综合指数影响的重要性依次为:煤层性质＞埋深＞直接顶岩性＞煤层厚度。其中煤层性质和埋深对应力分布的影响远大于后两个因素的影响,煤层厚度的影响最小。两者对护巷煤柱宽度综合指数的影响关系分别如图 11 和图 12 所示。

由图 11 可见,随着煤层性质的增强,煤柱宽度综合指数逐渐增加;由图 12 可知,增加煤层埋深,煤柱宽度综合指数呈减小的趋势。由此可以得出,煤柱宽度综合指数主要取决于煤层性质和埋深两个因素。

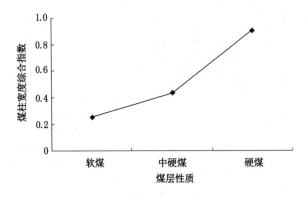

图 11　煤层性质与煤柱宽度综合指数关系

Fig.11　Synthetic factor of coal pillar with coal strength

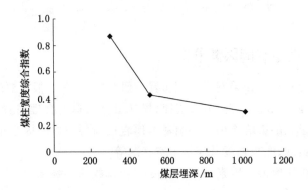

图 12　埋深与煤柱宽度综合指数关系

Fig.12　Synthetic factor of coal pillar with cover depth

5　新方法的不足和改进

煤柱宽度综合指数,是一定宽度煤柱稳定性和经济性的综合表征,综合指数越大,煤柱的宽度越合理,当综合指数为负值时,表明此宽度下的煤柱稳定性或经济性(煤炭采出率)偏低,在设计煤柱宽度时应该规避。在综合指数曲线中无煤柱开采是煤柱宽度综合指数的极限(正无穷大),此时煤柱的稳定性和经济性最高。综合指数主要以区段煤柱为研究对象,对房柱式开采、条带开采和煤柱工作面稳定性(煤柱回收)仍需进一步研究。

在煤柱宽度综合指数中,煤柱塑性区宽度的选取意义重大。本文主要采用了工作面充分采动后实体煤中塑性区宽度(回采塑性区),不同采动阶段塑性区宽度不同,比如掘巷塑性区、回采塑性区和多次回采塑性区;另外,借鉴 Wilson"弹性核"煤柱设计法,塑性区宽度可以选择一定的安全系数。

6　结　　论

(1)基于 Wilson 弹性核理论,将煤柱内垂直应力特征和塑性区分布规律相结合,以煤柱稳定性和经济性为目标,建立了"煤柱宽度综合指数"数学模型来量化煤柱合理宽度。

(2)煤柱宽度合理指数越大,煤柱的宽度可取性越强,当该指数为负值时,煤柱易发生动力型破坏。根据试验矿井生产地质条件确定合理的窄煤柱宽度为 5 m，并在工业性试验中得到应用成功。

（3）该方法主要适应于讨论护巷煤柱，对房柱式开采和煤柱工作面稳定性（煤柱回收）还需要进一步研究。同时塑性区宽度的选取需据实分析。

参考文献

[1] Feng G, Wang P, Chugh Y P. Stability of Gate Roads Next to an Irregular Yield Pillar: A Case Study[J]. Rock Mechanics & Rock Engineering, 2018:1-20.

[2] DOBROSKOK A A, LINKOV A M. Modeling of fluid flow, stress state and seismicity induced in rock by an instant pressure drop in a hydrofracture[J]. Journal of Mining Science, 2011, 47(1): 10-19.

[3] GALVIN J, HEBBLEWHITE B, SALAMON M. UNSW pillar strength determinations for Australian and South African conditions [C]. 37th US Rock Mechanics Symposium, Vail, Colorado, 1999: 63-71.

[4] DeMarco M J, Koehler J R, Maleki H. Gate road design considerations for mitigation of coal bumps in western U. S. longwall operations. In: Maleki H, Wopat P F, Repsher R C, Tuchman R J (eds) Proceedings: mechanics and mitigation of violent failure in coal and hard-rock mines. U. S. Department of the Interior, Bureau of Mines, SP 01-95, Spokane, WA, NTIS No. PB95211967,1995: 141-165.

[5] BIENIAWSKI Z T. Improved design of coal pillars for U. S. mining conditions[C]. Proc 1st conference on ground control in mining, West Virginia University, Morgantown, WV, 1983: 13-22.

[6] MARK C, HEASLEY K A, IANNACCHIONE A T, et al. Coal Pillar Mechanics and Design[C]. Proceedings of the Second International Workshop, Pittsburgh, PA, 1999: 145-154.

[7] Alejano L R, RamiíRez-Oyanguren P, Taboada J. Fdm predictive methodology for subsidence due to flat and inclined coal seam mining[J]. Int. J. Rock Mech. Min. Sci., 1999, 36(4):475-491.

[8] COLWELL M, Frith R, Mark C. Analysis of longwall tailgate serviceability (ALTS): A chain pillar design methodology for australian conditions[C]. Proceedings of the Second International Workshop on Coal Pillar Mechanics and Design, Vail, Colorado, 1999: 33-48.

[9] 宋选民,窦江海.浅埋煤层回采巷道合理煤柱宽度的实测研究[J].矿山压力与顶板管理,2003,20(03):31-33.
SONG Xuanmin, DOU Jianghai. Observation on rational coal-pillar width of mining tunnels of shallow coal seam[J]. Ground Pressure and Strata Control,2003,20(03): 31-33.

[10] 张开智,夏均民,蒋金泉.钻孔煤粉量变化规律在区段煤柱合理参数确定中的应用[J].岩石力学与工程学报,2004,23(8):1307-1310.

Zhang Kaizhi, Xia Junmin, Jiang Jinquan. Variation law of quantity of coal dust in drill hole and its application to determination of reasonable width of coal pillars[J]. Chinese Journal of Rock Mechanics and Engineering, 2004, 23(8):1307-1310.

[11] WILSON A H. A Hypothesis Concerning Pillar Stability[J]. The Mining Engineer (London), 1973(131): 409-417.

[12] WILSON A H. The Effect of Yield Zones on the Control of Ground[C]. Proceedings of the 6th International Strata Control Conference, Banff, Canada, 1977.

[13] 白矛,刘天泉. 条带开采中条带中条带尺寸的研究[J]. 煤炭学报, 1983, 4(12):19-26.
BAI Mao, LIU Tianquan. Study on pillar size and mining width for partial mining [J]. Journal of China Coal Society, 1983, 4(12):19-26.

[14] 侯朝炯,马念杰. 煤层巷道两帮煤体应力和极限平衡区的探讨[J]. 煤炭学报, 1989, 14(04):21-29.
HOU Chaojiong, MA Nianjie. Stress in in-seam roadway sides and limit equilibrium zone[J]. Journal of China Coal Society, 1989, 14(04):21-29.

[15] 翟所业,张开智. 煤柱中部弹性区的临近宽度[J]. 矿山压力与顶板管理, 2003, 20(4): 14-16.
ZHAI Suoye, ZHANG Kaizhi. The critical width of middle elastic area of the coal pillar to safeguard working[J]. Ground Pressure and Strata Control, 2003, 20(4):14-16.

[16] 谢广祥. 综放采场围岩三维力学特征[M]. 北京:煤炭工业出版社, 2007.

[17] 柏建彪,王卫军,侯朝炯,等. 综放沿空掘巷围岩控制机理及支护技术研究[J]. 煤炭学报, 2000, 25(5):478-481.
BAI Jianbiao, WANG Weijun, HOU Chaojiong, et al. Control mechanism and support technique about gateway driven along goaf in fully mechanized top coal caving face[J]. Journal of China Coal Society, 2000, 25(5):478-481.

[18] 韩承强,张开智,徐小兵,等. 区段小煤柱破坏规律及合理尺寸研究[J]. 采矿与安全工程学报, 2007(03):370-373.
HAN Chengqiang, ZHANG Kaizhi, XU Xiaobing, et al. Study on Failure Regularity and Reasonable Dimension of District Sublevel Small Coal Pillar[J]. Journal of Mining & Safety Engineering, 2007(03):370-373.

[19] JING L, HUDSON J A. Numerical methods in rock mechanics[J]. International Journal of Rock Mechanics and Mining Sciences, 2002, 39(4): 409-427.

[20] PENG S S. Coal Mine Ground Control[M]. Third Edition. Morgantown, WV, USA, 2008.

[21] Peng S S. Coal mine ground control[M]. 3rd Edition. Peng S S publisher, Morgantown, 2008.

[22] Li W, Bai J, Peng S, et al. Numerical modeling for yield pillar design: a case study [J]. Rock Mech. Rock Eng., 2005, 48(1):305-318.

[23] DEMARCO M J. Critical Pillar Concept in Yield-Pillar Based Longwall Gateroad Design[J]. Mining Engineering, 1996, 48(8):73-78.

[24] 张科学. 深部煤层群沿空掘巷护巷煤柱合理宽度的确定[J]. 煤炭学报,2011,36(S1):28-35.
ZHANG Kexue. Determining the reasonable width of chain pillar of deep coal seams roadway driving along next goaf[J]. Journal of China Coal Society,2011,36(S1):28-35.

[25] CHENG Y M,WANG J A,XIE G X,et al. Three-dimensional analysis of coal barrier pillars in tailgate area adjacent to the fully mechanized top caving mining face[J]. International Journal of Rock Mechanics and Mining Sciences,2010,47(8):137-138.

[26] 侯朝炯团队. 巷道围岩控制[M]. 徐州:中国矿业大学出版社,2012.
HOU Chaojiong,et al. Ground control of roadways[M]. Xuzhou:China University of Mining and Technology Press,2012.

拉伸荷载下盐岩力学及声发射特征研究

曾寅[1,2]，刘建锋[1,2]，邓朝福[1,2]，李志成[1,2]，向高[1,2]

(1. 四川大学 水力学与山区河流开发保护国家重点实验室，四川 成都，610065；
2. 四川大学 水利水电学院，四川 成都，610065)

摘　要：采用 MTS815 Flex Test GT 岩石力学试验系统及 PCI-Ⅱ 声发射测试系统，探讨了间接拉伸与直接拉伸加载方式对盐岩变形破坏的力学与声发射特征的影响。结果表明，纯盐岩间接强度约为 1.823 MPa，为其直接拉伸强度的 3.12 倍，但盐岩中含杂质时的直接拉伸强度显著增加，为纯盐岩直接拉伸强度的 1.6 倍。纯盐岩间接拉伸和直接拉伸下的峰后变形均呈缓慢下降特征，但含杂质盐岩直接拉伸峰后变形呈脆性特征，应力下降坡度转折点相比纯盐岩较明显；间接拉伸下峰值应力应变量约是直接拉伸下的 23.83 倍。研究揭示了间接拉伸下的声发射特征参数呈"低潮期-活跃期-次活跃期"变化特征，直接拉伸荷载下的声发射特征参数在 20%F 前处于一个较平静期，峰值应力后变形阶段声发射活跃度较高；间接拉伸与直接拉伸在卸载过程中声发射计数和能量变化特征不同，前者因受压缩加载应力影响，在卸载过程声发射相对活跃，而后者则无显著声发射特征。基于声发射定位点时空演化，揭示导致间接拉伸和直接拉伸损伤机理的差异；间接拉伸加载接触部位的压应力区，导致试样损伤出现较早，并沿加载破坏面发展和集聚，而直接拉伸下的损伤出现较晚，且沿破坏面的集聚程度不及间接拉伸。

关键词：盐岩；间接拉伸；直接拉伸；加卸载；声发射

Rock mechanics and acoustic emission characteristics under tensile loading of salt rock

ZENG Yin[1,2], LIU Jianfeng[1,2], DENG Chaofu[1,2], LI Zhicheng[1,2], XIANG Gao[1,2]

(1. *State key laboratory of Hydraulic and Mountain River Engineer, Sichuan Univ., Chengdu 610065, China*; 2. *College of Water Resource and Hydropower, Sichuan Univ., Chengdu 610065, China*)

Abstract: The MTS815 Flex Test GT Rock Mechanics Test System and PCI-Ⅱ Acoustic Emission (AE) Test System were used, The effects of indirect tensile and direct tensile loading on the mechanical and acoustic emission characteristics of salt rock deformation and failure were discussed. The results show that the indirect strength of pure salt rock is about 1.823 MPa, which is 3.12 times of its direct tensile strength, However, the direct tensile strength of the salt rock containing impurities is significantly increased, which is 1.6 times the direct tensile strength of pure salt rock. The post-peak deformation of pure salt rock under indirect tension and direct tension showed a slow decline, but the deformation after direct tensile peak of

---

基金项目：国家自然科学基金资助项目(51874202)；四川省青年基金资助项目(2017JQ0003)。
作者简介：曾寅(1994—)，男，岩土工程专业，主要从事岩石力学试验研究。E-mail：1026579434@qq.com。
通信作者：刘建锋(1979—)，男，博士，研究员，主要从事岩石力学与工程方面的研究。E-mail：liujf@scu.edu.cn。

impurity-containing salt rock was brittle, and the turning point of stress-decline slope was more obvious than that of pure salt rock.; The peak strain under indirect tension is approximately 23.83 times that of direct stretching. The study reveals that the characteristic parameters of acoustic emission under indirect tension are characterized by "low tide period-active period-secondary active period". The acoustic emission characteristic parameters under direct tensile load are in a quieter period before 20%F, In the stage of peak stress, the acoustic emission activity is higher in the later stage of deformation; Indirect stretching and direct stretching have different characteristics of acoustic emission count and energy change in the unloading process. The former is affected by the compressive loading stress and the acoustic emission is relatively active during unloading, while the latter has no significant acoustic emission characteristics. Based on the spatial and temporal evolution of acoustic emission sites, the differences in indirect tensile and direct tensile damage mechanisms are revealed. Indirect tensile loading of compressive stress zones at the contact site leads to earlier damage to the sample and develops and agglomerates along the loaded failure surface, However, the damage caused by direct stretching occurs later, and the degree of agglomeration along the failure surface is not as much as indirect stretching.

Keywords: Salt rock; Indirect tensile; Direct tensile; Loading and unloading; Acoustic emission

1 引 言

盐岩作为一种特殊性软岩,拥有极低渗透性、致密性、损伤自愈性等良好特性,是公认的能源地下储库的理想介质[1,2]。岩石抗拉强度是工程设计及建设需获知的重要参数之一,盐岩的拉伸特性也是盐穴能源储气库设计需获知关键依据之一。

虽然间接拉伸试验是研究岩石拉伸力学行为的主要手段[3,4],但直接拉伸试验结果更能有效反映岩石的拉伸力学行为[5-7]。研究表明[8-10],直接拉伸试验中岩石拉伸破坏过程力学特性与间接拉伸下存在较大的差异。通过自主研制直接拉伸试验装置,配合岩石力学试验机进行直接拉伸试验研究很有必要[11,12]。刘建锋等[13,14]利用自行研制的直接拉伸试验装置对层状盐岩进行了直接与间接拉伸单调加载变形破坏试验,分析了两种拉伸方式下的强度与声发射时空分布特征差异。对于盐岩循环荷载作用下力学特性研究,杨春和等[15]通过对盐岩进行单轴循环加卸载试验,研究了盐岩力学变形特征,有较多的学者[16,17]进行过盐岩单、三轴循环荷载下力学变形特性研究,但这些研究主要集中在盐岩循环荷载压缩破坏,对于拉伸循环荷载作用下研究相对较少,因而开展盐岩拉伸循环荷载下破坏力学特性研究很有必要。

本文将基于 MTS 岩石力学试验系统与自主研发的直接拉伸试验装置,进行盐岩拉伸加卸载条件下的变形破坏试验,揭示盐岩不同拉伸测试状态下的破坏力学特性与声发射特征差异。

2 试验设备及实验方案

2.1 试验设备

试验采用四川大学 MTS815 Flex Test GT 岩石力学试验系统与 PCI-II声发射系统(图1)。MTS815 试验系统拉伸荷载最大 2 300 kN,LVDT 为±2.5 mm,精度 0.5%RO。声发射监测采用 8 个 Micro30 型声发射传感器,频率为 100~600 kHz,放大器增益为 40 dB。

图 1 MTS 岩石力学试验系统

Fig. 1 MTS rock mechanics test system

2.2 试样制备及试验方案

盐岩样品取自某盐穴工程,室内参照《工程岩体试验方法标准》[18]要求,采用车床干车法加工。间接拉伸试件尺寸为 $\phi \times H = 90~\text{mm} \times 45~\text{mm}$,直接拉伸试件尺寸为 $\phi \times H = 100~\text{mm} \times 100~\text{mm}$。间接拉伸为纯盐岩,直接拉伸包括纯盐岩和杂质盐岩两种,每种测试试件均为 4 个。

间接拉伸循环荷载试验采用巴西劈裂方法进行加载,并参照国家规范标准[18]在试样两端加载部位处加置垫条,试验全程采用轴向位移(LVDT)控制,加载速率为 0.3 mm/min,在峰前荷载每隔 0.5 kN 进行一次加卸载循环,每次卸载到 0.3 kN 再进行加载。声发射采集系统的 8 个传感器在直径受拉面上下各均匀布置 4 个,并通过两个特制方形金属框固定在试样端部两侧,示意图见图 2(a),采用凡士林作为声发射传感器耦合剂,增强测试效果。

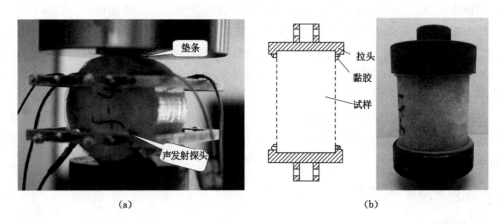

图 2 拉伸加载试样图

Fig. 2 Tensile loading specimen diagram

(a) 间接拉伸;(b) 直接拉伸

盐岩直接拉伸试验装置不同于间接拉伸,笔者为此研发了适配 MTS 岩石力学试验系统的岩石材料直接拉伸试验装置[19,20],该拉伸装置采用了液压限位支撑形式(图3)。试样的固定方式采用黏接法,利用拉头装置与工业黏胶将试样固定[图2(b)],有效保证了试样轴心受拉,避免拉伸过程中应力不均。直接拉伸试验全过程采用轴向位移(LVDT)控制,加载速率为 0.1 mm/min,间隔 0.5 kN 加卸载一次,峰值应力后停止循环加卸载,直至试样完全破坏。声发射传感器亦采用特制金属框固定在试样表面靠近上下端部位置,加载全过程实时同步监测。

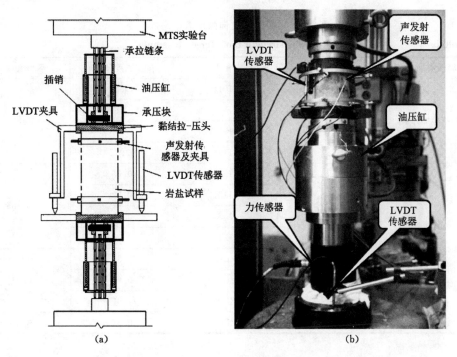

图 3 液压限位支撑式直接拉伸试验装置
Fig. 3 Hydraulic limit supported direct tensile testing device

3 试验结果及分析

3.1 强度与变形特征

基于笔者进行的盐岩拉伸荷载试验,对数据进行处理后得到各试样抗拉强度值(表1)。
其中纯盐岩间接拉伸下抗拉强度平均值为 1.823 MPa,纯盐岩与杂质盐岩直接拉伸下的抗拉强度平均值分别为 0.585 MPa、0.93 MPa。根据笔者已有的研究[21]表明纯盐岩单轴循环加卸载下抗压强度为 20.99 MPa,故而纯盐岩单轴循环荷载下抗压强度是间接拉伸循环荷载下抗拉强度的 11.5 倍,是直接拉伸下抗拉强度的 35.88 倍,是杂质盐岩直接拉伸下抗拉强度的 22.57 倍。在不含夹层面的情况下,杂质盐岩直接拉伸抗拉强度高于纯盐岩,是纯盐岩的 1.6 倍。图 4 为直接拉伸下杂质盐岩与纯盐岩破坏后形态。从破坏后宏观断面上来看,杂质盐岩中盐岩颗粒与硬质胶结的杂质颗粒紧密结合,改变了含杂质盐岩晶体颗粒之间的裂隙关系[22],在拉伸破坏过程中杂质颗粒能承受更大的强度,导致杂质盐岩直接拉伸抗拉强度相比纯盐岩有所提高。但在表1中含夹层盐岩直接拉伸抗拉强度明显低于其他杂质盐岩,主要是夹层部位易造成应力集中,拉伸过程中夹层部位最先断裂,破坏裂纹完全沿夹层断裂,故而导致抗拉强度骤降。纯盐岩直接拉伸强度相比间接拉伸强度较小,间接拉伸下强度是直接拉伸下的3.12倍。盐岩作为一种致密晶体颗粒组成的岩石,在直接拉伸应力状态下,微裂纹沿着各晶粒接触最薄弱部位扩展,而间接拉伸下沿着预定的破坏面加载,势必会造成最大的拉伸应力相对较高,故相比间接拉伸下,直接拉伸得到的抗拉强度更能反映盐岩真实抗拉强度特性[13]。

表 1　　各试样拉伸强度表

Table 1　　Tensile strength table for each sample

试验类别		编号	抗拉强度/MPa	平均值/MPa
间接拉伸加卸载	纯盐岩	F-1	1.562	1.823
		F-2	1.96	
		F-3	1.798	
		F-4	1.972	
直接拉伸加卸载	纯盐岩	L-1	0.65	0.585
		L-2	0.62	
		L-3	0.48	
		L-4	0.59	
	杂质盐岩	ZL-1	0.83	0.930
		ZL-2	0.95	
		ZL-3	1.01	
		ZL-4	0.32	含夹层

 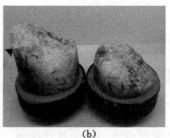

(a)　　　　　　　　　　　　(b)

图 4　直接拉伸荷载下试样破坏后

Fig. 4　Direct tension loading and unloading destruction after morphology

(a) 纯盐岩;(b) 杂质盐岩

图5为循环荷载条件下盐岩拉伸应力-应变曲线。从图5可得知,无论是直接拉伸还是间接拉伸应力下,在加卸载过程中,当应力超过上一级卸荷荷载时,应力应变曲线均会近似沿着原加载曲线上升,呈现岩石变形记忆现象[23]。在间接拉伸循环荷载条件下,峰值应力过后,试样仍能在一定程度的循环荷载下发生变形。直接拉伸应力状态下,加载曲线之间近

乎平行,卸载曲线之间亦近乎平行,随着循环加载应力的逐步增大,试样变形得到逐步增强,当达到峰值应力之时,纯盐岩应变量为 0.064%,而间接拉伸下峰值应力处应变量达到 1.525%,前者仅有后者的 4.2%。直接拉伸峰值应力后试样并不会马上失稳破坏,而是随着拉伸应力缓慢下降,出现一定的残余强度,即在直接拉伸应力下,存在类似于单轴压缩加载破坏下的劣化效应,在达到峰值应力过后,仍具有一定的抗拉承载力。

从图 5(c)可看出,相比纯盐岩,杂质盐岩在峰值应力后出现了明显的应力下降坡度转折点。笔者分析此是由于杂质盐岩在峰值应力过后,相对处于高拉应力,呈现较强脆性特征,由试样内部杂质颗粒与盐岩晶体颗粒拉伸破坏的过程,需要一个应力下降转折点起到缓冲作用,而纯盐岩处于一个相对低拉应力条件,呈现一个较缓慢的拉伸破坏过程,因而应力坡度下降转折点不明显。

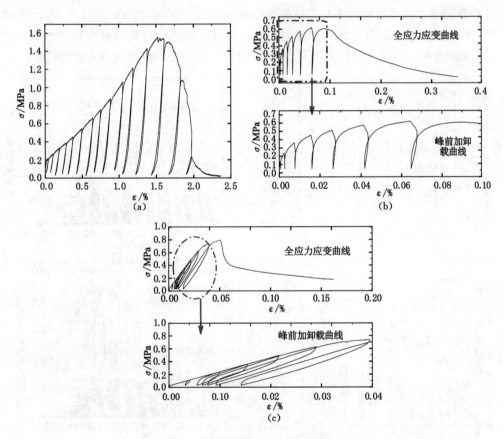

图 5 拉伸荷载下应力-应变曲线

Fig. 5 Tensile stress-strain curve under loading and unloading

(a) 间接拉伸(纯盐岩);(b) 直接拉伸(纯盐岩);(c) 直接拉伸(杂质盐岩)

3.2 声发射振铃计数与能量特征

从图 6 可看出,在间接拉伸加载过程中,声发射特征变化主要可以分为三个阶段:

(Ⅰ)低潮期:从初始加载至 80%F(F 为峰值应力)。此阶段声发射累计振铃计数与累计能量斜率较低,表明声发射活动处于一个较低的增幅期,声发射振铃计数率与能量率最大值分别仅有 230.13 次/s、143.25 mV/s。该阶段主要由于加载应力在递增的过程中,造成试样内部晶体颗粒相互挤压摩擦,出现微裂隙,产生少量声发射信号。

（Ⅱ）活跃期:从80%F至P80%F(P表示峰值应力后)。该阶段声发射累计振铃计数与累计能量斜率处于较高水平,声发射活动处于活跃期。循环加载应力接近峰值应力,试样内部晶体颗粒微裂隙急剧扩张、贯通形成宏观裂纹,声发射信号在这个阶段大量爆发,监测到声发射振铃计数率与能量率均是整个加载过程的最大值,分别为519.71次/s、347.57 mV/s。另一方面,在该阶段加载与卸载过程中出现较明显的Kaiser现象[24],表明试样在间接拉伸循环荷载中存在声发射记忆现象。

（Ⅲ）次活跃期:从P80%F至试验结束。随着峰值应力过后,声发射活跃期降低,但是依然在一个较高的水平,从声发射振铃计数率与能量率均可看出,其最大计数率与能量率为256次/s、195.8 mV/s。声发射累计振铃计数与累计能量增幅还维持在较高阶段,根据统计,该阶段声发射振铃累计技术与累计能量达到102.09(10^3次)与15.75(10^3 mV),分别约占整个加载过程的22%与18%。声发射振铃计数与能量累计数均超过或接近加载全过程的1/5,说明在间接拉伸峰值应力后阶段,试样仍承受着一定的拉伸应力造成的损伤。

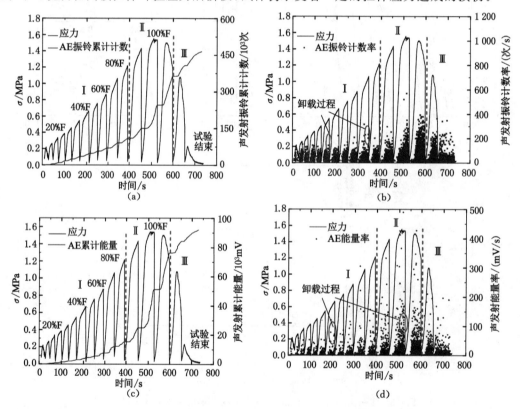

图6 间接拉伸荷载下声发射振铃计数与能量特征

Fig.6 AE ring count and energy characteristics under indirect tension loading and unloading

(a) 应力-时间-振铃累计计数；(b) 应力-时间-振铃计数率；
(c) 应力-时间-累计能量；(d) 应力-时间-能量率

从图7可看出,相比间接拉伸,直接拉伸下声发射特征参数累计曲线"台阶状"更为明显,全过程均体现Kaiser效应。直接拉伸下无论是声发射振铃计数率还是能量率均处于一

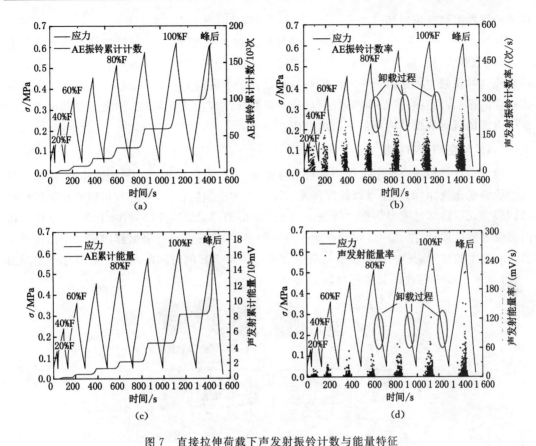

图 7 直接拉伸荷载下声发射振铃计数与能量特征

Fig. 7 AE ring counts and energy characteristics under direct tension loading and unloading

(a) 应力-时间-振铃累计计数;(b) 应力-时间-振铃计数率;

(c) 应力-时间-累计能量;(d) 应力-时间-能量率

个较低水平,全过程中声发射振铃计数率与能量率最大值仅为 362 次/s、215 mV/s,两者仅有间接拉伸最大值的 70% 与 62%,表明直接拉伸循环荷载过程中破坏是由较弱结构面发生,侧面体现出直接拉伸下抗拉强度更为接近实际拉伸破坏值。

直接拉伸循环荷载下,声发射在 20%F 前,声发射振铃计数率与能量率处于一个较平静的时期。岩石直接拉伸试验中,初始加载至某一加载应力段内声发射常出现平静期[9,10,13]。笔者分析是由于直接拉伸加载实验中,盐岩试样内部各结构面受到的拉伸应力呈均匀性,在加载应力 20%F 前,最薄弱的抗拉结构面并未"发现",当加载应力超过 20%F 时,抗拉能力最薄弱的结构面将会先"出现",并在拉伸应力下,微裂隙大量发育,声发射信号也开始出现。因而,盐岩在 20%F 前处于"找寻"最弱结构面阶段,通过声发射的监测可以在试样破坏前预判最弱结构面的空间位置。

当拉伸应力加载至峰值应力时,声发射活动更为活跃,此时试样最薄弱结构面发育的大量裂隙逐步汇聚,并贯通成宏观裂纹,声发射振铃计数率与能量率显著增加。随着拉伸峰值应力后的下降,声发射活动并未减弱,反而其活跃度达到最高阶段,声发射振铃计数率与能量率最大值即是在该阶段出现。根据声发射测试技术原理[25]可知,声发射活动表征了岩石破坏过程的释放能量剧烈程度,由此可知,在直接拉伸峰值应力过后,盐岩仍在承受着一定程度的缓慢张拉变形破坏。

从图 6(b)(d),图 7(b)(d) 间接与直接拉伸卸载过程可看出,在间接拉伸每次卸载过程中,均能监测到声发射信号,而在直接拉伸卸载过程中几乎不可见声发射信号活跃。笔者分析造成这种现象主要是由于在间接拉伸卸载过程中不可避免受到压缩应力对试样内部裂隙扩展性的影响,而在直接拉伸卸载过程中,不存在压缩应力的影响,拉伸应力卸载过程对于试样内部最薄弱抗拉结构面并不造成损伤。

3.3 声发射空间分布特征

通过对盐岩拉伸变形全过程声发射信号处理,不仅可以得到声发射振铃计数、能量等特征参数的变化规律,而且还可以通过声发射事件点的空间定位,再现还原出试样在两种拉伸加卸载方式下破坏过程损伤演化特征。图 8 表示应力增量 20%F(F 指峰值应力) 下各阶段内声发射事件累计空间分布,红色点指代声发射事件,每一个点代表试样内部一个损伤破裂信号所发生的位置,其中 40% 表示加载应力从 0 到 40%F 的时间段内声发射事件点空间分布,对应图 6、图 7 标注的加载应力段。

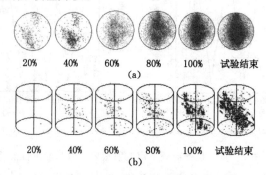

图 8　拉伸荷载下声发射空间分布特征

Fig. 8　Spatial distribution of AE under tensile cyclic loading

(a) 间接拉伸;(b) 直接拉伸

从图 8 可看出,间接拉伸下试样损伤破坏过程中,声发射主要沿预定加载面逐步汇聚,在加载应力的增大过程中,逐渐汇聚在同一直径面上,最后宏观裂纹也是沿着此破坏面发生。间接拉伸加载方式下,试样在加载部位面处会受到压缩应力集中效应,加载之初,内部形成细微裂隙和缺陷,声发射逐渐在该压缩应力区积聚、贯通,离加载点较边缘的部位,则声发射定位点较少。峰值应力后,加载应力虽有所降低,但声发射定位点在宏观破坏面上继续积聚,根据笔者对峰值应力后阶段声发射定位点统计发现,总共有声发射定位点 3 127 个,约占加载过程的 30%,侧面印证了盐岩在间接拉伸峰后仍承受着一定的承载变形能力。

相比间接拉伸加载方式,直接拉伸下声发射空间分布特征具有明显的差异。直接拉伸加载方式下,在加载应力 20%F 前,鲜有声发射事件信号点。在 20%F 后,试样内部抗拉强度最薄弱面出现,声发射定位点也在该部位大量形成、汇聚。峰值应力后阶段,声发射定位点也在最薄弱面抗拉面上积聚贯通成宏观裂纹面,试样发生破坏。根据笔者统计,直接拉伸试验全过程声发射定位点共有 3 780 个,相比间接拉伸,直接拉伸下声发射定位点集聚程度远远不及前者。分析认为主要是直接拉伸下试样内部仅受拉应力,在拉伸应力下沿着薄弱抗拉面发生损伤破坏;但在间接拉伸下,沿预定的加载应力线形成宏观破坏面,加载部位处压缩应力对内部晶体颗粒损伤影响较大,内部晶体颗粒形成较多的裂纹,造成声发射的活跃程度强于直接拉伸下。

4 结 论

(1) 通过盐岩拉伸加卸载试验,研究得到纯盐岩间接拉伸抗拉强度约为 1.823 MPa,直接拉伸下抗拉强度约为 0.585 MPa,相比单轴循环荷载抗压强度,拉伸循环荷载抗拉强度远远低于抗压强度。同时,盐岩直接拉伸抗拉强度受杂质与夹层影响。

(2) 研究表明盐岩间接与直接拉伸荷载下峰值应力后仍存在一定的变形破坏。间接拉伸下峰值应力应变量约是直接拉伸下的 23.83 倍,含杂质盐岩直接拉伸下峰值应力后变形呈现脆性特征,应力下降坡度转折点较明显。

(3) 间接拉伸和直接拉伸破坏过程的声发射计数和能量演化特征不同。间接拉伸下,声发射特征参数呈"低潮期-活跃期-次活跃期"变化特征,峰后声发射特征参数累计数均维持在一个较高的水平,约占全过程的1/5。直接拉伸下,声发射特征参数在 20%F 前处于一个较平静期,峰后声发射活跃度保持较高。间接拉伸卸载过程受压缩应力影响有声发射活动,而直接拉伸卸载过程并不出现较明显的损伤。

(4) 间接拉伸和直接拉伸破坏过程的声发射时空演化差异,表明两种拉伸方式导致的损伤破坏机理不同。间接拉伸下受加载接触部位压缩应力影响,声发射定位点从加载接触部位开始出现,并沿预定加载破坏面逐渐连通;直接拉伸下试件仅受拉应力,损伤出现较晚,声发射定位点从 20%F 后开始出现,声发射沿破坏面的集聚程度远不及间接拉伸。

参 考 文 献

[1] 杨春和,李银平,陈锋. 层状盐岩力学理论与工程[M]. 北京:科学出版社,2009:114-116.
YANG Chunhe,LI Yinping,CHEN Feng. Mechanics and engineering for laminated salt rock[M]. Beijing:Science Press,2009:114-116.

[2] 杨春和,李银平,屈丹安,等. 层状盐岩力学特性研究进展[J]. 力学进展,2008,38(04):484-494.
YANG Chunhe,LI Yinping,QU Danan,et al. Advances in researches of the mechanical behaviors of bedded salt rocks[J]. Advances in Mechanics,2008,38(4):484-494.

[3] 邓华锋,李建林,朱敏,等. 圆盘厚径比对岩石劈裂抗拉强度影响的试验研究[J]. 岩石力学与工程学报,2012(04):792-798.
DENG Hua-feng,LI Jian-lin,ZHU Min,et al. Research on effect of disc thickness-to-diameter ratio on splitting tensile strength of rock[J]. Chinese Journal of Rock Mechanics and Engineering,2012(04):792-798.

[4] 许金余,刘石,孙蕙香. 3 种岩石的平台巴西圆盘动态劈裂拉伸试验分析[J]. 岩石力学与工程学报,2014,33(S1):2814-2819.
Xu Jinyu,Liu Shi,Sun Huixiang. Analysis of dynamic split tensile tests of flattened brazilian disc of three rocks[J]. Chinese Journal of Rock Mechanics and Engineering,2014,33(supp.1):2814-2819.

[5] 王金星,王灵敏,杨小林.对岩石拉伸试验方法探讨[J].焦作工学院学报(自然科学版),2004,23(3):205-209.

WANG Jin-xing, WANG Ling-min, YANG Xiao-lin. Tensile test methods for rock materials[J]. Journal of Jiaozuo Institute of Technology (Natural Science), 2004, 23(3): 205-209.

[6] 喻勇.质疑岩石巴西圆盘拉伸强度试验[J].岩石力学与工程学报,2005,24(7):1150-1157.

YU Yong. Questioning the validity of the Brazilian test for determining tensile strength of rocks[J]. Chinese Journal of Rock Mechanics and Engineering, 2005, 24(7):1150-1157.

[7] ASTM C496—86. Standard test method for splitting tensile strength of cylindrical concrete specimens[S].

[8] 彭瑞东,鞠杨,谢和平.灰岩拉伸过程中细观结构演化的分形特征[J].岩土力学,2007,28(12):2479-2583.

PENG Rui-dong, JU Yang, XIE He-ping. Fractal characterization of meso-structural evolution during tension of limestone[J]. Rock and Soil Mechanics, 2007, 28(12):2479-2583.

[9] 李天一,刘建锋,陈亮,等.拉伸应力状态下花岗岩声发射特征研究[J].岩石力学与工程学报,2013,32(增2):3215-3221.

Li Tianyi, Liu Jianfeng, Chen Liang, et al. Acoustic Emission Characteristics Of Granite Under Tensile Loading[J]. Chinese Journal of Rock Mechanics and Engineering, 2013, 32(Supp2):3215-3221.

[10] 张泽天,刘建锋,王璐,等.煤的直接拉伸力学特性及声发射特征试验研究[J].煤炭学报,2013,38(6):960-965.

ZHANG Ze-tian, LIU Jian-feng, WANG Lu, et al. Mechanical properties and acoustic emission characteristics of coal under direct tensile loading conditions[J]. Journal of China Coal Society, 2013, 38(6):960-965.

[11] 余贤斌,王青蓉,李心一,等.岩石直接拉伸与压缩变形的试验研究[J].岩土力学,2008,29(1):18-22.

YU Xian-bin, WANG Qing-rong, LI Xin-yi, et al. Experimental research on deformation of rocks in direct tension and compression[J]. Rock and Soil Mechanics, 2008, 29(1):18-22.

[12] 余贤斌,谢强,李心一,等.岩石直接拉伸与压缩变形的循环加载实验与双模量本构模型[J].岩土工程学报,2005,27(9):988-993.

YU Xian-bin, XIE Qiang, LI Xin-yi, et al. Cycle loading tests of rock samples under direct tension and compression and bi-modular constitutive model[J]. Chinese Journal of Geotechnical Engineering, 2005, 27(9):988-993.

[13] 刘建锋,徐进,杨春和,等.盐岩拉伸破坏力学特性的试验研究[J].岩土工程学报,2011,33(4):580-586.

LIU Jianfeng, XU Jin, YANG Chunhe, et al. Mechanical characteristics of tensile failure of salt rock[J]. Chinese Journal of Geotechnical Engineering, 2011, 33(4):580-586.

[14] Liu J F, Pei J L, Ma K, et al. Damage evolution and fractal property of salt rock in tensile failure[J]. Underground Storage of CO_2 and Energy, 2010(7):105-112.

[15] 杨春和,马洪岭,刘建锋.循环加、卸载下盐岩变形特性试验研究[J].岩土力学,2009,30(12):3562-3568.
Yang Chunhe, Ma Hongling, Liu Jianfeng. Experimental study on deformation characteristics of rock salt under cyclic loading and unloading[J]. Rock and Soil Mechanics, 2009, 30(12):3562-3568.

[16] 高红波,梁卫国,徐素国,等.循环载荷作用下盐岩力学特性响应研究[J].岩石力学与工程学报,2011,30(增1):2617-2623.
Gao Hongbo, Liang Weiguo, Xu Suguo, et al. Response of rock mechanical properties under cyclic loading[J]. Chinese Journal of Rock Mechanics and Engineering, 2011, 30(S1):2617-2623.

[17] 马林建,刘新宇,许宏发,等.循环荷载作用下盐岩三轴变形和强度特性试验研究[J].岩石力学与工程学报,2013,32(4):849-856.
MA Lin-jian, LIU Xin-yu, XU Hong-fa, et al. Deformation and strength properties of rock salt subjected to triaxial compression with cyclic loading[J]. Chinese Journal of Rock Mechanics and Engineering, 2013, 32(4):849-856.

[18] 中华人民共和国电力工业部.GB/T 50266-1999 工程岩体试验方法标准[S].北京:中国计划出版社,1999.
The Electric Power Industry Ministry of the People's Republic of China. GB/T 50266-1999 Standard for tests method of engineering rock masses[S]. Beijing:China Planning Press,1999.

[19] LIU Jianfeng, XIE Heping, Ju Yang, et al. Device with position-limit spring for alternating tension-compression cyclic test[P]. USA:9488560B2,2016.

[20] 刘建锋,谢和平,鞠杨,等.弹簧组件限位拉压循环试验岩石试样固定装置[P]. ZL201510067094.9.

[21] 邓朝福,刘建锋,徐慧宁,等.单轴加卸载条件下岩盐变形过程能量和声发射特征研究[J].工程科学与技术,2017(S2):150-156.
Deng Chaofu, Liu Jianfeng, Xu Huining, et al. Study on the energy and acoustic emission characteristics of rock salt under uniaxial loading and unloading conditions [J]. Advanced Engineering Sciences, 2017(S2):150-156.

[22] 纪文栋,杨春和,刘伟,等.层状盐岩细观孔隙特性试验研究[J].岩石力学与工程学报,2013,32(10):2036-2044.
Ji Wen-dong Yang Chun-he, Liu Wei, et al. Experimental investigation on meso-pore structure properties of bedded salt rock[J]. Chinese Journal of Rock Mechanics and Engineering,2013,32(10):2036-2044.

［23］韩贝传.岩石的记忆功能及对岩体变形的影响［J］.工程地质学报,1998,6(4)：326-332.

Han Beichuan. The memory function of rock and its influence on rock mass deformation[J]. Journal df Engineering Geology,1998,6(4):326-332.

［24］徐速超,冯夏庭,陈炳瑞.矽卡岩单轴循环加卸载试验及声发射特性研究［J］.岩土力学,2009,30(10):2929-2934.

XU Su-chao, FENG Xia-ting, CHEN Bing-rui. Experimental study of skarn under uniaxial cyclic loading and unloading test and acoustic emission characteristics[J]. Rock and Soil Mechanics,2009,30(10)：2929-2934.

［25］秦四清,李造鼎,张倬元,等.岩石声发射技术概论［M］.成都：西南交通大学出版社,1993.

QIN Si-qing, LI Zao-ding, ZHANG Zhuo-yuan, et al. An introduction to acoustic emission techniques in rocks[M]. Chengdu：Southwest Jiaotong University,1993.

负压条件下瓦斯渗流实验及抽采模拟研究

李祥春[1,2,3]，**高佳星**[1,2]，**李安金**[1,2]，**聂百胜**[1,2]，**陈志峰**[4]

(1. 中国矿业大学(北京) 资源与安全工程学院, 北京, 100083;
2. 中国矿业大学(北京) 煤炭资源与安全开采国家重点实验室, 北京, 100083;
3. 河南省瓦斯地质与瓦斯治理重点实验室——省部共建国家重点实验室培育基地, 河南 焦作, 454000;
4. 新疆工程学院 安全科学与工程学院, 新疆 乌鲁木齐, 830023)

摘 要：通过加入负压调节系统的含瓦斯煤体假三轴应力加载实验，发现煤体的结构变化可分为四个阶段，即压实阶段、裂隙发展阶段、裂隙形成阶段和破碎阶段。分析了相同轴压或围压条件下负压变化对渗透率的影响，发现在裂隙发育阶段，随负压增大渗透率出现下降和上升两个阶段。结合实验结果建立了负压条件下的气固耦合模型，利用COMSOL模拟了瓦斯抽采过程中负压与孔径对有效抽采半径的影响，引入两个简单指标——压径比与双径比，作为提高抽采负压和增大抽采钻孔孔径对有效抽采半径的提升效果的参考依据，论文结论对提高瓦斯抽采的效率有一定的指导意义。

关键词：瓦斯渗流；负压；瓦斯抽采；耦合

Gas seepage experiment and extraction simulation under negative pressure condition

LI Xiangchun[1,2,3], GAO Jiaxing[1,2], LI Jinan[1,2],
NIE Baisheng[1,2], CHENG Zhifeng[4]

(1. School of Resources and Safety Engineering, China Univ. of Mining & Technol. (Beijing), Beijing 100083, China; 2. State Key Lab of Coal Resources and Safe Mining, Beijing 100083, China; 3. State Key Laboratory Cultivation Base for Gas Geology and Gas Control (Henan Polytechnic University), Jiaozuo 454000, China; 4. Department of Safety Engineering, Xinjiang Institute of Engineering, Urumqi 830000, China)

Abstract: Through the pseudo three axis stress loading experiment of the gas containing coal with negative pressure regulating system, it is found that the structural change of coal can be divided into four stages: compaction stage, crack development stage, crack formation stage and crushing stage. The effect of negative pressure on permeability is analyzed under the same axial compression or confining pressure. It is found that there are two stages of permeability change at the crack development stage: permeability decreases with the increase of negative pressure and permeability increases with the increase of negative pressure. Combined

---

基金项目：国家重点研发计划项目(2016YFC0600708)；中央高校基本科研业务费专项资金资助项目(2009QZ09)；河南省瓦斯地质与瓦斯治理重点实验室——省部共建国家重点实验室培育基地开放基金项目(WS2018B04)。
作者简介：李祥春(1979—)，男，博士，副教授，博士生导师，研究方向：瓦斯灾害防治。E-mail：chinalixc123@163.com。
通信作者：高佳星(1992—)，男，在读博士研究生，研究方向：瓦斯灾害防治。E-mail：18234089357@163.com。

with the experimental results, a gas-solid coupling model under negative pressure was established, and the effect of negative pressure and aperture on the effective extraction radius during gas extraction was simulated by COMSOL. The introduction of two simple indicators—Negative pressure difference/Effective extraction radius difference, Effective extraction radius difference/Borehole aperture difference. They as a reference for Improvement effect of effective extraction radius caused by Increase the negative pressure of gas extraction and increasing the diameter. The conclusion of the paper has certain guiding significance for improving the efficiency of gas extraction.

Keywords: gas seepage; negative pressure; gas extraction; coupling

瓦斯对于开采深部低透气性和高瓦斯含量的煤层,增透是瓦斯抽采要解决的首要问题。瓦斯事故给煤矿带来了巨大的威胁,每年造成数百人的死亡和巨大的经济损失[1],因此,加快瓦斯抽采的进度,提高瓦斯抽采的效率,使得煤层瓦斯压力降低,最大限度地预防由瓦斯气体造成的事故,对煤矿的安全生产和增加瓦斯的资源化利用有重要的现实意义[2]。

渗透率是评价煤层渗透性的定量指标,也是评价煤层瓦斯可采性的重要指标之一。大量学者研究分析了如有效应力、孔隙压缩系数、体积应力等对渗透率的影响,建立了一些模型来分析瓦斯渗流问题。Palmer 和 Mansoori[3]将煤体的弹性特性和由吸附引起的膨胀作用引入到渗透率模型,即 Palmer-Mansoori 模型,D. Jasinge[4]对澳大利亚的褐煤进行了实验研究,发现褐煤的渗透率与有效应力负指数拟合关系,并总结成经验公式。Liu[5,6]建立了新型的孔隙裂隙岩体的 TPHM 模型;周世宁院士[7-9]首先应用达西定律来解释瓦斯渗流问题;罗新荣[10]经过实验研究,提出 Darcy's law 的适用范围和用 Klinkenberg 效应修正的 Darcy 定律。在瓦斯渗流实验方面,Sommerton 等[11]发现应力场的存在使得煤体渗透性的发生了改变;该学者还对孔隙的压缩系数对渗透特性的影响进行了实验研究。F. Gu[12]等建立了煤层各向异性和不连续性的渗透率和孔隙率模型,并研究了其在耦合模拟中的应用;G. X. Wang[13]等改进了瓦斯回收和 CO_2 地质封堵的渗透率模型;G. Izadi[14]等研究了含离散裂隙流体渗流的渗透率演化规律;尹志光等[15]发现煤岩全应力-应变过程中,瓦斯流动速度变化的总体规律呈缓"U"型,且具有明显的应变滞后特性;蒋长宝等[16]建立了煤体与煤体内部损伤的变化规律之间的联系,也验证裂隙(瓦斯渗流的主要通道)会因轴压或围压的增大而压实闭合。实验研究还包括不同压力路径下含瓦斯煤体的渗流机理,以及在实验过程中加入含水率以及温度这些影响因子对渗流特性的影响[17-21]。利用三轴实验来模拟煤层的受力情况更接近实际情况,赵阳升,胡耀青[21-26]对原煤煤样进行了大量的假三轴瓦斯透性实验,综合探讨气体吸附解吸、轴压有效应力和围压有效应力,孔隙中的气体压力在渗透率变化中起到的作用,并拟合出相应的方程式。李祥春等[27]利用准三轴实验研究发现,煤体渗透率随体积应力的增加呈负指数规律衰减,随着温度的升高,煤体瓦斯的渗透率升高;杨天鸿等[28]根据煤体变形过程中各参数之间耦合关系,提出了含瓦斯煤岩变形直至破裂过程中的固气耦合模型;唐春安[29,30]应用数值模拟分析软件 RFPA,模拟出含瓦斯煤岩在瓦斯压力与地应力等共同作用下损伤发展直至破裂的全过程。

影响煤层瓦斯抽采效果的一个重要因素是抽采的负压。在负压产生的瓦斯压力梯度的作用下,瓦斯通过裂隙不断流向钻孔,从而达到瓦斯抽采的目的。程平远[31]等认为抽采负压的作用是给裂隙游离瓦斯流入钻孔提供动力,裂隙瓦斯流出后再形成基质与裂隙瓦斯压力差,从而达到抽采瓦斯的目的。尹光志等[32]进行了钻孔抽采瓦斯三维数值模拟,认为负压对钻孔瓦斯抽采的影响不明显。杨宏民[33]等对煤层瓦斯抽采合理孔口负压进行了研究，郑吉玉等[34]研究发现随着负压的增大，煤样渗透率不断增大，在负压较小时，负压对渗流的影响较大，当负压达到一定程度渗透率趋于稳定。轴压、围压和温度越大，煤样渗透率越小。

但是负压条件下的瓦斯渗流研究方面，数值模拟研究较多，实验研究较少，本文主要针对负压条件下瓦斯渗流实验过程煤体应力与瓦斯渗流进行实验研究。利用数值模拟研究，探究瓦斯抽采过程中气固耦合规律，为揭示抽采过程中瓦斯渗流机理、提高瓦斯抽采效率等提供理论基础，为保证我国煤炭安全生产奠定基础。

1 负压条件下瓦斯渗流实验

1.1 实验煤样

本实验研究煤样取自潞安集团余吾煤业大块原煤，通过钻孔取样机钻取直径为 50 mm 的煤柱，再用岩石切割将其长度切割成 100 mm，用双端面磨石机将煤样的两段打磨平整，将钻好的原煤煤样，用真空干燥箱干燥，干燥一般在 50 ℃的恒温箱内干燥 24 h，完成干燥后用放有干燥剂的密封袋密封。如图 1 所示。

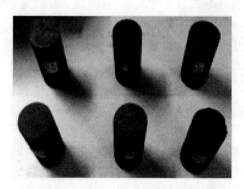

图 1 余吾原煤煤样

Fig.1 YuWu original coal samples

1.2 实验系统

本文为了使含瓦斯煤体的假三轴应力加载实验结果更加接近瓦斯抽采时煤岩体与瓦斯气体的耦合规律，在由轴压围压加载系统，三轴夹持器，恒温系统，流量采集系统，应力应变测试系统构成的加入了可调节负压大小并能使负压稳定在某一固定值的负压调节系统。整个实验的系统图如图 2 所示。

(1) 加载试验机

实验室用加载系统包括天辰试验机及其控制软件 TensonTest。通过 TensonTest 软件的各参数的设置实现不同加载路径的加载，同时具有较高的精度和稳定性。

(2) 三轴夹持器

瓦斯气体通过三轴夹持器腔体上方的进气孔通入腔体中，温控仪接口和围压控制口分布在腔体的侧面，通过这两接口连接温控仪和围压加载装置，控制温度和实验所需要的围压条件，气体通过腔体中煤样后从下端出气口出来。

(3) 恒温系统

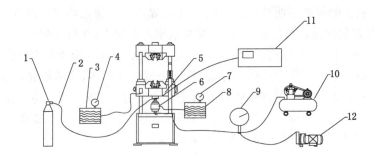

图 2　实验系统示意图

Fig.2　Experimental system schematic diagram

1——瓦斯气瓶;2——瓦斯输送管路;3——轴压泵;4——轴压控制仪表;5——压力机;
6——煤样罐;7——围压控制仪表;8——围压泵;9——流量计;10——负压控制系统;
11——应力应变采集系统;12——水环真空泵

实验时,为了保证实验在恒定的温度下进行,采用 ES-I 型温控仪,在一定的范围内可以实现温度的连续变化,具有较好的稳定性。

(4) 负压调节系统

实验时,为了保证负压恒定,应用数控显示仪,气动调节阀,压力变送器来完成,可以将出口负压调节为 0 kPa、10 kPa、20 kPa、30 kPa、40 kPa、50 kPa、60 kPa、70 kPa、80 kPa。

(5) 流量采集系统

流量采集系统是由七星华创 CS200 质量流量计和 D07 系列与相配套的流量积算仪及软件配合使用。CS200 和 D07 质量流量计采用新型恒功率传感器和温度补偿技术,因此具有高精度,快响应以及极地零漂及温漂等特点。CS200 质量流量计测量量程为 250 mL/min 和 D07 质量流量计的量程为 1.5 L/min 经过三轴夹持器排出夹持器排出的气体经过质量流量计,通过流量积算仪及相配套的软件就可以实现总流量及瞬时流量的记录与导出,实时监测流量变化,最终再通过计算就可以算出其渗透率。

(6) 应力应变测试系统

由压力机自带软件进行应力应变的测量,本实验配合电脑软件用其来测定煤样的轴向应变。

1.3　实验方案及结果分析

对传统假三轴加载渗流实验系统进行改进,加入了可调节负压大小并能使负压稳定在某一固定值的负压调节系统,使得实验系统可以测定不同负压下的渗透率规律,改进方法稳定可靠。实验测得的数据有瓦斯流量,轴向压力,围压,出气口压力,进气口压力以及应变。本文对实验结果分别从轴压对瓦斯渗流的影响,以及围压对渗流的影响,负压对渗流的影响三个方面进行分析。在三个方面的分析中着重研究负压对整个渗流过程的影响。

根据井下瓦斯抽采过程负压对抽采效果的影响,近似模拟井下瓦斯抽采过程中煤体渗流规律,通过控制轴向压力的变化来模拟支撑压力变化,用围压的变化来模拟水平应力变化,实验采用假三轴加载方法。采用设备为课题组自行研制的三轴负压-渗流实验装置,实验过程中选用瓦斯压力 0.4 MPa。

(1) 轴压加载对含瓦斯煤体渗透率的影响分析

对余吾矿煤样分别进行轴压分级加载实验,研究煤样破碎前不同负压下渗透率与轴压之间的关系。瓦斯压力为 0.4 MPa,研究瓦斯流量、轴向应力、应变以及时间的关系。如图 3 为瓦斯流量、轴向应力、应变与时间的关系曲线图。

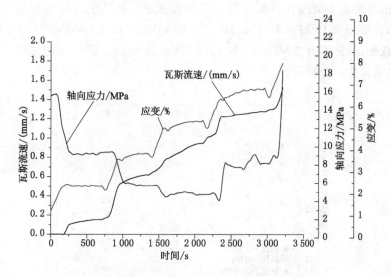

图 3 瓦斯流量、轴向应力、应变与时间的关系曲线图

Fig. 3 Relation diagram of gas flow, axial stress, strain and time

从图 3 可以看出瓦斯流速变化大致可分为四个阶段,瓦斯流速下降阶段(1~1 061 s),瓦斯流速稳定阶段(1 061~2 349 s),瓦斯流速上升阶段(2 349~3 054 s)和瓦斯流速急速上升阶段(3 054~3 209 s)。在轴向应力波动点都会伴随着瓦斯流量的波动,瓦斯气体流量的变化对应力的变化非常敏感,这说明渗透率和轴向应力之间的关系非常密切。通过应变的变化曲线将煤样的整体形态及孔隙裂隙形态的发展分为四个阶段,即压实阶段,裂隙发展阶段、裂隙形成阶段和破碎阶段如图 4 所示,压实阶段煤体中的原生裂隙逐渐消失,导致瓦

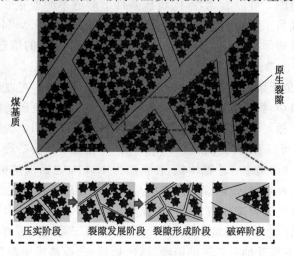

图 4 煤样受力变形模型图

Fig. 4 Model diagram of stress and deformation of coal sample

斯通道变窄,瓦斯流速持续降低;裂隙发育阶段开始在煤体内部形成新的裂隙但是还未能达到原生裂隙水平,瓦斯流速保持稳定低速;在裂隙形成阶段,新的裂隙接近原生裂隙水平,瓦斯流速开始上升;在破碎阶段,瓦斯流速急剧上升。

对瓦斯压力为 0.4 MPa 下煤样不同负压下渗透率与轴压之间的变化进行研究。在围压为 2 MPa,在负压为 10 kPa、20 kPa、30 kPa、40 kPa、50 kPa、60 kPa、70 kPa、80 kPa 的条件下分级加载轴压,分别为 2 MPa、6 MPa、10 MPa、14 MPa。研究上述条件下煤体渗透率的变化,实验结果如图 5 所示。

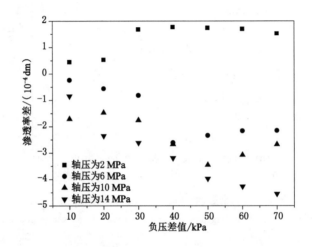

图 5 不同抽采负压下渗透率差值(轴压变化)

Fig. 5 Differences in permeability rates under different extraction negative pressures(axial pressure change)

由于不同负压情况下渗透率变化不明显,所以选择用 20 kPa、30 kPa、40 kPa、50 kPa、60 kPa、70 kPa、80 kPa 的渗透率分别减去 10 kPa 的渗透率,以此来研究负压变化条件下渗透率的变化情况。由图 5 可以看到,不同负压条件下渗透率的变化非常微小,数量级只有 10^{-4} dm,这说明提高负压所能增加的瓦斯抽采率非常有限。另外,可以看到轴压的变化对渗透率有着比较明显的影响,这也从另一方面验证了图 3 中呈现的信息是正确的,而且轴压的增大(2~14 MPa)使渗透率减小。

同理,研究不同负压情况下瓦斯流速的变化也采用上述方法,可以看到不同负压条件下瓦斯流速变化非常微小,数量级只有 10^{-2} mm/s,这也说明提高负压所能增加的瓦斯抽采率非常有限。图 6 中依然可看出轴压的变化对瓦斯流速有着比较明显的影响。

结合图 3~图 6 可以看到,在轴压为 2 MPa 时,煤体处于压实阶段,瓦斯流动通道较窄,随着负压的增加,煤体渗透率有小幅增加,在负压为 40 kPa 之后保持相对稳定,而瓦斯流速则随负压的增加非线性增加,此时煤体渗透率与瓦斯流速随负压增大的变化趋势相同。轴压为 6 MPa 时,煤体处于裂隙发育阶段,随着负压增加,瓦斯渗透率逐渐降低,在负压为 50 kPa 时达到最低值,之后又随负压增大。这应该是由于在裂隙发育阶段,瓦斯流动通道较窄,微小裂隙中的瓦斯受抽采负压的影响,迅速被抽出,煤基质中的气相瓦斯在压差作用下扩散进入裂隙,同时吸附态瓦斯迅速解吸成为气相瓦斯进一步扩散,此时渗透率相对较大。随着负压增大,上述过程持续进行,但基质中的瓦斯与吸附态瓦斯持续减少,而且煤体

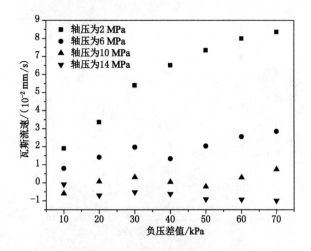

图 6 不同抽采负压下瓦斯流速差值(轴压变化)

Fig. 6 Differences in gas flow velocity under different extraction negative pressures(axial pressure change)

中瓦斯压力降低,根据有效应力效应,在一定程度上使煤体的裂隙与微孔隙变窄,从而导致煤体的渗透率持续降低,这一阶段称为渗透率下降阶段。之后由于负压的持续增加与轴压的作用下,煤体中未与外界连通的微孔隙开始破碎进而形成新的瓦斯流通通道,从而导致渗透率逐渐增大,这一阶段称为渗透率上升阶段。这就到导致瓦斯流速与渗透率的变化出现两个阶段,变化趋势相反阶段与变化趋势相同阶段,这是由于负压为 50 kPa 之前,随着负压的增加,瓦斯流速需要保持增大才能使瓦斯压力保持为 0.4 MPa,所以出现了瓦斯流速与渗透率变化趋势相反的现象,而在实际生产过程中,两者应该是相同的。轴压为 10 MPa 时的情形与 6 MPa 时相似,煤体也处于裂隙发育阶段,差别在于渗透率的最低点在负压为 60 kPa 时出现,发生了后移。这说明在轴压较大的条件下,需要更大的负压才能使未连通的微孔隙连通形成新的瓦斯流动通道,即轴压限制了负压作用下瓦斯流动新通道的产生。轴压为14 MPa 的渗透率变化情况可以证实这一点,高轴压条件下,渗透率随着负压的增加持续降低,没有出现 6 MPa 与 10 MPa 条件下的渗透率上升阶段。而轴压为 2 MPa 时,由于轴压较小,很容易进入了渗透率上升阶段,并保持平衡,这也从另一方面证实了高轴压限制了负压作用下瓦斯流动新通道的产生。

(2)围压加载对含瓦斯煤体渗透率的影响分析

对余吾矿煤样分别进行围压分级加载实验,研究煤样破碎前不同负压下渗透率与围压之间的关系。对瓦斯压力为 0.4 MPa 下煤样不同负压下渗透率与围压之间的变化进行研究。在轴压为 18 MPa,在负压为 10 kPa、20 kPa、30 kPa、40 kPa、50 kPa、60 kPa、70 kPa、80 kPa 的条件下分级加载围压,分别为 2 MPa、2.6 MPa、3.2 MPa、3.8 MPa、4.4 MPa。研究上述条件下煤体渗透率的变化,实验结果如图 7 所示。

由于不同负压情况下渗透率变化不明显,研究时依然采用差值负压。从图 7 可以发现不同围压条件下,随着负压的增大,渗透率在逐渐减小,尽管减幅非常小(数量级为 10^{-4} dm),但趋势却比较明显。这说明负压的增加所能增加的瓦斯抽采率非常有限,但负压的增加会降低渗透率的现象是客观存在的,而图 7 中渗透率的变化趋势与图 5 有所区别,没有出

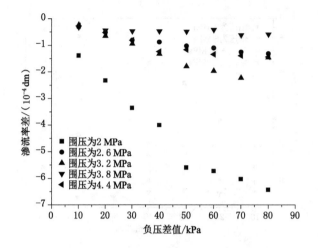

图7 不同抽采负压下渗透率差值(围压变化)

Fig. 7 Differences in permeability rates under different extraction negative pressures (confining pressure change)

现渗透率的波动的两个阶段。从图7和图8可以发现,在围压为2~4.4 MPa时,随着负压的增大渗透率持续下降,这是由于此时加载的轴压为18 MPa,煤样已处于破碎阶段,前文中分析认为高轴压限制了负压作用下瓦斯流动新通道的产生,所以此时渗透率的变化趋势与轴压为14 MPa时类似。而随围压的增大,渗透率也逐渐下降(图7中反映的只是同围压条件下渗透率随负压变化的差值),综合来看,在煤样破碎阶段,小围压(2~4.4 MPa)并不会改变煤样所处阶段,渗透率会随负压的增大而减小。同时在此阶段,围压的增加同样限制了负压作用下瓦斯流动新通道的产生。由此可见,煤体应当只有在裂隙发育阶段,瓦斯流动新通道才会在负压作用下产生。在围压为2~4.4 MPa时,随负压增大,瓦斯流速增大则是由于使瓦斯压力保持为0.4 MPa而造成的,综合形成了随负压增大渗透率与瓦斯流速变化趋

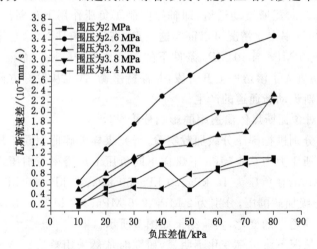

图8 不同抽采负压下瓦斯流速差值(围压变化)

Fig. 8 Differences in gas flow velocity under different extraction negative pressures (confining pressure change)

势相反的情形。

2 渗透率与有效应力之间的关系分析

分析图3、图5、图7可见,煤的渗透率与有效应力密切相关,有效应力增加,渗透率发生变化。目前常用的煤的渗透率计算公式 Louis 公式:

$$k = A\exp(B\sigma) \tag{1}$$

式中 k——渗透率,mD;

A,B——实验系数;

σ——有效应力,MPa。

从图9可以看出,不同负压下渗透率随有效应力的变化曲线变化趋势基本一致,不过会有细微差别,而这些差别是由于负压造成的。

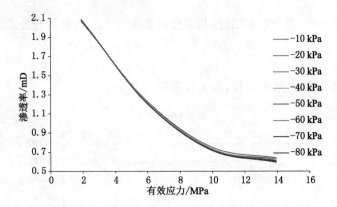

图9 不同抽采负压下渗透率随有效应力的变化曲线

Fig. 9 Curves of permeability versus effective stress under different extraction pressures

从图10可以发现拟合效果非常好,相似性系数为0.959,由此可以说明,实验结果符合 Louis 公式。所以对实验系数 A 和 B 与负压之间的关系进行拟合,结果如图11所示。

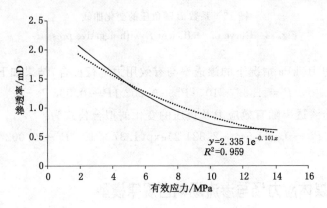

图10 渗透率随有效应力的拟合曲线

(抽采负压为 10 kPa)

Fig. 10 Fitting curve of permeability with effective stress(suction negative pressure is 10 kPa)

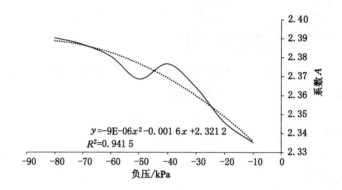

图 11 系数 A 随负压的变化曲线

Fig. 11 Curve of coefficient A with negative pressure

从图 11 可以看出,拟合效果很好,相关性系数为 0.941 5,其拟合公式为:

$$A = -9 \times 10^{-6} P^2 - 0.001\ 6P + 2.321\ 2 \tag{2}$$

式中 P——负压,kPa。

从图 12 可以看,拟合效果很好,相关性系数为 0.989。

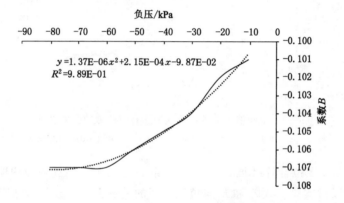

图 12 系数 B 随负压的变化曲线

Fig. 12 Curve of coefficient B with negative pressure

对抽采负压为 10 kPa 情况下的渗透率与有效用力进行拟合,结果如下:

$$B = 1.37 \times 10^{-6} P^2 + 0.002\ 1P - 0.098\ 7 \tag{3}$$

由此可以得出渗透率随有效应力及负压的变化的拟合公式为:

$$k = (-9 \times 10^{-6} P^2 - 0.001\ 6P + 2.321\ 2)\exp(1.37 \times 10^{-6} P^2 + 0.002\ 1P - 0.098\ 7) \tag{4}$$

3 负压条件下煤体应力场与渗流场耦合规律模型

3.1 模型假设

(1) 耦合系统的固相和气相分别为煤岩体和高纯瓦斯气体。

(2) 将煤岩体等效为双重孔隙介质模型,双重孔隙介质包括煤基质孔隙结构和煤基质间的裂隙结构,煤基质拥有较大的孔隙率,但其渗透率很小,基本可以忽略不计,因此是瓦斯气体的主要储存空间。而裂隙拥有很小的孔隙率和很大渗透率,因此是游离态瓦斯气体的主要渗流通道。由电镜扫描可以观测到煤体颗粒大致成方形,因此采用的双重介质模型为Warren-Root 正方形模型。

（3）通过气密性检测,可以得出煤样与煤样罐的胶套密封性很好,因此瓦斯气体的只会通过煤样内部流动,且假设含瓦斯煤样骨架为弹塑性体,拥有各向同性。

（4）充入煤样罐中的气体为单相饱和甲烷气体。

（5）应用达西定律计算煤样孔隙裂隙空间中的瓦斯气体流动。

（6）在煤样罐外有恒温装置,确保实验过程瓦斯渗流为等温渗流,因此模型中的渗流为等温渗流。

（7）采用朗缪尔方程解释实验过程中瓦斯的吸附解吸过程。

3.2 孔隙率变化模型

根据煤的双重孔隙介质假设,煤体中的孔隙和裂隙是为瓦斯提供了吸附表面和运移通道,同时孔隙率（包含孔隙和裂隙）也会受地应力场,瓦斯压力场和吸附膨胀导致的煤体结构变形的影响。因为孔隙率会受到多种因素的影响,所以是煤层渗流场和应场耦合纽带。

根据本节基本假设,只有单相饱和的瓦斯流体的含瓦斯煤体,通过将公式的计算和变形,得到综合作用下煤体孔隙率模型为：

$$\varphi = 1 - \frac{1-\varphi_0}{1+\varepsilon_v}\left\{1 - \frac{p_0-p}{K} + \frac{2a\rho_v RT}{3V_m K(1-\varphi_0)}[\ln(1+bp_0) - \ln(1+bp)]\right\} \quad (5)$$

式中 φ_0 ——煤体的初始孔隙率；

ε_v ——煤体体积应变；

p_0, p ——储层瓦斯压力；

K ——煤体体积模量；

ρ_v ——煤的密度,t/m^3；

V_m ——气体摩尔体积；

a ——单位质量的煤的最大吸附瓦斯量,m^3/t；

b ——Langmuir 甲烷吸附常数,Pa^{-1}；

R ——理想气体常数,$8.314 J/(mol·K)$；

T ——标准状态下绝对温度,273 K。

3.3 含瓦斯煤体瓦斯渗流方程

根据基本假设煤体瓦斯的流动规律符合达西定律,根据前人研究,气体在多孔介质中的渗流必须考虑 Klikenberg 效应,尤其是气体在低气压阶段在低渗透介质中的渗流,通过计算和变形得到含瓦斯煤体瓦斯渗流方程为：

$$\frac{(1-\varphi)}{1+\varepsilon_v}p\frac{\partial \varepsilon_v}{\partial t} + \left[\varphi + \frac{abcp_n}{(1+bp)^2} + \frac{(1-\varphi_0)p}{(1+\varepsilon_v)K} - \frac{2ab\rho_v RTp}{3V_m K(1+bp)(1+\varepsilon_v)}\right]\frac{\partial p}{\partial t} - \nabla \cdot \left[\frac{k}{\mu}\left(1+\frac{m}{p}\right)\rho \cdot \nabla p\right] = 0$$

(6)

式中 φ ——煤层孔隙率,%；

p ——瓦斯压力,Pa；

ρ_n ——标准状态下瓦斯密度,kg/m^3；

c——单位煤体吸附质量校正参数,kg/m^3;

ρ_v——煤体密度,kg/m^3;

k——煤层渗透率 m^2;

μ——瓦斯动力黏度系数,$Pa \cdot s$;

K——Klikenberg 系数,Pa。

3.4 煤体应力场方程

在含瓦斯煤体内的任一点取一平行微正六面体单元,不考虑煤体瓦斯渗流和变形运动的惯性力及瓦斯的体积力。考察的微元体处于应力平衡状态,考虑煤体骨架重力的影响,引入拉梅常数,经过整理运算,得到含瓦斯煤体变形控制方程张量表达式为:

$$G\sum_{j=1}^{3}\frac{\partial^2 \mu_j}{\partial x_j^2} + \frac{G}{1-2\nu}\sum_{j=1}^{3}\frac{\partial^2 \mu_j}{\partial x_j \partial x_i} - \frac{(3\lambda - 2G)}{3K}\frac{\partial P}{\partial x_i} - \frac{(3\lambda + 2G)2ab\rho_v RT}{9V_m K(1+bp)}\frac{\partial P}{\partial x_i} + \alpha\frac{\partial P}{\partial x_i} + F_i = 0$$

(7)

式中 ε_{ij}——煤体的应变张量;

μ——煤体位移;

F_i——体积力;

λ——拉梅常数;

G——剪切模量;

ν——泊松比。

3.5 气固耦合方程

根据本节以上分析及推导,结合的孔隙率和渗透率动态变化模型可知,煤体内气体流动的渗流场与煤岩体骨架变形的应力场通过多个耦合项相互耦合,将式(4)~式(7)联立建立瓦斯运移多场耦合偏微分方程组[35]:

$$\left.\begin{array}{l} \dfrac{(1-\varphi)}{1+\varepsilon_v}p\dfrac{\partial \varepsilon_v}{\partial t} + \left[\varphi + \dfrac{abcp_n}{(1+bp)^2} + \dfrac{(1-\varphi_0)p}{(1+\varepsilon_v)K} - \dfrac{2ab\rho_v RTp}{3V_m K(1+bp)(1+\varepsilon_v)}\right]\dfrac{\partial p}{\partial t} - \\ \nabla \cdot \left[\dfrac{k}{\mu}\left(1+\dfrac{m}{p}\right)p \cdot \nabla p\right] = 0 \\ G\sum_{j=1}^{3}\dfrac{\partial^2 \mu_j}{\partial x_j^2} + \dfrac{G}{1-2\nu}\sum_{j=1}^{3}\dfrac{\partial^2 \mu_j}{\partial x_j \partial x_i} - \dfrac{(3\lambda - 2G)}{3K}\dfrac{\partial P}{\partial x_i} - \dfrac{(3\lambda + 2G)2ab\rho_v RT}{9V_m K(1+bp)}\dfrac{\partial P}{\partial x_i} + \alpha\dfrac{\partial P}{\partial x_i} \\ + F_i = 0 \\ \varphi = 1 - \dfrac{1-\varphi_0}{1+\varepsilon_v}\left\{1 - \dfrac{p_0-p}{K} + \dfrac{2a\rho_v RT}{3V_m K(1-\varphi_0)}[\ln(1+bp_0) - \ln(1+bp)]\right\} \\ k = (-9\times10^{-6}P^2 - 0.0016P + 2.3212)e^{(1.37\times10^{-6}P^2 + 0.0021P - 0.0987)\sigma} \end{array}\right\}$$

(8)

4 瓦斯抽采过程数值模拟研究

4.1 物理几何模型构建及物理力学参数选取

选取余吾煤矿 3# 煤层为研究对象,根据现场实际情况,实验拟合结果以及理论分析结果建立二维平面几何物理模型。模型选取一个钻孔及其周围煤壁作为研究对象,因为 3# 煤层钻孔布置为每隔 4 m 布置一个抽采钻孔,煤层厚度为 6.34 m,钻孔半径为 0.094 m,钻孔高度为 1.8 m,倾角为 3°,钻孔长 160 m.因此设置的几何模型的长度为 4 m,高度为 6.34 m。并选取钻孔位于中心位置的断面进行分析;模型顶部根据余吾 3# 煤层的应力测试报告选取相似的均匀载荷。底部边界固定,左、右边界为水平应力边界。几何模型与网格模型图如图 13 所示。煤体和瓦斯气体各项物理力学参数如表 1 所示,钻孔周围煤体模拟的初始条件如表 2 所示。

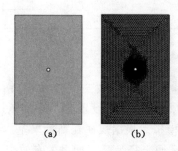

图 13　钻孔周围煤体几何模型与网格模型图

Fig. 13　Geometric model and mesh model of coal around borehole

根据余吾煤矿井下实际钻孔的分布,研究瓦斯抽采钻孔某一切面的煤体残余瓦斯压力和渗透率。本文绘制几何模型,其比例与实际矿井的比例相同,对此次模拟的几何模型采用较细网格划分。

表 1　　　　　　　　煤体和瓦斯气体各项物理力学参数

Table 1　　　　　The physical and mechanical parameters of coal and gas

参数名称	数值	单位
煤弹性模量	3 500	MPa
Klinkenberg 系数	0.251	Pa
Biot-Willis 系数	0.44	
泊松比	0.3	
煤层初始孔隙率	4.93	%
煤层透气性系数	0.524	$m^2/(MPa^2 \cdot d)$
煤体的密度	1.6	kg/m^3
瓦斯的动力黏度	1.08×10^{-5}	$Pa \cdot s$
瓦斯吸附体积常数 a	26.209 8	m^3/t
瓦斯吸附压力常数 b	1.683 6	MPa^{-1}
气体摩尔体积	22.4	L/mol
气体常数 R	8.314	$J/(mol \cdot K)$
热力学温度	298.15	K
瓦斯气体密度	0.714	kg/m^3

表 2　　　　　　　　　　　钻孔周围煤体模拟的初始条件
Table 2　　　　　The initial conditions of simulation of coal around the borehole

初始条件	数值	单位
初始瓦斯压力	1.2	MPa
煤体瓦斯含量	13	m^3/t
煤壁瓦斯涌出量	5.3	m^3/min
煤岩体垂直应力	14.96	MPa
煤岩体水平应力	6.66	MPa

4.2 数值模拟结果分析

本文对瓦斯抽采进行模拟,为了确定抽采过程中负压,钻孔孔径以及在钻孔斜下方补孔等不同因素中哪一种因素的影响最大,本分设置了不同的负压,不同孔径下的模拟,以及单独分析了在钻孔斜下方补孔的模拟。根据煤矿瓦斯抽采基本指标规定,抽采一定时间后,煤层瓦斯压力低于 0.74 MPa 的区域半径为瓦斯抽采的有效半径。在模拟的云图中会体现煤层瓦斯压力低于或等于 0.74 MPa 的等值线和标注出瓦斯抽采的有效半径。

实际抽采中,抽采的负压不得低于 13 kPa,并且某些煤矿的抽采负压达到了 60 kPa,因此本文选取了 6 个不同的抽采负压,分别为 13 kPa,22 kPa(余吾煤矿采用的抽采负压),31 kPa,38 kPa,45 kPa 和 52 kPa。其模拟结果如图 14 所示。

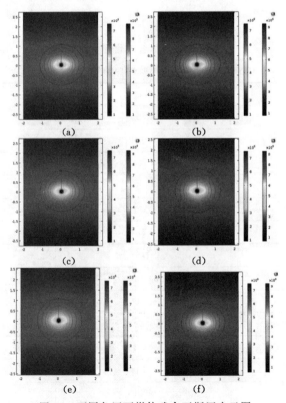

图 14　不同负压下煤体残余瓦斯压力云图

Fig.14　Nephogram of residual gas pressure of coal under different negative pressure
(a) 负压 13 kPa;(b) 负压 22 kPa;(c) 负压 31 kPa;(d) 负压 38 kPa;(e) 负压 45 kPa;(f) 负压 52 kPa

钻孔在大气压下自然释放的模拟结果如图 15 所示。

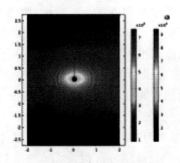

图 15　大气压下的煤体残余瓦斯压力云图

Fig. 15　Nephogram of residual gas pressure of coal under barometric pressure

表 3 为在抽采 90 d 后有效半径的模拟结果,可以看到负压比大气压下的抽采有效半径有比较明显的提升,但继续增加负压,抽采有效半径的提升非常有限,本文采取一种简单的指标来对提高负压能够得到的有效抽采半径做比较,简称为压径比,为无量纲量,记为 $\Delta_{(p_1 \sim p_2)}$ =抽采负压差/抽采有效半径差,p_1、p_2 为径压比中的抽采负压,表 4 是本次模拟中各负压段的压径比。

表 3　不同负压下的抽采有效半径

Table 3　Effective radius of extraction under different negative pressure

负压/kPa	抽采有效半径/m
0	0.910 85
13	1.102 32
22	1.102 4
31	1.102 49
38	1.102 56
45	1.102 63
52	1.102 7

表 4　压径比

Table 4　Negative pressure difference/Effective extraction radius difference

$\Delta_{(0\sim13\ kPa)}$	68	$\Delta_{(22\sim31\ kPa)}$	10^5
$\Delta_{(0\sim22\ kPa)}$	115	$\Delta_{(22\sim38\ kPa)}$	10^5
$\Delta_{(0\sim31\ kPa)}$	162	$\Delta_{(22\sim45\ kPa)}$	10^5
$\Delta_{(0\sim38\ kPa)}$	198	$\Delta_{(22\sim52\ kPa)}$	10^5
$\Delta_{(0\sim45\ kPa)}$	235	$\Delta_{(31\sim38\ kPa)}$	10^5
$\Delta_{(0\sim52\ kPa)}$	271	$\Delta_{(31\sim45\ kPa)}$	10^5
$\Delta_{(13\sim22\ kPa)}$	112 500	$\Delta_{(31\sim52\ kPa)}$	10^5
$\Delta_{(13\sim31\ kPa)}$	105 882	$\Delta_{(38\sim45\ kPa)}$	10^5
$\Delta_{(13\sim38\ kPa)}$	104 167	$\Delta_{(38\sim52\ kPa)}$	10^5
$\Delta_{(13\sim45\ kPa)}$	103 226	$\Delta_{(45\sim52\ kPa)}$	10^5
$\Delta_{(13\sim52\ kPa)}$	102 632		

由压径比的定义可知,压径比越大,说明增加负压可以提高的有效抽采半径越小,即性价比越低。由表 4 可以比较清晰地看到 $\Delta_{(0\sim13\ kPa)}$ 最小,说明由负压从 0 到 13 kPa 增加的有效抽采半径效果最好,而从 0 到 22 kPa、31 kPa、38 kPa、45 kPa、52 kPa 增加的有效抽采半径效果越来越差。而从后面的数据也可以发现,如果从一定负压增加到更好的负压,压径比非常大,数量级都在 10^5 左右,这也说明不同负压下,瓦斯抽采有效半径虽然不同,但是其差距非常小。综合来看负压对增加有效抽采半径有明显效果,但是持续增加负压对瓦斯提高瓦斯抽采的效果并不明显。

在实际抽采中,钻孔孔径经常为 94 mm,余吾采用的钻孔孔径也是 94 mm,同时其他矿井也有采用 75 mm,84 mm,97 mm,113 mm 以及 120 mm 的钻孔孔径,因此选用的 6 个抽采钻孔孔径分别为 75 mm,84 mm,94 mm,97 mm,113 mm 以及 120 mm。其模拟结果如图 16 所示。

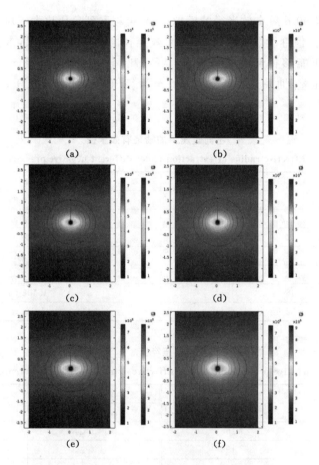

图 16 不同抽采孔径下的煤体残余瓦斯压力云图

Fig. 16 Nephogram of residual gas pressure of coal under different extraction aperture

(a) 孔径 75 mm;(b) 孔径 84 mm;(c) 孔径 94 mm;(d) 孔径 97 mm;
(e) 孔径 113 mm;(f) 孔径 120 mm

表 5 为在抽采 90 d 后不同钻孔孔径的有效半径模拟结果,结果中可以看出孔径对瓦斯抽采的效果较为明显,孔径越大,其抽采的有效半径越大,但是四周边角残余的瓦斯压力还

是很高,说明只增加钻孔孔径的方法并不是最优化的方法。

表 5 抽采 90 d 后不同钻孔孔径的有效抽采半径的模拟结果
Table 5　simulation results of effective drainage radius for different borehole diameters after 90 days' extraction

钻孔孔径/mm	有效抽采半径/m
75	0.839 848
84	1.066 24
94	1.102 4
97	1.110 06
113	1.168 49
120	1.190 39

本文利用一个简单指标对不同钻孔孔径的有效抽采半径做判断,简称为双径比,为无量纲量,记为 $\Gamma_{(r_1\sim r_2)}$ = 有效抽采半径差/钻孔孔径差(双径单位都以 mm 计算,其中 $r_1\sim r_2$ 表示钻孔孔径由 r_1 增加为 r_2),以此作为钻孔所能提升的有效抽采半径的参考指标。表 6 是本次模拟中各钻孔孔径分段的双径比。

表 6　双径比
Table 6　Effective extraction radius difference/Borehole aperture difference

$\Gamma_{(75\sim84)}$	25.15	$\Gamma_{(84\sim120)}$	3.45
$\Gamma_{(75\sim94)}$	13.82	$\Gamma_{(94\sim97)}$	2.55
$\Gamma_{(75\sim97)}$	12.28	$\Gamma_{(94\sim113)}$	3.48
$\Gamma_{(75\sim113)}$	8.65	$\Gamma_{(94\sim120)}$	3.38
$\Gamma_{(75\sim120)}$	7.79	$\Gamma_{(97\sim113)}$	3.65
$\Gamma_{(84\sim94)}$	3.62	$\Gamma_{(97\sim120)}$	3.49
$\Gamma_{(84\sim97)}$	3.37	$\Gamma_{(113\sim120)}$	3.13
$\Gamma_{(84\sim113)}$	3.53		

由双径比的定义可知,此比值越大代表提升钻孔孔径能够增加的有效抽采半径越大,从表 6 中的数据可以比较清晰地看出,钻孔孔径持续增加所提升的有效抽采半径并不是越来越大,$\Gamma_{(75\sim84)}$ 是 $\Gamma_{(84\sim94)}$ 的 6.95 倍,这也从另一方面证明单纯的增加钻孔孔径并不能增加抽采效率,应该利用多种指标来对钻孔孔径进行选取,利用双径比作为指标来看,本次模拟中选择 84 mm 的钻孔孔径是合适的。

图 17 为孔径为 94 mm,抽采负压为 22 kPa 下,不同抽采时间下距钻孔边缘不同位置的残余瓦斯压力图,从图中可以看出,在抽采时间为 90 d 时,瓦斯抽采有效半径为 1.102 4 m,而实际余吾煤矿的瓦斯抽采有效半径为 1.4 m,两者相差较小,说明模拟结果较为贴近实际抽采情况。

参照余吾现场为解决煤体四周残余瓦斯压力高的问题而采用的方法,决定在模拟的几何图形里,在瓦斯抽采钻孔的下方补钻一个瓦斯抽采钻孔。其模拟结果如图18所示。

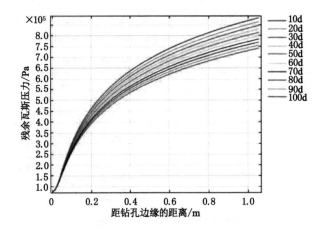

图 17　不同抽采时间下距钻孔边缘不同位置的残余瓦斯压力

Fig. 17　Residual gas pressure at different locations along the borehole under different pumping times

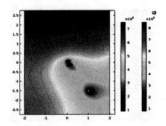

图 18　双钻孔残余瓦斯压力云图

Fig. 18　Nephogram analysis of double-borehole

从图18的结果可以看出,此方法确实解决了周围煤体残余瓦斯压力问题。同时也说明余吾所采的方法有效,并且可靠。但是根据两种钻孔渗透率云图的对比分析,如图19所示,双钻孔抽采的低渗透率区域要小于单一钻孔的低渗透率的区域,并且可以看出钻孔周围渗透率要低于其他区域,这是由于钻孔周围存在应力集中的缘故,因此,钻取抽采钻孔后,如果

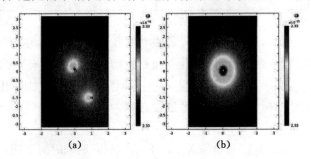

图 19　双钻孔渗透率云图与单钻孔渗透率云图对比分析

Fig. 19　Comparative analysis of double borehole permeability nephogram and single borehole permeability nephogram

(a) 双钻孔渗透率云图;(b) 单钻孔渗透率云图

不对其进行增透措施,其抽采并不能达到最佳效果。

此模拟结果虽然与余吾煤矿的实际测得结果不能完全对等,但是根据模拟结果所推出的问题,却与余吾煤矿抽采过程中所遇到的情况十分相似,就余吾现场报告可以看出,钻完抽采钻孔后,需要对钻孔进行水力冲孔,对钻孔进行预裂,根据第一节实验所得出的结论,煤体在应力的作用下,表现为四个阶段,即压实阶段,裂隙发展阶段、裂隙形成阶段和破碎阶段,而这四个阶段中,只有压实阶段,渗透率会随着应力的增加而减小,因此,矿井采用的水力冲孔,可以使钻孔周围的煤体不在处于压实阶段,防止应力集中造成的煤体渗透率的降低。

4.3 负压条件下瓦斯渗流率变化和瓦斯抽采的讨论

一般认为提高抽采负压可以相应提高煤体中瓦斯流动的压力差,从而达到提高煤层瓦斯抽采率的目的。但是抽采负压增加,不但提高了抽采设备的要求,而且漏风量也会随之增大,从而导致瓦斯抽采浓度降低。一些文献在实验室条件研究负压对渗透率的影响时发现随着负压的增大,煤样渗透率不断增大,轴压、围压和温度越大,煤样渗透率越小。在本文实验中分析了不同轴压和围压条件下,随着负压增加渗透率的变化情况,这是由于不同压力条件下煤样所处的阶段不同,导致瓦斯流动通道不同,渗透率随负压的变化情况不尽相同。

就本文实验而言,在轴压为 2 MPa 时,随负压的增加,煤样的渗透率在不断增加,这与一些文献中的结论一致。不过,在轴压 6 MPa 与 10 MPa 时,随负压增大渗透率出现下降和上升两个阶段,而在轴压为 14 MPa 时,随负压增大渗透率持续下降。所以在讨论负压对渗透率的影响时一定要说明煤样或者煤层所处的阶段。本文实验还发现,相同轴压或围压条件下,改变负压大小引起渗透率的变化非常微小,数据间差值的数量级只有 10^{-4} dm,这与一些文献中负压对钻孔瓦斯抽采的影响不明显的结论是一致的。

而就煤样所处阶段的渗透率随负压变化情况而言,压实阶段中渗透率随负压增加而增加,在裂隙发育阶段随负压增大渗透率出现下降和上升两个阶段,在裂隙形成阶段,负压增大渗透率持续下降。并且发现,高轴压限制了负压作用下瓦斯流动新通道的产生,轴压为 14 MPa 时渗透率随负压增大的变化证实了这一点。这与文献中轴压越大,煤样渗透率越小的结论有一定的相似性。但是本文实验中只在围压在 2 MPa 的条件下讨论了分级加载轴压下不同抽采负压条件下的渗透率变化,在轴压为 18 MPa 条件下讨论了分级加载围压下不同抽采负压条件下的渗透率变化,没有形成连续实验,这导致实验结果由一定的局限性,之后的研究需要做连续实验,以此得到更加全面的实验结论。

利用 COMSOL 对不同抽采负压、不同钻孔孔径条件下的瓦斯抽采进行模拟,得到对应的煤体残余瓦斯压力云图及对应的有效抽采半径。引入了两个简单指标作为提高抽采负压和增大抽采钻孔孔径对有效抽采半径的提升效果如何的参考依据,这给煤层瓦斯抽采过程中负压以及钻孔孔径的选取提供了一种思路。为了提高瓦斯抽采效率,可以在瓦斯抽采设计中加入类似的指标,当然需要在简单的指标中加入一些因子,如将增加抽采负压需要的能量,增加钻孔孔径增加的花费等加入到指标中,使指标更加贴近实际情况而成为一种能在瓦斯抽采设计中使用的指标,这也可以作为之后研究的改进方向。

毫无疑问,在研究负压对渗透率的影响时必须要明确煤体处于阶段,在制定瓦斯抽采和措施时需要综合更多科学合理的指标,这样才能更加真实的揭示抽采过程中瓦斯渗流机理,为提高瓦斯抽采效率等提供理论基础。

5 结　论

（1）通过应变的变化曲线将煤样的整体形态及孔隙裂隙形态的发展分为四个阶段，即压实阶段，裂隙发展阶段、裂隙形成阶段和破碎阶段。负压条件下瓦斯渗流实验表明，相同轴压或围压条件下，改变负压大小渗透率的变化非常微小，数据间差值的数量级只有 10^{-4} dm，对瓦斯流速影响也较小，数据间差值的数量级只有 10^{-2} mm/s，说明提高负压所能增加的瓦斯抽采率非常有限。

（2）在裂隙发育阶段，随负压增大渗透率出现下降和上升两个阶段，在渗透率下降阶段，随负压增大基质中的瓦斯与吸附态瓦斯持续减少导致渗透率逐渐降低；渗透率上升阶段煤样在负压的持续增加与轴压的作用下，煤体中未与外界连通的微孔隙开始破碎形成新的瓦斯流通通道，从而导致渗透率逐渐增大。同时发现在高轴压限制了负压作用下瓦斯流动新通道的产生。

（3）通过含瓦斯煤体负压条件下的渗流规律进行抽象简化，得到了推导气固耦合方程的基本假设，对 Langmuir 方程进行修正，将拟合出的方程进行整理，引入到耦合方程中，使耦合方程有更高的实用性。

（4）引入两个简单指标——压径比与双径比，作为提高抽采负压和增大抽采钻孔孔径对有效抽采半径的提升效果如何的参考依据。发现负压对增加有效抽采半径有明显效果，但是持续增加负压对瓦斯提高瓦斯抽采的效果并不明显，以及单纯的增加钻孔孔径并不能增加抽采效率。在瓦斯抽采过程数值模拟发现，距离钻孔越近，其渗透率的值越低，因此，可以看出如果改变钻孔周围煤体所处阶段，会非常有效地增加负压抽采的效果。因此，矿井所采用水力冲孔预裂的方式可有效地增加瓦斯的抽采效果。

参 考 文 献

[1] 陈晓坤,蔡灿凡,肖旸. 2005—2014 年我国煤矿瓦斯事故统计分析[J].煤矿安全,2016(02):224-226.

[2] Pan Rongkun, Cheng Yuanping, Yuan Liang, et al. Effect of bedding structural diversity of coal on permeability evolution and gas disasters control with coal mining [J]. Natural Hazards,2014,73(2):531-546.

[3] Palmer I, Mansoori J. How permeability depends on stress and pore pressure in coalbeds:a new model[C]. SPE Reservoir evaluation & engineering,1998:539-544.

[4] Jasinge D, Ranjith P G, Choi S K. Effects of effective stress changes on permeability of Latrobe valley brown coal[J]. Fuel,2011,90(3):1292-1300.

[5] Liu H H, Rutqvist J, Berryman J G. On the relationship between stress and elastic strain for porous and fractured rock[J]. Int. J. Rock Mech. Min. Sci. ,2009,46(2):289-296.

[6] Liu H H, Rutqvist J. A new coal-permeability model: Internal swelling stress and fracture matrix interaction[J]. Transport in Porous Media,2010,82(1):157-171.

[7] 周世宁,孙辑正.煤层瓦斯流动理论及其应用[J].煤炭学报,1965(1):24-37.

[8] 周世宁,林柏泉.煤层瓦斯赋存与流动理论[M].北京:煤炭工业出版社,1999.

[9] 周世宁.煤层透气性系数的测定和计算[J].中国矿业学院学报,1980(1):1-6.

[10] 罗新荣.煤层瓦斯运移物理模型与理论分析[J].中国矿业大学学报,1991(3):58-64.

[11] Sommerton W J,Soylemezoglu I M,Dudley R C. Effect of stress on permeability of coal[J]. Int. J. Rock Mech. Min. Sci. and Geomech. Abstr.,1975,12(2):129-145.

[12] Gu F,Chalaturnyk R J. Permeability and porosity models considering anisotropy and discontinuity of coalbeds and application in coupled simulation[J]. Journal of Petroleum Science and Engineering,2010,74(3-4):113-131.

[13] Wang G X,Massarotto P,Rudolph V. An improved permeability model of coal for coalbed methane recovery and CO_2 geo sequestration[J]. International Journal of Coal Geology,2009,77(1-2):127-136.

[14] Izadi G,Wang S,Elsworth D,et al. Permeability evolution of fluid-infiltrated coal containing discrete fractures[J]. International Journal of Coal Geology,2011,85(2):202-211.

[15] 尹志光,李广治,赵洪宝,等.煤岩全应力-应变过程中瓦斯流动特性试验研究[J].岩石力学与工程学报,2010,29(1):170-175.

[16] 蒋长宝,尹志光,李晓泉,等.突出煤型煤全应力-应变全程瓦斯渗流试验研究[J].岩石力学与工程学报,2010,29(2):3482-3487.

[17] 朱卓慧,冯涛,谢东海,等.不同应力路径下含瓦斯煤渗透特性的实验研究[J].采矿与安全工程学报,2012,29(4):570-574.

[18] Wang Guangrong,Xue Dongjie,Gao Hailian,et al. Study on permeability characteristics of coal rock in complete stress-strain process[J]. Journal of China Coal Society,2012,37(1):107-111.

[19] 马衍坤,王恩元,李忠辉,等.煤体瓦斯吸附渗流过程及声发射特性实验研究[J].煤炭学报,2012,37(4):641-646.

[20] 魏建平,王登科,位乐.两种典型受载含瓦斯煤样渗透特性的对比[J].煤炭学报,2013,38(1):93-99.

[21] 魏建平,位乐,王登科.含水率对含瓦斯煤的渗流特性影响试验研究[J].煤炭学报,2014,39(1):97-103.

[22] 胡耀青,赵阳升,魏锦平,等.三维应力作用下煤体瓦斯渗透规律实验研究[J].西安矿业学院学报,1996,16(4):308-311.

[23] 赵阳升,胡耀青.孔隙瓦斯作用下煤体有效应力规律的实验研究[J].岩土工程学报,1995,17(3):26-31.

[24] 赵阳升,胡耀青,杨栋,等.三维应力下吸附作用对煤岩体气体渗流规律影响的试验研究[J].岩石力学与工程学报,1999,18(6):651-653.

[25] Zhao Y S,Qing H Z,Bai Q Z. Mathematical model for solid-gas coupled problems on the methane flowing in coal scam[J]. Acta Mechanica Solida Sinica,1993,6(4):459-466.

[26] 赵阳升.煤体—瓦斯耦合数学模型及数值解法[J].岩石力学与工程学报,1994,13(3):229-239.

[27] 李祥春,聂百胜,刘芳彬,等.三轴应力作用下煤体渗流规律实验[J].天然气工业,2010(6):19-21.
[28] 杨天鸿,徐涛,刘建新,等.应力-损伤-渗流耦合模型及在深部煤层瓦斯卸压实践中的应用[J].岩石力学与工程学报,2005,24(16):2900-2905.
[29] 徐涛,唐春安,宋力.含瓦斯煤岩破裂过程流固耦合数值模拟[J].岩石力学与工程学报,2005,24(10):1667-1673.
[30] Tang C A,Yang T H,Tham L G,et al.. Coupled analysis of flow,stress and damage (FSD) in rock failure[J]. Int. J. Rock Mech. Min. Sci.,2002,39(4):477-489.
[31] 程远平,董骏,李伟,等.负压对瓦斯抽采的作用机制及在瓦斯资源化利用中的应用[J].煤炭学报,2017,42(6):1466-1472.
[32] 尹光志,李铭辉,李生舟,等.基于含瓦斯煤岩固气耦合模型的钻孔抽采瓦斯三维数值模拟[J].煤炭学报,2013,38(4):535-541.
[33] 杨宏民,沈涛,王兆丰.伏岩煤业3#煤层瓦斯抽采合理孔口负压研究[J].煤矿安全,2013,44(12):11-13.
[34] 郑吉玉,田坤云,王振江.负压对煤的瓦斯气体流动影响研究[J].煤炭技术,2016,35(3):175-177.
[35] 牛帅.煤体瓦斯运移多场耦合分析及应用[D].焦作:河南理工大学,2015.

静水压力下开挖卸荷快慢对硐室围岩变形-开裂影响的连续-非连续方法模拟

王学滨[1,2]，芦伟男[2]，白雪元[2]，张智慧[2]

(1. 辽宁工程技术大学 计算力学研究所，辽宁 阜新，123000；
2. 辽宁工程技术大学 力学与工程学院，辽宁 阜新，123000)

摘 要：为了模拟剪切开裂现象，在自主开发的基于拉格朗日元方法、变形体离散元方法及虚拟裂缝模型耦合的连续-非连续方法的基础上，引入了Ⅱ型断裂能，开展了静水压力条件下开挖卸荷快慢对硐室围岩模型变形-开裂过程影响的数值模拟研究，得到了下列结果。随着时步数目的增加，模型两条对角线与硐室表面交汇的4个位置首先发生破坏；然后，裂缝不断增多、长大，向围岩深处发展，一些单元脱离围岩，发生冒落。剪裂缝主要位于V形坑边缘，而拉裂缝位于V形坑内部，较分散。随着开挖卸荷时间的增加，围岩中V形坑的形态变得对称或未得到充分发展，拉、剪裂缝区段数目有降低的趋势。不同开挖卸荷时间时拉、剪裂缝区段数目随时步数目的演变规律具有类似性。

关键词：静水压力；开挖卸荷；剪切开裂；连续-非连续方法；硐室；围岩

Numerical simulation of effects of excavation unloading time on deformation-cracking processes of the chamber surrounding rock under the hydrostatic pressure

WANG Xuebin[1,2], LU Weinan[2], BAI Xueyuan[2], ZHANG Zhihui[2]

(1. *Institute of Computational Mechanics, Liaoning Technical University,
Fuxin 123000, China*;
2. *College of Mechanics and Engineering, Liaoning Technical University,
Fuxin 123000, China*)

Abstract: In order to model the shear fracture, the shear fracture energy is introduced into the developed continuum-discontinuum method in which the Lagrangian element method, deformational discrete element method and fictitious crack model are coupled. Effects of excavation unloading time on deformation-cracking processes of the chamber surrounding rock under the hydrostatic pressure are investigated. The following results are obtained. Firstly, cracking is observed in four positions in the vicinity of intersections between diagonal lines of the chamber surrounding rock and the chamber surface. Then, with an increase of the number of timesteps, the number of cracks increases, the length of cracks increases, cracks extend towards

---

基金项目：国家自然科学基金项目(51574144)；辽宁省百千万人才工程项目(2017)。
通信作者：王学滨(1975—)，男，黑龙江省双鸭山市人，教授，博士生导师，主要从事工程材料变形、破坏及稳定性研究。
Tel：13941824926，E-mail：wxbbb@263.net。

the depth of the chamber surrounding rock, and some elements separate from the chamber surrounding rock. Shear cracks are in the boundary of V-shaped notches, while scattering tensile cracks are in notches. With an increase of excavation unloading time, V-shaped notches become symmetrical or cannot develop fully, and the number of shear and tensile crack segments has a decreasing tendency. Evolution of the number of shear and tensile crack segments with timesteps is similar.

Keywords: hydrostatic pressure; excavation unloading; shear fracture; continuum-discontinuum method; chamber; surrounding rock

数值模拟研究是硐室围岩变形、破坏及稳定性研究的主要手段。目前，主要采用两种方法开展此问题研究。其一为连续方法，代表性的连续方法是有限元方法和有限差分方法。此类方法可以较好地描述硐室围岩的应力、应变及塑性区分布，但不适于硐室围岩开裂和坍塌的模拟。开裂会暴露出新的表面，开裂和坍塌会产生新的接触关系。采用连续方法难以处理这些问题。其二为非连续方法，代表性的非连续方法是离散元方法和不连续变形分析方法。此类方法可以较好地模拟开裂、运动和接触问题，但对于应力、应变的描述一般较为粗糙。在非连续方法中，引入黏结概念[1-2]可使模拟连续介质和开裂成为可能，但往往需要引入接触刚度，这会对应力、应变产生一定的影响。

为了弥补连续方法和非连续方法各自的缺陷，连续-非连续方法应运而生，正在快速发展[3-6]。王学滨发展了一种基于拉格朗日元方法、变形体离散元方法和虚拟裂缝模型耦合的二维连续-非连续方法[6]。该方法已经在巴西圆盘劈裂实验、采场岩层冒落等问题模拟方面初步展现了一定的优势[6-11]。文献[7]研究了集中加载、加载板加载和平台加载3种不同加载方式对巴西圆盘的宏观力学行为、变形-开裂过程及材料抗拉强度的影响。文献[8]对逐步卸荷条件下圆形硐室围岩中应力波传播、压缩位移控制加载条件下圆形及矩形硐室围岩的变形-开裂-垮塌过程进行了模拟，研究了卸荷时间及局部自适应阻尼系数对卸荷过程的影响。文献[9]通过在岩样左、右两侧预设V形缺口，对拉伸位移控制加载条件下岩样的变形-开裂过程进行了模拟。文献[10]对紧凑拉伸岩样的变形-开裂过程进行了模拟。文献[11]对采动诱发长壁开采水平岩层的开裂-冒落过程进行了模拟。然而，在上述二维连续-非连续方法中，仅引入了Ⅰ型断裂能，更适于处理拉裂问题，这使该方法的应用受到限制。

无论是在实验室中，还是在现场观测中，静水压力条件下硐室周边的均匀破坏极为少见，更多的是局部破坏，例如，出现1～4个V形坑[12-14]。V形坑的发生、发展和稳定伴随着岩爆的整个过程。硐室距离地表越深，硐室围岩所处应力状态越接近于静水压力。探讨静水压力条件下硐室围岩模型变形-开裂过程对于一些地质灾害的预防具有重要的理论和实际意义。另外，开挖卸荷快慢是人为可控的影响硐室围岩模型变形-开裂过程的主要因素之一，因此，其影响也是值得探讨的问题。

为了模拟剪切开裂现象，在自主开发的基于拉格朗日元方法、变形体离散元方法及虚拟裂缝模型耦合的连续-非连续方法的基础上，本文引入了Ⅱ型断裂能，开展了静水压力条件下开挖卸荷快慢对硐室围岩模型变形-开裂过程影响的数值模拟研究。

1 连续-非连续方法简介

1.1 原始方法简介

基于拉格朗日元方法、变形体离散元方法和虚拟裂缝模型耦合的二维连续-非连续方法[6]主要包括4个计算模块：应力、应变模块、节点分离模块、接触力求解模块和运动方程求解模块。

在应力、应变模块中，主要借鉴 FLAC 的基本原理，求解单元的应力和应变。

在节点分离模块中，采用带拉伸截断的莫尔-库仑强度准则判别节点是否分离。节点分离包括剪切分离和拉伸分离。对于上述两种分离，只引入Ⅰ型断裂能，不引入Ⅱ型断裂能。在节点发生剪切分离时，考虑了应力脆性跌落，即应力状态由初始抗剪强度参数决定的初始屈服面跌至由残余抗剪强度参数决定的残余屈服面上。在此过程中，最大主应力 σ_3（第3主应力），即围压，保持不变。

在接触力求解模块中，采用基于势的接触力计算方法求解单元的接触问题。

在运动方程求解模块中，借鉴 FLAC 的基本原理，采用中心差分方法求解节点的速度，进而求得节点的位移。

1.2 引入Ⅱ型断裂能

Ⅱ型断裂能，即为剪切断裂能，是剪切破坏过程中单位面积消耗的能量。

当节点发生拉伸分离时，首先，通过计算成对分离节点之间的法向张开度计算分离节点的法向黏聚力，对成对分离节点施加方向相反的法向力。法向力等于法向黏聚力与作用面积之积。法向黏聚力与法向张开度所围面积即为Ⅰ型断裂能。然后，计算裂缝面的平均剪应力和分离节点的切向滑移量，二者所围面积即为Ⅱ型断裂能，对成对分离节点施加切向力。

当节点发生剪切分离时，首先，通过计算成对分离节点之间的切向滑移量计算分离节点的切向黏聚力，对成对分离节点施加方向相反的切向力。切向力等于切向黏聚力与作用面积之积。切向黏聚力与切向滑移量所围面积即为Ⅱ型断裂能。然后，计算裂缝面的平均法向应力和分离节点的法向张开度，二者所围面积即为Ⅰ型断裂能，对成对分离节点施加法向力。

应当指出，在目前的连续-非连续方法中，开裂仅能沿单元边界进行，因此，裂缝面即为单元边界。

2 计算模型、结果及分析

2.1 计算模型、方案及参数

未开挖硐室的模型的高度和宽度均为 40 m，被剖分成 160×160 个正方形单元。模型的厚度取为正方形单元的边长，即 0.25 m。各种参数取值如下：密度 $\rho=2\,430$ kg/m³，弹性模量 $E=14$ GPa，泊松比 $\mu=0.27$，抗拉强度 $\sigma_t=5$ MPa，法向接触刚度 $K_n=10$ GPa，初始黏聚力 $c=12.9$ MPa，初始内摩擦角 $\varphi=17°$，摩擦系数 $f=0.1$，残余黏聚力 $c_r=2$ MPa，残余内摩擦角 $\varphi_r=11°$，Ⅰ型断裂能 $G_f^I=10.5$ N/m，Ⅱ型断裂能 $G_f^{II}=105$ N/m，局部自适应阻尼系数 0.2，时步长度为 $2.329\,37\times10^{-5}$ s。计算在平面应变、大变形条件下进行。在模型的底面上施加活动铰支座约束，在模型的左、右侧面和顶面施加 27 MPa 的压应力。开挖后的模型如图 1 所示。

计算过程包括 3 步。

第 1 步：对未开挖硐室的模型进行计算，直到达到静力平衡状态。此步骤花费的时步数

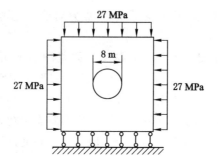

图 1 开挖后的模型

Fig. 1 Model after excavation

目为 12 000。

第 2 步：开挖圆形硐室，直径为 8 m。硐室的中心与模型的中心重合。采用由模型的中心向外逐圈删除单元的方式模拟硐室的开挖过程。应当指出，由于模型被剖分成正方形单元，因此，圆形硐室的表面呈锯齿状。

第 3 步：对开挖硐室后的围岩模型进行计算。

本文共采用 3 个计算方案。方案 1～方案 3 的硐室开挖卸荷时间 T 分别为 $9.317\ 48 \times 10^{-2}$ s（相当于 4 000 个时步数目）、$2.795\ 244 \times 10^{-1}$ s（相当于 12 000 个时步数目）及 $8.385\ 732 \times 10^{-1}$ s（相当于 36 000 个时步数目），以研究硐室开挖卸荷快慢的影响。

2.2 结果及分析

图 2～图 3、图 4～图 5 及图 6～图 7 分别给出了方案 1～方案 3 的剪裂缝、拉裂缝及 σ_3

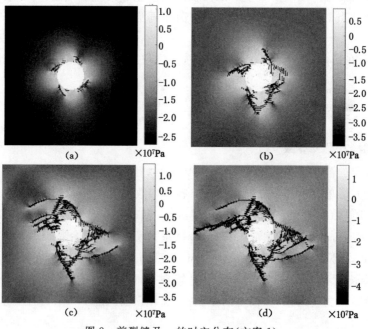

图 2 剪裂缝及 σ_3 的时空分布（方案 1）

Fig. 2 Spatiotemporal distributions of shear cracks and σ_3 (scheme 1)

(a) 时步数目＝16 000；(b) 时步数目＝56 000；(c) 时步数目＝136 000；(d) 时步数目＝176 000

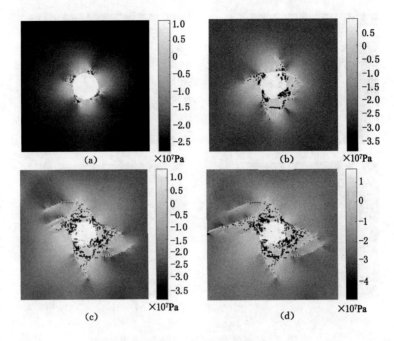

图 3 拉裂缝及 σ_3 的时空分布(方案 1)

Fig. 3 Spatiotemporal distributions of tensile cracks and σ_3 (scheme 1)

(a) 时步数目=16 000;(b) 时步数目=56 000;(c) 时步数目=136 000;(d) 时步数目=176 000

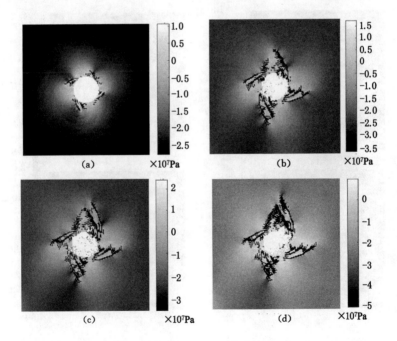

图 4 剪裂缝及 σ_3 的时空分布(方案 2)

Fig. 4 Spatiotemporal distributions of shear cracks and σ_3 (scheme 2)

(a) 时步数目=24 000;(b) 时步数目=64 000;(c) 时步数目=104 000;(d) 时步数目=184 000

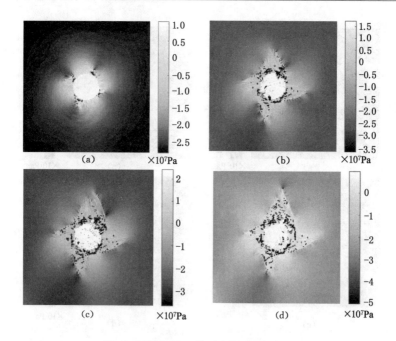

图 5 拉裂缝及 σ_3 的时空分布（方案 2）

Fig. 5 Spatiotemporal distributions of tensile cracks and σ_3 (scheme 2)

(a) 时步数目＝24 000；(b) 时步数目＝64 000；(c) 时步数目＝104 000；(d) 时步数目＝184 000

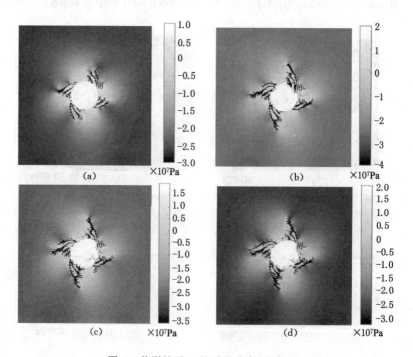

图 6 剪裂缝及 σ_3 的时空分布（方案 3）

Fig. 6 Spatiotemporal distributions of shear cracks and σ_3 (scheme 3)

(a) 时步数目＝48 000；(b) 时步数目＝88 000；(c) 时步数目＝168 000；(d) 时步数目＝248 000

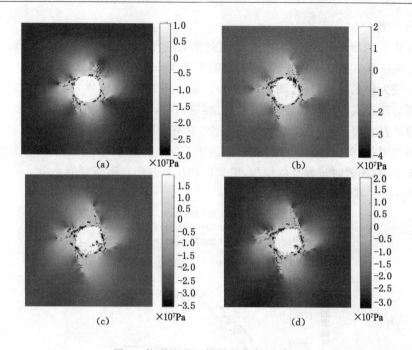

图 7 拉裂缝及 σ_3 的时空分布(方案 3)

Fig. 7 Spatiotemporal distributions of tensile cracks and σ_3 (scheme 3)

(a) 时步数目=48 000;(b) 时步数目=88 000;(c) 时步数目=168 000;(d) 时步数目=248 000

的时空分布规律,其中,负值代表压应力,黑色线段代表裂缝区段,两个单元之间的裂缝称之为 1 个裂缝区段。图 8～图 9 分别给出了剪裂缝区段数目-时步数目曲线和拉裂缝区段数目-时步数目曲线,其中,时步数目是从模型被建立时开始计起的,剪、拉裂缝区段数目是从硐室完全形成时开始计起的。由图 2～图 7 可以发现:

(1) 随着时步数目的增加,硐室周边 4 个位置首先发生破坏;然后,裂缝不断增多、长大,向围岩深处发展。

下面,以方案 1 为例详细介绍硐室围岩的 σ_3 的时空分布规律和破坏过程。当时步数目=16 000 时[图 2(a)和图 3(a)],硐室周边的 σ_3 较高,为拉应力;远离硐室位置(模型周边)的 σ_3 较低,为压应力;在硐室周边与模型周边之间,在硐室顶部、底部和左、右两帮,各形成 1 个半圆形的 σ_3 过渡区。此时,剪裂缝和拉裂缝出现在模型两条对角线与硐室表面多个交汇处附近。当时步数目=56 000 时[图 2(b)和图 3(b)],剪裂缝较弯曲,其最大长度接近硐室直径,在硐室右帮和底部,V 形坑较完整,而在硐室左帮和顶部,V 形坑并不完整。此时,在 V 形坑内部,σ_3 较高,为拉应力;V 形坑之外的 4 个 σ_3 过渡区尺寸较大,形状不规则。应当指出,此时,剪裂缝主要位于 V 形坑边缘,而拉裂缝位于 V 形坑内部,较分散;一些单元已脱离围岩,正在冒落。当时步数目=136 000 时[图 2(c)和图 3(c)],硐室顶部、底部及右帮的 V 形坑较清晰,由 V 形坑某些位置发展出了一些狭长、弯曲的剪裂缝。此时,V 形坑之外的 σ_3 过渡区形状极不规则,其外边界已接近模型周边;拉裂缝主要分布在 V 形坑内部;一些脱离围岩的单元已与硐室底部发生接触。在时步数目由 136 000 增至 176 000 的过程中[图 2(c～d)和图 3(c～d)],V 形坑进一步得到发展,由 V 形坑某些位置发展出的一些狭长、弯曲的剪裂缝已接近模型边界,硐室底部堆积了大量脱离围岩的单元;拉裂缝仍主要分布在 V 形坑内部。

（2）随着 T 的增加，硐室围岩中 V 形坑的形态变得对称或未得到充分发展。

对比方案 1 和方案 2 可以发现，方案 2 的 V 形坑较为对称，未观察到由 V 形坑某些位置发展出的较为狭长的剪裂缝（图 4 和图 5）。在方案 2 中，在硐室右肩位置出现一个由于单元脱离围岩而形成的较小尺寸的 V 形坑；相对而言，逆时针发展的剪裂缝多于顺时针发展的剪裂缝。在方案 3 中，V 形坑发育不完全，只有逆时针发展的剪裂缝（图 6）。而且，随着时步数目的增加，剪裂缝和拉裂缝的形态趋于不变，这一点不同于方案 1 和方案 2。在方案 1 和方案 2 中，随着时步数目的增加，剪裂缝和拉裂缝仍在不断发展之中。

由图 8～图 9 可以发现：

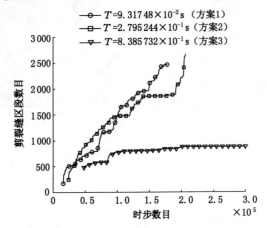

图 8　剪裂缝区段数目-时步数目曲线
Fig. 8　Evolution of shear crack segments with timesteps

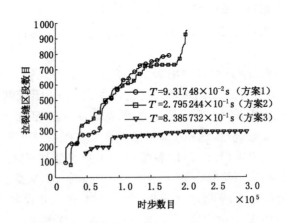

图 9　拉裂缝区段数目-时步数目曲线
Fig. 9　Evolution of tensile crack segments with timesteps

（1）随着时步数目的增加，方案 3 的剪裂缝区段数目首先缓慢增加；当时步数目＝84 000～92 800 时，剪裂缝区段数目快速增加；此后，剪裂缝区段数目缓慢增加直至趋于不变。随着时步数目的增加，方案 1 和方案 2 的剪裂缝区段数目总体上持续增加，二者会发生交叉。

(2) 不同 T 时拉、剪裂缝区段数目随着时步数目的演变规律具有类似性。应当指出,当硐室刚形成时,方案 3 的裂缝区段数目大于方案 1 和方案 2 的,而方案 1 和方案 2 的基本相同。

(3) 随着 T 的增加,拉、剪裂缝区段数目有降低的趋势。

3 讨论

(1) 关于硐室周边的不均匀破坏。根据经典弹塑性力学,静水压力条件下硐室周边将发生均匀的破坏。然而,无论是在实验室中,还是在现场观测中,硐室周边的均匀破坏极为少见,更多的是局部破坏,例如,出现 1～4 个 V 形坑[12-14]。基于连续方法的静水压力条件下硐室围岩剪切带演化规律的数值模拟研究[14]表明,对于有限尺寸的正方形模型,模型对角线与硐室周边相交处附近的剪应力对于 V 形坑的产生有重要作用。文献[12,14]和本文的模型都不是轴对称模型,模型中大部分区域会存在剪应力,除非在模型的对称线上。在本文中,由于模型的底部被施加了法向约束,所以,模型的对称线只有通过模型中心的垂直直线 1 条。在此直线上,剪应力为零。在通过模型中心的水平直线上,剪应力也接近于零。这样,在硐室周边,将有 4 个区域剪应力不为零。这些区域将首先出现开裂,进而发展成 V 形坑。

(2) 关于卸荷快慢的影响。卸荷快会在硐室围岩中激发应力波[15],从而会对硐室围岩产生较大的扰动,进而会对硐室围岩造成较大的破坏,这与目前的结果是一致的。然而,卸荷快会使破坏减轻的观点也存在。为了减少对地表结构的破坏程度,美国长壁工作面推进速度非常快,但是,德国长壁工作面却推进缓慢[16]。同样是为了保护地表结构,美、德两国采取完全相反的措施,这是由于美、德两国的煤矿开采条件存在显著的差异。美国的长壁工作面大部分都是单一煤层,而且煤层上覆岩层在采煤之前未受过扰动。德国多是多煤层,在采下部煤层时,由于先前对上部煤层的开采,上覆岩层已多次受过采煤的扰动。

4 结论

在自主开发的基于拉格朗日元方法、变形体离散元方法及虚拟裂缝模型耦合的连续-非连续方法的基础上,引入了 II 型断裂能,开展了静水压力条件下开挖卸荷快慢对硐室围岩模型变形-开裂过程影响的数值模拟研究,得到了下列结论:

(1) 随着时步数目的增加,模型两条对角线与硐室表面交汇的 4 个位置首先发生破坏;然后,裂缝不断增多、长大,向围岩深处发展,一些单元脱离围岩,发生冒落。剪裂缝主要位于 V 形坑边缘,而拉裂缝位于 V 形坑内部,较分散。

(2) 随着开挖卸荷时间的增加,硐室围岩中 V 形坑的形态变得对称或未得到充分发展,拉、剪裂缝区段数目有降低的趋势。不同开挖卸荷时间时拉、剪裂缝区段数目随着时步数目的演变规律具有类似性。

参考文献

[1] Potyondy D O, Cundall P A. A bonded-particle model for rock[J]. International Journal of Rock Mechanics and Mining Sciences, 2004, 41(8): 1329-1364.

[2] 魏巍,覃燕林,曹鹏,等. 粒状材料颗粒破碎过程分析[J]. 南水北调与水利科技, 2014, 12

(6):98-102.

WEI Wei, TAN Yanlin, CAO Peng, et al. Analysis on particle breakage processes of granular materials[J]. South-to-North Water Transfers and Water Science & Technology, 2014, 12(6):98-102.

[3] Lisjak A, Grasselli G. A review of discrete modeling techniques for fracturing processes in discontinuous rock masses[J]. Journal of Rock Mechanics and Geotechnical Engineering, 2014, 6(4):301-314.

[4] Mahabadi O K, Kaifosh P, Marschall P, et al. Three-dimensional FDEM numerical simulation of failure processes observed in Opalinus Clay laboratory samples[J]. Journal of Rock Mechanics and Geotechnical Engineering, 2014, 6(6):591-606.

[5] 严成增,孙冠华,郑宏,等.基于局部单元劈裂的FEM/DEM自适应分析方法[J].岩土力学,2014,35(7):2064-2070.

YAN Chengzeng, SUN Guanhua, ZHENG Hong, et al. Adaptive FEM/DEM analysis method based on local splitting elements[J]. Rock and Soil Mechanics, 2014, 35(7):2064-2070.

[6] 王学滨.拉格朗日元方法、变形体离散元方法及虚拟裂纹模型耦合的连续-非连续介质力学分析方法[R].北京:中国矿业大学(北京),2015.

WANG Xuebin. A method for continuum-discontinuum medium based on the coupled Lagrangian element method, deformational discrete element method and fictitious crack model[R]. Beijing:China University of Mining and Technology(Beijing), 2015.

[7] 郭翔,王学滨,白雪元,等.加载方式及抗拉强度对巴西圆盘试验影响的连续-非连续方法数值模拟[J].岩土力学,2017,38(1):214-220.

GUO Xiang, WANG Xuebin, BAI Xueyuan, et al. Numerical simulation of effects of loading types and tensile strengths on Brazilian disk test by use of a continuum-discontinuum method[J]. Rock and Soil Mechanics, 2017, 38(1):214-220.

[8] 王学滨,马冰,潘一山,等.巷道围岩卸荷应力波传播及垮塌过程模拟[J].中国矿业大学学报,2017,46(6):1259-1266.

WANG Xuebin, MA Bing, PAN Yishan, et al. Numerical simulation of the stress wave propagation and collapsing process of the tunnel surrounding rock[J]. Journal of China University of Mining & Technology, 2017, 46(6):1259-1266.

[9] 王学滨,陈忠元,郭瑞,等.预设V形缺口的单向拉伸岩样变形-开裂过程模拟——基于连续-非连续方法[J].防灾减灾工程学报,2018,38(2):209-215.

WANG Xuebin, CHEN Zhongyuan, GUO Rui, et al. Modeling of cracking processes of rock specimens with V-shaped notches in uniaxial tension——Based on a continuum-discontinuum method[J]. Journal of Disaster Prevention and Mitigation Engineering, 2018, 38(2):209-215.

[10] 王学滨,白雪元,祝铭泽.基于连续—非连续方法的地质体材料变形—拉裂过程模拟——以岩样紧凑拉伸试验为例[J].地质力学学报,2018,24(3):332-340.

WANG Xuebin, BAI Xueyuan, ZHU Mingze. Modeling of deformation-cracking processes of geomaterials in compact tension tests based on a continuum-discontinuum method[J]. Journal of geomechanics,2018,24(3):332-340.

[11] 王学滨,郭翔,芦伟男,等.单层采动诱发长壁开采水平岩层开裂、冒落过程模拟——基于连续-非连续方法[J].防灾减灾工程学报,2018,38(1):1-6.
WANG Xuebin, GUO Xiang, LU Weinan, et al. Modeling of cracking process of specimens in uniaxial tension and with prescribed positions permitted to crack using a continuum-discontinuum method[J]. Journal of Disaster Prevention and Mitigation Engineering,2018,38(1):1-6.

[12] 陆家佑,王昌明.根据岩爆反分析岩体应力研究[J].长江科学院院报,1994(3):27-30.
LU Jiayou, WANG Changming. Study on back analysis for stress of rock mass from information of rockbursts[J]. Journal of Yangtze River Scientific Research Institute,1994(03):27-30.

[13] Vardoulakis I, Sulem J, Goenot A. Borehole instabilities as Bifurcation Phenomena [J]. Int. J. Rock Mech. Min. Sci. & Geomech. Abstr. ,1988,25,159-170.

[14] 王学滨,潘一山,陶帅.不同尺寸的圆形隧洞剪切应变局部化过程模拟[J].中国地质灾害与防治学报,2009,20(4):101-108.
WANG Xuebin, PAN Yishan, TAO Shuai. Numerical simulation of shear strain localization processes along circular tunnels with different diameters[J]. The Chinese Journal of Geological Hazard and Control,2009,20(4):101-108.

[15] 董春亮,赵光明,李英明,等.深部圆形巷道开挖卸荷的围岩力学特征及破坏机理[J].采矿与安全工程学报,2017,34(3):511-526(518).
DONG Chunliang, ZHAO Guangming, LI Yingming, et al. Mechanical properties and failure mechanism of surrounding rocks in deep circular tunnel under excavation unloading[J]. Journal of Mining & Safety Engineering,2017,34(3):511-526(518).

[16] 彭赐灯.长壁开采[M].第二版.北京:科学出版社,2011.

极软弱地层双层锚固平衡拱结构形成机制研究

孟庆彬,韩立军,梅凤清,冯伟,周星

(中国矿业大学 深部岩土力学与地下工程国家重点实验室,江苏 徐州,221116)

摘　要:针对极软弱地层中巷道围岩变形大、支护困难等特点,基于锚杆、锚索支护理论与技术,提出了在极软弱地层中采用双层锚固平衡拱结构解决巷道围岩稳定控制问题。在分析极软弱地层双层锚固平衡拱几何特征的基础上,采用弹塑性理论研究了双层锚固拱结构的形成条件及结构承载力的影响因素,推导了锚固平衡拱结构参数的计算公式;并基于理论分析和 FLAC^{3D} 数值计算,分析了不同断面形状(断面矢高)、支护参数及预应力对锚固效应的影响规律,提出了形成合理的双层锚固平衡拱结构的控制参数。工程实践表明,采用"双层锚固平衡拱结构"可有效地控制极软弱地层中巷道围岩的大变形与破坏问题,维持了巷道围岩与支护结构的稳定及使用安全。

关键词:极软弱地层;双层锚固平衡拱结构;锚固效应;工程应用

Study on formation mechanism of anchorage structure of double balance arch in the very soft strata

MENG Qingbin, HAN Lijun, MEI Fengqing, FENG Wei, ZHOU Xing

(*State Key Laboratory for Geomechanics and Deep Underground Engineering, China University of Mining and Technology, Xuzhou 221116, China*)

Abstract: According to the characteristics of difficult supporting and large deformation in very soft layer, On the basis of combine support theory and technique, the double anchoring balanced arch supporting structure was presented. Based on the geometric features of double anchoring balanced arch supporting structure, the forming condition and bearing capacities were explored by using elastic-plastic theory, the calculating formula of supporting structure was obtained. By theoretical analysis and FLAC^{3D} numerical calculation to establish anchoring structural mechanics model, calculation formulas of anchor thickness of double anchoring balanced arch structure were derived. Practice shows, large deformation of soft rock in deep roadways could be controlled effectively by using the double anchoring balanced arch supporting structure, stability and service safety of surrounding rock and support structure could be kept effectively.

Keywords: very soft strata; anchor structure of double balance arch; anchor structure; engineering application

---

基金项目:国家自然科学基金资助项目(51704280,51574223);中国博士后科学基金资助项目(2015M580493,2017T100420);山东省土木工程防灾减灾重点实验室开放课题资助项目(CDPM2014KF03)。

通信作者:孟庆彬(1985—),男,山东菏泽人,博士/博士后,助理研究员,主要从事岩体加固理论与应用技术的研究。E-mail:mqb1985@126.com。

随着我国中东部地区煤炭资源逐步枯竭,矿井建设逐步向煤炭资源丰富的西部地区发展,在包括内蒙古、新疆在内的我国西部地区广泛分布着一类特殊岩层——弱胶结软岩地层[1-4],极软弱地层巷道围岩属于结构性较差的非均质散状赋存类型,具体表现为围岩结构性差(成岩时间晚,基本为半成岩或未成岩状态)、胶结性差(泥质胶结)、遇水泥化崩解(遇水短时间即呈泥化状)、强度低(单轴抗压强度一般小于 1 MPa)、围岩体自稳能力差(难以形成自承载结构)、开巷后围岩变形量大及围岩控制困难(锚杆、锚索等主动支护结构可锚性较差,易脱锚失效)。此外,由于巷道顶底板及两帮围岩变形相互影响,如果巷道围岩某位置变形过大,则势必会引起其他位置围岩的变形甚至破坏,形成恶性循环。因此,极软弱地层巷道支护是西部地区煤炭安全开采的技术难题,也是目前国内外学者和工程技术人员关注的重要工程问题。

虽然我国许多专家学者针对软弱围岩巷道断面形状优化[5-7]与支护技术[8-14]进行了一定的理论探索及工程实践,取得了一些研究成果,解决了许多工程难题,但目前的研究很少涉及极软弱地层巷道围岩稳定控制理论与支护技术。因此,本文针对极软弱地层巷道中难以形成有效的锚固结构、围岩变形大、维护困难等特点,以蒙东五间房矿区西一矿极软弱地层回采巷道为工程背景,研究极软弱地层回采巷道中应用锚网索联合支护形成的锚固结构的几何特征、结构效应及形成机制,探讨极软弱地层回采巷道围岩稳定控制技术,力图为解决极软弱地层煤巷支护难题提供合理的技术途径。

1 双层锚固平衡拱结构的几何特征

基于锚杆、锚索的锚固作用机理[15-17],提出了极软弱地层巷道采用锚杆与锚索构成的双层锚固平衡拱结构(图 1)[18,19]。其主要内容如下:

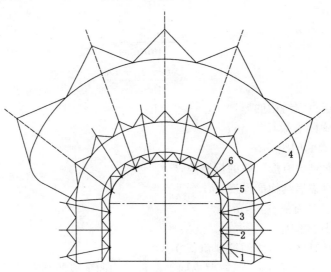

图 1 双层锚固平衡拱结构示意图

Fig.1 Structural figure of double balance arch

1——高强预应力锚杆;2——钢筋网;3——锚杆托盘;
4——低松弛预应力锚索;5——锚索托盘;6——钢筋托梁

(1) 巷道开挖后,及时对巷道周边松动破碎围岩体进行锚网喷初次支护,可对破碎围岩产生径向约束作用,形成内层压缩拱结构;锚杆间的相互挤压作用保证了内层锚固拱的形成与稳定,由锚杆形成的压缩拱称为内层锚固平衡拱。

(2) 采用锚索进行加强支护,若锚索的间排距合理,则由锚索压缩锥的叠加作用亦可形成外层压缩拱结构,可对内层锚固平衡拱结构起到了强化作用,并对内层锚固结构施加了径向约束力,提高了内层锚固结构的承载力。

(3) 锚索作为加强支护可锚固到更深部岩层中(稳定或破碎岩层),锚索作用形成的外层锚固结构与锚杆形成的内层锚固结构相互叠加形成双层锚固平衡拱结构,并可增加预应力锚固层的厚度。锚索一般较长,可在较大范围内对巷道松动围岩产生挤压作用,既可分担部分巷道围岩压力,又可提高围岩的稳定性与自承载力。

(4) 钢筋网、钢筋梯、喷层等构件可防止锚杆锚固区域内的破碎岩块掉落,随着锚固区岩层扩容、离层的增大,钢带和钢筋托梁的受力逐渐增加,通过构件对锚杆间的围岩施以径向约束,可阻止围岩产生进一步的扩容和离层,增加了内层锚固结构的整体性和承载力。

2 双层锚固平衡拱的结构效应分析

2.1 双层锚固平衡拱结构形成的几何条件

"组合拱"理论是解释锚杆控制巷道围岩稳定的主要理论依据,但其未考虑到锚索对组合拱厚度的影响;一般而言,锚索可施加较大的预紧力,可对岩层中的层理、节理等裂隙面形成挤压效应,增加了不连续面间的摩擦力,扩大了锚杆锚固形成的内层锚固结构厚度,提高了锚固组合拱的整体强度。本文约定,由锚杆作用形成的锚固结构称为内层锚固结构,其厚度记为 b_2;由锚索作用形成的锚固结构为外层锚固结构,其厚度记为 b_1,由组合拱理论[15-20]可得:

$$\begin{cases} b_1 = \dfrac{L_a \tan \alpha_a - D_a}{\tan \alpha_a} \\ b_2 = \dfrac{L_b \tan \alpha_b - D_b}{\tan \alpha_b} \end{cases} \quad (1)$$

式中　b_1——外层锚固结构的厚度,m;

　　　L_a——锚索的有效长度,m;

　　　D_a——锚索的间距,m;

　　　α_a——锚索对破裂围岩体的控制角,(°);

　　　b_2——内层锚固结构的厚度,m;

　　　L_b——锚杆的有效长度,m;

　　　D_b——锚杆的间距,m;

　　　α_b——锚杆对破裂岩体的控制角,(°)。

对锚杆和锚索支护系统而言,双层锚固平衡拱结构的形成的几何条件为:(1) 锚杆在径向约束作用下,当间排距合理时,由径向应力约束作用形成的压缩锥可形成连续的压缩层,即内层锚固结构;(2) 由锚索径向约束作用形成的外层锚固结构的厚度要大于内层锚固结构的厚度,即:

$$\dfrac{L_a \tan \alpha_a - D_a}{\tan \alpha_a} > \dfrac{L_b \tan \alpha_b - D_b}{\tan \alpha_b} > 0 \quad (2)$$

双层锚固平衡拱结构形成的几何条件为：

$$\begin{cases} D_b < L_b \tan \alpha_b \\ D_a < (L_a - L_b) \tan \alpha_a + \dfrac{D_b \tan \alpha_a}{\tan \alpha_b} \end{cases} \quad (3)$$

2.2 巷道锚固结构承载力的影响因素分析

在锚杆约束围岩形成内层锚固平衡拱结构后，锚索可继续压缩内层锚固拱及其外部的松动围岩，假定锚索的约束力均匀作用于锚固拱的内表面，且锚固拱的外表面受围岩压力的作用；由于锚索一般能够伸入较深的岩层中，在内层锚固平衡拱外侧形成外层锚固拱结构，即最终所形成的双层锚固平衡拱结构的支护力与厚度大于锚杆"组合拱"的厚度。双层锚固平衡拱的实质是锚杆约束围岩形成的内层锚固拱与锚索约束围岩形成的外层锚固拱叠加而成的双层锚固拱结构。双层锚固平衡拱的承载力主要与锚杆约束阻力、破裂岩石性质、拱的厚度和拱的形状等有关，即：

$$P_c = f(p_s, \varphi_b, b_0, F_c, S_c) \quad (4)$$

式中 P_c——双层锚固平衡拱的承载力，N/m；

p_s——锚杆、锚索约束阻力的合力，N/m；

φ_b——岩石的内摩擦角，(°)；

b_0——锚固拱的厚度，m；

F_c——锚固拱的截面；

S_c——锚固拱的形状。

锚杆（索）约束阻力一般采用锚杆（索）最大拉拔力表示，对于端锚结构，锚杆（索）的约束阻力与最大拉拔力的关系[18,20]：

$$P_s = \dfrac{Q_b}{D_b^2} + \dfrac{Q_a}{D_a^2} \quad (5)$$

式中 P_s——锚杆（索）约束阻力，kN；

Q_a——锚索最大拉拔力，kN；

Q_b——锚杆最大拉拔力，kN。

参考有关组合拱厚度的相似模型试验结果，可初步得到组合拱的厚度与锚杆（索）的长度、间排距关系式为[18,21]：

$$b_0 = (L_a - D_a)k_1 + (L_b - D_b)k_2 \quad (6)$$

式中 k_1——锚索群锚时对锚固拱厚度的影响系数，$k_1 = L_a D_b / (L_a D_b + L_b D_a)$；

k_2——锚杆群锚时对锚固拱厚度的影响系数，$k_2 = L_b D_a / (L_a D_b + L_b D_a)$。

根据 Mohr-Coulomb 强度准则，可求得沿巷道轴向单位长度上锚固拱承载力的表达式为：

$$N = b_0 K_r P_s + \dfrac{1}{2} b_0^2 k \quad (7)$$

式中 k——径向应力增加斜率；

K_r——称为支护力放大系数。

联立式(5)、(6)、(7)，可获得双层锚固平衡拱结构承载力与锚杆（索）支护参数和围岩参数的关系式为：

$$N = \frac{k}{2}\left[(L_a - D_a)k_1 + (L_b - D_b)k_2\right]^2 +$$
$$K_r\left[(L_a - D_a)k_1 + (L_b - D_b)k_2\right]\left(\frac{Q_b}{D_b^2} + \frac{Q_a}{D_a^2}\right) \tag{8}$$

2.3 双层锚固平衡拱结构的厚度计算

根据结构力学原理,考虑到结构的对称性(图2),由平衡方程可得:

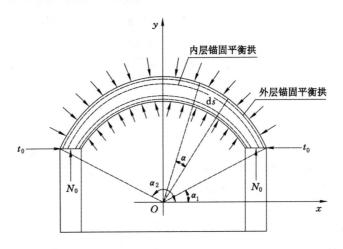

图 2 双层锚固平衡拱结构力学模型

Fig. 2 Structure mechanical model figure of double balance arch

$$2N_0 - \int_{\alpha_1}^{\alpha_2} q\mathrm{d}s\sin\alpha = 0 \tag{9}$$

可求得:

$$N_0 = \frac{q(R + b_0)}{2}(\cos\alpha_1 - \cos\alpha_2) \tag{10}$$

式中 N_0——环向轴力,kN;

R——巷道半径,m;

q——均布外载密度,N/m;

α_1、α_2——矢跨比控制角,(°);

$\mathrm{d}s$——单元体的宽度范围,m。

由于在极软弱岩层中,围岩中径向应力的增加斜率 k 很小,设 $k=0$,则有:

$$N = K_r\left[(L_a - D_a)k_1 + (L_b - D_b)k_2\right]\left(\frac{Q_b}{D_b^2} + \frac{Q_a}{D_a^2}\right) = mN_0 \tag{11}$$

式中 m——应力改变系数。

联立式(6)、(10)、(11),可得到双层锚固平衡拱的厚度为:

$$b_{0m} = \frac{mRqD_a^2D_b^2(\cos\alpha_1 - \cos\alpha_2)}{2K_r(Q_bD_a^2 + Q_aD_b^2) - qmD_a^2D_b^2(\cos\alpha_1 - \cos\alpha_2)} \tag{12}$$

由锚固结构承载力影响因素与公式(12)可知,双层锚固平衡拱的厚度与围岩荷载、巷道矢跨比、锚杆(索)间距、锚杆(索)最大拉拔力、围岩体的内摩擦角等因素有关。

3 双层锚固平衡拱结构的形成机制研究

3.1 巷道断面矢高对锚固结构效应的影响分析

煤矿巷道常用的断面形状为矩形断面、切圆拱断面、直墙拱形断面等,其中矩形断面的断面利用率最高,但其稳定性较差;直墙拱形断面稳定性较好,巷道围岩受力也较为合理,但断面利用率较低,切圆拱断面的受力状态和断面利用率介于二者之间[18,19]。采用 FLAC^{3D} 模拟分析巷道矢高变化对锚杆、锚索预应力扩散范围及围岩应力状态的影响,以探究断面矢跨比对双层锚固平衡拱结构形成的影响规律和确定巷道较为合理的矢高(拱高)。

(1) 数值模型的建立

建立矢跨比为 1/2(直墙拱形断面)、1/4、1/6、1/8、1/10、0(矩形断面)等 6 组断面形状的数值模型,模型的尺寸为 60 m(长度)×60 m(宽度)×30 m(厚度),巷道尺寸为 5.2 m(宽度)×3.6 m(高度)。锚网索支护参数同下文 4.1,锚杆(索)采用 cable 单元,围岩体物理力学参数见表 1[18]。本数值计算约束周边位移,为揭示锚杆(索)施加预应力后在围岩中产生的支护应力场,因其与原岩应力场在数值上相差 2~3 数量级,故在数值模拟时不考虑原岩应力场[22-24]。

表 1 围岩物理与力学参数
Table 1 Surrounding rock parameters

名称	煤层	顶板岩层	底板岩层
弹性模量/MPa	480.68	458.50	301.70
泊松比	0.18	0.29	0.28
天然抗压强度/MPa	8.00	5.89	5.64
抗拉强度/MPa	0.55	0.53	0.46
体积模量/MPa	250.35	363.89	228.56
剪切模量/MPa	203.68	177.70	117.85
内摩擦角/(°)	19.42	36.50	34.62
黏聚力/MPa	2.56	1.79	0.70
密度/(kg/m³)	1 460	2 114	2 068

(2) 锚杆和锚索应力扩散效应分析

由图 3 可知,当巷道的矢跨比为 1/8~1/2 时,在巷道顶部范围内,锚杆对围岩产生了明显的压缩作用,压缩区域的外轮廓线呈现拱形状;对于锚索而言,在锚索的自由段(靠近巷道一侧的端头部位)产生了较为明显的高压应力区(采用张拉设备对锚索施加的预应力,通过安装在锚索自由端端头的锁具和托盘对巷道围岩产生挤压作用),随着锚索锚固范围的增加(离巷道表面越远),锚索对围岩的约束作用减弱;由于锚杆、锚索应力扩散的叠加作用,锚杆作用范围内围岩应力较大,形成了内层锚固平衡拱结构;在锚杆、锚索锚固段间,由于锚索的应力扩散较为明显,因此在两者锚固段间形成了外层锚固平衡拱结构;由于内、外层锚固平衡拱结构间并没有比较明显的界限,所以锚杆、锚索应力场相互叠加形成双层锚固平衡拱结构。

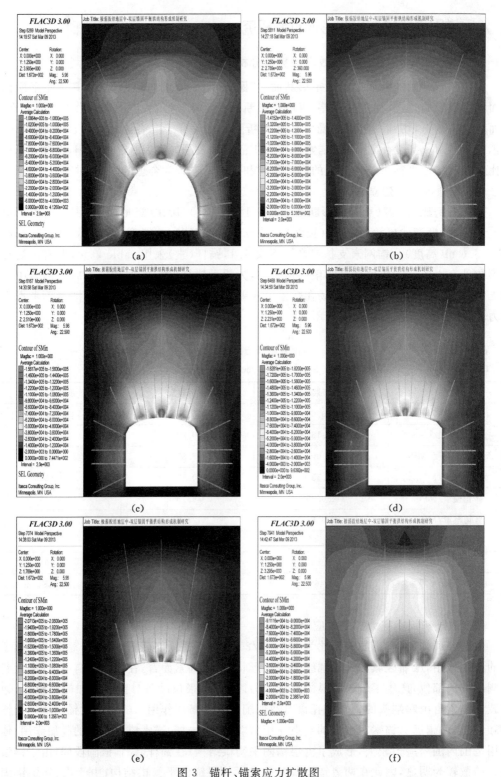

图 3 锚杆、锚索应力扩散图

Fig. 3 Stress diffusion of bolt and cable

(a) 矢跨比 1/2;(b) 矢跨比 1/4;(c) 矢跨比 1/6;(d) 矢跨比 1/8;(e) 矢跨比 1/10;(f) 矢跨比 0

当巷道矢跨比由 1/2 至 1/8 逐渐减小的过程中,锚索自由段靠近巷道侧部位围岩应力明显减小;这是由于当巷道矢跨比逐渐减小时,巷道拱顶逐渐由弧线向直线过渡,造成锚索预应力损失,使得外层锚索对围岩径向约束作用逐渐减弱,表现为外层锚固拱结构效应减弱。

3.2 支护结构对锚固结构效应的影响分析

（1）锚杆的作用效应

当采用单根锚杆支护后,也会在锚杆两端形成一定的压缩锥,但所形成压缩锥的范围较小。当采用群锚结构时,若锚杆的间排距和预应力合理,则各锚杆形成的压缩锥相互叠加,最终可在锚杆锚固范围内形成连续的压缩带,压缩区范围内围岩强度有所增强。当巷道的断面形状为拱形时,均匀压缩带形成了压缩拱结构,最终可在锚杆锚固范围内形成压缩状的锚固拱结构。在巷道拱顶中心位置布设测线以监测反映巷道围岩中的应力分布状态,锚杆所形成的锚固拱效应如图 4 所示。

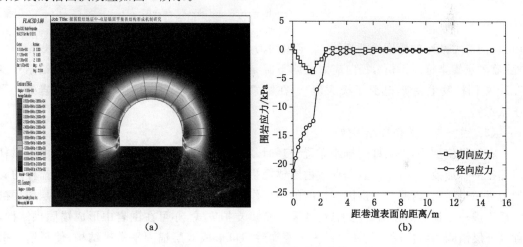

图 4 锚杆作用效应及应力场曲线

Fig. 4 Effect of anchor rod and stress curve

由图 4 可知,由于锚杆间相互挤压及对围岩径向压密作用,使得在巷道围岩中形成的径向应力远大于切向应力。径向应力为负值（围岩处于受压状态）,随离围岩表面距离的增大,径向应力逐渐减小;在锚杆锚固范围内,径向应力变化较为明显,当监测点超出锚杆的锚固范围时,径向应力变化不大;切向应力在锚杆锚固范围内处于受压状态,超出锚杆的锚固范围则变为受拉状态。在锚杆 2/3 长度处（监测点横坐标 1.5 m）,径向和切向应力发生突变,切向应力达到了最大值,径向应力在锚杆长度 2/3 范围外急剧下降。因此,当采用单一的锚杆支护作用时,巷道围岩应力场的主要扩散范围在锚杆长度的 2/3 之内。

（2）锚索的作用效应

锚索可在巷道顶部较大范围内形成应力扩散区,与预应力锚杆相比,其亦具有压缩围岩形成锚固体的特性;此外,锚索还可将其锚杆支护结构锚固在上部稳定的围岩之中,起到加强支护的作用;预应力锚索可调动上部围岩的承载力,激发支护结构与围岩共同作用的潜能。锚索所形成的锚固拱效应如图 5 所示。

随离围岩表面距离的增大,径向应力的数值呈减小趋势,在顶板深度 0～4.5 m 范围

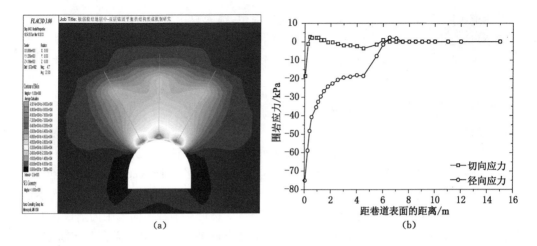

图 5 锚索作用效应及应力场曲线

Fig.5 Effect of anchor rope and stress curve

内,径向应力减小的幅度逐渐变缓,稳定值约为 −18.1 kPa,在监测范围超出锚索长度 7/10 处(监测点横坐标 4.5 m),径向应力急剧下降;切向应力在经过初期的波动后,在锚索长度 7/10 处同样发生突变(达到了极大值)。在锚索锚固范围内,径向应力为负值(围岩处于受压状态)。

(3) 锚杆与锚索联合作用效应

在锚杆施加后,当锚杆间排距布置合理时,在锚固范围内可形成挤压加固圈,由于锚杆压缩锥的叠加效应,挤压加固圈形成内层锚固拱结构;待围岩产生一定的变形后,锚杆的支护力进一步增强,内层锚固拱因锚杆间的挤压作用使得锚固层厚度进一步增强,内层锚固拱承载力进一步提升。在锚索施加后,若采用密集支护方式,亦可在围岩中形成锚固拱结构,称为外层锚固拱结构,故内、外层锚固拱叠加后即可形成双层锚固平衡拱结构,使得锚固拱厚度和承载力都得到了增强。此外,由于锚索支护力的影响范围广,可将双层锚固平衡拱结构组合成一个整体,同时亦可充分调动深部围岩的承载力,起到将围岩应力向围岩深部传递的作用,体现了支护结构与围岩共同承载的原理。锚杆和锚索共同作用效应如图 6 所示。

通过对锚杆、锚索单独作用及共同作用下巷道围岩应力场的比较分析,锚杆和锚索共同作用的应力场是两者单独作用的应力场的几何叠加,即巷道围岩应力场的损失较小。在锚杆锚固长度范围内(0~2.4 m),锚杆单独作用时切向应力为负值(围岩处于受压状态),且数值相对较大。锚索单独作用时,切向应力基本为正值(围岩处于受拉状态),且数值相对较小。当锚杆和锚索联合作用时,在锚杆锚固范围内,切向应力场是锚杆起主要作用产生的;在锚杆锚固范围之外、锚索锚固范围之内(2.4~6.3 m),锚杆作用产生的切向应力为正值(围岩处于受拉状态),且数值相对较小;锚索作用形成的切向应力为负值(围岩处于受压状态),且数值相对较大。

因此,当锚杆和锚索联合作用时,锚杆径向与切向应力产生了 2 次突变,存在 2 个拐点。第 1 次突变位置发生在 2/3 锚杆长度处,主要是由锚杆的锚固作用引起的,第 2 次突变发生在 7/10 锚索长度处,主要是由于锚索的锚固作用引起的。

(4) 锚索布置方式对锚固结构效应的影响分析

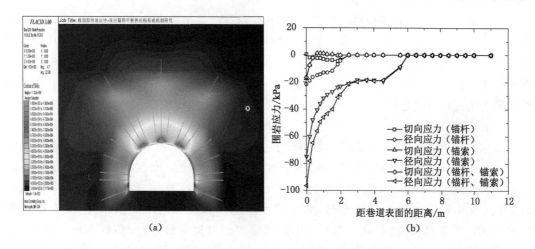

图 6 锚杆与锚索共同作用效应及应力场曲线

Fig. 6 Combined effect of anchor rod and anchor rope and stress curve

在锚杆参数和锚索排距不变的情况下,锚索布置方式(布置 3～6 根锚索)对巷道围岩体的锚固效应的影响如图 7 所示。

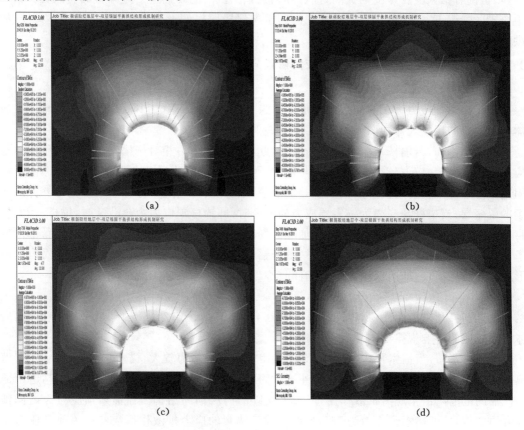

图 7 锚索布置方式对锚固效应的影响

Fig. 7 Anchorage effect of cable arrangement mode

(a) 3 根锚索;(b) 4 根锚索;(c) 5 根锚索;(d) 6 根锚索

由图 7 可知,当在巷道顶板布置 3 根锚索时,尽管锚索由于预应力作用产生的压力锥相互叠加形成了连续的压缩层,但由于锚索支护密度小,未在巷道顶部较大范围内形成双层锚固结构;当在巷道顶板布置 5 或 6 根锚索时,由于支护密度增大,锚索与锚杆锚固作用相互叠加形成了较为明显的双层锚固结构,锚固结构的厚度及范围明显增大。因此,在巷道顶板增加锚索的支护密度对锚固效应的影响较明显,增加锚索密度有助于双层锚固平衡拱结构的形成。

锚索布置方式的变化对径向应力的影响主要表现在整个锚索锚固范围内,而对切向应力的影响则表现在锚杆锚固段与锚索锚固段间。在径向应力监测过程中出现了反常情况,即布置 3 根锚索时在巷道顶部产生的径向应力大于布置 4 根锚索时巷道顶板的径向应力。这是因为布置 3 根锚索时,为了满足巷道顶板锚索间距要求,锚索位置相对集中地布置在巷道顶板处,造成顶部围岩中径向应力相对集中分布;而布置 4 根锚索时,锚索可均匀的布置在巷道顶板处,应力扩散较为均匀,故出现了顶板围岩中径向应力不增反降的情况。当布置 5 或 6 根锚索时,随着锚索支护密度的增大,径向应力的数值有所提高。

对于切向应力而言,在锚杆锚固范围内,锚索支护密度的变化对围岩切向应力的影响不明显;但随着锚索支护密度的提高,锚杆和锚索锚固段间的切向应力逐渐提高。故在锚索间距合理的情况下,采用密集型锚索支护有利于外层锚固拱的形成及对锚杆形成的内层锚固拱的叠加效应的增强。

3.3 锚杆(索)预应力对锚固结构效应的影响分析

(1) 锚杆预应力的影响分析

设计锚索预应力为 120 kN,锚杆预应力为 20 kN、40 kN、60 kN、80 kN 时对巷道围岩体锚固效应的影响规律如图 8 所示。

由图 8 可知,当锚杆预应力为 20 kN 时,此时锚杆预应力相对于锚索较小。由于锚索预应力较大、锚杆预应力较小,造成锚杆与锚索所形成的应力场不相协调,在巷道顶部锚杆锚固范围内未形成明显的内外层锚固拱叠加效应。且此时在巷道帮部未形成范围足够大的连续的压缩层;当锚杆预应力为锚索预应力的 1/2 时,即锚杆预应力为 60 kN 时,锚杆、锚索各自应力扩散作用在顶部形成明显的叠加应力场,叠加区范围和厚度较大,并在巷道帮部形成了连续的压缩层;当锚杆预应力较大时,由于锚杆应力扩散明显,巷道周边锚杆锚固范围内形成了连续压缩带,即内层锚固拱,并有利于双层锚固平衡拱结构的形成。

(2) 锚索预应力的影响分析

设计锚杆预应力为 40 kN,锚索预应力为 40 kN、80 kN、120 kN、160 kN 时对巷道围岩体锚固效应的影响规律如图 9 所示。

由图 9 可知,当锚索预应力为 40 kN 时,由于锚杆预应力较大、锚索预应力较小,在巷道顶部锚索锚固范围内仍未形成明显的外层锚固拱和有效的双层锚固叠加区,而在巷道帮部锚杆锚固范围内由于应力扩散作用,形成了连续的内层锚固结构。当锚索预应力为锚杆预应力的 2 倍时,即锚索预应力为 80 kN 时,锚杆、锚索的应力扩散作用在巷道顶部形成明显的叠加应力场,叠加区的范围和厚度较大。在巷道帮部,由于应力的叠加作用,连续的压缩层较为明显,形成了较为稳定的帮部结构。当锚索预应力较大时,由于锚索应力扩散明显,巷道周边锚索锚固范围内形成了连续的压缩带,即外层锚固拱,而锚杆由于未形成内层锚固

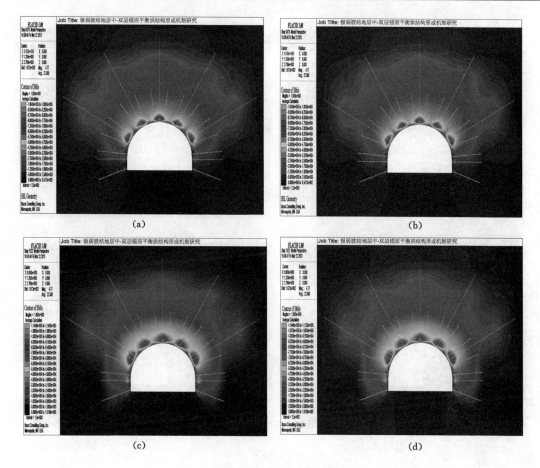

图 8 锚杆预应力对锚固效应的影响

Fig. 8 Anchorage effect of bolt prestress

(a) 20 kN;(b) 40 kN;(c) 60 kN;(d) 80 kN

结构,故仍未形成有效的锚固叠加区。

(3) 锚杆与锚索预应力耦合效应分析

当锚杆预应力过大时[图 10(a)],在巷道周边锚杆锚固范围内形成了均匀的压缩层,即内层锚固结构和稳定的帮部结构,但因锚索预应力较小,在巷道顶板未能形成较大范围的外层锚固拱结构,故很难与锚杆预应力场叠加形成双层锚固平衡拱结构。当锚杆和锚索预应力协调时[图 10(b)],锚杆、锚索锚固作用形成的内外层锚固结构易在巷道顶板处叠加,形成双层锚固平衡拱结构,且帮部同样可形成均匀连续的压缩层。当锚索预应力过大时[图 10(c)],锚索在巷道顶部形成了范围足够大的外层锚固结构,但由于锚杆未能够发挥其预应力支护的作用,故造成锚杆与锚索所产生的应力场未能够相互叠加,而未形成双层锚固结构。

因此,在采用密集型锚索支护方式下,当锚索预应力为锚杆预应力的 2～3 倍时,锚杆、锚索形成的锚固拱结构可在巷道顶板叠加形成双层锚固平衡拱结构,亦形成了稳定的帮部结构。即表明,锚杆和锚索的预应力仅在耦合的情况下才能发挥支护结构的协同作用。

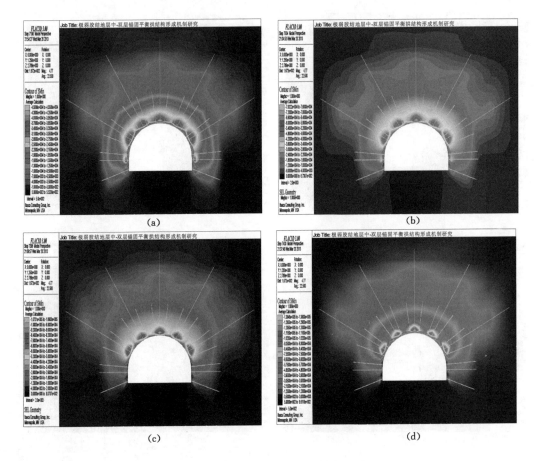

图 9 锚索预应力对锚固效应的影响
Fig. 9 Anchorage effect of cable prestress
(a) 40 kN;(b) 80 kN;(c) 120 kN;(d) 160 kN

4 工程实践

4.1 煤巷锚网索耦合支护技术

结合国内外煤巷锚网支护理论与技术,基于双层锚固平衡拱结构效应分析,提出了极弱胶结地层中西一矿1320工作面回采巷道采用锚网索耦合支护技术方案[19,25],支护结构如图11所示。

锚杆规格为 ϕ20 mm×2 400 mm,间、排距为 700 mm×700 mm,预紧力不低于 50 kN;锚索规格为 ϕ17.8 mm×5 100 mm(6 000 mm),间、排距为 1 600 mm×2 100 mm,按 3-2-3 的方式布置,预紧力不低于 150 kN;钢筋托梁由 ϕ14 mm 的圆钢焊接,全断面使用;金属网采用 ϕ6.5 mm 冷拔铁丝编制的菱形网,网格为 50 mm×50 mm;底拱混凝土厚度为 300 mm,墙角混凝土厚度为 200 mm,强度等级为 C35。

4.2 现场监测与结果分析

(1)围岩收敛变形监测结果分析

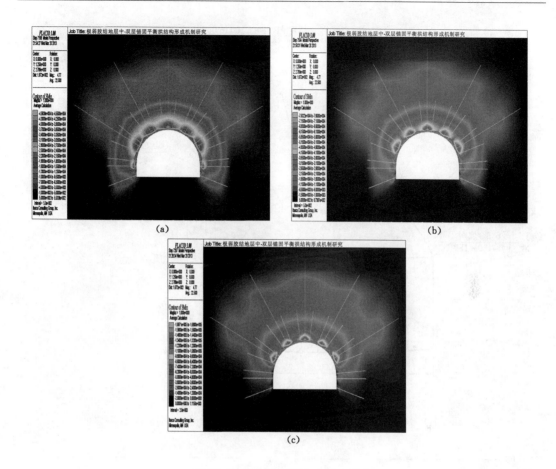

图 10　锚杆与锚索预应力耦合效应

Fig. 10　Coupling effect of pre-stress of bolt and anchor cable

(a) 锚杆预应力过大；(b) 锚杆与锚索预应力耦合；(c) 锚索预应力过大

由图 12(a) 可知，1302 回风巷在埋深 169.1～224.1 m 处，两帮最大收敛变形量为 49.93 mm，顶板最大下沉量为 36.41 mm；在埋深 254.4～281.5 m 处，两帮最大收敛变形量为 46.99 mm，顶板最大下沉量为 36.02 mm。由图 12(b) 分析可知，1302 运输巷在埋深 166.7～248.31 m 处，两帮最大收敛变形量为 48.57 mm，顶板最大下沉量为 32.95 mm；在埋深 251.19～270.79 m 处，两帮最大收敛变形量为 51.43 mm，顶板最大下沉量为 36.19 mm。监测结果表明，巷道两帮收敛变形量大于顶板下沉量，说明所提出的双层锚固平衡拱结构能有效地控制巷道顶板岩层的下沉与离层变形。

(2) 锚杆受力监测结果分析

由图 13 可知，在锚杆安装后 20 d 内，锚杆锚固力出现小幅度波动现象，这与巷道收敛变形急剧增长周期是一致的。在 98 d 后，锚杆受力基本进入稳定状态，和锚索几乎同步，即表明了锚杆、锚索形成的内外层锚固结构共同作用的特点。左帮、左拱肩、拱顶、右拱肩、右帮锚杆受力稳定的数值依次为 54.19 kN、58.08 kN、64.54 kN、52.81 kN、59.80 kN。锚杆受力较小，表明锚索形成的外层锚固结构承担了较多的围岩荷载。

(3) 锚索受力监测结果分析

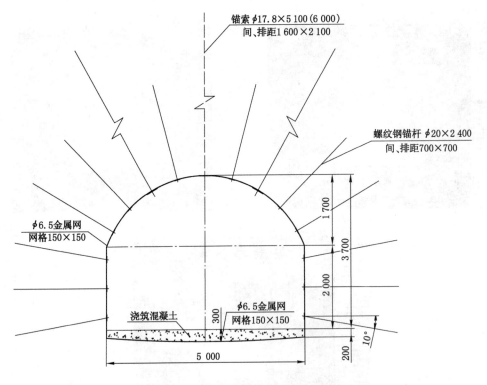

图 11 极软弱地层双层锚固平衡拱支护结构

Fig. 11 Anchorage structure of double balance arch in very soft strata

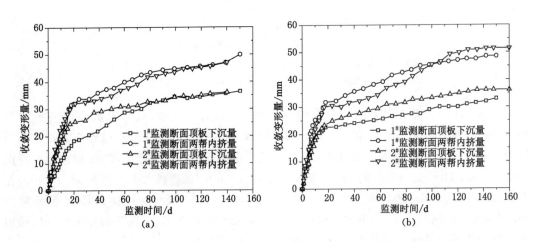

图 12 巷道围岩表面位移监测曲线

Fig. 12 Displacement monitoring curve of the rock surface

(a) 回风巷；(b) 运输巷

由图 14 可知，左拱肩、拱顶、右拱肩锚索受力稳定的数值依次为 150.93 kN、173.74 kN、125.65 kN，尚未达到锚索的破断强度，未出现锚索破断现象，表明锚索可将荷载传递到围岩深处，体现了巷道围岩与支护结构共同承载、相互作用的原理。

锚杆与锚索受力监测结果表明，锚杆、锚索受力均处于正常状态，内层锚固拱结构和外

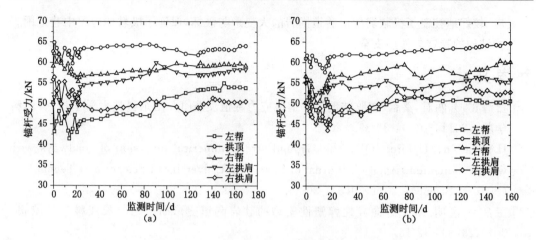

图 13 锚杆受力监测

Fig.13 The load-bearing monitor of bolt

(a) 1# 监测断面;(b) 2# 监测断面

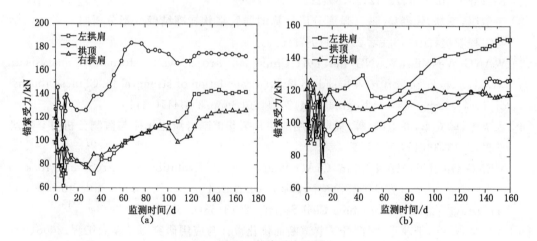

图 14 锚索受力监测

Fig.14 The load-bearing monitor of cable

(a) 1# 监测断面;(b) 2# 监测断面

层锚固拱结构共同作用,提高了支护结构的承载力,有效地控制了煤巷围岩变形。此外,锚索受力相对较大,体现了锚索支护力的传递作用,锚索既可提高内层锚固结构的承载力,又可调动深部围岩的承载力,使内层锚固拱、外层锚固拱、深部围岩三者共同承担支护荷载。

5 结 论

(1) 基于锚杆和锚索锚固机理,分析了极软弱地层巷道顶板双层锚固平衡拱的几何特征,提出了双层锚固拱结构形成条件,推导出了双层锚固平衡拱结构锚固厚度的计算公式。

(2) 通过理论与数值分析,揭示了断面矢跨比、支护结构、预应力等对煤巷围岩体锚固结构效应的影响规律,当采用锚索密集型布置方式、锚索预应力为锚杆预应力的 2~3 倍及矢跨比为 1/8~1/2 的切圆拱形断面有利于双层锚固平衡拱结构的形成。

(3) 工程实践表明,采用"双层锚固平衡拱"支护理念,可充分发挥锚杆与锚索的主动支护效应,较好地控制了极软弱地层巷道围岩的大变形与破坏,保证了极软弱地层中巷道围岩及支护结构的稳定与使用安全。

参 考 文 献

[1] 贾海宾,苏丽君,秦哲.弱胶结地层巷道地应力数值反演[J].山东科技大学学报(自然科学版),2011,30(5):30-35.
JIA Haibin,SU Lijun,QIN Zhe. Ground stress numerical inversion of roadways with weakly cemented strata[J]. Journal of Shandong University of Science and Technology (Natural Science),2011,30(5):30-35.

[2] 王磊,王渭明.基于各向异性弹塑性模型的软岩巷道变形研究[J].长江科学院院报,2012,29(8):12-16.
WANG Lei,WANG Weiming. Analysis of soft rock roadway deformation based on anisotropic elasto-plastic constitutive model[J]. Journal of Yangtze River Scientific Research Institute,2012,29(8):12-16.

[3] 王渭明,王磊,代春泉.基于强度分层计算的弱胶结软岩冻结壁变形分析[J].岩石力学与工程学报,2011,30(增2):4110-4116.
WANG Weiming,WANG Lei,DAI Chunquan. Frozen wall deformation analysis in weakly cemented soft rock based on layered calculation of strength[J]. Chinese Journal of Rock Mechanics and Engineering,2011,30(Supp. 2):4110-4116.

[4] 孟庆彬,韩立军,乔卫国,等.极弱胶结地层开拓巷道围岩演化规律与监测分析[J].煤炭学报,2013,38(4):572-579.
MENG Qingbin,HAN Lijun,QIAO Weiguo,et al. Evolution of surrounding rock in pioneering roadway with very weakly cemented strata through monitoring and analyzing[J]. Journal of China Coal Society,2013,38(4):572-579.

[5] 孙广义,陈刚,于蒲喜.深部开采巷道断面优化设计与应用研究[J].煤炭工程,2008(9):57-60.
SUN Guangyi,CHEN Gang,YU Puxi. The optimized design and application research of roadway cross-section in deep mining[J]. Coal Engineering,2008(9):57-60.

[6] 贾陈海.大断面高地应力煤巷断面形状选择的探讨[J].江西煤炭科技,2008(3):72-74.
JIA Chenhai. Study on selection of cross-sectional shape in roadway with large cross section and high strata stress[J]. Jiangxi Coal Science & Technology,2008(3):72-74.

[7] 李思峰.急倾斜煤层断面形状及支护方式的选择及应用[J].煤炭技术,2009,28(5):69-70.
LI Sifeng. Selection and application of section form and support way in steep coal seam [J]. Coal Technology,2009,28(5):69-70.

[8] 康红普,王金华,林健.煤矿巷道锚杆支护应用实例分析[J].岩石力学与工程学报,2010,29(4):649-664.
KANG Hongpu,WANG Jinhua,LIN Jian. Case studies of rock bolting in coal mine roadway[J]. Chinese Journal of Rock Mechanics and Engineering, 2010, 29(4):

649-664.

[9] 张国锋,于世波,李国峰,等.巨厚煤层三软回采巷道恒阻让压互补支护研究[J].岩石力学与工程学报,2011,30(8):1619-1626.
ZHANG Guofeng, YU Shibo, LI Guofeng, et al. Research on complementary supporting system of constant resistance with load release for three-soft mining roadway in extremely thick coal seam[J]. Chinese Journal of Rock Mechanics and Engineering,2005,24(21):3959-3964.

[10] 张农,高明仕.煤巷高强预应力锚杆支护技术与应用[J].中国矿业大学学报,2004,33(5):524-527.
ZHANG Nong, GAO Mingshi. High strength and pretension bolting support of coal roadway and its application[J]. Journal of China University of Mining and Technology,2004,33(5):524-527.

[11] 陈登红,华心祝,李英明,等.回采巷道围岩分类治理模式及关键技术研究[J].岩石力学与工程学报,2012,31(11):2240-2247.
CHEN Denghong, HUA Xinzhu, LI Yingming, et al. Study of key technologies and management modes for classifying surrounding rocks of gateway[J]. Chinese Journal of Rock Mechanics and Engineering,2012,31(11):2240-2247.

[12] 王金华.全煤巷道锚杆锚索联合支护机理与效果分析[J].煤炭学报,2012,37(1):1-7.
WANG Jinhua. Analysis on mechanism and effect of rock bolts and cables in gate road with coal seam as roof[J]. Journal of China Coal Society,2012,37(1):1-7.

[13] 王琦,李术才,李为腾,等.让压型锚索箱梁支护系统组合构件耦合性能分析及应用[J].岩土力学,2012,33(11):3374-3384.
WANG Qi, LI Shucai, LI Weiteng, et al. Analysis of combination components coupling of pressure relief anchor box beam support system and application[J]. Rock and Soil Mechanics,2012,33(11):3374-3384.

[14] 何满潮,齐干,程骋,等.深部复合顶板煤巷变形破坏机制及耦合支护设计[J].岩石力学与工程学报,2007,26(5):987-993.
HE Manchao, QI Gan, CHENG Cheng, et al. Deformation and damage mechanisms and coupling support design in deep coal roadway with compound roof[J]. Chinese Journal of Rock Mechanics and Engineering,2007,26(5):987-993.

[15] 侯朝炯,郭励生,勾攀峰.煤巷锚杆支护[M].徐州:中国矿业大学出版社,1999:7-10.
HOU Chaojiong, GUO Lisheng, GOU Panfeng. Bolt supporting in coal roadway[M]. Xuzhou: China University of Mining and Technology Press,1999:7-10.

[16] 何满潮,孙晓明.中国煤矿软岩巷道工程支护设计与施工指南[M].北京:科学出版社,2004:56-65.
HE Manchao, SUN Xiaoming. Supporting design and construction guide of soft rock engineering in china's coal mine[M]. Beijing: Science Press,2004:56-65.

[17] 孟庆彬,韩立军,乔卫国,等.极弱胶结地层煤巷锚网索耦合支护效应研究及应用[J].采矿与安全工程学报,2016,33(5):770-778.

MENG Qingbin,HAN Lijun,QIAO Weiguo,et al. Supporting effect and application of bolt-net-anchor coupling support under extremely weak cementation formation[J]. Journal of Mining & Safety Engineering,2016,33(5):770-778.

[18] 梅凤清. 极软弱地层锚固结构形成机制与承载特性研究[D]. 徐州：中国矿业大学,2013.

MEI Fengqing. Study on formation mechanism and bearing behavior of anchor structure in very soft strata[D]. Xuzhou:China University of Mining and Technology,2013.

[19] 孟庆彬,韩立军,浦海,等. 极弱胶结地层煤巷支护体系与监控分析[J]. 煤炭学报,2016,41(1):234-245.

MENG Qingbin,HAN Lijun,PU Hai,et al. Research and monitoring analysis of coal roadway bolting system in very weakly cemented stratum[J]. Journal of China Coal Society,2016,41(1):234-245.

[20] 余伟健,高谦,朱川曲. 深部软弱围岩叠加拱承载体强度理论及应用研究[J]. 岩石力学与工程学报,2010,29(10):2134-2142.

YU Weijian,GAO Qian,ZHU Chuanqu. Study of strength theory and application of overlap arch bearing body for deep soft surrounding rock[J]. Chinese Journal of Rock Mechanics and Engineering,2010,29(10):2134-2142.

[21] 杨大伟. 软岩巷道中围岩变形控制机理及锚固结构稳定性研究[D]. 沈阳：东北大学,2011.

YANG Dawei. Surrounding rock deformation control mechanism and anchorage structure stability study in soft rock roadway[D]. Shen Yang:Northeastern University,2011.

[22] 张镇,康红普,王金华. 煤巷锚杆-锚索支护的预应力协调作用分析[J]. 煤炭学报,2010,35(6):881-886.

ZHANG Zhen,KANG Hongpu,WANG Jinhua. Pre-tensioned stress coordination function analysis of bolt-cable anchor support in coal roadway[J]. Journal of China Coal Society,2010,35(6):881-886.

[23] 康红普,王金华. 煤巷锚杆支护理论与成套技术[M]. 北京：煤炭工业出版社,2007：114-138.

KANG Hongpu,WANG Jinghua. Rock bolting theory and complete technology for coal roadways[M]. Beijing:China Coal Industry Publishing House,2007:114-138.

[24] MENG Qingbin,HAN Lijun,SUN Jingwu,et al. Experimental study on the bolt-cable combined supporting technology for the extraction roadways in weakly cemented strata[J]. International Journal of Mining Science and Technology,2015,25(1):113-119.

[25] MENG Qingbin,HAN Lijun,QIAO Weiguo,et al. Support technology for mine roadways in extreme weakly cemented strata and its application[J]. International Journal of Mining Science and Technology,2014,24(2):157-164.

Deformation and stress analysis of surface subsidence at the Jingerquan Mine

DING Kuo[1,2], MA Fengshan[2], ZHAO Haijun[2],
GUO Jie[2], ZHU Jianjun[3], TIAN Maohua[3]

(1. Key Laboratory of Shale Gas and Geoengineering, Institute of Geology and Geophysics, Chinese Academy of Sciences, Beijing 100029, China;
2. School of Civil Engineering, North China University of Technology, Beijing 100144, China;
3. Hami Jingerquan Mining Co. Ltd., Hami 839000, China)

Abstract: Over the past several years, the Jingerquan Mine has experienced a long-term and progressive surface fracturing and deformation which is directly related to the block caving method and then we start our investigation based on this background. The trace of fracturing was based on a series of damage mapping activities and the deformation data was collected by GPS from 2013 to 2016. In this paper, following an introduction to the engineering geology and mechanical properties of the rock mass, emphasis was put on an analysis of the fissures, deformation and stress of surface subsidence. Results reveal the diversity magnitude and structural features of surface movement and ground fissures. In addition, the time dependent behavior is comprehended and the subsidence zone reflects different types of time-displacement curve-regressive phase, steady phase and progressive phase, all these achievements indicate the complexity and diversity of the subsidence zone. On the other hand, stress calculation which inspired from the mechanical model of the cracking of hole wall is carried out, it is meaningful to understand the relation between fracture features, displacement vectors and horizontal stress.

Keywords: block caving; surface deformation; ground fissures; stress analysis

1 Introduction

The Jingerquan nickel mine is located in the east border of Hami, Xinjiang Province, in West China. Over the last 10 years, the mine has experienced a long-term and progressive deformation and fracturing in the surrounding rock and result in a significant surface subsidence eventually. This is directly related to the block caving method which is one of the most effective underground mining techniques coupled with enormous environmental impacts.

Block caving is initiated by blasting an extensive horizontal panel, stress redistribution and gravity combine to trigger progressive fractures in the vicinity of the mining area. The ground surface does not subside immediately until caving thinned the overlying cap rock

---

Supported by the Natural Science Foundation of China (41372323) and the Xinjiang Corps Thirteenth Divisions of State-owned Asset Management.
Corresponding author: DING Kuo, Tel:010-82998590, E-mail:309520428@qq.com.

and cannot transfer the load effectively. Caving progressively extends upwards with a further extraction, causing significant surface depression or subsidence eventually, above the goaf and adjacent areas. The complex temporal-spatial process is controlled by geological structures, dip angle of orebody, joint orientation and fault location, and so on[1,2].

In general, two types of deformation zones can be observed on the ground surface-continuous and discontinuous deformation zones[3]. Block caving mining is typically characterized by discontinuous subsidence. Discontinuous deformation zone is characterized by irregular failure surface and large disturbances affected regions, Features such as tension cracks, steps and chimney caves normally appear in this area[4,5]. Continuous deformation zone is a relatively regular and smooth lowering of the ground surface and shows only elastic deformation or continuous non-elastic strain[6].

According to Benko et al.[7], surface subsidence above longwall operations where the topography is relatively flat tends to be symmetric while surface subsidence above mines where a slope is present shows a pattern of greater subsidence developing in the upper part of the slope. Woo et al.[8] showed that caving-induced surface deformations tend to be discontinuous and asymmetric due to large movements around the cave controlled by geologic structures, rock mass heterogeneity and topographic effects. Villegas et al.[9] proved the time-dependent movement of the hangingwall surface is described based on the surveying data and three different stages of the time-displacement behavior is identified. Stephenson et al.[10] performed a detailed analysis of the subsidence by three-dimension physical modelling, concluded that the failure mode is a combination of topping and shear failure. Jennings[11] pioneered detailed step-path analyses of rock slopes with the development of a limit equilibrium approach that incorporated shear failure along joints, shear through intact rock, and tensile failure of rock bridges. Brummer et al.[12] indicated a plausible evolution path of the footwall fracturing that was subsequently described by conceptual numerical models created in 3DEC. Vyazmensky et al.[13,14] evaluate the significance of geological structure on surface subsidence combined FEM/DEM-DFN (discrete fracture network) modeling.

In this paper, the main objective is improving the understanding of subsidence phenomena and figuring the mechanism of block caving by a particular case study of Jingerquan. On the other hand, little attention has been paid to tectonic and stress patterns of the subsidence area, which is significant to fracture propagation, hence there is a genuine need for an inclusive study on stress redistribution of surface subsidence.

2 Basic conditions of Jingerquan mine

The Jingerquan mine, located in the Xinjiang province in northeast China, is a typical metal mine with steep dip angle. The ore deposit is approximately 2.0 km long, 700 meters wide and extends to more than 500 meters underground, the strike of the orebody is almost

330° and dips southwardly with dip angle range from 35° in the deep excavating orebody to about 65° which located in shallow orebody excavated before. The mining activities at Jingerquan commenced in 2009, a total of 2 Mt ore was excavated and roughly 800 thousands of square meters mined-out areas was produced up to 2015. In 2012 which being mined with block caving mining method for about 3 years, large scale ground movement and ground fissures appeared, as a result, we start our fundamental investigation in 2013.

The topography in the mining area is simple, the terrain is relatively flat and shows a low surface altitude difference. The strata of the mine are igneous rocks composed mainly of dunite, pyroxenite, anorthosite and amphibolite. Ore-bearing ultrabasic rocks were intruded into the Precambrian stratum as irregular dikes. The ore body lies in the central and lower part of the ultrabasic rock mass, which the rich ore is located in the middle of the ore body and the lean ore is at the periphery of the rich ore.

The mechanical properties and corresponding parameters of the rock mass are the fundament of stability evaluation of engineering rock mass. A large quantity of laboratory tests on various kinds of rock mass have been conducted in Jingerquan mine, the results of tests are shown in Fig. 1 It can be seen from the text result that the principal strengths of rock mass improve sharply with the increasing of mining depth, except a slight decline of poisson ratio from plane 965 to plane 981. On the other hand, the features of strong alteration and weathering widely develop in the superficial and shallow of rock mass, which greatly decrease the mechanical properties of the rock mass.

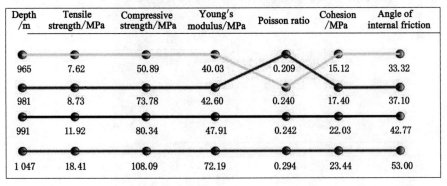

Depth /m	Tensile strength/MPa	Compressive strength/MPa	Young's modulus/MPa	Poisson ratio	Cohesion /MPa	Angle of internal friction
965	7.62	50.89	40.03	0.209	15.12	33.32
981	8.73	73.78	42.60	0.240	17.40	37.10
991	11.92	80.34	47.91	0.242	22.03	42.77
1 047	18.41	108.09	72.19	0.294	23.44	53.00

Fig. 1 Mechanical properties of different depth

Beside the engineering geology and rock mechanical properties, tectonic stress is another prominent factor inducing a large scale deformation of ground surface. In terms of Jingerquan mine, field tests and measurements have been taken on the virgin stresses by the stress releasing technique. According to the in situ stress results, the distribution characteristics of the virgin stresses in the Jingerquan mine area are as follows:

• The horizontal stress is the maximum principal stress of the mine area which trends NE with value of 34.9 MPa.

• The maximum principal stress is basically parallel to the regional tectonic stress field in Xinjiang region.

• The vertical stress of rock mass with value of 18.5 MPa is the intermediate principal stress, which exceed the weight of overlying strata.

• The shear stress reaches 8.55 MPa which is liable to damage the roadway roof.

3 Monitoring

3.1 Fissures

In 2013, a brief visual inspection and surface mapping was conducted and the mapping was firstly focused on the surface fissures. The investigation results showed that all of the visual surface deformation and failure situated in the hanging rocks, a crater subsidence formed over a large area, the overburden of the center part of crater moves vertically downward and the adjacent moves towards the center and downwards at the same time, those movements have enough cumulative effect to form fissures. The fissures can be divided into two groups according to their distribution characteristics: Fissure Zone No. 1 and Fissure Zone No. 2-shown in Fig. 2.

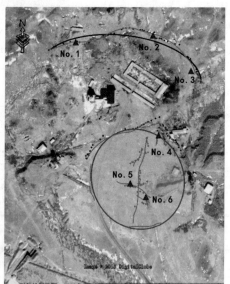

Fig. 2 Schematic diagram of surface cracks, red lines represent visible fissures, blue triangle represent the monitoring points around the typical fissures

Fissure Zone No. 1 is located above the goaf and two kinds of fissures are visible compression fissures and tension fissures, as shown in Fig. 3. The main fissures of this area are compression fissures of X-shaped which the average extended length is approximately 50~100 meters. The angle of X-shape is controlled by maximum principal stress and minimum principal stress which formed by squeezing action of the crater. There are many branch fissures parallel to the main compression fissures in this zone. The tension fissures often appear around the crater and steps are likely formed with the continuous and intensive subsidence. Fissure Zone No. 2, formed by tension stress, is located 300 m northward away from Fissure Zone No. 1. It has a length of 30~40 m average and a width of 10~20 mm. The fissures in the zone trend to be connected. Fissures have been often associated with geological discontinuities and faults and the extending orientation of the fissure zones appears to be approximately parallel to the trend of the ore body.

Fig. 3 Scene photographs of fissure No. 4 and No. 5

The measurement results by the fissure gauges revealed that the relative displacement of the two sidewalls of a ground fissure usually appears to be of three-dimensional features. The relative displacement of monitoring points around fissure No. 4 kept inconsistent and increased continuously with the mining deep, it means fissure No. 4 is still in activity. Similarly, the data of monitoring points around fissure No. 5 and No. 6 shows that fissure No. 5 and No. 6 appeared a tendency of expansion, while it's much gentler than fissure No. 4. We can deduct from results above that the deep extension section below Fissure Zone No. 1 had wrecked and separated from each other. According to the monitoring data, the monitoring points around fissure No. 1, No. 2 and No. 3 moved to subsidence center as a whole and the width of fissures remained stable in general. We could deduce from above that the deep extension section below Fissure Zone No. 2 keep a state of bending instead of wrecking.

3.2 GPS

In 2013, a monitoring network of 101 stations, which are marked bars mounted in cylindrical concrete bases buried 0.5 m into the ground, are distributed along the hanging wall in eight surveying lines oriented north-south (94,95,96,97,98,99,100,101) and monitoring data are updated every four months. Besides, seven additional monitoring stations are applied around the subsidence center for safety reasons.

As shown in the vertical displacement contours-Fig. 4, the vertical subsidence zone was an irregular cycle which similar to tubular-shaped and the affected area extended 300 meters along the dip direction of the ore body, 350 m along the strike. By the latest monitoring data of 2016. 4, the maximum vertical displacement reached 1 650. 6 mm. According to the comparison of each stage of subsidence curve, the magnitude of the subsidence increase with the long-term and large-scale repeated excavation, however, the location of the subsidence center are not changed obviously in spite of the center of the orebody moved with the mining depth. It means that the ground surface subsidence center was mainly decided by the geologic weaknesses in the rock mass underground mainly.

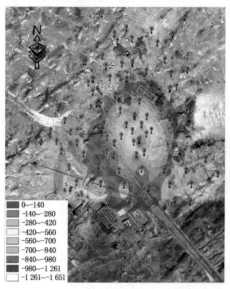

Fig. 4　Accumulative vertical displacement from 2013. 7~2016. 4, on the base map of Google Earth(the red pointer represent the monitoring points)

In terms of horizontal displacements-shown in Fig. 5, the horizontal displacement zone was an irregular cycle similarly, the maximum horizontal displacement reached 473. 99 mm which was less than vertical displacement, it means the vertical displacement was more prominent but the horizontal displacement can't be ignore simultaneously. On the other hand, the subsidence center sink continuously during the excavation, we can speculated that new air faces appeared and horizontal structure stress released coincidentally which manifested as tensile cracks. Consequently, the horizon deformation zone expanded to north by the influence of tectonic joints and fracture zones. The horizontal displacement of expanded area was roughly 20~30 mm from 2013. 10 to 2016. 4.

3.3　Time-dependent deformation

The typical time-dependent deformation goes through three different stages-regressive, steady and progressive-shown in Fig. 6. During the regressive stage, the deformation starts with dilation and relaxation of the rock mass, there is immediate

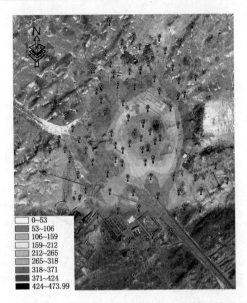

Fig. 5 Accumulative horizontal displacement from 2013.7~2016.4, on the base map of Google Earth(the red pointer represent the monitoring points)

correlation between the rate of displacement and the variation in the production. Continuation of mining, the rock adjacent to the collapsed rubble column is resisted by the stress of the rubbles, the time displacement curve enter a creeping motion named steady state which ranges at a constant velocity. With the large-scale stress adjustment and redistribution, the balance of the stress between the rubbles may be broke under the disturbance, starting the progressive state which the time-dependent deformation at acceleration rate to the status of collapse.

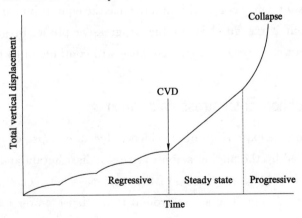

Fig. 6 Typical time-displacement curve

As to the study area, we can't acquire the integrated deformation curve by a single monitoring station, for the reason of a relatively short time for monitoring. The survey monitoring system was initiated in 2013 which conspicuous cracks and crater was visible at that time, regressive state cannot be detected anymore in central subsidence region. On the other hand, the subsidence zone shows a characteristic of complexity and diversity according to the long-term observation and monitoring, it means the monitoring stations which diversity in space reflect different types of time-displacement curve, we describe the time-displacement curve respectively, as shown in Fig. 7.

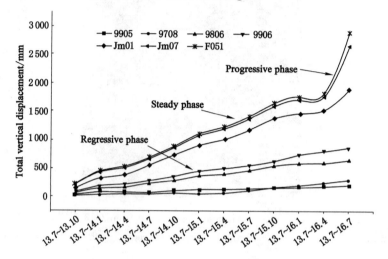

Fig. 7　Time-displacement curve of different monitoring points

The rate of curve is the most conspicuous different between regressive phase and steady state. Generally, the movement rate of the regressive phase is 50~200 mm/year and 600~1 000 mm/year in steady phase, the change of phase between regressive and steady is defined as the critical vertical displacement (CVD). If the range of CVD values is 0.8~1.0 m, the first fissures appear and the curve steps in the steady phase afterwards[9]. In extreme cases, the subsidence curve will mutate once the equilibrium arch is broke and then the time-displacement curve go through the progressive phase. In general, the value of displacement saltation is 1.8~2.0 m, the collapse will continue until the next equilibrium status.

4　Methodology of horizontal stress calculation

Crater is the surface expression of subsidence due to collapse of support pillar where large voids are created by the mining activity, stress redistribution and gravity combine to trigger progressive fracturing and caving of the rock mass between surface collapse crater and underground mined area. The stress around the collapse crater has varied during the conversion from rock-intact to rock-fractured downwards and we put emphasis on the horizontal stress on ground surface in this paper.

Before the excavation, the horizontal stress of the in-situ stress can be expressed by tectonic stress and horizontal component caused by gravity stress.

$$\sigma_1 = \sigma_1^k + \sigma_H = \sigma_1^k + k\gamma_0 h = \sigma_1^k + \frac{\mu}{1-\mu}\gamma_0 h \tag{1}$$

$$\sigma_2 = \sigma_2^k + \sigma_H = \sigma_2^k + k\gamma_0 h = \sigma_2^k + \frac{\mu}{1-\mu}\gamma_0 h \tag{2}$$

In which σ_1 and σ_2 are the horizontal stress of the in-situ stress, σ_1^k and σ_2^k are horizontal tectonic stress, σ_H is the horizontal component caused by gravity stress, k is side pressure coefficient, μ is Poisson ratio, γ_0 is rock bulk density, h is depth to surface ground.

Due to the formation of collapse crater, the stress point to the wall of the collapse crater is defined as σ_0, therefore the stress around the collapse crater have changed, compared to the in-situ stress.

$$\nabla\sigma_1 = \sigma_1 - \sigma_0, \nabla\sigma_2 = \sigma_2 - \sigma_0 \tag{3}$$

In which $\nabla\sigma_1$ and $\nabla\sigma_2$ are the horizontal stress following the collapse.

The horizontal slice around the surface crater is in a state of plane stress, a large quantity of experimental research shows that the stress state is in accordance with plane solution of elastic theory, by a condition of the depth of mine-out area is much greater than the diameter of collapse crater. The radial stress ranges from in-situ stress away from the collapse crater to lateral stress of the wall of the collapse crater, the normal shear-stress concentrate and reach the maximum value.

The stress calculation model of Jingerquan is inspired from the mechanical model of the cracking of hole wall, we take a hypothesis that the caving zone is a cavity filled with debris of rock mass, hence $\sigma_0 = 0$, $\sigma_H = 0$ on the surface ground of caving zone, the formula (1) and (2) turn into formula (3) as below:

$$\nabla\sigma_1 = \sigma_1^k \quad \nabla\sigma_2 = \sigma_2^k \tag{4}$$

We obtain the correlation formula between radial and tangential displacement u, v and horizontal tectonic stress σ_1^k, σ_2^k based on the solution of G-Kirsch, as follows:

$$u = \frac{1+\mu}{2E}\left[(\sigma_1^k + \sigma_2^k)\left(r + \frac{R^2}{r} - 2\mu r\right) + (\sigma_1^k - \sigma_2^k)\left(r + 4\frac{R^2}{r} - 4\mu\frac{R^2}{r} - \frac{R^4}{r^3}\right)\cos 2\theta\right] \tag{5}$$

$$v = \frac{1+\mu}{2E}\left[(\sigma_2^k - \sigma_1^k)\left(r + \frac{2R^3}{r} - 8\mu\frac{R^2}{r} + \frac{R^4}{r^3}\right)\right]\sin 2\theta \tag{6}$$

In which R is the radius of collapse crater, r and θ are the polar coordinates of the monitoring points, E and μ are the elastic modulus and poisson ratio of rock mass.

The formula of surrounding rock displacement before excavation is as follows:

$$u_0 = \frac{1+\mu}{2E}r\left[(\sigma_2^k + \sigma_1^k)(1-2\mu) + (\sigma_2^k - \sigma_1^k)\cos 2\theta\right] \tag{7}$$

$$v_0 = \frac{1+\mu}{2E}r(\sigma_2^k - \sigma_1^k)\sin 2\theta \tag{8}$$

The final formula of displacement in the subsidence area is as follows:

$$u_r = u - u_0 = \frac{1+\mu}{2E}\left[(\sigma_1^k + \sigma_2^k)\frac{R^2}{r} + (\sigma_1^k - \sigma_2^k)\left(4\frac{R^2}{r} - 4\mu\frac{R^2}{r} - \frac{R^4}{r^3}\right)\cos 2\theta\right] \tag{9}$$

$$v_\theta = v - v_0 = \frac{\sigma_1^k - \sigma_2^k}{2E}(1+\mu)\left[\frac{R^4}{r^3} - 2\frac{(1-\mu)R^2}{(1+\mu)r}\right]\sin 2\theta \tag{10}$$

5 Results and discussion

It is a normal practice to analyze the surface displacements, because the displacements can be acquired directly by GPS or InSAR, thus less effort is done to reflect on stress reaction coupled with continuous ore extraction, which is the key factor of quantity and direction of displacement vector, as well as the fracture distribution. A horizontal stress calculation based on formula (9) and (10) is conducted by MATLAB and the input parameters are u, v, r, θ. r, θ represent the position of the monitoring points in polar coordinate system, which the zero point of the polar coordinate system is the subsidence center according to the previous monitoring results, u, v represent the radial-displacement and tangential-displacement surveyed by GPS from 2013～2016. The radial-stress and tangential-stress are calculated and the detailed features can be summarized in the following.

Fig. 8 presents the quantity and distribution of radial-stress and shows a completely distinct trend compared with distribution of displacement. The extension can be divided into two regions parallel to the strike of the orebody, located in the north and south of the subsidence center. The maximum radial-stress of south region is 31.77 MPa and concentrates on two monitoring points on both sides of a compression fissure of X-shaped located in Fissure Zone No.1, the radial-stress of other area ranges from 2.9 MPa to 6.7 MPa and extends to the mining direction. The north region is unconnected and located in Fissure Zone No.1 and Fissure Zone No.2 respectively, the maximum radial-stress of north region is 11.43 MPa located Fissure Zone No.1, the radial-stress lies in Fissure Zone No.2 range from 2.1 MPa to 3.6 MPa.

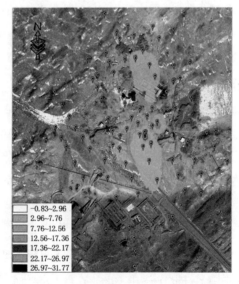

Fig. 8 Schematic diagram of radial-stress, positive value represents
the direction pointed to the subsidence center

As follow from Fig. 9, the distribution of tangential-stress shows a similar trend with the radial-stress, two regions of north and south and parallel to the strike of the orebody. While the bigger stress of 30.31 MPa lies in the north region compared with the south region, and the north region just extends a narrow area compared with the radial-stress, we cannot observe the concentration of tangential-stress in Fissure Zone No. 2. The maximum tangential-stress of south region is 9.92 MPa and extends to the mining direction which is similar to the radial-stress.

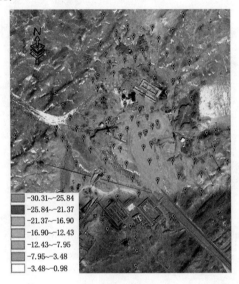

Fig. 9　Schematic diagram of tangential-stress, positive value represents the direction of clockwise

In order to understand the relation between horizontal stress, fissures and displacement direction fully, we draw the stress distribution graph as shown in Fig. 10. Fissure No. 5 and No. 6 located in the two wings of X-shaped fissure, corresponding to the concentration of radial-stress. The X-shaped is irregular, the further wing is about 100 meters and in north-south direction which is almost parallel to the spread of radial-stress, the other extends just about 40 meters. Fissure No. 1, No. 2 and No. 3 are tensile cracks and related to the radial-stress in north region, these fissures existed before we begin our investigation and monitoring in 2013, the maximum radial-stress of these fissures is 3.6 MPa and less than the ultimate tensile strength, it means that the maximum radial-stress reached the ultimate tensile strength at some time in the past and reduced rapidly following the appearance of tensile fissures. Additionally, we cannot observe tensile fissures in the south of the subsidence center corresponding with Fissure No. 1, No. 2 and No. 3 at present, but the maximum of tangential-stress in this area is 6.7 MPa and we can deduce that this area is the potential breakage with tensile failure coupled with continuous ore extraction. As to fissure No. 4, there are numerous secondary fissures parallel to fissure No. 4 which is unique in the subsidence area, these secondary fissures situated around the concentration region of tangential-stress,affected and dominated by tangential-stress.

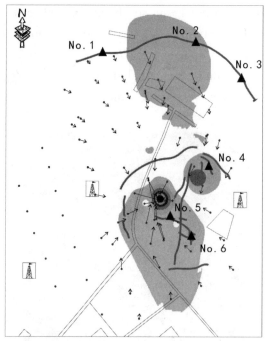

Fig. 10　Schematic diagram of the relationship between radial stress, surface fissures and displacement vectors

Prophase research have proved that displacement vector pointed to the underground mine and changed with the distance from the monitoring points to the center of the ore body (Haijun Zhao,2013). The conclusion emphasized the effort of radial-stress while the influence of tangential-stress is ignored. As shown in Fig. 11, the direction of displacement vectors doesn't point to the subsidence center absolutely but deflect to the direction of the tangential-stress, especially in the area of stress concentration. On the other hand, it does not mean that the stress focus on all angle of the subsidence center, the position of stress concentration is controlled and influenced by the joint orientation and faults.

6　Conclusions

The long-term GPS monitoring of ground deformation revealed the spatial distribution characteristic and the reaction to the large-scale excavation, the maximum vertical displacement reached 1 650. 6 mm and the maximum horizontal displacement reached 473. 99 mm in the past four years. As to the fissures, two groups are divided according to the distribution characteristics, one of them is located above the goaf and constituted of compression fissures of X-shaped, the other are tension fissures and lie on the border of subsidence area. Moreover, the failure of deep extension section is speculated by the relative displacement of the fissures.

The time-dependent deformation goes through three different stages-regressive, steady

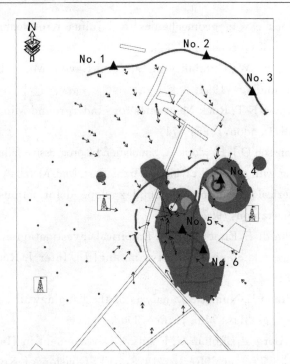

Fig. 11 Schematic diagram of the relationship between tangential stress, surface fissures and displacement vectors

and progressive and represent different failure mechanism respectively, it could be an important tool to estimate the movement rate of different regions. We classified monitoring points by different subsidence stages and strengthen the understanding of the whole surface subsidence.

Inspired from the mechanical model of the cracking of hole wall, we acquired the calculation formula of horizontal stress on ground surface based on a serious of hypothesis, and then the stress distribution of surface subsidence accompanied with underground mining is distinct. Based on the achievement above, we got a more reasonable explanation of the displacement vectors and fissures-X-shaped fissures, tensile fissures and secondary fissures.

Although the long-term GPS monitoring of ground deformation is a laborious, this research illustrates that it is beneficial for a deeper understanding of the deformation features and fracture mechanism. Furthermore, the mine construction engineering can be arranged rationally to avoid these disasters.

References

[1] Laubscher D. Block caving manual. Report for the international caving study[M]. JKMRC and Itasca Consulting Group, Brisbane, 2000.

[2] Van AsA, Davison J, Moss A. Subsidence definitions for block caving mines[J]. Technical Report, 2003:59.

[3] Brown E T. Block caving geomechanics[R]. Julius Kruttschnitt Mineral Research Centre, University of Queensland, 2003.

[4] Herdocia A. Hanging Wall Stability of Sublevel Caving Mines in Sweden[D]. Luleå University of Technology, 1991.

[5] Brady B H G, Brown E T. Rock Mechanics for Underground Mining[M]. 2nd Edition. Chapman & Hall, London, 1995: 571.

[6] Singh U K, Stephansson O J, Herdocia A. Simulation of progressive failure in hanging-wall and footwall for mining with sub-level caving[J]. Trans. Inst. Min. Metall A, 1993(102): 188-194.

[7] Benko B. Numerical modelling of complex slope deformations[D]. University of Saskatchewan, Saskatoon, 1997.

[8] Woon K S, Eberhardt E, Elmo D, et al. Empirical investigation and characterization of surface subsidence related to block cave mining[J]. Inter J. Rock Mech. & Mining Sci., 2013(61): 31-42.

[9] Villegas T, Nordlund E. Numerical analysis of the hangingwall failure at Kiirunavaara mine[C]. Proceedings Mass Min., Luleå, 2008.

[10] Stephansson O, Borg T, Bkblom G. Fracture Development in Hanging Wall of North Kiruna Mine[C]. Technical Report 1978: 51T, Division of Rock Mechanics, Luleå University of Technology, Sweden, 1978.

[11] Jennings J E. A mathematical theory for the calculation of the stability of slopes in open cast mines[C]. In: Proceedings of the Symposium on the Theoretical Background to the Planning of Open Pit Mines, Johannesburg, 1970: 87-102.

[12] Brummer R K, Li H, Moss A. The transition from open pit to underground mining: an unusual slope failure mechanism at Palabora[C]. The South African Institute of Mining and Metallurgy, 2006: 411-420.

[13] Vyazmensky A, Stead D, Elmo D, et al. Numerical analysis of block caving-induced instability in large open pit slopes: A finite element discrete element approach[J]. Rock Mech. Rock Eng., 2010(43): 21-39.

[14] Vyazmensky A, Elmo D, Stead D. Role of rock mass fabric and faulting in the development of block caving induced surface subsidence[J]. Rock Mech. Rock Eng., 2010(43): 533-556.

[15] Haijun Zhao, Fengshan Ma, Yamin Zhang, et al. Monitoring and Analysis of the Mining-Induced Ground Movement in the Longshou Mine, China[J]. Rock Mech. Rock Eng., 2013(46): 207-211.

光纤光栅煤矿安全智能监测系统

方新秋,吴刚,梁敏富,薛广哲,马盟,陈宁宁

(中国矿业大学 深部煤炭资源开采教育部重点实验室 矿业工程学院,江苏 徐州,221116)

摘 要:基于光纤光栅的煤矿安全智能监测系统是保证煤矿生产正常进行的重要技术措施。光纤传感技术具有安全防爆、抗电磁干扰、抗腐蚀、防水、体积小、灵活方便等诸多优点,尤其适用于煤矿井下的恶劣环境。文章介绍了光纤光栅的煤矿安全智能监测系统的整体架构与工作原理,对光纤光栅安全智能监测系统的软硬件开发、现场布置及安装方法进行了论述。实践表明:光纤光栅的煤矿安全智能监测系统有效解决了井下恶劣多变环境及高瓦斯矿井的安全在线监测和实时预警等关键技术问题,实现了高瓦斯矿井安全监测的无人化与智能化,对于高效矿井的减员提效、灾害预测预报、安全生产等方面具有重要意义。

关键词:智能监测;光纤光栅;传感技术;软件开发;监测设备

FBG intelligent monitoring system of coal mine safety

FANG Xinqiu, WU Gang, LIANG Minfu, XUE Guangzhe, MA Meng, CHEN Ningning

(Key Laboratory of Deep Coal Resource Mining, Ministry of Education of China, School of Mines, China University of Mining & Technology, Xuzhou 221116, China)

Abstract: The intelligent monitoring system for coal mine safety based on Fiber Bragg grating(FBG) is an important technical measure to ensure the production of coal mine normally. FBG sensing technology has many advantages, such as safety explosion prevention, electromagnetic interference resistance, corrosion resistance, waterproof, small size, flexible and convenient, and so on. It is especially suitable for the harsh environment in the coal mine. The whole structure and working principle of the FBG intelligent monitoring system of coal mine safety is introduced, and the software and hardware development, the field work and the installation method is discussed. This paper shows that the FBG intelligent monitoring system of coal mine safety can effectively solve the security of online monitoring and real-time warning and other key technical problems under the changeful environment and high gas coal mine and achieve unmanned and intelligent monitoring for high gas coal mine safety, It has important significance for mine efficiency, forecasting disaster and safety production etc.

Keywords: intelligent monitoring; FBG; sensor technology; software development; monitoring equipments

目前,我国很多矿区的安全监测多数属于人工监测和电子监测,能够部分实现安全的在线监测,鉴于煤矿井下复杂多变的恶劣环境及高瓦斯等开采技术条件的限制,导致煤矿安全

基金项目:中央高校基本科研业务费专项资金资助项目(2017CXNL01)。
作者简介:方新秋(1974—),男,浙江省永康市人,教授,博导,主要从事煤矿智能开采、矿山智能监测等方面的研究。Tel:0516-83590577,E-mail:xinqiufang@163.com。

在线监测的技术手段和监测设备水平低下,在采掘巷道和工作面仍需要现场作业人员进行安全监测,使得安全监测数据滞后、监测误差较大,从某种意义上讲还未真正实现煤矿井下的安全在线传输与实时监测,这同样是影响煤矿安全高效开采的技术瓶颈。光纤传感技术是20世纪70年代中期发展起来的一门新技术,在光纤传感技术的基础上开发了光纤传感器,并通过集成创新形成各种系统,应用于相关行业的监测和预警中。光纤传感器监测技术是利用外界因素使光在光纤中传播时光强、相位、偏振态和波长等特征参量发生变化,从而对外界因素进行监测和信号传输的技术。光纤传感器安全防爆、抗电磁干扰、抗腐蚀、防水、体积小、灵活方便,尤其适用于煤矿井下的恶劣环境。近年来光纤传感监测技术开始在煤矿中推广,在复杂地层、地质构造及水文地质条件下的煤矿围岩和支护结构中进行监测,能够准确监测巷道顶板的变形、破坏,进而选择巷道治理的最佳时机,为巷道顶板变形监测和隐患预警提供了一种稳定可靠的技术方案,光纤光栅传感技术的安全在线监测技术实现了由传统人工监测、电子监测向光纤监测的高新技术变革,推动了煤炭行业整体安全监测科学技术水平的进步,对煤矿安全高效开采及开采智能监测提供了一条新的技术途径[1]。

1 光纤光栅的煤矿安全智能监测系统概述

1.1 光纤光栅矿压在线监测系统整体架构

基于光纤光栅的煤矿安全智能监测系统是一套功能齐全、扩展方便、高性价比的煤矿安全智能实时动态监测系统,基于最为先进的光电子集成技术,采用模块化设计,系统将计算机技术、光纤通信与数据处理技术、传感器技术和煤矿安全技术融为一体,具有结构紧凑、功耗低、性能高、系统稳定和便于维护等优点[2]。系统的整体架构如图1所示。

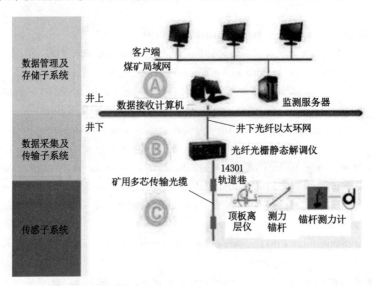

图1 光纤光栅的煤矿安全智能监测系统整体架构

Fig.1 Overall architecture of the intelligent monitoring system for coal mine safety based on FBG

基于光纤光栅的煤矿安全智能监测系统由传感子系统、数据采集与传输子系统和数据理与存储子系统三部分组成。

（1）传感子系统（图1所示C位置）

传感子系统是由各种基于光纤光栅传感技术的传感设备组成。

（2）数据采集与传输子系统（图1所示B位置）

数据采集与传输子系统主要由传感光缆、光纤光栅信号解调仪和配套设备组成。

（3）数据管理与存储子系统（图1所示A位置）

数据管理与存储子系统主要由数据接收计算机、监测服务器和客户端组成。

1.2 光纤光栅的煤矿安全智能监测系统工作原理

基于光纤光栅的煤矿安全智能监测系统将计算机技术、光纤通信与数据处理技术、传感器技术和巷道围岩安全技术融为一体，整个系统由地面数据接收计算机、用于数据共享的煤矿局域网、对波长进行解调的光纤光栅静态解调仪、可24 h连续不间断工作的监测服务器、地面各科室的客户端、井下传输光缆、传输光纤、各类光纤光栅传感器及与其他的连接接口组成，具体的工作原理如图2所示。

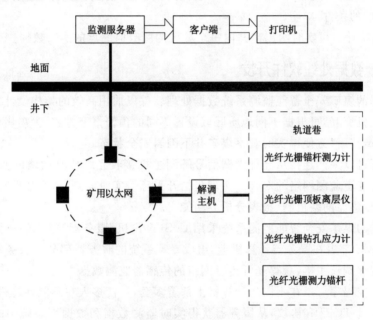

图2 光纤光栅的煤矿安全智能监测系统原理图

Fig.2 Schematic diagram of the intelligent monitoring system for coal mine safety based on FBG

当系统工作时，通过光纤光栅锚杆测力计、光纤光栅顶板离层仪、光纤光栅钻孔应力计、光纤光栅测力锚杆等不同功能的光纤光栅传感器，将被测的轨道巷试验段顶板及两帮锚杆载荷、顶板离层、围岩应力、锚杆杆体应力等不同形式的物理量信号转变成便于记录及再处理的光信号（波长信号），通过光波信号的变化间接测量围岩稳定参数，然后通过传输光纤、跳线和多芯传输光缆将光纤光栅传感设备采集的围岩物理参数传输到光纤光栅信号解调仪，再通过光纤光栅信号解调仪将光信号解调为数字信号，利用南三采区变电所的以太网将数据上传至地面监测服务器，利用服务器—客户端网络将数据传输到生产科计算机，构成一套合理的基于光纤光栅的煤矿安全智能监测系统。

1.3 光纤光栅的煤矿安全智能监测系统的功能

（1）实时显示。本系统能够自动对多种井下光纤光栅传感器所在区域进行实时监测，计算机实时动态显示监测数据。

（2）异常预警。本系统能够自动感知现场监测数据的异常波动，在灾害事故发生前及时预警。

（3）监测点定位。本系统以电子地图方式实时显示矿井内各个位置监测点的编号、当前数据以及实际地理位置，方面管理人员操作和维护。

（4）状态查询。各个监测点的安全数据和报警信息都保存到本系统的大容量储存器，系统按照时间将数据分为历史信息、实时信息，管理操作人员可以根据不同的需求动态调整监测点的监测时间间隔。管理操作人员可查看各监测点的历史变化曲线，为决策和维护提供数据支持。

（5）报警设定。管理操作人员可对本系统的报警触发条件进行设定，以适用各种不同工况、不同区域的条件。

（6）数据统计。可实时给出矿井内每个分区的所有监测点的实时数据。

2 光纤光栅数据处理软件开发

光纤光栅的煤矿安全智能监测系统数据处理软件以通用高效的架构设计来管理监测数据和友好的人机交互，可根据不同地质区域设置不同的预警报警方式，为煤炭企业监控工作人员、科研人员使用，方便得实时观察煤矿井下围岩安全状态。

本系统通过以太网连接从光纤光栅信号解调主机接收光纤传感网络的监测数据，以实现与光纤传感巷道监测网络的对接。实现包括开采环境实时监测、监测数据存储管理、数据分析与预警、传感器网络控制、报表分析等功能[3]。

光纤光栅的煤矿安全智能监测系统采用 C/S(客户机/服务器)模式的设计思路。服务器程序和 DBMS 部署在服务器计算机上，组成专家系统的服务器部分。服务器通过以太网连接光纤光栅解调仪主机，获得光纤传感网络的传感器监测数据[4,5]。

客户端监控程序可以在不同的计算机上部署多套，可使多人在不同计算机上同时监测。客户端程序通过 TCP/IP 协议，从服务器获得实时监测数据和数据库系统中的数据来进行监测和分析，系统总体架构如图 3 所示。

监测系统的软件主界面如图 4 所示。共包括实时监控模块、系统管理模块、数据查询模块、综合报表信息模块、传感器详细信息模块、离层详细信息模块、综合测站信息模块和传感器布置总体显示模块。

3 光纤光栅传感设备硬件简介

3.1 光纤光栅锚杆测力计

本系统采用的光纤光栅锚杆测力计，主要用来测量和监测巷道锚杆所受的载荷应力大小和预应力的损失情况。在传统的锚杆测力计的设计思想上，设计出光纤光栅锚杆测力计，摒除了传统非在线监测的锚杆测力计普遍具有测点分散、监测不连续、监测信息不完整且滞

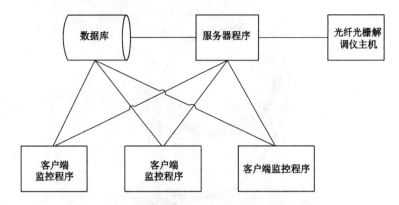

图 3 光纤光栅的煤矿安全智能监测系统软件架构
Fig. 3 Software architecture of the intelligent monitoring system for coal mine safety based on FBG

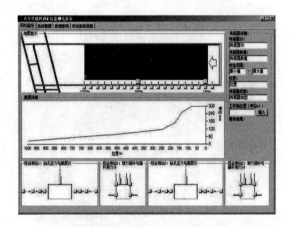

图 4 监测系统的软件主界面
Fig. 4 Software main interface of monitoring system

后的缺点,同时兼顾了可现场直接读数据的优点,便于现场读数,结构简单,安装便捷,其结构与工艺适于普遍推广应用[6-8]。光纤光栅锚杆测力计的实物图如图5所示。

3.2 光纤光栅顶板离层仪

系统采用的光纤光栅顶板离层仪,是将巷道顶板离层时对光纤的作用,引起光纤光栅反射波长的发生变化,只要测出这些参数随外界因素的变化关系,就可以通过光纤光栅波长的变化来检测外界离层变化,克服了传统顶板离层仪现场人工读数数据滞后、误差大等缺点。这一技术的应用对测量锚杆支护巷道锚固区内外顶板离层大小,以及评价锚杆支护效果和巷道安全程度具有重要意义[9-12]。光纤光栅顶板离层仪的实物图如图6所示。

3.3 光纤光栅钻孔应力计

光纤光栅钻孔应力计采用的是基于光纤式的一维应力应变测量技术,测量的是煤体或岩体垂直载荷应力的大小。光纤光栅钻孔应力计的传感原理为受应力作用钻孔周围的煤体或岩体产生破坏变形,将应力传递到应变体上产生横向变形,置于光纤光栅钻孔应力计内的光纤光栅传感器将变形量转换成光波信号,经过光纤光栅信号解调仪解调为数字信号,传输

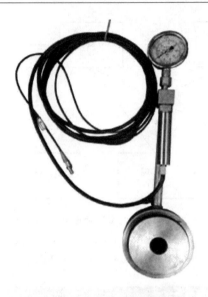

图 5 光纤光栅锚杆测力计
Fig. 5 Dynamometer for bolt based on FBG

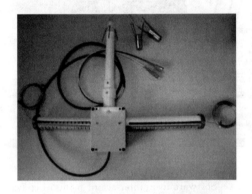

图 6 光纤光栅顶板离层仪实物图
Fig. 6 Physical map of roof separation monitoring sensor based on FBG

至数据接收计算机动态显示煤岩体应力的大小。随着工作面空间和时间关系的不同及工作面开采的不断推进,用于监测因采动影响的煤矿井下煤层或岩层内部应力变化、工作面超前支承压力峰值分布位置与显著影响范围,分析其在相对稳定和显著运动过程中的变化规律,是研究采场动压作用规律的重要手段之一。光纤光栅钻孔应力计实物图如图 7 所示。

3.4 光纤光栅测力锚杆

系统采用的光纤光栅测力锚杆,是针对传统的测力锚杆测量精度不高,抗干扰能力不强,电阻应变片易受潮湿、电磁等恶劣工作环境的影响设计的。这种光纤光栅测力锚杆结构简单、测量精度高、抗震性能好、工作稳定可靠,安装方便、测试快捷、便于远距离传输和自动化监测,其传输信号为光信号,不受电磁干扰,允许长距离传输。光纤光栅测力锚杆是本安型产品,适合应用于危险环境中。主要用来测量锚杆受力大小和分布状况,进行锚杆支护质量监测,对巷道锚杆支护的安全施工及参数设计优化具有重要意义。具体的实物图如图 8 所示。

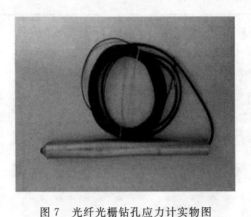

图 7 光纤光栅钻孔应力计实物图
Fig. 7　Physical map of borehole stress-meter based on FBG

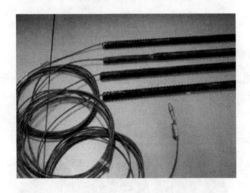

图 8 光纤光栅测力锚杆实物图
Fig. 8　Physical map of force-measuring bolt based on FBG

4 光纤光栅的煤矿安全智能监测系统的布置与安装

以某轨道巷为研究对象,针对轨道巷顶板压力、离层、锚杆受力、围岩应力等影响巷道围岩稳定性的参量因素监测手段落后、数据传输滞后、监测误差大、利用程度低、传感元件寿命短、测量易受环境影响,大部分甚至还是靠人工观测,不能实现实时自动监测及数据传输,不能进行准确、全面预测和预警,且传感仪器现场提供电源工作,在高瓦斯矿井使用会增加安全隐患。在这种背景下,本项目基于光纤光栅传感技术,提出合理的光纤光栅的煤矿安全智能监测系统的技术方案,最终形成一套成熟完善的基于光纤光栅的煤矿安全智能监测系统,实现煤矿井下的高效安全生产,并可进一步推广使用[13-15]。

4.1 测站的布置

轨道巷的矿压监测包括巷道顶板及两帮锚杆载荷监测、顶板离层监测、煤岩体应力监测、锚杆杆体应力分布特征监测等内容,所以每个综合测站包括一个锚杆载荷监测断面、一个煤岩体应力监测断面和一个锚杆杆体应力分布特征监测断面,测点布置示意图如图 9 所示。

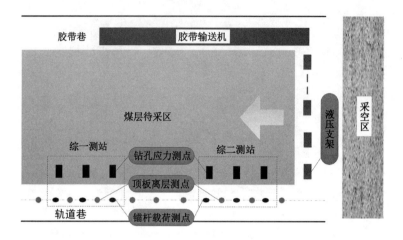

图 9 传感设备布置示意图

Fig. 9 Layout of sensing equipment

4.2 测站中光栅传感设备的安装

(1) 光纤光栅锚杆测力计安装在综合测站内,根据轨道巷的开采状况、煤层赋存特点、地质特征,将光纤光栅锚杆测力计合理布置在顶板和两帮,光纤光栅锚杆测力计的安装示意图如图 10 所示。

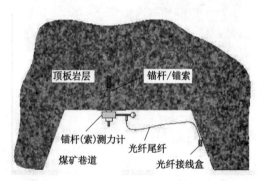

图 10 光纤光栅锚杆测力计安装示意图

Fig. 10 Installation diagram of dynamometer for bolt based on FBG

(2) 在轨道巷(距离巷口 200 m 开始实施)每隔 50 m 布置 1 个光纤光栅顶板离层仪,根据现场未采煤的长度、采煤进度及生产计划,合理布置顶板离层仪的监测点,光纤光栅顶板离层仪安装示意图如图 11 所示。

(3) 光纤光栅钻孔应力计安装在轨道巷的综合测站内,根据轨道巷的开采状况、煤层赋存特点、地质特征,合理布置在巷道的两帮,光纤光栅钻孔应力计的安装示意图如图 12 所示。

(4) 光纤光栅测力锚杆安装在综合测站内,布置轨道巷的两帮,根据煤层赋存特点和地质特征,合理布置监测点,轨道巷光纤光栅测力锚杆的安装示意图如图 13 所示。

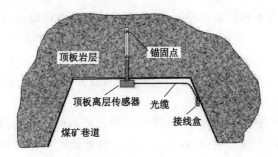

图 11 光纤光栅顶板离层仪的安装示意图
Fig. 11 Installation diagram of roof separation monitoring sensor based on FBG

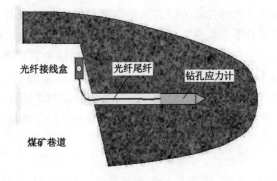

图 12 光纤光栅钻孔应力计的安装示意图
Fig. 12 Installation diagram of borehole stress-meter based on FBG

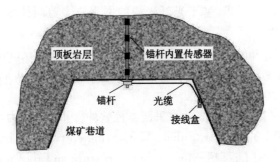

图 13 光纤光栅测力锚杆的安装示意图
Fig. 13 Installation diagram of force-measuring bolt based on FBG

5 结 论

本研究以某轨道巷为工程技术背景,围绕研究目标,经过理论分析、实验室实验、数值模拟和现场监测相结合的方法,分析了光纤光栅传感理论及特性,开发了光纤光栅数据处理软件,研制了一批光纤光栅智能监测设备,设计并建立了基于光纤传感技术的高精度、本质安全型煤矿安全智能监测系统,取得的研究成果主要包括以下几个方面:

(1) 开发了光纤光栅的煤矿安全智能监测系统数据处理软件,其通用高效的架构设计来管理监测数据和具有友好的人机交互界面,根据不同地质区域设置不同的预警报警方式,为煤炭企业监控工作人员、科研人员使用,方便得实时观察煤矿井下围岩安全状态。本系统通过以太网连接从光纤光栅信号解调主机接收光纤传感网络的监测数据,实现与光纤传感巷道监测网络的对接,包括开采环境实时监测、监测数据存储管理、数据分析与预警、传感器网络控制、报表分析等功能。

(2)研制了光纤光栅锚杆测力计、光纤光栅顶板离层仪、光纤光栅钻孔应力计、光纤光栅测力锚杆等矿用光纤光栅监测设备,并获得了光纤智能监测相关的国家授权发明专利16项,软件著作权4项。对矿用光纤光栅监测设备进行力学分析可以得出本系统所监测的物理量与波长变化呈线性关系,通过标定测试实验对其进行了验证,结果表明设计的矿用光纤光栅传感器波长变化与物理量有很好的正相关,且监测灵敏度高、误差小。

(3)在某轨道巷进行了光纤光栅的煤矿安全智能监测系统的现场实践,通过对光纤光栅的煤矿安全智能监测系统井下安装施工、地面安装施工、软件调试,实现了光纤光栅的煤矿安全智能监测系统的井下监测与传感采集系统及井上数据存储与处理系统的正常运行,并得到了全部的矿压监测数据,从而验证了光纤光栅的煤矿安全智能监测系统的可靠性和可行性,对监测数据进行整理分析,得到了巷道的矿压显现规律,同时检测了锚杆支护质量。结果表明基于光纤光栅的煤矿安全智能监测系统能够较准确的得到巷道矿压数据,达到了预期的效果。

参考文献

[1] 方新秋,何杰,郭敏江,等.煤矿无人工作面开采技术研究[J].科技导报,2008(09).
FANG Xinqiu,HE Jie,GUO Minjiang,et al. Research on manless working face mining technology in coal mine[J]. Science & Technology Review,2008(09).

[2] 李虎威,方新秋,梁敏富,等.基于光纤光栅的围岩应力监测技术研究[J].工矿自动化,2015,41(11):17-20.
LI Huwei,FANG Xinqiu,LIANG Minfu,et al. Research on monitoring technology of surrounding rock stress based on fiber grating[J]. Industry and Mine Automation,2015,41(11):17-20.

[3] 方新秋,瞿群迪,何富连.高产高效综采面监控软件[J].矿山压力与顶板管理,1998(3).
FANG Xinqiu,QU Qundi,HE Fulian. Monitoring software of high yield and efficiency fully mechanized coal face[J]. Mine ground pressure and roof management,1998(3).

[4] 方新秋,何富连,梁袁.综采面支架-围岩保障系统软件设计[J].矿山压力与顶板管理,1997(3):59-61.
FANG Xinqiu, HE Fulian, LIANG Yuan. Support system of software design of support-surrounding rock[J]. Mine ground pressure and roof management,1997(3):59-61.

[5] 方新秋,万德钧,王庆.离散元法在分析综放采场矿压中的应用[J].湘潭矿业学院学报,2003(4):11-14.
FANG Xinqiu,WAN Dejun,WANG Qing. DEM application of mine ground pressure analysis in full-mechanized caving mining field[J]. Journal of Xiangtan Mining Institute,2003(4):11-14.

[6] 林传年,刘泉声,高玮,等.光纤传感技术在锚杆轴力监测中的应用[J].岩土力学,2008,629(11):3161-3164.
LIN Cuannian,LIU Quansheng,GAO Wei,et al. Application of optical fiber sensing technology in axial load monitoring of bolt[J]. Rock and Soil Mechanics,2008,629(11):3161-3164.

[7] 鞠文君.煤巷锚杆支护监测仪器的开发与应用[J].煤矿开采,2004,9(3):52-54.
JU Wenjun. Development and application of monitoring equipment of bolt support in coal roadway[J]. Coal Mining Technology,2004,9(3):52-54.

[8] 冯仁俊.基于光纤光栅测试的全长锚固锚杆实验研究[D].西安:西安科技大学,2006.

[9] 方新秋,吴刚,梁敏富,等.基于光纤光栅传感的煤矿采空区顶板应变监测系统及方法,ZL103528732A[P].2014.
FANG Xinqiu,WU Gang,LIANG Minfu,et al. Strain monitoring system and method of gob roof that based on FBG[P]. 2014.

[10] 孔恒,马念杰.基于顶板离层监测的锚固巷道稳定性控制[J].中国安全科学学报,2002,12(3):55-58.
KONG Heng,MA Nianjie. Stability control of anchoring roadway that based on roof Separation Monitoring[J]. China Safety Science Journal,2002,12(3):55-58.

[11] 王云海.顶板离层仪的工作原理及应用研究[J].煤炭与化工,2013(03):100-101.
WANG Yunhai. Working mechanism and application research of roof separation indicator[J]. Coal and chemical,2013(03):100-101.

[12] 倪正华,方新秋,李嘉薇.基于光纤光栅的顶板离层监测系统[J].仪表技术与传感器,2013(2).
NI Zhenghua,FANG Xinqiu,LI Jiawei. Monitoring system of roof separation based on FBG[J]. Instrument Technique and Sensor,2013(2).

[13] 詹亚歌,吴华,裴金诚,等.高精度准分布式光纤光栅传感系统的研究[J].光电子·激光,2008,19(6):758-762.
ZHAN Yage,WU Hua,PEI Jincheng,et al. Research on high precision and quasi-distributed FBG sensor system[J]. Journal of Optoelectronics Laser,2008,19(6):758-762.

[14] 柴敬,邱标,李毅,等.钻孔植入光纤Bragg光栅检测岩层变形的模拟实验[J].采矿与安全工程学报,2012,29(1):44-47.
CAI Jing,QIU Biao,LI Yi,et al. Simulation Experiment of Embedded Fiber Bragg Grating Monitoring in Rock Deformation Through Borehole[J]. Journal of Mining & Safety Engineering,2012,29(1):44-47.

[15] 莫淑华,张晓晔,高农.光纤传感技术在力学测试中的发展与应用[J].哈尔滨师范大学自然科学学报,2000,16(2):51-55.
MO Shuhua,ZHANG Xiaohua,GAO Nong. Development and application of FBJ in mechanical testing[J]. Natural Science Journal of Harbin Normal University,2000,16(2):51-55.

Repairing technology based on analysis of deformation or failure on main roadways in Shuanglong Coal Mine

YUN Dongfeng[1,2], WANG Zhen[1], WU Yongping[1,2],
REN Fengtian[1], XING Taijun[3], CAO Mingshi[3]

(1. School of Energy, Xi'an University of Science and Technology, Xi'an 710054, China;
2. Key Laboratory of Western Mine Exploitation and Hazard Prevention Ministry of Education, Xi'an University of Science and Technology, Xi'an 710054, China;
3. Shanxi Shuanglong Coal Development Co. Ltd., Huangling 727307, China)

Abstract: During mining process of Shuanglong coal mine of Shannxi Coal Chemical Firm, the main roof transportation belt roadway and the main coal ventilation rail roadway supported by bolt-shotcrete all suffered severe deformation or failure, shotcrete layer cracking, the roof hanging bag, the lateral wall bulging, the floor heave and other ground pressure phenomenon were frequent, resulting in the normal transportation and ventilation of the mine roadways are disturbed, which are forced to repair main roadways. According to the current situation of layout and production in Shuanglong coal main, the theoretical analysis of the system, numerical calculation, physical similar simulation experiment, and on-site observation of the deformation or failure of two main roadways show that the main reasons for the deformation of the main roadways are the superposition of mining pressure on both sides of the main roadway, the swelling of the soft rock with water, the interlacing of the surface peaks and valleys, transportation roadway arranged in the roof, and other compound effects. Based on the results of deformation analysis of main roadways, the maintenance technology program was determined to use the form of "anchored bolt and cable ＋ metal mesh ＋ W steel strip ＋ shotcrete" after removing the cracked shotcrete layer, the net bag and loose rock, the section of serious deformation supplements the U-shaped steel semi-circular arch retractable supports to strengthen the support. On-site tests showed that the maintenance technology scheme can effectively control the further deformation or failure of the main roadways, which is safe, reliable, economical and reasonable, and has a good effect, thus ensuring the smoothness of the mine transportation and ventilation system.

Keywords: main entry; soft rock; bolt-shotcrete; coal pillar; support; repair

  Main transportation roadway and main ventilation roadway are necessary passages for transportation, pedestrians, ventilation and other systems in the process of underground mining coal, and they have a long service life, therefore, ensuring their smooth flow is a prerequisite for the safe and efficient production of mine[1-2]. With the development of coal mining technology, mature theories have been formed in the deformation or failure mechanism of roadways, coal pillars, support methods, soft rock support and so on[3-6], but the problems of coal mining engineering often due to the complexity of objective geological conditions and human factors disturbance, which makes a certain theoretical technology has some limitations for the solution of complex engineering problems.

Corresponding author: Tel: 18192091825, E-mail: 2597893810@qq.com.

Main roadways of bolt-shotcrete support have been rated as high-quality model project about Shuanglong coal mine. As the strips progresses and the gob area expands, main transportation roadway and main ventilation roadway continue deforming, and local tops and the lateral walls cracking, shedding, and the phenomenon of the floor heave are significant, resulting in narrowing of main roadways sections, hindered pedestrians, transportation, and impeded ventilation of mines, which causes great potential safe hazards and is forced to stop producing for maintenance of main roadways. For this reason, the transportation roadway is arranged along the roof instead of being arranged in coal seam, deformation or failure of Shuanglong coal mine roadways is shown in Fig. 1. However, in Shuanglong coal mine, the coal, the roof and the floor rock are relatively soft, especially, the coal and mudstone are easily swelling with water, after repeated maintenance, the effect is not satisfactory, which directly affects the normal production and seriously limits the development of the mine. Therefore, aiming at the deformation or failure of the main roadways in Shuanglong coal mine, that the reasons for the deformation and destruction of the Shuanglong coal mine are studied and put forward an economically reliable and reasonable remediation plan are urgent problems to guarantee the normal production of mine.

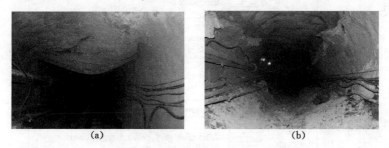

Fig. 1 Deformation or failure on main roadways
(a) the roof hanging bag; (b) shotcrete layer cracking

1 Engineering situation

Shuanglong coal minefield is located in the south of Shuanglong Town, Huangling County, Shaanxi Province, and there are no villages in the mine field. Mine field is a simple irregular syncline structure with an inclination of 2°~5°. The coal seam is classed into fat gas coal by the type of coal; the floor is seat rock and mudstone; the roof is gray black mudstone and fine siltstone interbed; the roof, the floor and coal seam are softer. The hydrogeology within the mine field is not very complicated and has little impact on coal mining, so the mine field hydrogeology is mostly classed into a crack that is a simple type. The mine 855 main transportation roadway is located in the roof fine siltstone interbed layer, and the 850 main ventilation roadway is located in the No. 2 coal seam. 855 transportation roadway and 850 ventilation roadway are repaired with bolt-shotcrete support. The bolt adopts $\phi 18\times 2\,000$ mm round steel resin bolt, and the distance between the bolts is 800 mm × 800 mm; Laying 12$^\#$ diamond iron wire netting, specifications 10 m × 1 m, shotcrete thickness of 100 mm; at the same time, the constructed anchor cable strengthens the support, and the anchor cable adopts $\phi 15.24\times 7\,300$ mm steel strand, and the distance between anchor cables is 1 200 mm × 3 000 mm. Shuanglong coal mine integrated strata histogram shown in Fig. 2.

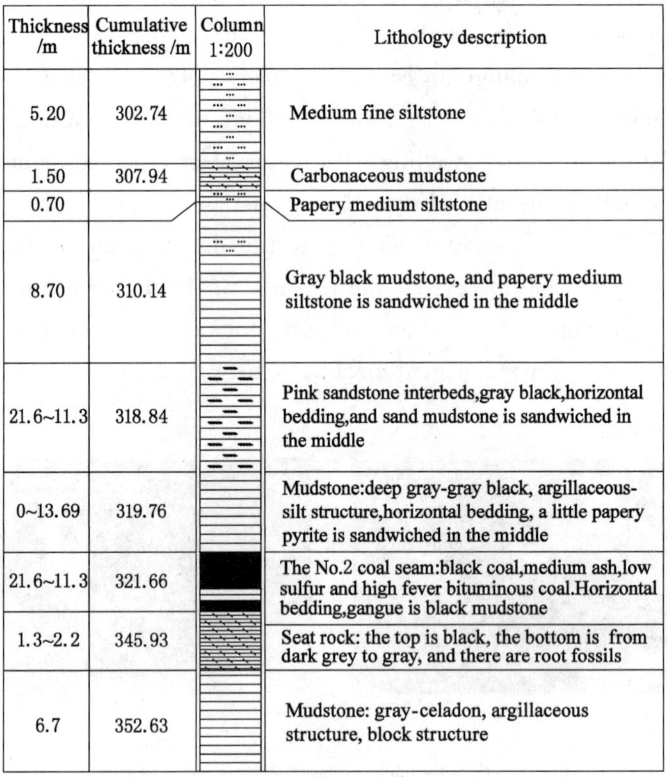

Thickness /m	Cumulative thickness /m	Column 1:200	Lithology description
5.20	302.74		Medium fine siltstone
1.50	307.94		Carbonaceous mudstone
0.70			Papery medium siltstone
8.70	310.14		Gray black mudstone, and papery medium siltstone is sandwiched in the middle
21.6~11.3	318.84		Pink sandstone interbeds, gray black, horizontal bedding, and sand mudstone is sandwiched in the middle
0~13.69	319.76		Mudstone: deep gray-gray black, argillaceous-silt structure, horizontal bedding, a little papery pyrite is sandwiched in the middle
21.6~11.3	321.66		The No.2 coal seam: black coal, medium ash, low sulfur and high fever bituminous coal. Horizontal bedding, gangue is black mudstone
1.3~2.2	345.93		Seat rock: the top is black, the bottom is from dark grey to gray, and there are root fossils
6.7	352.63		Mudstone: gray-celadon, argillaceous structure, block structure

Fig. 2　Shuanglong coal mine integrated strata histogram

2　Analysis of deformation or failure on main roadways

In order to ensure the normal mining of the Shuanglong coal mine, the research team has studied the destruction mechanism of the roadways and the stress variation characteristics and the movement of the high stress about surrounding rock through on-site investigation, experimental research, theoretical analysis, and numerical calculation. Based on the research of literature[7-8], this paper comprehensively considers the various types of factors that main roadways encounter severe deformation or failure, and lays the foundation for the maintenance of main roadways and the mitigation of continuous deformation.

2.1　The superposition of abutment pressure

Shuanglong coal mine is arranged with an advancing strips layout, 1401~1405 fully-mechanized coal faces are arranged on the right side of main roadways, and 1304 full-mechanized coal face is arranged on the left side of main roadway, and the direction of advancement of the 1304 full-mechanized coal face is from the outside to the inside and parallel to the transportation roadway. As the face is mined forward, the rear roof rock will fall over a large area, forming a high abutment pressure that can move with the face progresses in the front and the rear of the face, and eventually abutment pressure on the pillar of main roadways. In particular, with the formation of gob on both sides of main roadways, the abutment pressures of the two sides of the main roadways are superimposed, and the "island" effect of the coal pillars is obvious. Two sides layout of strips of Shuanglong coal mine 855 transportation roadway and 850 ventilation roadway as shown in Fig. 3.

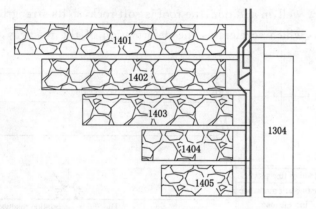

Fig. 3 Layout of strips about the main roadways of Shuanglong coal mine

The results of the literature show that the internal pressure of an island pillar is an arch-shaped distribution and the vertical pressure at the center of the pillar is the largest[9]. After Shuanglong coal mine was disturbed by surrounding rock pressure such as roadway excavation, mining, repeated support, etc. , a certain width of plastic deformation zone appeared on both sides of the coal pillars, and the carrying capacity decreases, so the actual protected coal pillars became smaller, only the central part of the coal pillars body serves to protect the roadways. In addition, the coal quality of Shuanglong coal mine is softer, and the characteristics of advancing mining will increase the creep effect of the overlying strata on the pillars of the roadway, in the long run, the carrying capacity of the coal pillars will drop sharply, which will aggravate the deformation of the roadways. The superposition of abutment pressure in the coal pillar of the roadway is as shown in Fig. 4.

The main ventilation roadway is arranged in the coal seam, the transportation roadway is adjusted to the roof rock due to the floor heave affects the normal production. The superposition of the mining pressure has a greater impact on the deformation or failure of coal roadway, especially the soft rock roadway, Shuanglong coal mine roadway layout as shown in Fig. 5. The transportation roadway is arranged in the rock strata of the roof, and

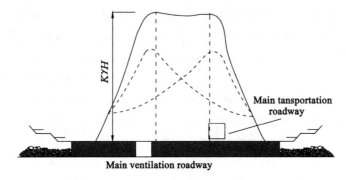

Fig. 4　The superposition of abutment pressure of the coal pillar

the plastic deformation zone on both sides of the pillars body that can't support the overlying loads very well, in addition, the roof is soft rock, so its strength and poor carrying performance is low, which accelerates the deformation of the roadway, so maintenance is more difficult.

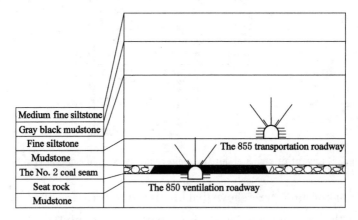

Fig. 5　Shuanglong coal mine roadway layout

Due to the strong deformation of the roadway caused by the abutment pressure, the size of pillar is an important factor affecting the roadway stability[10]. The strips on both sides of the roadways are mined at different time. One side is mined first and then the other side is mined, which causes the mining pressure to form the pressure superposition on the pillar; The pressure of the overlying strata on both sides of the gob of is transferred to the middle of the coal pillars, so coal pillars at the front of the face will be double affected by the overburden pressure and the lateral pressure of the coal pillars, which will cause further deformation or failure of pillars. The main roadways experienced multiple disturbances that aggravated the expansion of the plastic zone of the coal pillars and increased the difficulty of roadways maintenance. For coal quality, the roof and floor rock are softer than ordinary coal seam conditions, especially in the Shuanglong coal mines where the interlacing of peaks and valleys on the surface, the special roadway location, the mining and layout are not conducive to the stability of the roadways, so the width of the protection pillar should be appropriately increased to ensure the stability of surrounding rock of main roadways.

2.2 The swelling of the mudstone with water

In Shuanglong coal main, the floor heave, the roof and the lateral walls appear the tuck net that both occur at ventilation roadway and transport roadway. After excavating floor and dismantling the tuck net, the roadways disaster phenomenon occur repeatedly, which has seriously affected the normal mining of the mine. The research team conducts rock sample processing on the more complete rock blocks that can be processed on site in key Laboratory of western mine exploitation and hazard prevention ministry of education, and coal samples are sampled standardly and then tested with the press in preparation apparatus room, incomplete rock sample tested with point load tester. On-site observations and tests show that the roof and floor are thick mudstones, and the strata beddings are widely developed, the swelling of rock with water not only reduces the strength of rock, but also deteriorates the carrying capacity of rock. In particular, the strength of floor mudstone is weakened due to the soaking of floor in the gob water and production water, and roadway pillars are under higher pressure, on the one hand, which leads to the pressure is directly transmitted to the floor rock body, causing the floor heave of roadway; on the other hand, soft coal and soft rock have poor carrying capacity, thus the roof and the lateral walls are separated to form a tuck net under heavy pressure.

2.3 The interlacing of the surface peaks and valleys

The Shuanglong coal mine is located in the mountainous area, and the interlacing of the surface peaks and valleys above the gob, so the caving of overburden strata transfers the surface (1402～1403 strips have been mined). There is a wide range of landslides and collapses whose area is about 300 m^2, and the height of the landslide is about 4～8 m on both sides of the surface valley. The landslides occur on the ground surface between the 1402 face and the 1403 face, which is 300 m from main roadway pillars, and the collapsed area corresponds to a serious deformation of main ventilation roadway. The surface collapse and landslides of Shuanglong coal mine are shown in Fig. 6 below.

(a)　　　　　　　　　　　(b)

Fig. 6　Hillside collapse on the ground surface

(a) Hillside collapse on the left;(b) Hillside collapse on the right

Due to the completion of mining when the 1403 face advances about 600 m, and the overlying strata fractures in the middle of the advance direction of face and intersected with the surface, resulting in huge loads that are from caving zone and fractured zone above the gob exerts the pillars. In particular, the fractured and reversed (anticlockwise rotation) movements of the cantilevered overlying strata are applied to the coal pillars with huge loads, so 30~40 m coal pillars cannot bear the overlying load, which exacerbates the deformation or failure of main roadways. Judging from the time and space of landslides and deformation or failure of roadway that have been discovered, the overburden migration here is still active.

2.4 The effect of large mining height

The mining height of the face of the Shuanglong coal mine is about 2 m, and the mining height is not large, however, due to the bulking coefficient of thick mudstone after falling is very small, in particular, the mudstone is difficult to support the overburden pressure and is compressed after being softened by water, resulting in range of the caved zone increases, which forms the effects of large mining heights, even transfers to ground surface. In Shuanglong coal mine, the interlacing of the surface peaks and valleys, uneven pressure act on the main roadways, leading to pillars are under higher pressure, which causes strength of pillars are weakened, and aggravates the deformation or failure of the roadways.

From the above analysis, it can be known that the most fundamental reason for the deformation or failure of main roadways in the Shuanglong coal mine is unreasonable mining layout, and Shuanglong coal mine is equipped with an advancing layout of strips, which makes compressive time of main roadway pillars prolong and increase maintenance of main roadways. In the process of mining, factitious factors and unfavorable objective geological conditions make the deformation or failure of the main roadway intensify. Before mining and disturbing, the swelling of mudstone with water is the main factor of deformation or failure of main roadways; The mining pressure formed by the mining of the face and the large mining height effect caused by the collapse of the overlying strata in the gob are the main factors of the deformation or failure of the roadway; the interlacing of the surface peaks and valleys in the Shuanglong coal mine, and the objective factors such as the roof and floor and the soft coal quality accompany the end of mining of the coal seam where the main roadways are arranged; the mining layout, the small size of the coal pillars, the spillage of the production water, the special roadway location, the repeated overhaul of roadways and other factitious factors further exacerbate the deformation or failure of the main roadways.

There are many complicated and interrelated reasons for the deformation or failure of the main roadways in the Shuanglong coal mine, and their relationship between them is intricate and interrelated. When the coal seam, the roof and the floor are soft and the strength is low, the range of the loose surrounding rock is larger (6~8 m) and the carrying capacity is lower. The support of anchor-shotcrete net + anchor cable only temporarily controls the large-scale separation of the roof and can not longer meet the requirements of maintenance, thus, with the coal seam mining, the creep effects of surrounding rock on both sides of the roadway are obvious.

3 Physical similar simulation experiment

3.1 Experiment procedure

The experiment uses a 5 m similar simulation experiment frame, the size of 5 000 mm (length)×1 600 mm (height)×200 mm (width)[11], the geometric ratio of 1 ∶ 100. The main testing methods are: using 108-channel pressure data acquisition system and CL-YB-114 pressure sensor to monitor the pressure distribution of coal pillars, and adopting CM-1L-10 static resistance strain gauge and the newly developed small ratio roadway simulation testing device [12] monitor the surrounding rock pressure changes in main roadways. Before the start of the experiment, testing and excavating the roadway, adding weight to the top of the model, installing a small ratio roadway similar simulation testing device.

First, the 1403 face was mined, and open-off cut is mined at a distance of 30 cm from the left border, which the length of the open-off cut was 10 cm, and it is advanced from left to right, each time advances 5 cm. After the face is advanced 170 cm, reaching the stop line that was 40 cm from the center of main ventilation roadway. The collapse pattern of the overlying strata in the time that 1 403 face reaches to the stop line is shown in Fig. 7(a).

Then the 1304 face is mined, and the 1304 face is on the right side of the roadway. Excavate Open-off cut at 10 cm from the right border, open-off cut length 10 cm. When the face advances 30 cm to the left, the roof is falling. When the face is advanced 95 cm, the reading of the testing device on main ventilation roadway increased significantly, and the height of the roof fell by 30 cm. When the face advances 105 cm, the saw blade is snapped off during the coal mining process, and the model shows dropping dregs phenomenon, and the height of the roof fell by 52 cm. The face is advanced to the stop line when face is advanced 170 cm. The collapse pattern of the overlying strata in the time that 1304 face reaches to the stop line is shown in Fig. 7(b).

In order to simulate the large mining effect caused by the roof where strength of the mudstone falls and then is weaken, the floor rock layer of the left gob of the model is removed, during this process, the roof began to drop slowly, and the pressure of main roadways increases obviously. The rock mass above the roadway is an inverted trapezoid, as shown in Fig. 8 below.

3.2 Characteristics of pressure variation above overlying strata

After the face is mined and overlying strata has stabilized, collecting data from pressure sensor. The curve of the overburden pressure after the 1403 face is mined is

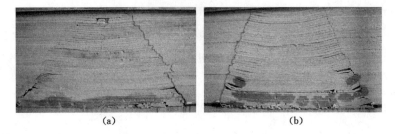

Fig. 7　The collapse pattern of the overlying strata

(a) The collapse pattern of the overlying strata in the time that；(b) The collapse pattern of the overlying strata in the time that 1403 face reaches to the stop line 1304 face reaches to the stop line

Fig. 8　The rock mass above the roadway is an inverted trapezoid

shown in Fig. 9 below.

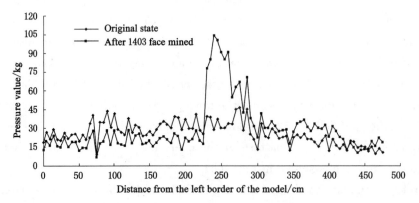

Fig. 9　The curve of the overburden pressure after the 1403 face mined

As can be seen from Fig. 9, after the mining of the 1403 face, the value of the pressure sensor in the gob is greatly increased, the maximum pressure is 104 kg, and the increase is 63 kg. The abutment pressure influence range is approximately 70 cm in front of the gob, and two roadways are within this range, which indicates that the roadways have been under heavy pressure at this time.

The curve of the overburden pressure after the 1304 face is mined is shown in Fig. 10 below.

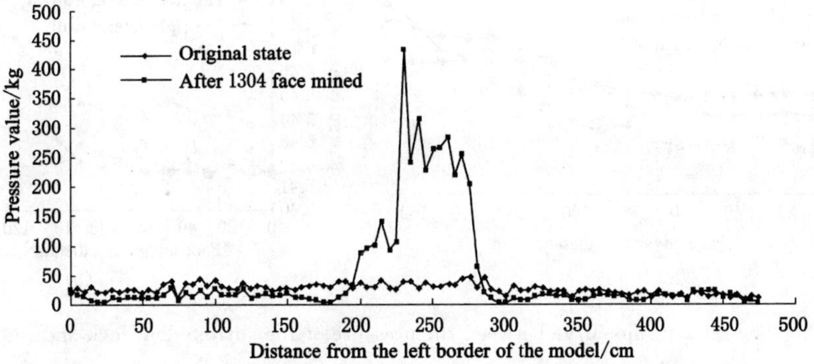

Fig. 10 The curve of the overburden pressure after the 1304 face mined

It can be seen from Fig. 10 that after the 1304 face is mined, the abutment pressure of the gob on both sides of the main roadways is superimposed on the pillars of the roadway, and the number of sensors below the pillars increases sharply, and the maximum pressure is 432 kg, and the increase is 400 kg, indicating main roadways are affected seriously by dynamic pressure.

3.3 Variation characteristics of surrounding rock pressure about main roadways

The relationship curve between the pressure on the roof and the lateral walls of main roadways and the advancing distance of the face is shown in Fig. 11 and Fig. 12.

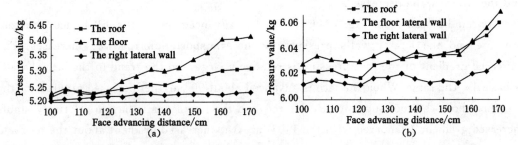

Fig. 11 The relationship curve between the pressure of the surrounding rock and the lateral walls of main roadways and the advancing distance of the face about the main roadways
(a) Main ventilation roadway; (b) Main transportation roadway

As can be seen from Fig. 11(a), with the advancement of the 1403 face, the pressure value of the roof and the lateral walls of the ventilation roadway increased, and the largest increase rate occurs in the top. When the face is advanced to 115 cm, the top pressure change rate is significantly accelerated, the increase of pressure on the left is more pronounced than that on the right.

As can be seen from Fig. 11(b), with the advancement of the 1403 face, the pressure values of the roof and the lateral walls of the transportation roadway all increased, and the increase rate is basically the same. Compared with the ventilation roadway, the increase

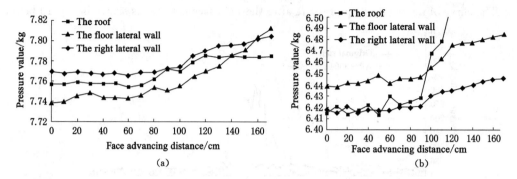

Fig. 12 The relationship curve between the pressure of the surrounding rock and the lateral walls of main roadways and the advancing distance of the face about the main roadways
(a) Main ventilation roadway; (b) Main transportation roadway

rate is significantly smaller. It can be seen that the mining of the 1403 face has a certain influence on the transportation roadways, but it is less affected than the ventilation roadways. When the face was advanced to 155 cm, the influence began to increase.

As can be seen from Fig. 12(a), with the advancement of the 1304 face, the roof pressure and the lateral walls pressure increase in main ventilation roadway. The increased gradient of the roof pressure is relatively large, and the trend in the lateral walls is basically the same. The growth of pressure accelerated when the face is advanced about 80 cm. Compared with the 1403 face, the increased gradient of the pressure about the roof and the lateral walls decreased.

As can be seen from Fig. 12(b), with the advancement of the 1304 face, the roof pressure and the lateral walls pressure of the transportation roadway increases. The increased gradient of the roof pressure is relatively large, and the trend in the lateral walls is basically the same. When the face is advanced about 90 cm, the increased gradient of the pressure about the roof suddenly accelerated, and the lateral walls have a relatively stable increased gradient. Compared with the 1403 face, the increased gradient about the roof and the two lateral walls has increased. Thus, it can be seen that the mining impact of the 1304 face on the transportation roadway is even more significant.

The physical similarity simulation experiment was used to analyze the influence of the 1403 and 1304 faces on the surrounding rock stability of main roadways, and the relationship between the pressure on the surrounding rock of main roadways and the face advanced distance, which further confirmed unreasonable mining layout is the main reason leading to the deformation of main roadways.

4 Mine pressure observation

Observing stations are set according to the different deformation or failure stages of the main roadways, three observation stations are set for main transportation roadways, and two observation lines are set for each station. The first observation station of main transportation roadway is located at the fifth single prop on the side of the gob, and the second observation station is located at 30 m from the coal mining face; the third station is located 30 m from the second station. The second and third stations moves forward in turn with advancing of the coal mining face. The observation station set up in the main ventilation roadway is basically the same as the transportation roadway. The displacement of rock strata at different depths of the roof was observed by the roof separation indicator, and the convergence or variation of roadway section was measured by the measuring rod.

The observational work lasts for 38 days. The observation results about the surface displacement of the transportation roadway are shown in Fig. 13 ((1) shows NO. 1 observation lines of NO. 1 observation stations). From the data, the roadway has a large amount of convergence, the maximum moving distance of the roof is 442 mm, the maximum moving distance of the floor is 394 mm, and the maximum moving distance of the lateral walls is 341 mm.

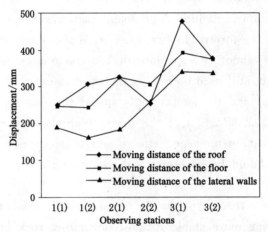

Fig. 13 The observation results about the surface displacement of the transportation roadway

It can also be seen from the Fig. 13 that the moving distance measured at Station No. 3 is much larger than stations No. 1 and No. 2, because No. 3 station is in the stage of mining influence, and under advancing abutment pressure caused by mining, the rock pressure is redistributed again and the deformation of the surrounding rock is significantly increased. With the advancement of the face, the deformation speed of surrounding rock of the roadway gradually increases, which indicates that the mining pressure has an important influence on main roadway deformation. In addition, the moving distance of the roadway roof is obviously larger than the lateral walls, and the main position of the deformation of the roadway is the top, so the roof should be strengthened the support.

5 Roadways maintenance project

5.1 Problem of the old maintenance project

The old maintenance project is mainly to passively drop factitiously loose mudstone layers of the top, cutting off the original rock bolt or anchor cable, leaving more stable rock, and then re-anchoring rock bolts and anchor cables, hanging the barbed wire, and spraying a thin layer of cement mortar, finally, implementing test of flat iron steel strip net under anchor net at local, using the technology of excavating floor to treat the floor heave. The main problems of this maintenance project are that roadway cross section increase sharply, large amount of slag discharge, high difficulty in maintenance, slow speed, high cost, shallow surrounding rock not anchored well, and only temporary reinforcement to expand the roadway cross section, which has not solved the problem of large deformation or failure of roadway and difficulty in maintenance from the roots, in the long run, the maintenance effect is poor.

5.2 Roadways maintenance new project

Based on the analysis of the cause of the deformation or failure about the main roadway, the similar simulation experiment, and the mine pressure on-site observation and the degree of deformation or failure about main roadways, the roadway deformation or failure is classed three categories. The first category is that roadway has been deformed or destroyed severely and cannot be used normally; the second category is that roadway is damaged moderately but still used normally; the third category is that roadway damaged slightly, that is, the top and the lateral walls appear tiny cracks. Roadway maintenance combined with the actual situation of the mine, comprehensive consideration from the technical feasibility, cost, maintenance effects and other aspects shows that the first category of roadway should be governed, and the second and third category should be prevented.

The first category of roadway is mainly to passively drop factitiously loose mudstone rocks of the top, leaving more stable rocks, re-anchoring rock bolt + anchor cable + hanging-net + W steel strip (spraying shotcrete at local), and besides, adding retractable support at large spans; at fracture development zone, using steel strip to support, lengthening of the rock bolt or anchor cable range, preferably full-length anchoring.

The second category of roadway should first dispose the tuck net and the shed shotcrete layer, and then add the rock bolts + anchor cables+ hung-net + spraying, at the same time, add props or retractable supports in the larger deformation zone for strengthening the support. At the surface fractures of main roadways, adding W steel strips to support, similarly, lengthening of the rock bolt or anchor cable range, preferably full-length anchoring. As much as possible not to drop factitiously loose rock layers, transforming passivity into initiative to reduce the amounts of maintenance. The maintenance of roadways is mainly to prevent, and the governance is supplement.

When necessary, the third category of roadway should first pry off the local ruptured spray layer, tuck net and loose rock block, then lay the rough wire mesh and the W steel strip, and re-anchor rock bolt and the anchor, at the same time, observe the degree of local roadway deformation, considering erects U-shaped retractable as supplement support, prevention is the main.

Combined with the actual situation of the Shuanglong coal mine, deciding delays the production of 1304 face, and finally mines 1304 face, and advances forward from the inside out to avoid the influence of the "island" effect on the ventilation and transportation roadway, which fundamentally eliminates the damage to the main roadways caused by the superposition of abutment pressure. As much as possible to reduce the amounts of maintenance, at the same time, properly intercept water, conduct water, and drain water. In particular, it is necessary to enhance the drainage of main roadways and minimize water soaking to the floor.

5.3 Analysis of the stability of main roadways after erecting shed support

(1) Displacement characteristics

It can be seen from Fig. 14 that the displacement of the surrounding rock of the 850 main ventilation roadway and the 855 main transportation roadway after erecting shed support is effectively controlled. The displacement of the lateral wall of main ventilation roadway is about 16 cm, the displacement of the roadway roof and floor is also relatively small, around 8 cm; the displacement of the top and the lateral wall of main transportation roadway is about 6 cm, the displacement of the left side and the bottom floor is also small, about 4 cm, the displacement of the overlying strata on the top of the roadway is about 6cm; which explain that erecting shed plays a very good supporting role.

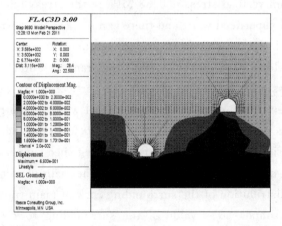

Fig. 14 Displacement distribution of the main roadway after erecting shed

(2) Destruction field characteristics

Fig. 15 shows that the plastic failure zone of surrounding rock about the roadway is significantly reduced after erecting shed support. Sheared failure occurred in the lateral walls of main ventilation roadway, and the deformation is not obvious; the plastic failure range of the floor is small, which shows that the floor heave of main ventilation roadway is well controlled; The roadway roof has both tensile failure and sheared failure, and the roadway roof-corner and floor-corner occur sheared failure and have a wide range. The surrounding rock of main transportation roadway occur mainly sheared failure, the plastic failure of the roadway roof is small, and the failure zone of the floor and the left lateral wall is large.

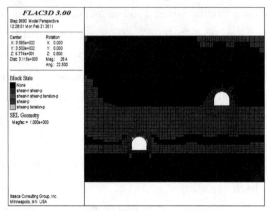

Fig. 15 Destruction field distribution of the main roadway after erecting shed

(3) Stress characteristics

Fig. 16 below shows the stress characteristics of main roadways after erecting shed support. Fig. 16(a) shows that the vertical stress distribution changes significantly after erecting shed support, and the vertical stress value significantly decreases. The range of stress release zone is significantly reduced in the roof and floor of main ventilation roadway, the maximum vertical stress is 0.38 MPa in the stress release zone, and stress is released in the lateral superficial walls and there is no stress concentration zone in the deep filed, and the stress distribution is more uniform around the anchor cable surround; the range of the stress release zone on the floor of the transportation roadway is larger than the range of the stress release zone on the top, the vertical stress value is 0.38 MPa, and there are stress concentration zones on the two lateral walls of the roadway, however, the stress concentration zone is significantly smaller than before, and the vertical stress value is reduced to 14 MPa; Fig. 16(b) shows that after erecting shed support, the horizontal stress distribution variation of the surrounding rock about the main ventilation roadway is more obvious; the stress distribution of the surrounding rock of main transportation roadway is smaller, but the horizontal stress value of the surrounding rock about the roadway is generally higher than before.

6 Effect analysis of main roadways maintenance

The 855 transportation roadway and 850 ventilation roadway of Shuanglong Coal Mine are supported by "rock bolts + anchor cables + hung-net + spraying" at the two ends along trending direction of main roadways where the two roadways are deformed slightly, and it is same with the two ends along trending direction of main roadways in the minor and slightly severe sections of the deformation. In the fall section, the use of "rock bolt + anchor cable + hanging-net + retractable support" support.

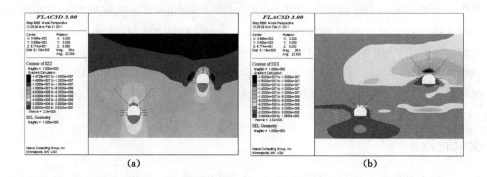

Fig. 16 Stress characteristics of the main roadway after erecting shed
(a) Horizontal stress distribution of the main roadway after erecting shed;
(b) Vertical stress distribution of the main roadway after erecting shed

(1) On-site observation

The observation shows that within 100 days after the maintenance, moving distance of the lateral walls about is was approaching 25.3~68.84 mm, and the moving distance of the top is 17~81.3 mm. In the following 100 days, the moving distance of lateral walls is closer to 3.12~17.05 mm, and the top moved closer to only 0~1.48 mm, achieving the ideal support effect.

(2) Economic benefit analysis

Main roadways maintenance project is not only technically feasible but also economically reasonable. The following is an analysis of the economic benefits of the new roadway maintenance project.

The price of rock bolt used in Shuanglong Coal Mine is 45 yuan each and 15 pieces for every meter of roadway. Anchor cable 100 yuan each, and 0.83 piece s per meter of roadway. Metal net 3 yuan per square meter, and 8 m² per meter of roadway. 1 m roadway spraying 1cm thick concrete takes 20 yuan, spraying thickness 0.94 cm in main roadways of shuanglong coal mine. W steel strips are worth 50 yuan each and 0.8 piece for every meter of roadway. 29U steel retractable support cost 2 800 yuan each, and this repair uses 100 pairs. The cost of different supporting materials for roadway per meter is shown in Table 1.

Table 1　　The cost of different supporting materials for roadway per meter

Supporting materials	Unit price	Quantity	Total (yuan)
Rock bolt	45	15	675
Metal net	3	8	83
Anchor cable	100	0.83	24
Beton	20	0.94	19.6
W steel strip	50	0.8	40
U steel retractable support	2 800	0.5	1 400

After calculation, the cost of roadway support before maintenance is 840 yuan per meter, and the support cost after maintenance is 2 240 yuan per meter, and the labor cost of 9 700 yuan per meter. Although the one-time investment is large, it greatly reduces the number of maintenance, the cost of maintenance per meter can be saved about 10 540 yuan per year. It can be seen that the economic benefit about the new project of maintenance is remarkable.

7 Conclusions

(1) The research on the deformation or failure of the main roadway in Shuanglong coal mine shows that the fundamental cause of the failure is the unreasonable the mining layout. Before mining disturbances, the swelling of mudstone with water is the main factor for the deformation or failure of main roadways. In the process of face mining, the abutment pressure formed by mining is the main factor of roadways deformation or failure, and factitious factors further exacerbate the deformation or failure of the roadways.

(2) Through physical similarity simulation experiments, the influence of the 1403 and 1304 faces on the stability of the surrounding rock of main roadways were comparatively analyzed, Summarizing the relationship between the pressure on the surrounding rock of main roadways and the advancing distance of the faces, which further confirmed unreasonable mining layout is the main reason leading to the deformation of large roadways.

(3) Based on the results of research and analysis of the deformation or failure of main roadways, It is proposed to erect U steel semi-circular arch retractable supports to strengthen the support on the basis of the support of "anchor bolt and cable + metal mesh + W steel strip + shotcrete", at the same time, adjust the current mining layout. After the numerical simulation and the new on-site experiment, the test show new project is economical and reliable, and the maintenance effect is good, so the continuous deformation or failure of the two main roadways are effectively is controlled well, ensuring smooth flow of coal mine transportation and ventilation system.

Reference

[1] Wang Lei, Zhao Jianzhong, LI Chuansen, et al. Study on roadway floor heave mechanism and preventive measure under mining influence[J]. Coal Science and Technology,2014,42(S1):16-18.

[2] Meng Yi. Damage evolution law and control technology of soft rock roadway under mining influence[J]. Coal Engineering,2017,49(11):50-54.

[3] Yuan Yue, Wang Weijun, Yuan Chao, et al. Large deformation failure mechanism of surrounding rock for gateroad under dynamic pressure in deep coal mine[J]. Journal of China Coal Society,2016,41(12):2940-2950.

[4] Zhi Guanghui, Zheng Bin. Protective coal pillar set method and development tendency

in China[J]. Coal Engineering,2009,41(06):9-11.
[5] Kang Hongpu, Wang Jinhua, Lin Jian. Study and applications of roadway support techniques for coal mines[J]. Journal of China Coal Society,2010,35(11):1809-1814.
[6] He Manchao. Progress and challenges of soft rock engineering in depth[J]. Journal of China Coal Society,2014,39(08):1409-1417.
[7] Yun Dongfeng, Wang Zhiyu, SU Puzheng, et al. Experiment Research on Repairing Technology of Soft-rock Main Roadway in Shuanglong Colliery[J]. Coal mining Technology,2013,18(05):63-65.
[8] Yun Dongfeng, Wang Zhiyu, SU Puzheng, et al. Numerical Simulation of Soft Main Roadway Repairing Technique in Shuanglong Coal Mine[J]. Safe in Coal Mines, 2013,44(08):39-41.
[9] Niu Zongtao, Guo Jun. Study on roadway layout and support technique in deep isolated coal pillar[J]. Coal Science and Technology,2017,45(10):54-59.
[10] Dong Fangting. Support theory based on broken rock zone in surround rock[J]. Proc. Of the 2nd Int. Symp. On Min. Tech. And Sci. ,1991(10):1130-1137.
[11] Yun Dongfeng, Wang Chenyang, Zhou Ying, et al. Study on Mine Strata Pressure Behavior Law of Coal Mining Face in Deep Mining[J]. Coal Engineering,2010,38(05):52-54.
[12] Yun Dongfeng, Su Puzheng, Du Qiang, et al. Testing device of similar simulation experiment about small ratio roadway[P]. China Patent:CN202255708U,2012-05-30.

Study on the height of fractured zone in overburden at the high-intensity longwall mining panel

GUO Wenbing[1,2], ZHAO Gaobo[1], LOU Gaozhong[1], WANG Shuren[3]

(1. *School of Energy Science and Engineering, Henan Polytechnic University, Jiaozuo 454003, China*; 2. *Coal Production Safety Collaborative Innovation Center in Henan Province, Henan Polytechnic University, Jiaozuo 454003, China*; 3. *International Joint Research Laboratory for Underground Space Development and Disaster Prevention of Henan Province, Henan Polytechnic University, Jiaozuo 454003, China*)

Abstract: The height of fractured zone (HFZ) at the high-intensity longwall mining panel plays a vital role in the safety analysis of coal mining under bodies of water. The processes of overburden failure transfer (OFT) were analyzed, which could be divided into the development stage and the termination stage. Through theoretical analysis, the limited suspension-distance and limited overhanging distance were proposed to judge the damage of each stratum. The mechanical models of strata suspended integrity and overhanging stability were established. A theoretical method to predict the HFZ at the high-intensity longwall mining panel was put forward based on the processes of OFT. Taking a high-intensity longwall mining panel (No. 11915 panel) as an example, the theoretical method proposed, the engineering analogy and the empirical formulas in the Regulation were used to predict the HFZ. The results show that the theoretical result is consistent with the engineering analogies' result and empirical formulas' result. The rationality and reliability of the theoretical method proposed is verified.

Keywords: fractured zone; overburden failure; high-intensity longwall mining panel; strata movement

1 Introduction

The height of fractured zone (HFZ) due to the exploitation of coal is a key parameter for the safety analysis of coal mining under a body of water[1-2]. Accordingly, it is necessary to predict the HFZ under the corresponding mining conditions before coal mining[3-5].

The HFZ caused by longwall mining under the traditional general geological mining conditions is sufficient. In China, the traditional calculation method of HFZ is empirical formulas given by the Regulation[6]. In addition, there are many research achievements on the HFZ in China. Xu et al. collected measured values of the HFZ with thick coal seam caving coal mining faces, and adopted a method of mathematical statistics regression analysis to obtain empirical formulas for calculating the HFZ[7]. Xu et al. accessed the influence of the roof key strata position on the HFZ by combining theoretical analysis with engineering detection, and a method for predicting the HFZ was put forward[8-9]. Gao et al.

Supported by the National Natural Science Foundation of China (51774111), Henan Province Science and Technology Innovation Outstanding Talent Fund (184200510003).

Corresponding author: ZHAO Gaobo, Tel: 18336835692, E-mail: zgb. hpu@foxmail. com.

investigated the relationship between overburden water-conducting fractures and the tensile deformation of rock stratum[10]. Based on this, Huang et al. proposed a calculated method to predict the HFZ[11]. Majdi et al. argued that the short-term the HFZ ranges from 6.5 to 24 times the mining height, it is 11.5 to 46.5 times the mining height in the long term[12]. Wu et al. proposed a method of determining the HFZ using the radial basis function neural network (RBFNN) model in MATLAB[13]. These pioneer works have offered a good insight into the prediction methods of the HFZ under the traditional geological mining conditions.

However, with the improvement of the coal mining technology and equipment in China, the high-intensity mining panels are increasing. The theoretical prediction method of the HFZ at the high-intensity longwall mining panel is poor. In this study, overburden failure transfer (OFT) was presented. A theoretical method to predict HFZ at the high-intensity longwall mining panel was put forward. The theoretical method proposed, the engineering analogy were used to predict the HFZ, and the results were compared with the predicted value by empirical formulas in the Regulation.

2 Analysis of overburden failure transfer

It is usual to cause some unique behaviors of ground pressure at the high-intensity longwall mining panel. For example, the fracture zone caused by high-intensity longwall mining sometimes could go straight the surface. There are only "two zones" failure mode above the mined-area above the high-intensity longwall mining panel, namely, the caving zone and the fractured zone, which different from the "three zones" mode due to the traditional underground coal mining[14]. The issues of surface subsidence, soil erosion, desertification and environmental pollution caused by high-intensity mining are more severe in eco-environmental fragile areas[15]. Therefore, it is vital to study the development mechanism of overburden failure and the HFZ at the high-intensity longwall mining panel.

2.1 Processes of overburden failure transfer

As a panel of sufficient width and length is excavated, the overlying rock strata are disturbed[16]. Before extracting a panel, overburden is in a state of primary rock stress; after setup entry, the stratum 1 is in a state of suspended integrity [as shown in Fig. 1(a)]. Afterwards, with advanced distance increasing, the stratum 1 is in a state of suspended rupture. Finally, this stratum is in collapse, the collapsed stratum 1 is in a state of overhanging stability. As advanced distance increases, the stratum 1 is in the state of overhanging rupture. At last, this stratum is broken [as shown in Fig. 1(b)].

Above all, each stratum is in the state of primary rock stress before extracting a panel, after which experiences suspended in collapse with the progression of coal face, and then the stratum is broken. When the failure of overburden transfers to the stratum n, each of strata 1 to n has failed, where the overburden failure height (H_f) is:

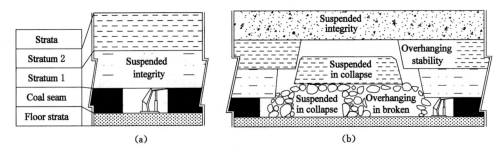

Fig. 1   The processes of OFT above the goaf

$$H_f = m + \sum_{i=1}^{n} h_i \tag{1}$$

Where m is the mining thickness; h_i is the thickness of stratum i.

2.2   Stages division of overburden failure transfer

According to above analysis, failed strata 1 to n would fill the goaf due to the accumulation of advanced distance. The separation height between the stratum $n-1$ and the stratum n is:

$$H_{s(n-1,n)} = m - \sum_{i=1}^{n} [h_i(K_i - 1)] \tag{2}$$

Where K_i is the volume expansion coefficient of failed stratum i.

When $H_{s(n-1,n)} > 0$, OFT is in the OFT development stage. At this moment, the failure of overburden transfers to the stratum $n-1$. Overburden failure would continue to transfer to upper strata and $H_{s(n-1,n)}$ reduces if the advanced distance increases.

When $H_{s(n-1,n)} < 0$ (the negative $H_{s(n-1,n)}$ would be revised as 0), the failed stratum $n-1$ fails to transfer to the stratum n, and the process of transferring ends up at stratum $n-1$. Then OFT steps into the OFT termination stage. Therefore, the processes of OFT could be divided two stages: the OFT development stage and the OFT termination stage.

3   Development mechanism of the HFZ

3.1   Failure criteria

According to above analysis and Fig. 1, after a period of extracting time, when the suspension-distance (D_s) of stratum 1 peaks at its limited suspension-distance (D_{smax}), this layer would experience suspended rupture, instability, and be suspended in collapse. After collapsing, the stratum 2 is in the state of suspended integrity. With the progression of the panel distance, when the suspension-distance of stratum 2 reaches its limited suspension-distance, this layer would also experience suspended rupture, instability, and be suspended in collapse. Therefore, destruction would transfer from the stratum 1 to the stratum 2 as the panel advance distance increases.

When the failure of overburden transfers to the stratum 2 which is suspended in collapse, one section of stratum 2 is fixed inside rock mass. The opposite section of stratum

2 is in the state of overhanging stability. As the advanced distance continues to increase, the overhanging distance (D_o) of stratum 2 reaches its limited overhanging distance (D_{omax}), and this layer experiences overhanging rupture, instability, and the overhang is broken.

Therefore, the height of overburden failure could be obtained through judging whether each stratum above goaf reaches its D_{smax} and D_{omax} or not. In order to calculate D_{smax} and D_{omax}, the mechanical models of strata suspended integrity and overhanging stability are built.

3.2 Mechanical model of strata suspended integrity

When the stratum i ($i=1,2,3,\cdots,n-1$) is in the state of suspended integrity, two opposite sections of stratum i are fixed inside the rock mass, and thus the suspended integrity mechanical model is built, as shown in Fig. 2.

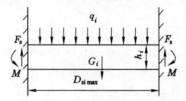

Fig. 2 Mechanical model of strata suspended integrity

As shown in Fig. 2, G_i is the gravity of stratum i of the suspended integrity part; D_{simax} is the limited suspension-distance of stratum i; q_i is the load collection degree given by strata above the stratum i of the suspended integrity part; M is the bending moment; F_s is the shear force. According to the strength criterion of the clamped beam, D_{simax} could be derived as:

$$D_{simax} = h_i \sqrt{\frac{2R_T}{q_i + k_i}} \tag{3}$$

Where E_i is the elastic modulus of stratum i; k_i is the gravity load collection degree of stratum i. $k_i = \gamma_i h_i$, where γ_i is the bulk density of stratum i; R_T is the limited tensile strength.

3.3 Mechanical model of strata overhanging stability

According to this analysis, after the stratum i is in collapse, one section of the stratum i is fixed inside the rock mass, and the opposite section is in the state of overhanging stability, and thus the overhanging stability mechanical model is established, as shown in Fig. 3.

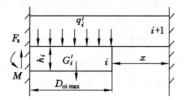

Fig. 3 Mechanical model of strata overhanging stability

According to Fig. 3, G'_i is the gravity of stratum i of the overhanging stability part; x is the suspension-distance of stratum $i+1$ after stratum i is in collapse; D_{oimax} is the limited overhang-distance of stratum i; q'_i is the load collection degree given by stratum i of the overhanging stability part. D_{simax} could be derived as:

$$D_{oimax} = h_i \sqrt{\frac{R_T}{3(q_i + k_i)}} \tag{4}$$

The judging formulas of strata suspended in collapse and overhanging being broken could be obtained using Eq. 4. The HFZ above the goaf is affected by many factors, such as the advanced distance, thickness of each stratum and coal seam, limited tensile strength of strata, elastic modulus, volume expansion coefficient, and so on.

4 Example analysis and engineering analogy

4.1 General situation

The No. 11915 longwall mining panel at Gequan Coal Mine in the Jizhong energy Group was taken as an example. The vertical depth of the 9# coal seam is 175 m below ground. The average mining thickness is about 6.5 m. The dip of the 9# coal seam is about 16°, and the length of the working face is about 950 m, and the dip length is about 80 m. The immediate roof is Daqing limestone, which thickness is about 5 m. Overlying strata are mainly sandstone. Fig. 4 shows the stratigraphic columns of overlying strata above 9# coal seam.

Fan defined the high-intensity mining as a mining method characterized by a large mining area, large mining face width, with a high extracted speed[17]. Guo and Huang, on the basis of previous studies, defined high-intensity mining as a high-yield, high-efficiency coal mining method[18]. Furthermore, there are: a thick coal seam (more than 3.5 m); fully mechanized mining of all height at once (fully mechanized top coal caving mining or high mining height support); large panel width (more than 200 m); rapid extracted velocity (more than 5 m a day); high output in a panel (usually ranges from 5~10 million tons a year-minimum is 3 million tons a year); small depth-thickness-ratio of coal seam (less than 100); serious failure of overburden and surface.

According to above definitions, the No. 11915 longwall panel has the characteristics of high-intensity mining, such as large mining thickness, large panel size, high production and efficiency, and thus it belongs to the high-intensity longwall mining panel.

4.2 Calculation of HFZ

The volume expansion coefficient of the immediate roof above the goaf of the No. 11915 panel was 1.160, which was measured in the field. According to some research results[19-20], other upper overlying strata residual volume expansion coefficients (K) could be calculated by the following Eq:

$$K = K_d - 0.017\ln h \quad (h < 100 \text{ m}) \tag{5}$$

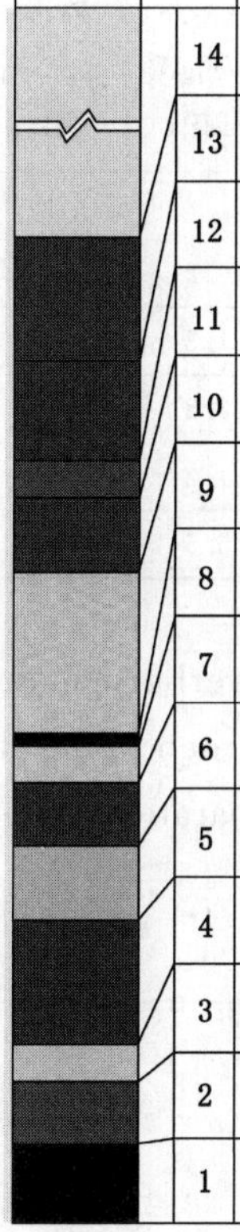

Column		Lithology	Thickness /m	Distance form 9# coal seam roof/m
	14	Alluvium soil	102	175
	13	Medium fine sandstone	10	73
	12	Siltstone	8	63
	11	Fuqing limestone	3	55
	10	Medium sandstone	6	52
	9	Pelitic siltstone	13	46
	8	7# coal seam	1	33
	7	Pelitic siltstone	3	32
	6	Medium fine sandstone	5	29
	5	Interbedding of fine siltystone	6	24
	4	Siltstone	10	18
	3	Mudstone	3	8
	2	Daqing limestone	5	5
	1	9# coal seam	6.5	0

Fig. 4　Stratigraphic columns

Where K_d is the volume expansion coefficient of immediate roof; h is the height between a overlying stratum and the coal seam.

D_{smax} and D_{omax} of each stratum above the No. 11915 panel can be calculated through combining Eqs (3) and (4) and Fig. 4. The results of D_{smax} and D_{omax} are shown in Table 1.

Table 1　　　　　　　　Physico-mechanical parameters of rocks

	Lithology	Thickness /m	Bulk density /(kN/m³)	Volume expansion coefficient	Tensile strength /MPa	Elastic modulus /GPa	D_{smax} /m	D_{omax} /m
12	Medium fine sandstone	10	24.10	1.087	6.2	30	60.28	24.61
11	Siltstone	8	24.10	1.090	2.0	23	34.91	14.25
10	Fuqing limestone	3	28.00	1.092	6.1	35	33.45	13.66
9	Medium sandstone	6	24.10	1.093	4.6	32	35.05	14.31
8	Pelitic siltstone	13	24.10	1.095	1.2	19	29.55	12.06
7	7# coal seam	1	13.00	1.101	0.05	4.2	2.77	1.13

Continued Table 1

Lithology		Thickness /m	Bulk density /(kN/m³)	Volume expansion coefficient	Tensile strength /MPa	Elastic modulus /GPa	D_{smax} /m	D_{omax} /m
6	Pelitic siltstone	3	24.10	1.101	1.2	19	15.89	6.49
5	Medium fine sandstone	5	24.10	1.103	4.2	30	25.68	10.48
4	Interbedding of fine sandstone	6	25.00	1.106	1.3	20	16.62	6.79
3	Siltstone	10	24.10	1.111	2.8	23	35.32	14.42
2	Mudstone	3	24.30	1.125	1.2	18	21.80	8.90
1	Daqing limestone	5	28.00	1.160	5.5	35	31.31	12.78

According above results, the failure processes of overburden during the No. 11915 panel extraction can be obtained. And the curve of overburden failure height with different advanced distance is shown in Fig. 6. The curve of separation height with overburden failure height is obtained, as shown in Fig 5.

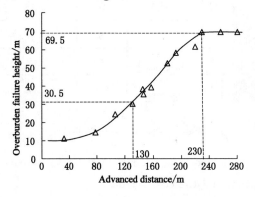

Fig. 5 Curve of overburden failure height vs. different advanced distance

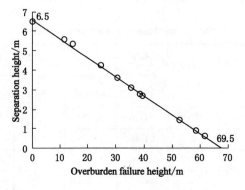

Fig. 6 Curve of separation height vs. overburden failure height

According to Fig. 5, with the increasing of advanced distance, the overburden failure height increases. When the advanced distance is 130 m, the OFT is in the stage Ⅰ: the development stage, and the overlying strata failure height is 30.5 m.

When the advanced distance reaches 230 m, the separation height between stratum 11 and stratum 12 decreases at −0.12 m, and the separation height would be revised as 0 based on the reality. Therefore, the failed stratum 11 fails to transfer to the stratum 12. The lower strata of stratum 12 are in state of failure and are full of mined-out areas.

According to Fig 6, when the No. 11915 panel is in the state of setup entry, the separation height between the immediate roof and floor is a maximum value of 6.5 m. The separation height between strata decreases linearly.

The intersection of the curve and the horizontal axis in Fig. 6 marks the maximum height of the overburden failure height (69.5 m). The HFZ, in this paper, is taken as the equivalent to the height of overburden failure. Therefore, the HFZ of the No. 11915 high-intensity longwall mining panel is 69.5 m by theoretical method proposed.

4.3 Engineering analogy

In order to adopt an engineering analogy method to predict the HFZ of the No. 11915 panel, part of the in-situ measurements similar to the geological and mining conditions of this panel were collected and arranged[21], as shown in Table 2. Lithological character is mid-hard, and the mining method belongs to the fully mechanized top caving.

Table 2　　Sample data about height of fractured water-conducting zone

Coal mine	Panel	Mining thickness/m	Dip angle/(°)	Mining depth/m	HFZ/m	Observation method
Kongzhuang	7 192	5.0	23	212	61.1	Borehole televiewer
Xieqiao	1 121(3)	6.0	13	534	67.9	Surface borehole
Renlou	7 212	4.7	17	311	56.0	Underground borehole
Nantun	6 310	5.8	5	368	70.7	Underground borehole
Nantun	9 301	5.3	15	541	67.5	Geophysical prospecting
Xinglongzhuang	1 301	6.4	9	428	72.9	Underground borehole
Xinglongzhuang	5 306	7.1	5	410	74.4	Underground borehole
Yaoqiao	7 507	4.9	5	372	63.6	Underground borehole

According to the sample data in Table 2, the factors influencing the HFZ include the mining thickness, lithological character, mining methods, the dip angle of coal seam, and the mining depth. The differences between the No. 11915 panel and above part panels in Table 2 are mainly the thickness of mining, the dip angle of coal seam, and the depth of mining.

In order to analyze the relationship between the HFZ and the mining thickness, a scatter diagram is obtained, as shown in Fig. 7.

As seen from Fig. 7, a fitting formula to predict the HFZ could be obtained as follows:

$$H = 39.833 \ln m - 1.850\ 2 \qquad (6)$$

Where H is the HFZ.

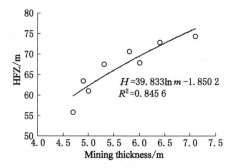

Fig. 7 Curve of HFZ vs. mining thickness

From Eq. 6, The HFZ, at the No. 11915 panel of Gequan coal mine with a 6.5 m mining thickness, can be predicted, and the height value is 72.7 m by the engineering analogy method.

The methods of the theoretical proposed and the engineering analogy are further compared with the predicted value by the empirical formulas in the Regulation. The results of comparing these three methods are shown in Table 3.

Table 3 Comparisons of HFZ calculation results with different methods

Methods	Formulas	HFZ/m
Theoretical proposed	$H = m + \sum_{i=1}^{n} h_i$	69.5
Engineering analogy	$H = 39.833\ln m - 1.8502$	72.7
Regulation (Mid-hard)	$H = 20\sqrt{\sum m} + 10$	61.0

According to the Table 3, the predicted HFZ result using the theoretical method proposed is in close agreement with the engineering analogies' result and empirical formulas' result. Therefore, the rationality of predicting the HFZ at the high-intensity longwall mining panel based on the theoretical method proposed is verified.

5 Conclusions

The processes of OFT are analyzed and can be divided into the development stage and the termination stage. Through theoretical analysis, the limited suspension-distance and limited overhanging distance of each stratum are proposed to judge the damage of each layer. The mechanical models of strata suspended integrity and overhanging stability are established. A theoretical method to predict the HFZ at the high-intensity longwall mining panel is put forward based on the processes of OFT.

Based on a high-intensity longwall mining panel (No. 11915 panel) in Gequan coal mine, the theoretical method proposed, the engineering analogy were used to predict the HFZ, and the results were compared with the predicted value by empirical formulas in the Regulation. The results show that the theoretical result is consistent with the engineering analogies' result and empirical formulas' result. The rationality of the theoretical method is verified.

References

[1] Guo W B, Shao Q, Yin S X, et al. Analysis of the Security of Mining Under a Reservoir [J]. J. Min. Safety Eng., 2006, 23(3): 324-328.

[2] Guo W B, Zou Y F, Hou Q L. Fractured zone height of longwall mining and its effects on the overburden aquifers[J]. Int. J. Min. Sci. Technol., 2012, 22(5): 603-606.

[3] Trueman R, Lyman G, Cocker A. Longwall roof control through a fundamental understanding of shield-strata interaction[J]. Int. J. Rock. Min. Sci., 2009, 46(2): 371-380.

[4] Cheng G W, Chen C X, Ma T H, et al. A case study on the strata movement mechanism and surface deformation regulation in Chengchao underground iron mine[J]. Rock Mech. Rock Eng., 2017, 50(4): 1011-1032.

[5] Adhikary D P, Guo H. Modelling of longwall mining-induced strata permeability change[J]. Rock Mech. Rock Eng., 2015, 48(1): 345-359.

[6] State Bureau of Coal Industry. Specification for coal pillar retention and coal mining in buildings, water bodies, railways and main wells and alley[M]. Beijing: Coal Industry Press, 2017.

[7] Xu Y C, Li J C, Liu S Q, et al. Calculation formula of "Two-zone" height of overlying strata and its adaptability analysis[J]. Coal Min. Technol., 2011, 16(2): 4-7+11.

[8] Xu J L, Wang X Z, Liu W T, et al. Effects of primary key stratum location on height of water flowing fracture zone[J]. Chinese J. Rock Mech. Eng., 2009, 28(2): 380-385.

[9] Xu J L, Zhu W B, Wang X Z. New method to predict the height of fractured water-conducting zone by location of key strata[J]. J. China Coal Society, 2012, 37(5): 762-769.

[10] Gao Y F, Huang W P, Liu G L, et al. The relationship between permeable fractured zone and rock stratum tensile deformation[J]. J. Min. Safety Eng., 2012, 29(3): 301-306.

[11] Huang W P, Gao Y F, Wang B, et al. Evolution rule and development height of permeable fractured zone under combined-strata structure[J]. J. Min. Safety Eng., 2017, 34(2): 330-335.

[12] Majdi A, Hassani F P, Nasiri M Y. Prediction of the height of destressed zone above the mined panel roof in longwall coal mining[J]. Int. J. Coal Geol., 2012(98): 62-72.

[13] Wu Q, Shen J J, Liu W T, et al. A RBFNN-based method for the prediction of the developed height of a water-conductive fractured zone for fully mechanized mining

with sublevel caving[J]. Arab. J. Geosci. ,2017, 10(7):172.

[14] Wang J C, Wang Z H. Stability of main roof structure during the first weighting in shallow high-intensity mining face with thin bedrock[J]. J. Min. Safety Eng. ,2015, 32(2): 175-182.

[15] Xu N Z, Gao C, Ni X Z, et al. Study on surface cracks law of fully-mechanized top coal caving mining in shallow buried depth and extra thick seam[J]. Coal Sci. Technol. , 2015, 43(12): 124-128.

[16] Wang X, Meng F B. Statistical analysis of large accidents in China's coal mines in 2016[J]. Nat. Hazards, 2018, 92(1): 311-325.

[17] Fan L M. On coal mining intensity and geo-hazard in Yulin-Shenmu-Fugu mine area [J]. China Coal, 2014, 40(5): 52-55.

[18] Guo W B, Wang Y G. The definition of high-intensity mining based on green coal mining and its index system[J]. J. Min. Safety Eng. ,2017, 34(4): 616-623.

[19] Miao X X, Mao X B, Hu G W, et al. Research on broken expend and express and press solid characteristics of rocks and coals[J]. J. Exp. Mech. ,1997(3): 64-70.

[20] Guo G L, Miao X X, Zhang Z N. Research on rupture rock mass deformation characteristics of longwall goals[J]. Sci. Technol Eng. ,2002(5): 44-47.

[21] Hu B N, Zhang H X, Shen B H. Guidelines for the retention and coal mining of coal pillars in buildings, water bodies, railways and main wells and alley[M]. Beijing: Coal Industry Press, 2017.

断层破碎带区域巷道围岩差异性分类及关键控制对策

赵启峰[1,2]，**张农**[2]，**李桂臣**[2]，**彭瑞**[1]，**郑思达**[3]，**郭玉**[2]，**殷帅峰**[1]

(1. 华北科技学院 矿井灾害防治河北省重点实验室，河北 三河，065201；
2. 中国矿业大学 深部煤炭资源开采教育部重点实验室 矿业工程学院，江苏 徐州，221116；
3. 皖北煤电集团公司 祁东煤矿，安徽 宿州，234000)

摘 要：针对断层破碎带区域煤岩穿层巷道围岩失稳垮冒技术难题，以祁东煤矿 6_138 回风巷为研究对象，在断层破碎带不同区域煤(岩)体物理力学参数测试基础上，采用岩体基本质量指标 BQ 理论计算、围岩松动圈实测、钻孔电视探测和数值模拟相结合的研究方法，将断层破碎带不同地段巷道围岩类型划分为 3 类：亚稳定型(Ⅲ类)、失稳渐变-趋稳型(Ⅳ类)和失稳渐变-垮冒型(Ⅴ类)。继而依据围岩差异性分类控制标准，采用理论计算、数值模拟和工程类比方法，确定不同类型围岩分类控制技术方案，即：亚稳定型(Ⅲ类)围岩继续采用原有支护方式，失稳渐变-趋稳型(Ⅳ类)采用优化后的"新型水泥基锚固剂高强度全长锚固锚杆＋高强度大延伸率锚索"支护方式，失稳渐变-垮冒型(Ⅴ类)围岩在"优化"基础上，辅加"底角锚杆＋间歇式注浆"补强加固对策，并确定各项技术参数。工业性实践结果表明，采用分类控制技术方案后，原Ⅴ类围岩的顶底板收敛率由 9.7% 下降至 6.5%，两帮收敛率由 8.6% 下降至 6.2%，即由失稳垮冒型(Ⅴ类)转变为渐变-趋稳型(Ⅳ类)，提高了破碎煤岩体的自身承载能力及抗变形能力，确保了断层破碎带不同区域巷道掘进施工安全。

关键词：断层破碎带；分类控制；巷道变形；补强支护；数值模拟

Study on differentiation classification and control countermeasures of roadway surrounding rock through the fault fracture zone

ZHAO Qifeng[1,2], ZHANG Nong[2], LI Guichen[2],
PENG Rui[1], ZHENG Sida[3], GUO Yu[2], YIN Shuaifeng[1]

(1. Key laboratory of Mine Disaster Prevention and Control of Hebei Province,
North China Institute of Science and Technology, Sanhe 065201, China;
2. School of Mines, Key Laboratory of Deep Coal Resource Mining, Ministry of Education of
China, China University of Mining and Technology, Xuzhou 221116, China;
3. Qidong Coal Mine, Wanbei Coal and Electric Power Group Corporation, Suzhou 234000, China)

基金项目：国家重点研发计划专项资助项目(2017YFC0603001)；中央高校基本科研业务费资助项目(3142017087，3142015087)；国家自然科学基金资助项目(51804119,51574224,51604114)；河北省高等学校科学技术研究资助项目(QN2018302)。

作者简介：赵启峰(1982—)，男，山东省枣庄市人，副教授，博士研究生，从事巷道围岩控制方面的研究。Tel：15132665168，E-mail：mineqfz@sina.com。

Abstract: According to the technical problems of failure collapse of roadway surrounding rock pass through the fault fracture zone in No. $6_1 38$ return airway of Qidong Coal Mine, the classification research methods by using mechanical parameters testing, classification BQ value theoretical calculation, loose circle field testing and numerical simulation are put forward. The roadway surrounding rocks pass through the fault fracture zone are divided into three types, namely the sub-stable type (Ⅲ), instability-to-gradual stabilization type (Ⅳ) and instability-to-gradual collapse type (Ⅴ). The classification control supporting countermeasures during excavation are determined by using classification standard of different surrounding rock, theoretical calculation, numerical simulation and engineering analogy methods. The supporting countermeasures of sub-stable type (Ⅲ) surrounding rock is the original supporting, the instability-to-gradual stabilization type (Ⅳ) surrounding rock adopts the optimization of "high strength and large elongation rate of anchors ＋", and the instability-to-gradual collapse type (Ⅴ) surrounding rock adopts reinforcement combined supporting, that is: high strength and large elongation rate of anchors, floor bolts and dense intermittent short grouting. The industrial practice shows that using the classification control countermeasures of roadway surrounding rock through the fault fracture zone, the convergence rate of roof-to-floor decreased from 9.7％ to 6.5％, the convergence rate of both sides decreased from 8.6％ to 6.2％ of the original type Ⅴ surrounding rock. The instability-to-gradual collapse type (Ⅴ) is transformed into the instability-to-gradual stabilization type (Ⅳ). The classification control countermeasures ensure the safety of roadway excavating through the fault fracture zone.

Keywords: fault fracture zone; classification control; roadway surrounding rock deformation; reinforced support; numerical simulation

我国华北、两淮等矿区一直是煤矿资源开采的重要基地,但因煤系地层赋存条件限制,断层发育,回采巷道大多经历断层破碎带内煤岩穿层掘进。因断层切割,煤(岩)体裂隙发育,结构面广布,微观结构间联结被削弱,围岩强度衰减;同时,断层破碎带内常遭受裂隙水影响,加剧了围岩软化、膨胀和风化程度,而且造成锚杆(索)孔淋水,削弱支护体锚固性能,导致支护失效、承载结构失稳垮冒[1-2]。皖北煤电祁东矿、开滦集团林南仓矿、中煤平朔井工三矿等在断层破碎带区域巷道掘进工程实践中,均出现过不同程度的承载结构失稳垮冒事故,严重威胁矿井安全生产。目前在上述矿区有针对性地开展断层破碎带巷道围岩差异性分类控制研究工作还非常欠缺,但国内外学者针对破碎围岩支护失效机理及稳定性控制方面进行过诸多深入研究,已有研究成果将为本课题研究提供借鉴和指导[3-5]。刘泉声等[6]提出了顾北矿巷道穿越 F_{104} 断层破碎带区域岩体加固技术,采用分段-间断-重复注浆策略,实现对目标区域充填密实度最大化。王克忠等[7]结合地下厂房洞群开挖过程,研究大断面硐室围岩断层破碎带变形破坏及分次支护耦合作用机制,提出强柔性支护技术。王炯等[8]采用数值分析和工程试验研究穿层巷道围岩非对称变形破坏机理及控制对策,提出锚网索＋底角锚杆的非对称耦合支护对策。康红普等[9]提出分段锚固,即在外段未凝固前施加给定预紧力,实现全长(加长)锚固,解决了端锚压缩区范围小及黏锚力小的问题。张农等[10]研究得出软岩巷道因围岩松散软弱破碎及裂隙水渗流诱发支护失效而导致失稳冒顶事故机理。王卫军等[11]研究指出软弱破碎直接顶锚网索施工中的变形和安装对锚固剂的破坏是导致巷道失稳的主因。马立强等[12]研究了遇断层破碎带预掘巷道群围岩控制技术、布置方式及支护参数。宋瑞刚等[13]研究了断层破碎带围岩的突发失稳破坏,建立了穿越断层破碎带围岩失稳尖点突变模型。韦四江等[14]采用高强锚固系统控制复杂软弱巷道围岩,并研究了预紧力锚杆作用下锚固体的形成因素。蒋康前等[15]研究了刘庄煤矿断层破碎带软岩巷道围岩破坏特征,提出分步耦合支护对策。郝育喜等[16]分析了旗山矿断层破碎带穿层巷道破坏特征,提出注浆＋非对称锚网索＋底角锚杆耦合支护控制对策。

本文在前人研究基础上,以祁东煤矿 6_138 回风巷作为研究对象,该巷道是皖北煤电祁东矿 6_138 回风巷是典型的穿越断层破碎带掘进巷道,不同地段围岩变形差异性显著,巷道稳态-亚稳态-失稳多重差异状态交替呈现,现有常规意义上的软岩分类控制机制已不能完全指导该类型巷道安全高效掘进。鉴于此,本文针对性开展断层破碎带区域巷道围岩差异性分类,并提出针对性的围岩控制对策。

1 工程概况

皖北煤电祁东矿 6_138 回风巷埋深 580 m,其南部布置有东翼胶带机巷、轨道大巷,东部为运输上山,西部为三采区回采边界,北部为待掘的 6_138 机巷,如图 1 所示。该巷基本顶为深灰色细砂岩,厚度 0.6～7.8 m,平均 4.6 m;直接顶为灰色-深灰色泥岩,破碎,厚度 3.7～6.9 m,平均 5.3 m;直接底为灰色-深灰色泥岩,泥质结构,含植物根茎化石,厚度 2.3～5.9 m,平均 4.0 m;基本底为浅灰色中砂岩,厚度 5.2～14.6 m,平均 12.31 m。该回风巷布置于 6_1 煤层中,该煤层厚度 0.6～3.8 m,平均 2.7 m,倾角 4°～17°,平均 14°。由于煤层倾角起伏变化大,又遭受多组断层切割(地堑结构),局部出现上抬、下沉或缺失,导致该回风巷 HF1～HF11、HF17～HF35 标记区间为煤巷段,HF11～HF17 标记区间为岩巷段(图 1),断层切割破碎带煤(岩)体遭受剪切与拉伸破坏、垂直节理与风化裂隙发育,其完整性遭到破坏,巷道存在失稳垮冒风险。

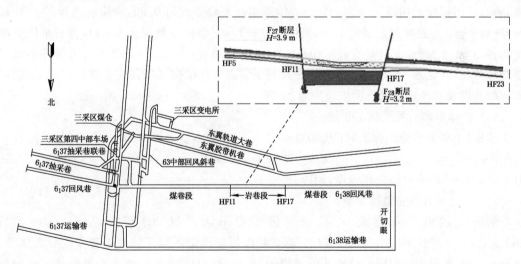

图 1 6_138 回风巷布置示意图(含断层破碎带地质剖面)

Fig.1 Schematic diagram of No. 6_138 return airway

2 断层破碎带区域围岩差异性分类

2.1 围岩差异性分类依据

(1)煤(岩)体物理力学参数测试

选取 6_138 回风巷断层破碎带不同区域顶板钻取岩芯,在煤炭资源与安全开采国家重点实验室进行物理力学参数及岩体完整性测试,如图 2 所示。

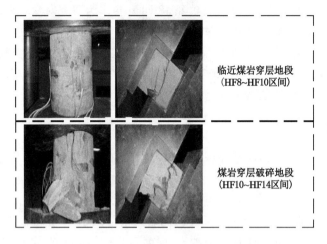

图 2 不同地段岩石力学参数测试

Fig. 2 Rock mechanics parameter test of different segments

测试结果表明:6_138 回风巷断层破碎带掘进全过程不同区域围岩强度差异性显著(表 1):① HF5~HF8 区间,巷道迎头距离断层切割面较远,岩石抗压强度平均值为 38 MPa,岩体完整性系数平均为 0.68,岩体较完整;② HF8~HF10 区间,巷道逐渐靠近断层切割地段,岩石抗压强度平均值为 28.8 MPa,岩体完整性系数平均为 0.52,岩体较破碎;③ HF10~HF14 区间,巷道穿越断层切割面,岩体松散破碎(完整性系数仅为 0.34),岩石抗压强度平均值为 22.4 MPa,尤其是直接顶(0~3.0 m)和直接底(0~2.2 m)范围内力学参数衰减至与煤体相当(见表 2),该空间范围恰好隶属于锚固承载结构体所需支护维稳空间,给断层破碎带巷道安全掘进造成重大安全隐患。

(2)岩体基本质量指标 BQ 值

依据《工程岩体分级标准》(GB 50218—94),岩体基本质量指标如式(1)。

$$BQ = 90 + 3\sigma_c + 250K_v \tag{1}$$

其中　σ_c——岩石单轴抗压强度,MPa;

　　　K_v——岩体完整性系数。

将表 1 数据代入公式,可得断层破碎带不同区域 BQ 值:$BQ_{(HF5\sim HF8)} = 374$,$BQ_{(HF8\sim HF10)} = 306$,$BQ_{(HF10\sim HF14)} = 242$。据此将断层破碎带不同区域围岩初步划分为 3 类:HF5~HF8 区间为远离断层破碎地段Ⅲ类围岩、HF8~HF10 区间为临近断层破碎地段Ⅳ类围岩、HF10~HF14 区间为穿越断层破碎地段Ⅴ类围岩。

(3)不同地段围岩松动圈实测

采用 BA-Ⅱ型松动圈测试仪实测不同地段围岩松动圈发育范围[13]。在上述 3 类围岩地段分别测试,每个地段各布置 2 个测站,垂直顶板、两帮及底板打钻,钻头 $\phi 42$ mm,孔深 4 000 mm。由测试结果(图 3)可知:Ⅲ类、Ⅳ类、Ⅴ类围岩松动圈分别为 1.4 m、1.8 m 和 2.2 m,均属于大松动圈范畴。大松动圈时应按照锚固区内形成某种结构(梁、层、拱、壳)采用加固拱理论设计支护参数[4]。

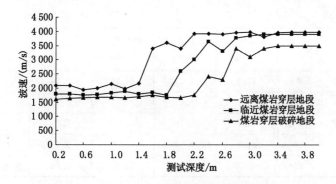

图 3 不同地段巷道围岩松动圈测试

Fig. 3 Loose circle detection of different part surrounding rock

2.2 围岩差异性类别

根据岩石物理力学参数测试、岩体基本质量指标 BQ 值、松动圈等测试结果,断层破碎带不同区域围岩类型划分为 3 类,即:亚稳定型(Ⅰ)、失稳渐变-趋稳型(Ⅱ)、失稳渐变-垮冒型(Ⅲ),见表 3 和表 4。6$_1$38 回风巷顶底板收敛率、两帮收敛率和塑性区扩展率在各断层间的变化情况见图 4。

表 1 断层破碎带不同区域煤岩体强度和完整性差异性对比

Table 1 Differences compared on rock mass strength and integrity among different segments of roadway around fault fracture zones

层位名称	煤岩穿层破碎地段(HF10~HF14 区间)				临近煤岩穿层地段(HF8~HF10 区间)			
	抗压强度/MPa	均值/MPa	完整性系数	均值	抗压强度/MPa	均值/MPa	完整性系数	均值
基本顶 0~4.6 m	32.91		0.48		41.5		0.71	
直接顶 3.0~5.3 m	22.4		0.38		28.3		0.58	
直接顶 0~3.0 m	15.1		0.16		17.2		0.28	
煤 0~2.7 m	13.5	22.4	—	0.34	13.3	28.8	—	0.52
直接底 0~2.2 m	14.7		0.17		16.7		0.31	
直接底 2.2~4.0 m	21.3		0.42		27.8		0.57	
基本底 0~12.31 m	36.87		0.45		56.8		0.69	

注:HF5~HF8 区间全部为煤巷段,前期已有围岩物理力学参数(矿地测科提供),岩体单轴抗压强度平均值为 38 MPa,岩体完整性系数平均为 0.68。

表 2 穿越断层破碎带煤岩体物理力学参数(HF10~HF14 区间)

Table 2 Rock physical and mechanical properties of the fault fracture zone (HF10~HF14)

类别	层位名称	抗压强度/MPa	抗拉强度/MPa	弹性模量/GPa	泊松比	内聚力/MPa	内摩擦角/(°)
标准试件	基本顶 0~4.6 m	32.91	2.28	25.02	0.37	1.28	29
标准试件	直接顶 3.0~5.3 m	22.4	1.58	21.02	0.31	0.78	34

续表 2

类别	层位名称	抗压强度/MPa	抗拉强度/MPa	弹性模量/GPa	泊松比	内聚力/MPa	内摩擦角/(°)
非标准试件	直接顶 0～3.0 m	15.1	1.21	14.53	0.28	0.57	31
非标准试件	煤 0～2.7 m	13.5	1.01	14.14	0.27	0.42	20
非标准试件	直接底 0～2.2 m	14.7	1.18	14.13	0.29	0.51	31
标准试件	直接底 2.2～4.0 m	21.3	1.99	20.04	0.20	0.75	33
标准试件	基本底 0～12.31 m	36.87	5.24	28.20	0.28	1.83	31

表 3 断层破碎带不同区域围岩类型划分

Table 3 Type classification of surrounding rock of different segments around fault fracture zones

巷道段类别	顶底板收敛率/%	两帮收敛率/%	塑性区扩展/%	工程地质描述	支护措施
亚稳定型	<3.7	<3.5	<5.7	距离煤岩交界断层切割面较远，围岩较完整，围岩变形和塑性区扩展均较小	原有支护下保持自稳，无须实施补强加固措施
失稳渐变-趋稳型	3.7～7.0	3.5～6.5	5.7～9.8	由亚稳定态向失稳倾向态转变，围岩破碎程度增大，呈现失稳倾向，但围岩变形和塑性区均在可控范围内	优化原有支护参数
失稳渐变-垮冒型	>7.0	>6.5	>9.8	临近并穿越断层切割面，围岩松散破碎，围岩变形和塑性区均急剧增大，掘巷后短时间内随即出现失稳垮冒、底鼓等破坏现象	及时采取补强加固措施

表 4 6_138 回风巷断层破碎带不同区域巷道围岩稳定性分类

Table 4 Stability classification of roadway surrounding rock of different segments through fault fracture zones

调查地点	岩体基本质量指标(BQ)	松动圈实测值	主要地压形态	巷道围岩类别	稳定性类别
HF5～HF8 标记区间	374	1.4 m	变形	煤巷段(亚稳定)	I
HF8～HF10 标记区间	306	1.8 m	变形-散体	煤巷段(失稳渐变趋稳)	II
HF10～HF14 标记区间	242	2.2 m	散体	煤岩穿层段(失稳垮冒)	III

3 围岩差异性支护研究

根据不同区域巷道围岩分类结果，提出针对性的分类控制对策：① 亚稳定型围岩依靠原支护保持围岩自稳，无须实施补强加固措施，如图 5(a)所示；② 失稳渐变-趋稳型围岩将

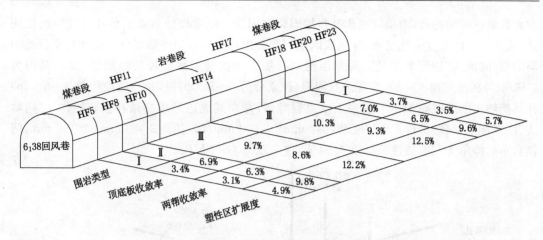

图 4 不同地段巷道围岩变形率及塑性区扩展对比

Fig. 4 Differences compared on surrounding rock deformation and plastic zones expansion among different segments

原有支护设计中的普通树脂药卷端锚锚杆优化为新型锚固剂全长锚固锚杆,原有普通锚索优化为高强度大延伸率锚索,如图 5(b)所示;③ 失稳渐变-垮冒型围岩在"优化"基础上,辅加补强加固措施,即"新型锚固剂全长锚固锚杆＋高强度大延伸率锚索＋底角锚杆＋间歇式注浆",如图 5(c)所示。

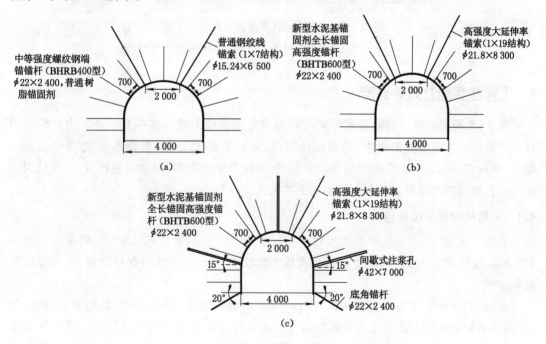

图 5 断层破碎带不同区域巷道分类控制支护设计图

Fig. 5 Supporting design schemes of classification control countermeasures through fault fracture zones

(a) 亚稳定型;(b) 失稳渐变-趋稳型;(c) 失稳渐变-垮冒型

间歇式注浆见图 6。针对失稳渐变-垮冒型围岩破碎严重、松动圈范围大、煤岩体裂隙发育胶结性差的特点,需采取注浆补强加固裂隙围岩。但常规注浆工艺单孔持续一次性注浆,浆岩结石体强度强化受限,且漏浆跑浆问题突出,为此,采用能够形成高强度网络骨架的马丽散"间歇式"注浆新工艺。课题组前期开展了多次马丽散凝胶特性测试(凝胶时间为2′19.45″)和注浆加固实验,发现多次间隔(注浆分为3~5个时段,每个时段停歇3~5 min)注浆能够在断层破碎带中形成多个渗流裂隙面,并在其周边形成黏结补强介质体。"间歇式"注浆技术参数:每排3个注浆孔,$\phi 42$ mm×7 000 mm,间、排距2 000 mm×2 100 mm,两侧肩角处仰角15°,有效注浆深度5.64 m,注浆压力4~6 MPa。

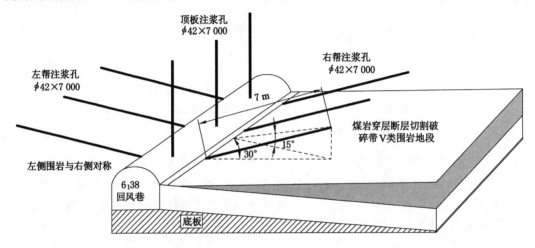

图 6　间歇式注浆工艺示意图

Fig. 6　Schematic diagram of intermittent grouting technology

4　工业性实施及效果监测

在 6_138 回风巷掘进过断层破碎带不同地段分别实施上述分类控制对策。为检验不同技术方案围岩稳定性控制效果,选取亚稳定型、失稳渐变-趋稳型、失稳渐变-垮冒型围岩地段各 30 m、50 m、100 m 跟踪观测围岩变形、支护体力学性态等,并在失稳渐变-垮冒型围岩使用钻孔电视探测间歇式注浆封堵加固裂隙效果[19]。

4.1　不同地段围岩位移监测

如图 7(a)所示,亚稳定型巷道顶底板移近量为 122 mm(收敛率 3.5%)、两帮移近量为 132 mm(收敛率 3.3%),围岩变形量和收敛率均较小,在原支护下可保持自稳,无须实施补强加固措施。

如图 7(b)所示,HF8~HF10 和 HF18~HF20 地段原属于失稳渐变-趋稳型,采取优化后"新型水泥基锚固剂高强度全长锚固锚杆+高强度大延伸率锚索"支护后,顶底板移近量减少至 170 mm(收敛率 4.9%)、两帮移近量减少至 182 mm(收敛率 4.6%),由表 1 分类标准可知:采用优化后的支护对策后围岩类别虽然依旧属于失稳渐变-趋稳型,但呈现出整体趋稳状态。

如图 7(c)所示,HF10~HF18 地段原属于失稳渐变-垮冒型,采取"新型水泥基锚固剂高强度全长锚固锚杆+高强度大延伸率锚索+底角锚杆+间歇式注浆"联合控制对策后,顶

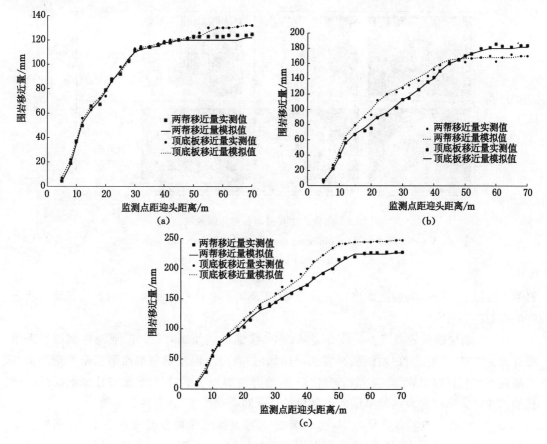

图 7 不同区域分类控制围岩变形监测
Fig.7 Monitoring curves of roadway surrounding rock
deformation of classification control countermeasures
(a)亚稳定型;(b)失稳渐变-趋稳型;(c)失稳渐变-垮冒型

底板移近量减少至 227 mm(收敛率 6.5%)、两帮移近量减少至 248 mm(收敛率 6.2%),表明该地段围岩类别由失稳渐变-垮冒型转变为失稳渐变-趋稳型,确保了过断层破碎带最危险区域巷道掘进施工安全。

4.2 注浆效果监测

采用 YTJ20 型岩层钻孔电视探测仪检测"间歇式"注浆对失稳渐变-垮冒型围岩裂隙封堵加固效果。钻孔成像截图显示(图 8):注浆前孔周环形、纵向裂隙均发育,孔周围岩破碎,完整性差;注浆后钻孔孔壁较平整,白色迹线为注浆后马丽散充填原有裂隙,原破碎煤岩体胶结成整体结构,断层破碎带Ⅴ类围岩完整性得到强化。

5 结 论

(1)断层破碎带不同区域围岩类型划分为 3 类,即:亚稳定型、失稳渐变-趋稳型和失稳渐变-垮冒型。亚稳定型围岩可依靠原支护方式保持自稳;失稳渐变-趋稳型围岩虽不会开挖即失稳垮冒,但若不及时优化原有支护,围岩将由局部失稳向失稳垮冒劣变;失稳渐变-垮

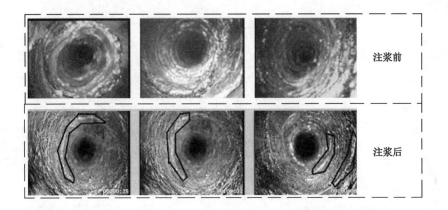

图 8 间歇式注浆封堵裂隙加固效果

Fig. 8 Photos with plugging crack and strengthen fracture zones effect of intermittent grouting technology

冒型围岩掘巷后短时间内随即出现失稳垮冒,需及时采取补强加固对策,确保失稳垮冒状态向渐变趋稳态转变。

(2) 在断层破碎带围岩差异性分类基础上,实施分类控制对策,即:亚稳定型围岩采用现有支护方式;失稳渐变-趋稳型围岩采用优化的"新型水泥基锚固剂高强度全长锚固锚杆＋高强度大延伸率锚索"支护方式;失稳渐变-垮冒型围岩采用"新型水泥基锚固剂高强度全长锚固锚杆＋高强度大延伸率锚索＋底角锚杆＋间歇式注浆"联合控制对策。

(3) 工业性分段实践效果表明,失稳渐变-趋稳型围岩采取优化支护参数后,围岩变形呈现整体趋稳状态;失稳渐变-垮冒型围岩采用"优化＋补强"的联合控制对策后,围岩状态由失稳渐变-垮冒型转变为失稳渐变-趋稳型,确保了断层破碎带最危险区域巷道掘进施工安全。

参考文献

[1] 王宏伟,姜耀东,赵毅鑫,等.软弱破碎围岩高强高预紧力支护技术与应用[J].采矿与安全工程学报,2012,29(4):475-479.
WANG Hongwei, JIANG Yaodong, ZHAO Yixin, et al. Application of support technology with high strength and high pretension stress for weak-broken rocks[J]. Journal of Mining & Safety Engineering,2012,29(4):475-479.

[2] 詹平.高应力破碎围岩巷道控制机理及技术研究[D].北京:中国矿业大学(北京),2012.
ZHAN Ping. Study on technology and control mechanism in high-stress fractured surrounding rock roadway[J]. Beijing: China University of Mining & Technology (Beijing),2012.

[3] 余伟健,高谦,韩阳,等.非线性耦合围岩分类技术及其在金川矿区的应用[J].岩土工程学报,2008,30(5):663-669.
YU Weijian, GAO Qian, HAN Yang, et al. Non-linear coupling classification technique of surrounding rock mass and its application in Jingchuan Mine[J]. Chinese Journal of

Geotechnical Engineering,2008,30(5):663-669.

[4] 康红普.我国煤矿巷道锚杆支护技术发展60年及展望[J].中国矿业大学学报,2016,45(6):1071-1081.

KANG Hongpu. Sixty years development and prospects of rock bolting technology for underground coal mine roadways in China[J]. Journal of China University of Mining & Technology,2016,45(6):1071-1081.

[5] 柏建彪,王襄禹,贾明魁,等.深部软岩巷道支护原理及应用[J].岩土工程学报,2008,30(5):632-635.

BAI Jianbiao, WANG Xiangyu, JIA Mingkui, et al. Theory and application of supporting in deep soft roadways[J]. Chinese Journal of Geotechnical Engineering, 2008,30(5):632-635.

[6] 刘泉声,卢超波,卢海峰,等.断层破碎带深部区域地表预注浆加固应用与分析[J].岩石力学与工程学报,2013,32(Supp.2):3688-3695.

LIU Quansheng, LU Chaobo, LU Haifeng, et al. Application and analysis of ground surface pre-grouting strengthening deep fault fracture zone[J]. Chinese Journal of Rock Mechanics and Engineering,2013,32(Supp.2):3688-3695.

[7] 王克忠,李仲奎,王玉培,等.大型地下硐室断层破碎带变形特征及强柔性支护机制研究[J].岩石力学与工程学报,2013,32(12):2455-2462.

WANG Kezhong, LI Zhongkui, WANG Yupei, et al. Study of strong flexible supporting mechanism and deformation characteristics for fracture zone in large underground caverns[J]. Chinese Journal of Rock Mechanics and Engineering,2013,32(12):2455-2462.

[8] 王炯,郭志飚,蔡峰,等.深部穿层巷道非对称变形机理及控制对策研究[J].采矿与安全工程学报,2014,31(1):28-33.

WANG Jiong, GUO Zhibiao, CAI Feng, et al. Study on the asymmetric deformation mechanism and control countermeasures of deep layers roadway[J]. Journal of Mining & Safety Engineering,2014,31(1):28-33.

[9] 康红普,姜铁明,高富强.预应力在锚杆支护中的作用[J].煤炭学报,2008,33(8):680-685.

KANG Hongpu, JIANG Tieming, GAO Fuqiang. Effect of pretensioned stress to rock bolting[J]. Journal of China Coal Society,2008,33(8):680-685.

[10] 张农,李桂臣,许兴亮.泥质巷道围岩控制理论与实践[M].徐州:中国矿业大学出版社,2011.

[11] 王卫军,郭罡业,朱永建,等.高应力软岩巷道围岩塑性区恶性扩展过程及其控制[J].煤炭学报,2015,40(12):2748-2754.

WANG Weijun, GUO Gangye, ZHU Yongjian, et al. Malignant development process of plastic zone and control technology of high stress and soft rock roadway[J]. Journal of China Coal Society,2015,40(12):2748-2754.

[12] 马立强,余伊河,金志远,等.大倾角综放面预掘巷道群快速过断层技术[J].采矿与安

全工程学报,2015,32(1):84-89.

MA Liqiang,YU Yihe,JIN Zhiyuan,et al. Fast pushing through fault of the pre-driven roadway groups in fully mechanized top-coal caving face with big dip angle[J]. Journal of Mining & Safety Engineering,2015,32(1):84-89.

[13] 宋瑞刚,张顶立,文明.穿越断层破碎带深埋隧道围岩失稳的突变理论分析[J].土木工程学报,2015,48(S1):289-292.

SONG Ruigang,ZHANG Dingli,WEN Ming. The cusp catastrophe theory analysis for instability of deep-buried tunnels surrounding rock through fault fracture zone[J]. China Civil Engineering Journal,2015,48(S1):289-292.

[14] 韦四江,李宝富.预紧力锚杆作用下锚固体的形成与失稳模式[J].煤炭学报,2013,38(12):2126-2131.

WEI Sijiang,LI Baofu. Anchor bolt body formation and instability mode under the influence of anchoring pretension[J]. Journal of China Coal Society,2013,38(12):2126-2131.

[15] 蒋康前,汪良海.深井穿断层破碎带软岩巷道围岩控制技术研究[J].煤炭工程,2014,46(3):42-44.

JIANG Kangqian,WANG Lianghai. Study on construction and control technology of mine soft rock roadway passing through broken zone of fault[J]. Coal Engineering,2014,46(3):42-44.

[16] 郝育喜,王炯,王浩,等.深井断层破碎带穿层软岩巷道锚网索耦合控制对策[J].采矿与安全工程学报,2016,33(2):231-237.

HAO Yuxi,WANG Jiong,WANG Hao,et al. Coupled bolt-mesh-anchor supporting technology for deep fault fracture zones throughout layers soft rock roadway[J]. Journal of Mining & Safety Engineering,2016,33(2):231-237.

近距离巨厚坚硬岩层破断失稳特征及分区控制

赵通[1],刘长友[2]

(1. 太原理工大学 采矿工艺研究所,山西 太原,030024;
2. 中国矿业大学 矿业工程学院 煤炭资源与安全开采国家重点实验室,江苏 徐州,221116)

摘 要:为实现近距离巨厚坚硬岩层下厚煤层的安全高效开采,根据朱仙庄煤矿北翼采区煤岩赋存特征和工作面生产技术条件,建立了厚硬岩层下开采三维模型,分析了厚硬岩层厚度和直接顶充填系数对厚硬岩层稳定性的影响规律以及支架—围岩相互作用特征。研究发现厚硬岩层厚度影响破断块体尺寸和力学结构形成,直接顶充填系数影响断裂块体下沉空间、破断位置和破断尺寸;根据矿压显现程度和顶板控制难易程度,把回采区域划分为顶板强烈来压区、顶板较强烈来压区和顶板来压不明显区,根据顶板来压强烈程度的不同,提出了厚硬岩层下煤层开采岩层分区控制的原则和方法。

关键词:近距离;巨厚坚硬岩层;破断失稳特征;分区控制

Roof fracture and instability characteristics and hierarchical control study of thick coal seams mining for super thick and hard rock in close distance

ZHAO Tong[1], LIU Changyou[2]

(1. *Mining Technology Institute, Taiyuan University of Technology, Taiyuan 030024, China*;
2. *School of Mines, State Key Laboratory of Coal Resources & Safe Mining, China University of Mining & Technology, Xuzhou 221116, China*)

Abstract: To achieve safe and efficient mining of thick coal seams for super Thick and Hard Rock (THR) in close distance, with the consideration of the occurrence characteristics of coal and rock in the mining area of the north-wing of Zhuxianzhuang Coal Mine and the production technical conditions, a three-dimensional model of mining under THR was established, and the influence of thickness of the THR and filling coefficient of immediate roof on the stability of THR and the interaction characteristics of support-surrounding rock were analyzed. Previous research has found that the thickness of THR affects the size and mechanical structure of the fractured block. The filling coefficient of immediate roof directly determines the sinking space of the fractured block, which affects its breaking position and size. Based on the degree of strata behaviors and the difficulty of roof control, the mining area is further divided into three zones: strong strata behaviors zone, less-strong strata behaviors zone, and strata behaviors mitigation zone, with the first

---

基金项目:国家自然科学基金资助项目(51574220)。
作者简介:赵通(1987—),男,山东省金乡人,博士,讲师,从事岩石力学与岩层控制、安全高效开采、深部开采方面的研究。
通信作者:刘长友(1965—),男,山东省东营市人,博士,教授,博士生导师,从事矿山压力与岩层控制等方面的研究。
Tel:15950686350,E-mail:ztcumt@cumt.edu.cn。

two as the roof control zone. Based on the research results, the principle of roof hierarchical-control under THR is put forward, and a hierarchical roof coordination control method based on pre-control of THR is proposed.

Keywords: in a close distance; super-thick and hard rock; fracture and instability characteristics; hierarchical control

我国煤层赋存条件复杂多样,煤层上方赋存坚硬、厚度较大岩层的煤矿占30%左右[1]。硬厚岩层条件下,顶板的破断特征、垮落运移规律、载荷作用机制以及支架围岩关系等将发生显著变异,淮北矿区海孜煤矿Ⅱ1026工作面、义马矿区千秋煤矿21141工作面、新汶矿区华丰煤矿1405工作面和同煤安平煤矿8117工作面等生产实践中存在砾岩及岩浆岩大面积悬顶、突然断裂后整体运动引发工作面支架立柱油缸变形、压爆、支架大面积压死等严重问题[2-4],极大影响岩层的有效控制和采面的安全生产。因此,研究近距离巨厚坚硬岩层的破断失稳规律和致灾机理,掌握厚硬顶板覆岩活动的特点和支架与围岩相互作用的本质,提出合理的顶板分区控制技术,成为该条件下岩层控制的关键。

目前对厚硬岩层控制的研究,国内外相关研究成果主要体现在巨厚坚硬顶板失稳致灾机理、巨厚坚硬顶板破断结构以及顶板控制技术等方面[5-10]。已有研究中高位厚硬岩层的岩层下界与煤层距离多大于50 m,存在整层厚度较小或多个分层情况。由于朱仙庄煤矿巨厚坚硬砾岩层位较低,其下0~40 m厚直接顶整体厚度小且变化大,部分砾岩直接与煤层接触,加之8号特厚煤层开采形成的较大采空空间对巨厚坚硬岩层的复杂影响机理不清等,因而将带来覆岩活动的复杂性和对生产影响的严重性,还需要进一步深入研究。

因此,本文以淮北朱仙庄煤矿北翼采区"五含"巨厚坚硬岩层赋存条件为研究背景,采用数值模拟和理论分析方法对厚硬岩层破断失稳特征与分区控制技术研究,为该类煤层的安全高效开采奠定基础,提供技术保障。

1 工作面地质及生产技术条件

淮北朱仙庄煤矿井田北翼采区8煤顶板分布侏罗纪砾岩岩溶含水层(简称"五含")(图1),

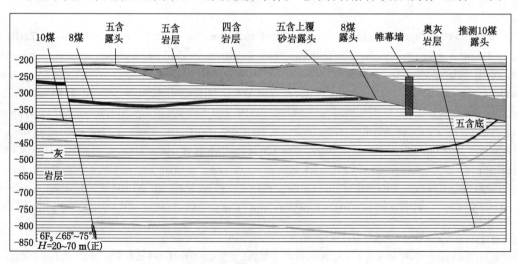

图1 五含岩层与煤层赋存关系剖面示意图(1∶1000)

Fig.1 The space relationship between thick and hard conglomerate and coal seam(1∶1000)

该岩层在平面上呈长舌状分布,倾角15°～25°,厚度0～102 m,平均50～65 m,与煤层平均间距小于40 m,剥蚀面大致与8号煤层斜交,实测五含岩层单轴抗压强度最大达到67.79 MPa。朱仙庄煤矿880工作面回采8号煤,煤厚7.02～13.50 m,平均10.03 m,采用走向长壁综采放顶煤采煤法。工作面面长165 m,走向长度400 m。

由关键层理论计算可知,五含砾岩厚度大、强度高、完整性好,作为基本顶,见图2;厚硬砾岩下直接顶岩性为泥岩、砂质泥岩,强度低且厚度较小,随工作面推进即垮落充填于采空区,作为直接顶。朱仙庄煤矿直接顶厚度小且变化大,为更直观描述直接顶厚度变化及与矿压显现关系,定义直接顶厚度h_i与煤层采高h_m的比值N($N = h_i/h_m$)为直接顶充填系数。

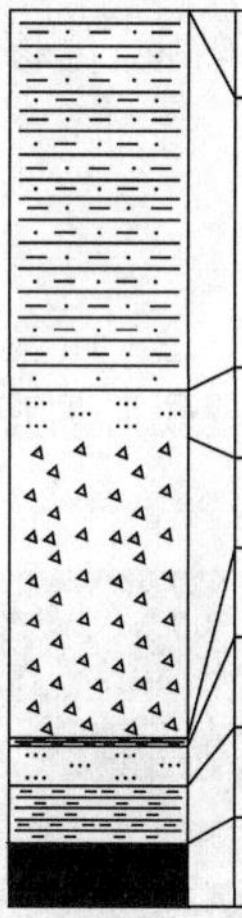

厚度/m	埋深/m	岩性	备注
65.0	246.96	泥岩	上覆岩层
4.80	311.96	细砂岩	
60.00	371.96	砾岩(五含)	基本顶
1.20	373.16	泥岩	
4.20	377.36	砂岩	直接顶
6.36	383.72	泥岩	
10.03	393.75	8号煤	煤

图2 朱仙庄煤矿岩层柱状图

Fig.2 Rock stratum histogram of Zhuxianzhuang Coal Mine

2 厚硬岩层下开采三维模型的建立

3DEC软件考虑裂隙等不连续面控制作用,构建的采场覆岩由块体单元组成,相邻单元可以接触和分离,其相互作用满足牛顿定律[11],能模拟剪切错动和脱开等破坏形式[12],准确表征大位移、大变形等覆岩运动演化特征。

2.1 模型构建和参数设定

根据朱仙庄矿880工作面生产地质条件,建立三维模型,研究直接顶充填系数和厚硬岩层厚度变化时近距离巨厚坚硬岩层破断特征和应力场分布规律。模型中岩块和节理本构关系为摩尔-库伦塑性模型[13]。在模型上表面施加均布载荷,形成应力边界条件;左、右和下表面均为零位移边界条件,具体约束边界条件为:

(1)模型左、右表面的位移矢量和速度矢量均为0,作为水平位移约束边界。

(2)模型下表面为全约束边界,即水平和竖直方向均固定。

(3)模型上表面为自由边界,模型岩层以外覆岩以均布载荷形式垂直施加在上表面边界。

模型尺寸:走向长度×倾向长度×高度=400 m×170 m×160 m,块体单元尺寸划分(长×宽×高):煤层0.25 m×3.3 m×0.5 m,直接顶岩层0.5 m×5 m×1 m,厚硬岩层0.5 m×5 m×5 m,沿推进方向模型两侧各留50 m煤柱。

2.2 模拟方案

建立近距离巨厚坚硬岩层下厚煤层开采的采场模型,研究厚硬岩层厚度和直接顶充填系数等影响因素对近距离巨厚坚硬岩层稳定性的影响机制。

(1) 厚硬岩层厚度对近距离巨厚坚硬岩层稳定性的影响

选定煤层采高 10 m、直接顶岩层厚度 10 m,直接顶充填系数为 1,模拟厚硬岩层厚度分别为 20 m、40 m 和 60 m 条件下,围岩破断特征、位移场和应力场分布规律。

(2) 直接顶充填系数对近距离巨厚坚硬岩层稳定性的影响

选定煤层采高 10 m、厚硬砾岩岩层厚度 60 m,分析直接顶充填系数 $N=1$、2 和 3,对应的直接顶岩层厚度分别为 10 m、20 m 和 30 m 条件下,围岩破断特征、位移场和应力场分布规律。

3 模拟结果分析

3.1 厚硬岩层厚度对近距离巨厚坚硬岩层稳定性的影响

当采高和直接顶厚度均为 10 m,直接顶充填系数为 1,厚硬砾岩厚度分别为 20 m、40 m 和 60 m 时,巨厚坚硬岩层破断特征见图 3。

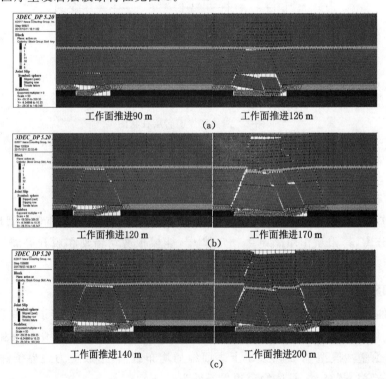

图 3 不同厚度砾岩下开采(初次和周期来压阶段)围岩破断运移特征

Fig. 3 Breakage and migration characteristics of surrounding rocks under conglomerates of different thickness (First weighting and periodic weighting stages)

(a) 上、下砾岩分层厚度分别为 40 m 和 20 m;(b) 上、下砾岩分层厚度分别为 20 m 和 40 m;(c) 整层砾岩厚度为 60 m

分析不同厚度砾岩下开采围岩破断特征可知,厚硬岩层在初次和周期来压时易发生整层破断,整层断裂岩层向工作面侧回转,形成不等高断裂体,周期破断块体沿已垮岩块的接触面向煤壁滑落失稳。砾岩层厚度为20 m时,其初次、周期破断步距分别为90 m和36 m;砾岩层厚度为40 m时,其初次、周期破断步距分别为120 m和50 m;砾岩层厚度为60 m时,其初次、周期破断步距分别为140 m和60 m。

图4中周期破断的块体以工作面前方煤壁为支点向采空区回转滑动,形成了应力升高区,砾岩厚度分别为20 m、40 m和60 m时,工作面煤壁前方应力峰值分别为10 MPa、12.5 MPa和15 MPa,峰值点分别距工作面8 m、10 m和16 m。由于厚度越小厚硬岩层周期破断形成块体的块度越小,越容易发生滑落失稳,使得作用在工作面前方煤壁上的应力越小,煤壁发生塑性破坏程度越小。

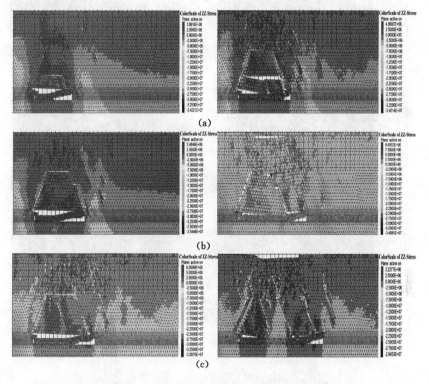

图4　不同厚度砾岩下开采(初次和周期来压阶段)围岩应力场分布

Fig. 4　Stress field distribution of surrounding rocks under conglomerates of different thickness (First weighting and periodic weighting stages)

(a)上、下分层厚度分别为40 m和20 m;(b)上、下分层厚度分别为20 m和40 m;(c)五含岩层厚度为60 m

由上述分析可知,当直接顶厚度和采高相同,即采场直接顶充填系数一定时,厚硬岩层厚度越大,其初次来压和周期来压步距越大、破断生成块体的块度越大,失稳时对围岩扰动越大。厚硬岩层初次破断一般发生整层破断,破断后一般整体垮落,但垮落后容易出现不等高断裂体;周期破断时,厚硬岩层厚度越小破断块体的块度越小,难以形成铰接结构,受到水平约束力越小,越容易滑落失稳。

厚硬岩层初次破断前,在工作面侧煤壁出现应力集中,砾岩厚度越大,应力峰值越大,峰值点距煤壁越远。厚硬岩层周期破断时在工作面前方煤壁形成应力集中,破断块度越小,滑动失稳时作用在工作面前方煤壁的应力越小。

当存在上薄、下厚砾岩分层时,上分层随下分层发生同步破断和运动,具有相同的破断步距、垂直位移和应力；当赋存条件为上厚、下薄时，两分层独立运动，下分层首先发生破断和运动，具有较小的来压步距及较大的垂直位移和应力，工作面会出现大小来压现象。

3.2 直接顶充填系数对近距离巨厚坚硬岩层破断运移的影响机制

当采高为 10 m、厚硬砾岩厚度为 60 m，直接顶充填系数 N 分别为：1、2 和 3，直接顶厚度分别为 10 m、20 m 和 30 m 时，巨厚坚硬岩层破断特征见图 5。

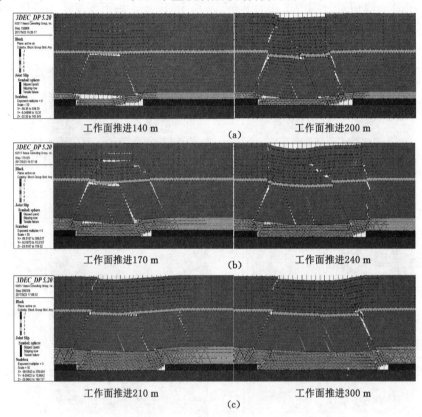

图 5 不同直接顶充填系数条件下开采围岩破断运移特征
Fig. 5 Breakage and migration characteristics of surrounding rocks under different immediate roof filling coefficients
(a) 充填系数 $N=1$；(b) 充填系数 $N=2$；(c) 充填系数 $N=3$

对图 5 不同直接顶充填系数条件下覆岩破断运移特征分析，可知直接顶充填系数为 1、2 和 3 时，厚硬岩层初次、周期来压步距分别为 140 m 和 60 m，170 m 和 70 m，210 m 和 90 m。充填系数越大，直接顶厚度越大，其垮落后对采空区充填程度越好，使得基本顶块体垮落运移空间越小、破断运移特征越不明显，工作面来压时覆岩破坏高度和破坏范围越小、矿压显现越缓和。

对图 6 不同直接顶充填系数条件下采场围岩应力分析，厚硬岩层初次破断前，在开切眼和工作面煤壁形成了应力集中，其中工作面侧煤壁前方 10~80 m 范围为应力增高区；直接顶充填系数越大，应力峰值越小，直接顶充填系数为 1、2 和 3 时，工作面侧煤壁上应力峰值分别为 20 MPa、15 MPa 和 10 MPa，峰值点分别距工作面 20 m、16 m 和 12 m。破断块体边缘应力较大,说明初次垮落前破断块体对周围岩体产生挤压作用,引起围岩应力集中。

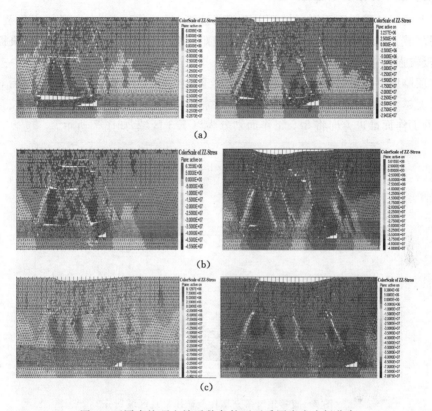

图 6 不同直接顶充填系数条件下开采围岩应力场分布

Fig. 6 Stress field distribution of surrounding rocks under different immediate roof filling coefficients

(a) 充填系数 $N=1$;(b) 充填系数 $N=2$;(c) 充填系数 $N=3$

周期破断的块体以工作面煤壁为支点向采空区回转滑动,在煤壁出现应力集中,直接顶充填系数越大,破断块体垮落下沉中受到支撑力越大,在工作面前方煤壁应力越小,直接顶充填系数为1、2和3时,工作面煤壁上应力峰值分别为 15 MPa、10 MPa 和 5 MPa,峰值点分别距工作面 14 m、12 m 和 8 m。

由上述分析可知,当厚硬岩层厚度相同,而直接顶充填系数增大时,厚硬岩层初次来压和周期来压步距越大、发生破断越困难,采场围岩破坏范围越小。厚硬岩层初次和周期破断,在工作面煤壁形成了应力升高区,直接顶充填系数越大,应力峰值越小,峰值点距工作面越近。

综上所述,岩层厚度对巨厚坚硬岩层稳定性和工作面矿压显现程度有显著影响,其影响机制为:不同坚硬岩层厚度及分布形式将产生不同的破断失稳形式和破断块体尺寸,形成不同的力学结构,从而产生不同的载荷作用方式和来压显现程度;直接顶充填系数(直接顶厚度和采煤高度比值)的影响机制则表现在对坚硬岩层破断失稳的限制程度不同,影响厚硬岩层破断位置、破断尺寸和断裂块体下沉空间,从而改变载荷的作用位置和大小。

3.3 厚硬岩层下开采顶板控制分区

根据上述厚硬岩层下不同区域开采引起的破断特征、位移场、应力场的变化及其特征,以及不同区域开采的矿压显现程度和顶板控制难易程度,可把回采区域划分为顶板强烈来压区、顶板较强烈来压区和顶板来压不明显区。

直接顶厚度直接影响厚硬基本顶岩层垮落下沉空间,直接顶厚度越大、垮落后对采空区的充填程度越大,基本顶破断形成稳定结构的概率越大,工作面来压强度越小。按直接顶充填系数,沿工作面推进方向将880工作面划分为以下3区域,如图7所示。

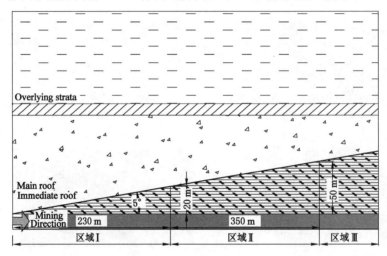

图 7 直接顶垮落充填性及矿压显现区域划分

Fig. 7 Immediate roof collapse filling and strata behavior area division

区域Ⅰ:$N<2$,直接顶厚度较小甚至缺失,此时厚硬顶板近距离赋存在煤层之上,部分与8煤层露头直接接触。结合朱仙庄矿参数,得到区域Ⅰ范围内工作面区间为0~230 m、直接顶厚度0~20 m。在区域Ⅰ内回采时,基本顶最小下沉空间大于4 m,巨厚坚硬顶板的悬露、破断和垮落运移对工作面的影响非常强烈,该回采区间称为顶板强烈来压区。

区域Ⅱ:$2<N<5$,直接顶厚度较大但垮落后不能全部充满采空区,结合朱仙庄矿参数,得到区域Ⅱ范围内工作面区间为230~580 m、直接顶厚度20~50 m。巨厚坚硬岩层破断垮落后受到采空区垮落岩石较有力支撑,回采时基本顶的失稳对工作面的影响程度为严重,该回采区间称为顶板较强烈来压区。

区域Ⅲ:$N>5$,直接顶厚度很大,垮落后可以密实充满采空区,结合朱仙庄矿参数,得到区域Ⅲ范围内工作面走向长度>580 m、直接顶厚度>50 m。回采后采空区被有效充填,巨厚坚硬岩石受到采空区垮落岩石有力支撑且几乎没有下沉空间,基本顶不发生破断或失稳垮落对工作面无明显影响,该回采区间称为顶板来压不明显区。

4 厚硬岩层下开采分区控制原则及方法

4.1 厚硬岩层下开采的支架—围岩相互作用特征

综合分析可知,对工作面产生显著影响的是厚硬顶板直接赋存在煤层之上和直接顶充填系数较小的情况下,即顶板强烈来压区和顶板较强烈来压区。

(1) 顶板强烈来压区的支架—围岩相互作用特征

顶板强烈来压区的直接顶充填系数 $N<2$,厚硬顶板直接赋存。支架与顶板岩层是相互作用、动态变化的关系。在厚硬顶板直接赋存煤层或直接顶较薄的条件下,采空区垮落冒矸对厚硬顶板的失稳运动影响很小,因此,厚硬顶板的断裂失稳主要取决于其自身的强度、赋存厚度和其上作用载荷的大小。

工作面在此阶段因为厚硬顶板的断裂失稳呈现不同的状况,其支架与围岩关系将呈现不同特征:① 初次来压阶段,由于砾岩岩层强度高、完整性好,初次破断后将形成长度大、厚度高的巨大块体,断裂岩体间水平约束力较小,自稳能力较弱,发生滑落失稳,而采空区矸石的支撑作用缺失,近距离巨厚坚硬岩层破断失稳后在支架上方产生巨大冲击动压,将造成支架围岩体系的极不稳定性;② 周期来压阶段,厚硬岩层受相邻岩层挤压作用,形成悬空的类"砌体梁"铰接结构,将有部分载荷作用在支架上。

(2) 顶板较强烈来压区的支架—围岩相互作用特征

顶板较强烈来压区的直接顶充填系数 $2<N<5$,直接顶厚度较薄。在此条件下,采空区有一定厚度的直接顶垮落,在工作面推进方向将形成煤壁-支架-采空区冒落矸石的不等高支撑体系,厚硬顶板的垮落失稳在一定程度上将受到冒落矸石的支撑作用,支架的载荷及动压影响将小于顶板强烈来压区的情况。但厚硬顶板断裂岩块仍将形成类"砌体梁"铰接结构。由于直接顶较薄,厚硬顶板的失稳仍有较大的回转空间,因此,来压较为强烈,支架仍将承受较强的动载荷作用。

结合不同回采区厚硬岩层破断失稳的特征及矿压显现规律,提出以下分区控制原则。

① 厚硬岩层下开采顶板的控制范围以顶板强烈来压区和顶板较强烈来压区为主,加大采场支架工作阻力无法根本解决顶板的安全控制问题。

② 对厚硬岩层采取高效的辅控措施以预控制,实现中、低位坚硬顶板及时切断回转、垮落充填,减小厚硬岩层块体下沉空间和作用在支架上的载荷,是降低采场矿压强度的有效手段。

③ 提出以坚硬顶板深孔承压爆破预裂为前提、支架合理选型为中心的协同控制方法和技术,实现分区控制。

4.2 近距离巨厚坚硬岩层顶板的爆破弱化控制方法

研究结果表明,在基本顶岩层厚硬、直接顶较薄、充填系数较小条件下,从支架选型角度难以实现顶板安全控制[14],必须进行近距离巨厚坚硬顶板的预控制,实现顶板爆破弱化[15],因此提出厚硬顶板深孔承压爆破弱化技术[16]的技术方案。

① 基本参数:根据 880 工作面条件及承压爆破预期效果,确定深孔内充水承压大小为 2.0 MPa;爆破钻孔水平转角分别设定为 15°与 20°[17],装药孔间距设为 7.0 m,设置大孔径导向孔直径为 100 mm,见图 8。

② 主要设备:地质钻机、钻杆、切槽钻头、$\phi 60$ mm 三翼钻头与圈钻头、注水泵、单向阀、封孔装置(带注液管和排液管)等。

③ 炸药及参数:三级煤矿许用乳化炸药,药卷规格为 $\phi 60 \times 500$ mm,1.5 ± 0.1 kg/卷,炸药密度为 1 194.265 kg/m³,炸药爆速:2 800 m/s。8# 普通瞬发电雷管,煤矿许用导爆索,以泰安、黑索金炸药为药芯、用棉线和塑料编织丝等作包缠物,并以塑料为防潮层组成,规格

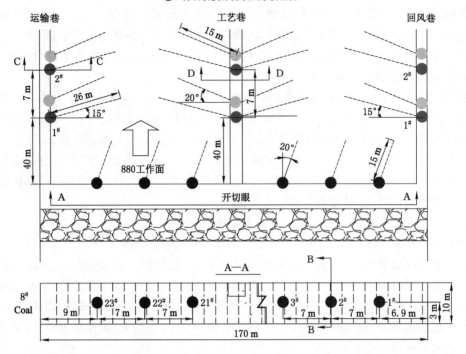

图 8 朱仙庄煤矿 880 工作面厚硬岩层深孔承压爆破钻孔布置图
Fig. 8 Deep hole confined blasting drilling arrangement of thick and hard rock in the working face 880 of Zhuxianzhuang Coal Mine

为 $\phi 5.2 \sim 5.5$ mm,爆速$\geqslant 6\,000$ m/s。

④ 工艺流程:根据施工目的和作业内容,提出近距离巨厚坚硬岩层承压爆破工艺流程如下:深孔打钻→装药→封孔→注承压水→爆破→效果检测。

5 结 论

(1)厚硬岩层厚度变化影响了破断块体尺寸和力学结构形成,直接顶充填系数直接决定断裂块体下沉空间,从而影响了其破断位置和破断尺寸,改变了工作面上作用载荷的位置和大小。

(2)根据厚硬岩层下不同开采区域顶板的破断运移和应力场的变化特征,以及不同区域开采的矿压显现程度和顶板控制难易程度,把回采区域划分为顶板强烈来压区、顶板较强烈来压区和顶板来压不明显区。

(3)顶板强烈来压区和顶板较强烈来压区对支架-围岩关系具有显著影响,是顶板控制的重点区域。

(4)近距离巨厚坚硬岩层控制的原则和方法是分区域顶板预控与支架协同控制相结合,采用坚硬顶板深孔承压爆破预裂控制为核心。

参 考 文 献

[1] 蒋金泉,代进,王普,等.上覆硬厚岩层破断运动及断顶控制[J].岩土力学,2014,35(S1):264-270.
JIANG Jinquan, DAI Jin, WANG Pu, et al. Overlying hard and thick strata breaking movement and broken-roof control[J]. Rock and Soil Mechanics, 2014, 35(S1): 264-270.

[2] 程占博,孔德中,杨敬虎.综放工作面厚硬顶板破断特征与支架工作阻力确定[J].矿业科学学报,2016,1(2):172-180.
CHENG Zhanbo, KONG Dezhong, YANG Jinghu. The breaking characteristics of thick-hard roof and determination of support capacity in fully mechanized caving face[J]. Journal of Mining Science and Technology, 2016, 1(2): 172-180.

[3] 张培鹏,蒋力帅,刘绪峰,等.高位硬厚岩层采动覆岩结构演化特征及致灾规律[J].采矿与安全工程学报,2017,34(05):852-860.
ZHANG Peipeng, JIANG Lishuai, LIU Xufeng, et al. Mining-induced overlying strata structure evolution characteristics and disaster-triggering under high level hard thick strata[J]. Journal of Mining & Safety Engineering, 2017, 34(05): 852-860.

[4] GU Shitan, JIANG Bangyou, PAN Yue, et al. Bending Moment Characteristics of Hard Roof before First Breaking of Roof Beam Considering Coal Seam Hardening[J]. Shock and Vibration, 2018(22).

[5] 杨敬轩,鲁岩,刘长友,等.坚硬厚层顶板条件下岩层破断及工作面矿压显现特征分析[J].采矿与安全工程学报,2013,30(2):211-217.
YANG Jingxuan, LU Yan, LIU Changyou, et al. Analysis on the rock failure and strata behavior characteristics under the condition of hard and thick roof[J]. Journal of Mining & Safety Engineering, 2013, 30(2): 211-217.

[6] ZHAO Tong, LIU Changyou, YETILMEZSOY Kaan, et al. Realization and engineering application of hydraulic support optimization in residual coal remining[J]. Journal of Intelligent & Fuzzy Systems, 2017(32): 2207-2219.

[7] 蒋金泉,张培鹏,聂礼生,等.高位硬厚岩层破断规律及其动力响应分析[J].岩石力学与工程学报,2014,33(7):1365-1374.
JIANG Jinquan, ZHANG Peipeng, NIE Lisheng, et al. Fracturing and dynamic response of high and thick stratas of hard rocks[J]. Chinese Journal of Rock Mechanics and Engineering, 2014, 33(7): 1365-1374.

[8] 祝捷,张敏,唐俊,等.顶板断裂瞬间煤体稳定性的动力学分析及数值模拟[J].煤炭学报,2014,39(2):253-257.
ZHU Jie, ZHANG Min, TANG Jun, et al. Dynamics analysis and numerical simulation on coal stability at the moment of roof fracture[J]. Journal of China Coal Society, 2014, 39(2): 253-257.

[9] 于斌,杨敬轩,高瑞.大同矿区双系煤层开采远近场协同控顶机理与技术[J].中国矿业

大学学报,2018,47(03):486-493.

YU Bin, YANG Jingxuan, Gao Rui. Mechanism and technology of roof collaborative controlling in the process of Jurassic and carboniferous coal mining in Datong mining area[J]. Journal of China University of Mining & Technology,2018,47(03):486-493.

[10] 李化敏,蒋东杰,李东印.特厚煤层大采高综放工作面矿压及顶板破断特征[J].煤炭学报,2014,39(10):1956-1960.

LI Huamin, JIANG Dongjie, LI Dongyin. Analysis of ground pressure and roof movement in fully- mechanized top coal caving with large mining height in ultra-thick seam[J]. Journal of China Coal Society, 2014,39(10):1956-1960.

[11] 潘俊锋,齐庆新,毛德兵,等.冲击性顶板运动及其应力演化特征的3DEC模拟研究[J].岩石力学与工程学报,2007,26(S1):3546-3552.

PAN Junfeng, QI Qingxin, MAO Debing, et al. Study on movement and stress evolutionary process of impacted roof with 3DEC[J]. Chinese Journal of Rock Mechanics and Engineering,2007,26(S1):3546-3552.

[12] 郝金鹏.陈家沟煤矿特厚煤层综放采场矿压特征及顶煤冒放性研究[D].太原:太原理工大学,2016.

[13] 曹胜根,姜海军,王福海,等.采场上覆坚硬岩层破断的数值模拟研究[J].采矿与安全工程学报,2013,30(2):205-210.

CAO Shenggen, JIANG Haijun, WANG Fuhai, et al. Numerical simulation of overlying hard strata rupture in a coal face[J]. Journal of Mining & Safety Engineering,2013,30(2):205-210.

[14] ZHAO Tong, LIU Changyou, YETILMEZSOY Kaan, et al. Fractural structure of thick hard roof stratum using long beam theory and numerical modeling[J]. Environmental Earth Sciences, 2017(76): 751.

[15] LIU Changyou, YANG Jingxuan, YU Bin. Rock-breaking mechanism and experimental analysis of confined blasting of borehole surrounding rock[J]. International Journal of Mining Science and Technology,2017,27(05):795-801.

[16] 杨敬轩,刘长友,于斌.围岩孔裂隙充水承压爆破过程分析[J].中国矿业大学学报,2017,46(05):1024-1032.

[17] YANG Jingxuan, LIU Changyou, YU Bin. Blasting process analysis of confined water filled in surrounding rock pores and fissures[J]. Journal of China University of Mining & Technology,2017,46(05):1024-1032.

[18] HE Hu, DOU Linming, FAN Jun, et al. Deep-hole directional fracturing of thick hard roof for rockburst prevention[J]. Tunnelling and Underground Space Technology,2012(32):34-43.

深井软岩下山巷道群非对称破坏机理与控制研究

刘帅[1,2],杨科[2],唐春安[1,3]

(1. 东北大学 资源与土木工程学院,辽宁 沈阳,110819;
2. 安徽理工大学 深部煤矿采动响应与灾害防控安徽省重点实验室,安徽 淮南,232001;
3. 大连理工大学 岩石破裂与失稳研究所,辽宁 大连,116024)

摘 要:受邻近巷道或工作面回采影响,巷道群中单一巷道通常处于复杂应力环境中并呈现非对称变形特点。以皖北煤电公司刘桥一矿Ⅱ66下山巷道群为支护工程实践,采用现场调查、实验室实验、理论分析、数值模拟以及工业性试验相结合的方法,探讨了巷道群非对称变形破坏机理,并针对各巷道提出相应优化支护方案。研究表明:(1)巷道群中后掘进巷道打破了先掘巷道围岩应力平衡,应力重新分布,在达到新的平衡过程中,围岩变形加剧。(2)巷道群围岩岩性软弱强度低,煤样CT扫描发现存在较多光滑板状裂隙,基质被裂隙网络分割和包围。X射线衍射实验显示,轨道下山底板为弱膨胀性岩石。(3)Ⅱ66采区最大水平主应力为31.87 MPa,侧压系数1.5,非均匀系数1.61。巷道群最优布置方位角为N269°E或N197°E,实际走向与最大主应力方向夹角为71°,巷道断面承受较高水平应力作用。(4)巷道群开拓延深后,二次应力场在巷间围岩叠加,辅助下山呈非对称变形。4煤回采对巷道群没有影响,6煤西翼工作面回采致使各巷道顶板和右帮位移急剧增大,停采线距离应由60 m改为120 m。(5)提出中空组合锚杆(索)分区注浆全断面控制对策,增加了控底措施,并对应力集中及变形的始发部位加密支护,同时以U型钢棚进行结构补强。工程实践表明,优化后的支护方案使围岩变形得到了有效控制,围岩完整程度显著提高。

关键词:深井;软岩;巷道群;非对称变形;全断面支护

Research on the asymmetric failure mechanism and control of downhill roadway group of soft rock in deep mine

LIU Shuai[1,2], YANG Ke[2], TANG Chunan[1,3]

(1. School of Resources and Civil Engineering, Northeastern University, Shenyang 110819, China; 2. Anhui Province Key Laboratory of Mining Response and Disaster Prevention and Control in Deep Coal Mine, Anhui University of Science and Technology, Huainan 232001, China; 3. Institute for Rock Instability and Seismicity Research, Dalian University of Technology, Dalian 116024, China)

Abstract: Single roadway in the roadway group is usually in the complex stress environment and presents the characteristics of asymmetric deformation under the influence of adjacent roadway or mining in the working

---

基金项目:安徽省重点研究与开发项目(1704a0802129)。
作者简介:刘帅(1990—),男,河南省郑州市人,现为博士研究生。E-mail:lsjz1990@163.com。
通信作者:杨科(1979—),男,博士,教授,主要从事矿山压力与岩层控制、煤与瓦斯共采方面的教学与研究工作。E-mail:yksp2003@163.com。

face. The mechanism of asymmetric deformation failure of roadway group is discussed and the corresponding optimization scheme of support is proposed for each roadway by means of field investigation, laboratory experiment, theoretical analysis, numerical simulation and industrial test on the basis of support at the Ⅱ66 downhill roadway group in Liuqiao No.1 coal mine of Wanbei coal and electricity company. The results show that (1) The later excavation roadway breaks the stress balance of surrounding rocks in the former roadway at the roadway group with the stress redistribution. the deformation of surrounding rock is intensified in the process of reaching a new equilibrium. (2) The surrounding rock of roadway group shows weak lithology and low strength. CT scan of coal sample showed that there were many smooth plate cracks and matrix was divided and surrounded by fracture network. X-ray diffraction experiment shows that the floor of dip track roadway is weak expansibility rock. (3) The maximum horizontal principal stress is 31.87 MPa with lateral pressure coefficient of 1.5 and non-uniform coefficient of 1.61 in Ⅱ66 mining area. The optimal azimuth of roadway group arrangement is N269°E or N197°E. The Angle between the actual direction and the direction of maximum principal stress is 71°. The horizontal stress of roadway section bearing is relatively high. (4) The secondary stress field is superimposed in the surrounding rocks between roadways and asymmetric deformation occurred in auxiliary downhill roadway after the development stretching of roadway group. The mining of No.4 coal seam has no influence on roadway group. The displacement of roof and right side of roadway increased sharply after the mining of western working face in No.6 coal seam and The distance of stop-line should be changed from 60 m to 120 m. (5) The control strategy of whole section that zoning grouting with Hollow combined bolt (cable) was proposed and measures to control the floor have been added. The intensive support was applied in initial part of stress concentration and deformation and the structure of u-shaped steel shed is reinforced at the same time. Practical application shows that the deformation of surrounding rock was effectively controlled and the integrity of surrounding rock was significantly improved by the optimized support scheme.

Keywords: deep mine; soft rock; downhill roadway group; asymmetric deformation; full section support

随着国民经济的快速发展,煤炭资源需求也在急剧增大。由于综合机械化开采水平的提高和开采强度的不断增大,浅部资源日益枯竭,东部大部分矿区都相继进入深部资源开采状态[1]。与浅部开采相比,深部煤(岩)体处于高渗透压力、高地应力、高地温梯度和强烈采掘扰动影响下的复杂耦合环境中,伴随发生的软岩非线性塑性大变形现象突出[2]。近几年来,对于深井软岩巷道工程问题,国内外学者从围岩性质、围岩赋存状态、围岩变形特点、支护作用等角度出发开展了大量研究,积累了丰富的理论和工程实践经验[3-6]。针对深井软岩巷道"小破坏、大位移"的变形特点,在控制技术方面目前主要采用锚杆(索)、围岩注浆以及钢筋混凝土砌碹等加固形式,通过分步联合支护来实现高强度耦合或刚柔耦合动态加固[7];在支护理论方面主要有主动有控卸压控制原理和深部软弱围岩叠加拱承载体强度理论等[8-9];而在支护设计方面,已有成果多是从巷道(硐室)群中单一巷道稳定性出发,研究各自的围岩变形规律及支护对策,难以从总体上阐明巷道群围岩破裂失稳机理及支护结构的力学特性[10]。

工程实践表明,受邻近巷道或工作面回采影响,巷道群中单一巷道通常处于复杂应力环境中并呈现非对称变形特点[11]。采用传统方法进行支护设计时,由于较少考虑巷道周围采掘工程分布及其时空关系,不了解围岩应力分布特点及变形的关键部位,往往采取均称支护手段。这种不考虑地质生产条件的设计方案往往造成过度支护、增加成本或难以对巷道变形的关键部位实施有效控制,出现非对称破坏进而导致支护结构的整体失稳。本文以刘桥一矿Ⅱ66下山巷道群为支护工程实践,采用现场调查、实验室实验、理论分析、数值模拟和工业性试验相结合的方法,探讨了巷道群非对称变形破坏机理,并针对各巷道提出了相应优化支护方案,为类似条件下山巷道群强矿压控制提供了借鉴。

1 下山巷道群工程概况及变形破坏特征

1.1 工程概况

刘桥一矿是皖北煤电公司主力生产矿井,采用立井多水平集中大巷布置方式。主采煤层为4煤和6煤,层间距60~80 m,煤层倾角10°~15°,每层单独布置准备巷道,现为"两综一炮"生产格局。其Ⅱ46采区、Ⅱ66采区主采工作面埋深已达800 m以上,并向-1 000 m水平开拓延深。本文以刘桥一矿Ⅱ66下山巷道群为研究对象,该巷道群由回风下山、轨道下山、辅助下山和运输下山组成。因运输下山与其他三条下山距离较远,受采掘活动影响较小,不在此次研究范围。三条下山平行布置其开拓延深顺序为:先掘轨道下山后掘回风下山和辅助下山。巷道群及周围采掘工程空间布置如图1所示。

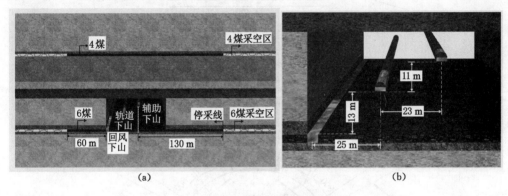

图1 巷道群及周围采掘工程空间布置

Fig.1 Layout of roadway group and surrounding working faces

(a) 巷道群及周围采掘工程空间布置;(b) 各巷道空间相对位置

回风下山布置在6煤中,沿煤层顶板掘进。直接顶为黑色泥岩,平均厚度2 m,泥质胶结,含植物化石,裂隙较为发育,遇水泥化;基本顶为浅色细砂岩,平均厚度6 m,细粒结构,钙质胶结并含有一条煤线;直接底为深灰色粉砂岩,平均厚度2.6 m,平行层理发育。辅助下山距离6煤顶板24 m左右,主要位于细砂岩中,部分区段穿层掘进,穿层岩性包括粉砂岩、泥质细砂岩、粉砂质泥岩、铝质泥岩等。现场观测显示,巷道两帮及顶板围岩完整性较好,底板极为软弱破碎。轨道下山距离6煤顶板13 m左右,围岩主要以细砂岩为主,同样为穿层掘进,穿层岩性与辅助下山相似。

1.2 原支护形式

回风下山为矩形断面,4.4 m×3.0 m(宽×高),部分区段采用工字钢对棚+帮部锚索支护(不作介绍),主要区段采用锚网索对称支护,设计参数如图2(a)所示。顶板选用左旋无纵筋螺纹钢锚杆,帮部选用右旋等强螺纹钢锚杆,间、排距800 mm×800 mm,钢筋梁连接;锚索选用 $\phi17.8$ mm、$L6300$ mm 的1860钢绞线,间、排距1 600 mm×2 400 mm,配加 M

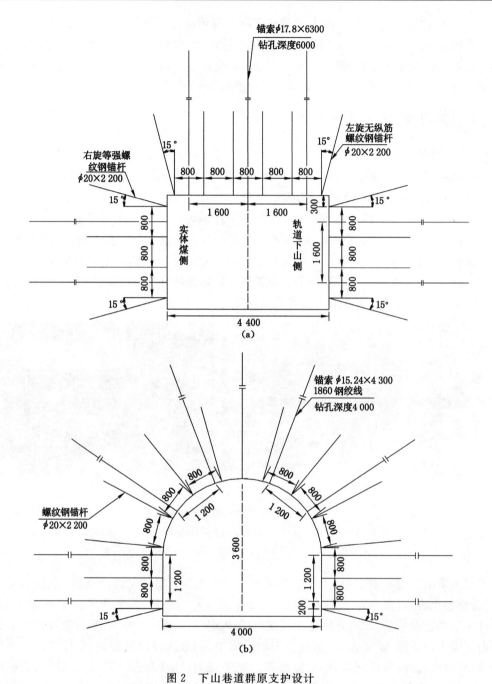

图 2 下山巷道群原支护设计

Fig.2 Former support designs of roadway group

(a)回风下山原支护断面;(b)轨道下山与辅助下山原支护断面

型钢带和钢笆网。

轨道下山和辅助下山为直墙半圆拱断面,宽 4.0 m,中线高 3.6 m,采用锚网索喷+U型钢联合支护,设计参数如图 2(b)所示。选用 U29 型钢作为基本支护,棚距 600 mm,3 节拱形式,搭接长度 500 mm。锚杆为螺纹钢锚杆,间、排距 800 mm×800 mm,钢筋梯子梁连接,锚索间、排距 1 200 mm×1 600 mm。

1.3 巷道群围岩变形破坏特征

采用原支护方案,巷道群掘出不久即产生严重变形。相继出现顶板离层冒落、垮帮、底鼓等现象,部分 U 型钢支架呈压扭状且时有锚杆、锚索被拉断或剪断现象发生(图 3)。现场观察与实测分析发现,巷道群中后掘巷道打破了先掘巷道围岩应力平衡,使应力重新分布,在达到新的平衡过程中,围岩变形加剧。返修过程中也出现上述情形,三条下山相互扰动影响显著。观测发现,巷道群同时受工作面回采影响,停采线前方长期存在的超前支承压力加剧了围岩失稳破坏。复杂的地质生产条件所引发的非均匀应力场致使三条下山围岩变形整体呈非对称性,具体特征如下:

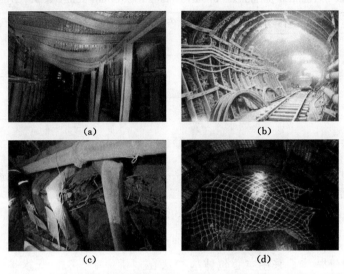

图 3 下山巷道群变形破坏特征

Fig. 3 Characteristics of roadway group deformation

(a) 回风下山;(b) 轨道下山;(c) 轨道下山;(d) 辅助下山

(1) 回风下山顶板大范围沉降,棚梁压弯,右侧肩窝完全压平,网兜现象突出;两帮强烈内移,局部区域达到 500~1 000 mm,锚索扭曲、钢带断裂;底鼓强烈,最大位移量达 800 mm。

(2) 轨道下山 U 型钢棚完全扭曲,部分区段拱部压平,顶板沉降偏于辅助下山侧;巷道底鼓严重并带动巷帮内移,右侧帮脚位移明显偏大,部分风水管路拖架挤压断裂。

(3) 辅助下山顶板出现台阶下沉,锚索退锚,锚固体整体沉降,部分支架右侧肩窝出现 U 型钢棚压弯内凸现象,形成不对称尖顶;底鼓严重且右帮强烈内移,水沟被局部填实。

2 下山巷道群变形失稳机理分析

巷道围岩稳定性与自身地质成因及应力场分布密切相关,是岩石强度、岩体完整性及结构面状态、原岩应力场、生产地质条件以及支护结构力学特性等综合作用的结果。

2.1 围岩物理力学特性

回风下山布置在煤层中,直接顶为泥岩;轨道下山和回风下山虽布置在细砂岩中,但围岩完整性较差,而且部分区段穿层掘进,穿层岩性软弱且节理裂隙发育。首先进行了现场取芯并开展了室内力学实验,力学参数如表 1 所示。结果显示,下山巷道群顶、底板物理力学强度较低,摩尔库伦破裂线均在煤层强度包络线附近。对煤样进行 CT 扫描并三维重构(图 4)。从图中可以看出,煤中存在较多光滑板状裂隙,基质被裂隙网络分割和包围,微裂隙的存在大大削弱了煤的物理力学强度。对轨道下山底板软岩黏土矿物开展了 X 射线衍射实验。结果显示,底板岩样以高岭石、石英为主,还有部分伊利石、绿泥石和长石等矿物,不含蒙脱石,为弱膨胀性岩石。

表 1　　　　　　　　下山巷道群围岩力学参数
Table 1　　The mechanical parameters of surrounding rock of downhill roadway group

岩层	密度/(kg/m³)	体积模量/GPa	剪切模量/GPa	内摩擦角/(°)	黏聚力/MPa
细砂岩	2 600	4.5	3.5	25	1.6
粉砂岩	2 550	4.0	3.0	23	1.4
煤	1 390	1.9	0.93	18	0.8
铝质泥岩	2 200	2.7	1.4	21	0.85
泥岩	2 200	2.5	1.3	20	0.83

综上可知,下山巷道群围岩节理裂隙发育,岩性软弱强度低,部分遇水膨胀力学性质劣化,围岩难以承受深部较高的二次应力场。

2.2 地应力场分析

地应力是所有地下工程变形失稳的原动力,掌握详实的原岩应力场分布特征是进行科学设计与支护的前提[12]。采用空心包体应变法在刘桥一矿−810 m 水平变电所泵房开展了原岩应力测试,测试结果如表 2 所示。

表 2　　　　　　　　Ⅱ66 采区主应力大小与方向
Table 2　　The magnitude and direction of principal stress in the Ⅱ66 mining area

水平主应力 σ_H/MPa	水平主应力 σ_h/MPa	铅直应力 σ_v/MPa	主应力 σ_H 方位/(°)	最大主应力倾角/(°)
31.87	19.77	21.25	N233°E	13.4

由测试结果可知Ⅱ66 采区原岩应力场有如下特点:(1) 通常认为应力值在 18~30 MPa 时为高应力区[13],Ⅱ66 采区最大应力为 31.87 MPa,表明该区域整体处于高原岩应力场。(2) 最大主应力和最小主应力近似为水平应力,侧压系数 λ 为 1.5,说明该区域为典型的构造应力场。(3) 水平地应力非均匀系数 ζ($\zeta = \sigma_H/\sigma_h$)为 1.61,说明该区域地应力具有明显的方向性,这将对巷道布置与支护产生重要影响。

进行巷道布置时,必须考虑巷道轴线与最大水平主应力之间的夹角,最优夹角应使作用在巷道围岩边界上的法向应力比值 σ_n/σ_v(σ_n 为作用在巷道两侧的水平法向应力)等于 1[14]。在 σ_{Hv} 型应力场中,根据任意斜面上法向应力与主应力的关系,可确定最优夹角 α_0 为:

$$\alpha_0 = \frac{1}{2}\arccos\frac{\sigma_H + \sigma_h - 2\sigma_v}{\sigma_H - \sigma_h} \tag{1}$$

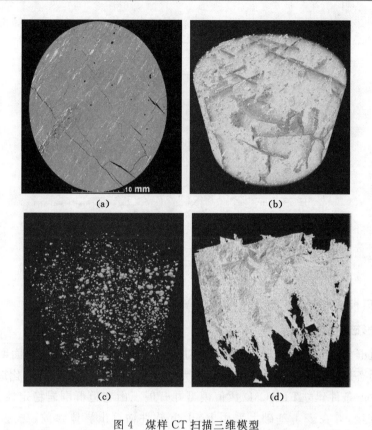

图 4 煤样 CT 扫描三维模型

Fig. 4 The 3D model of coal sample resulted by CT scanning

(a) 内部裂隙切片图；(b) 三维结构展示；(c) 孔隙空间分布；(d) 裂隙空间分布

将表 2 数据带入式(1)可得：$\alpha_0 = 36°$，即巷道最优布置方位角为 N269°E 或 N197°E。Ⅱ66 采区下山巷道群轴线方位为 N162°E，与最大主应力夹角为 71°。夹角较大致使巷道断面承受较高水平应力作用，同时非对称原岩应力场使围岩变形偏于巷道一侧。

2.3 支护结构稳定性分析

现有支护措施，无论是 U 型钢、工字钢等棚式被动支护，还是锚网索等主动支护，均是在围岩表面或是浅部区域形成具有自稳能力的承载结构。在进行支护设计时，大多注重提高支护体的强度、刚度而忽略承载结构的稳定性。已有研究表明，U 型断面顶板拱结构承载能力远高于两帮梁结构，而拱结构的承载能力又受梁结构制约，一旦梁结构产生破坏，拱结构承载能力随之急剧降低[15]。轨道下山和辅助下山均为直墙半圆拱断面，其支护结构可抽象为可动铰支座的二铰拱模型，铰链处具有三个自由度，该结构抗侧压能力较差。在进行支护设计时应将模型中的可动铰支座近似转化为固定铰支座(图 5)。

下山巷道群原支护方案虽进行了两帮底脚加固，但底板未采取控底措施。W. J. Gale 研究提出，深井水平应力通常大于铅直应力，巷道开挖引起应力重新分布时，铅直应力向两帮转移，水平应力向顶底板转移；铅直应力主导两帮破坏，水平应力主要影响顶底板岩层[16]。Ⅱ66 采区最大主应力为水平应力，与巷道群轴线夹角 71°，巷道顶底板岩层剪切错动风险较大。原支护方案未采取控底措施使得底板成为巷道失稳突破口，底鼓严重进而带

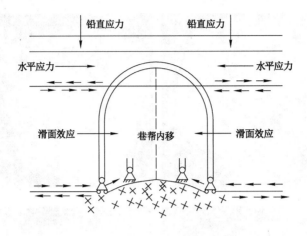

图 5　可缩性支架失稳模式
Fig.5　Instability mode of yieldable support

动帮墙失稳、顶板拱结构承载能力降低,最终导致支护结构整体失稳。

2.4　围岩变形破坏模拟分析

下山巷道群先后受到巷间掘进扰动、4 煤层工作面回采和 6 煤层工作面回采影响,其变形破坏是一动态过程。为明确巷道群围岩应力、变形以及塑性区的动态演化过程,基于工程地质概况与生产条件建立 FLAC3D 数值模型并开展下山巷道群围岩稳定性分析。采用摩尔库伦强度准则,并在表 1 基础上采用 RocLab 软件估算出岩体参数。模拟过程:模型建立—原岩应力加载平衡—各条下山分步开挖与支护—4 煤工作面回采—6 煤工作面回采。

(1) 下山巷道群掘进期间围岩稳定性分析

图 6 为下山巷道群掘进期间围岩垂直应力等值线图。从图中可以看出,轨道下山掘出以后,垂直应力在巷道周围呈椭圆对称分布,最大值为 36.1 MPa。围岩由浅入深形成应力降低区、增高区、原岩应力区。回风下山开挖后,其二次应力场与轨道下山二次应力场在巷间围岩叠加,应力增高区范围显著增大。同时,轨道下山右侧控制应力峰值升高,达到 40 MPa。辅助下山开挖后,巷道群巷间应力叠加更为显著,控制应力峰值分布范围均有所扩大,应力集中程度进一步提高且整体呈非对称分布。

图 7 和图 8 为下山巷道群掘进期间围岩表面位移演化过程与最终分布形态。从图中可以看出,轨道下山掘出以后,其最大变形量出现在巷道底板达到 290 mm,围岩呈对称变形。回风下山开挖后,轨道下山变形量有所增大,此时右肩变形大于左肩,底鼓变形稍偏于左侧。辅助下山开挖后,轨道下山变形继续发展,除肩窝位移外变形整体呈对称分布。回风下山顶板下沉较为剧烈,其最大位移量为 460 mm,两帮及底板位移量在 420 mm 左右,由图 7(b)可以看出,辅助下山开挖对回风下山变形影响较小,此时回风下山变形稍偏于左侧。辅助下山最大变形量为 320 mm,右帮和右肩位移明显大于巷道左侧,巷道非对称变形显著。

图 9 为下山巷道群开拓延深结束围岩塑性破坏分布图。红色区域为剪切破坏,紫色区域为张拉破坏。回风下山布置在煤层中,塑性破坏范围最大,顶底板张拉破坏较为严重,与前述支护结构稳定性分析结论一致。各下山巷道扰动影响显著,辅助下山围岩塑性破坏范围偏于右侧,呈非对称性。

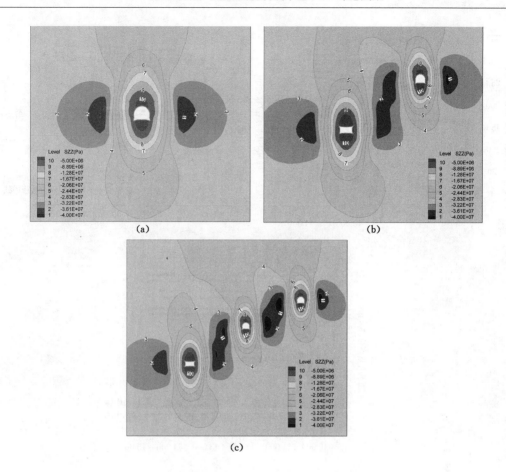

图 6 巷道群掘进期间围岩垂直应力等值线图

Fig. 6 Contour map of vertical stress of roadway group during excavation

(a) 轨道下山开挖；(b) 回风下山开挖；(c) 辅助下山开挖

(2) 采动影响下巷道群围岩稳定性分析

图 10 为Ⅱ46、Ⅱ66 采区 4 煤、6 煤工作面回采完成后下山巷道群围岩垂直应力等值线图。Ⅱ46 西翼工作面停采线距回风下山垂直距离约 85 m，水平距离约 60 m，东翼停采线与辅助下山水平距离约 130 m。对比图 10(a)与图 6(c)发现，4 煤回采后巷道群垂直应力场无明显变化，说明由于距离较远，4 煤保护煤柱底板集中应力未对巷道群产生扰动影响。对比图 10(b)与图 6(c)发现，由于 6 煤工作面与下山巷道群近似在同一层位，巷道群受 6 煤工作面回采影响显著。围岩应力升高区范围急剧增大，控制应力峰值从 40 MPa 增加到 48 MPa，增大 20%。峰值应力转移至巷间围岩中部，两侧均为峰后破坏区，不存在弹性应力升高区。

图 11 为工作面回采期间下山巷道群围岩表面位移曲线。从图中可以看出，4 煤回采对巷道群围岩变形影响较小，各巷道表面位移发生微小变化。从图 11(b)和 11(d)可以看出，6 煤回采后，巷道群顶板和右帮变形急剧增大，位移增大幅度在 87%～95%，底板和左帮位移增幅有限。不合理的 6 煤工作面停采线设计，加剧了巷道群变形失稳，同时使围岩变形呈现显著非对称性。巷道群受采动影响最终围岩变形形态如图 12 所示。

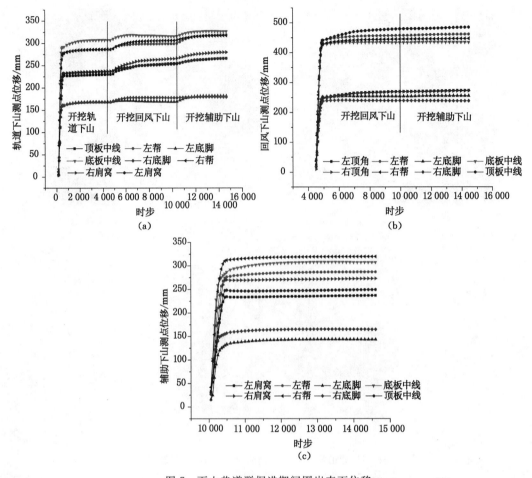

图 7 下山巷道群掘进期间围岩表面位移

Fig.7 Surface displacement of surrounding rock of roadway group during excavation

(a) 轨道下山表面位移；(b) 回风下山表面位移；(c) 辅助下山表面位移

图 13 为 4 煤、6 煤工作面回采完成后巷道群围岩塑性区分布状况。对比图 13(a)与图 9 可以看出，4 煤回采对巷道群影响较小。而 6 煤回采完成后，巷道群围岩塑性破坏单元数由 9 922 增长到 59 869，涨幅超过 5 倍；塑性区体积由 $6.92×10^4$ m³ 增长到 $6.52×10^5$ m³，涨幅超过 8 倍，巷间围岩几乎全部处于塑性破坏状态。由此说明，原设计方案下 6 煤采动影响是造成巷道群大面积失稳破坏的主要原因。矿压观测显示，6 煤工作面回采过程中超前支承压力影响范围在 120 m 左右，而回风下山距Ⅱ66 西翼工作面停采线仅 60 m。在支护优化设计前，必须改变原有的停采线设计方案。

3 支护方案优化

3.1 支护对策

根据现场调查及数值分析结果，确定支护方案。

(1) 深井高地应力环境中的软岩巷道，无论采取何种支护方式，围岩都不可避免地进入塑性状态。浅部围岩采用"高预应力锚杆＋M 型钢带＋金属网＋混凝土喷层"支护，使原来

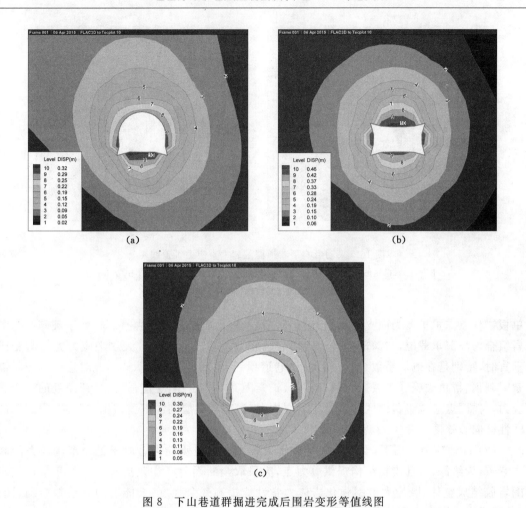

图 8 下山巷道群掘进完成后围岩变形等值线图

Fig. 8 Contour map of surrounding rock deformation of roadway group after excavation

(a) 轨道下山变形;(b) 回风下山变形;(c) 辅助下山变形

图 9 下山巷道群掘进完成后围岩塑性区分布

Fig. 9 Plastic zone distribution of surrounding rock of roadways after excavation

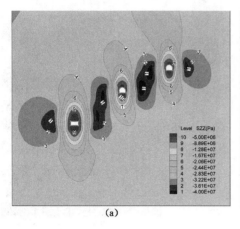 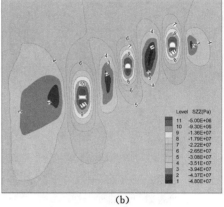

图 10 采动影响下巷道群围岩垂直应力等值线图

Fig. 10 Contour map of vertical stress of roadways affected by mining

(a) 4 煤工作面回采后；(b) 6 煤工作面回采后

单根锚杆点支护扩展为面支护，防止破碎区围岩垮落、离层、滑动、内挤。将处于破碎状态围岩组合为浅部承载层，与深部弹性应力增高区联系。巷道轴向与最大水平应力方向斜交或垂直时，特别是在侧压系数 $\lambda > 1$ 情况下，巷道顶、帮角塑性范围最大[17]，此时应采用倾斜锚索限制顶、帮角破碎区扩展。与锚杆相比，锚索因其柔性更佳，更能适应顶、帮角处的集中剪应力。锚固点应位于塑性区外的稳定岩层中，将浅部承载层与深部稳定围岩联结，限制深部塑性区向破碎区转变。

（2）高地应力作用下，支护的关键在于发挥围岩的自承能力。对于松动范围较大的软岩巷道，为给深、浅部支护结构提供着力点，应采取全断面分区注浆措施。通过注浆将破碎围岩胶结成整体，既能大范围强化岩体力学性能提高残余强度，又能利用浆液封堵围岩裂隙，降低水对围岩强度的弱化作用。

（3）针对围岩非对称变形，应加强关键部位控制以增强承载结构稳定性。笔者认为，应力集中程度高以及原支护方案下变形破坏始发位置均为需加强控制的关键部位。原支护形式未采取控底措施，底板率先破坏后，带动 U 型钢等支护结构失稳并失去承载能力，因此必须对巷道底板进行支护。通过打底板锚杆（索）和分区注浆方式，避免底板率先破坏发生底鼓。同时，对 U 型钢支架棚腿进行锁棚，以改善抗侧压能力差的特点，提高拱形支架稳定性。

3.2 支护方案优化

（1）回风下山支护方案优化

基于冒落拱理论确定锚固支护参数。巷道掘出后，两帮最大破坏深度 C 为[18]：

$$C = \left(\frac{K_{cx} \gamma H B}{10^4 f_c} - 1 \right) h \tan \frac{90 - \varphi}{2} \tag{2}$$

式中 K_{cx}——巷道周边挤压应力集中系数；

γ——上覆岩层平均重度，kN/m^3；

H——巷道埋深，m；

B——表征采动影响程度的无因次参数；

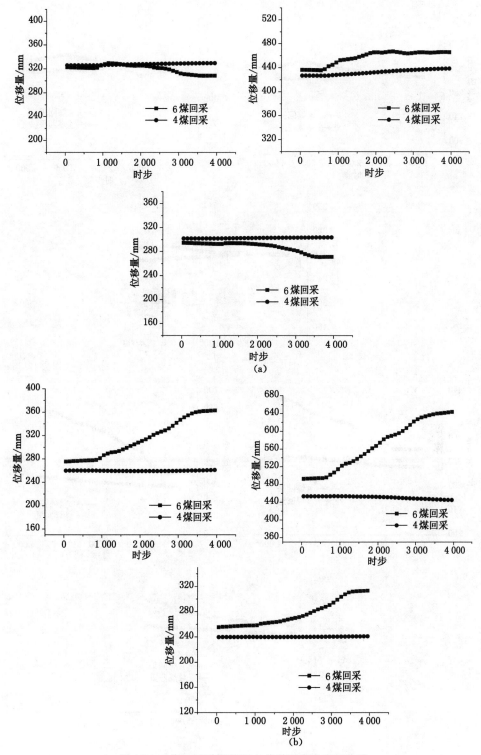

图 11 工作面回采期间下山巷道群围岩表面位移

Fig. 11 Surface displacement of surrounding rock of roadways during mining

(a) 底板;(b) 顶板;(c) 左帮;(d) 右帮

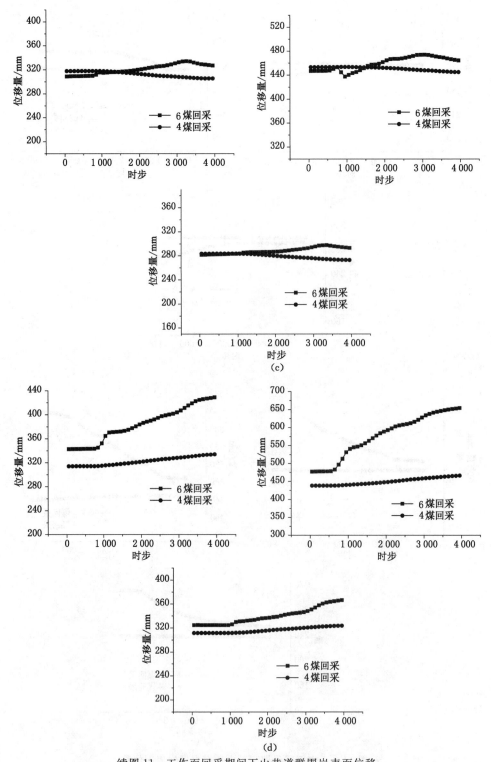

续图 11 工作面回采期间下山巷道群围岩表面位移

Continued Fig. 11 Surface displacement of surrounding rock of roadways during mining

(a) 底板;(b) 顶板;(c) 左帮;(d) 右帮

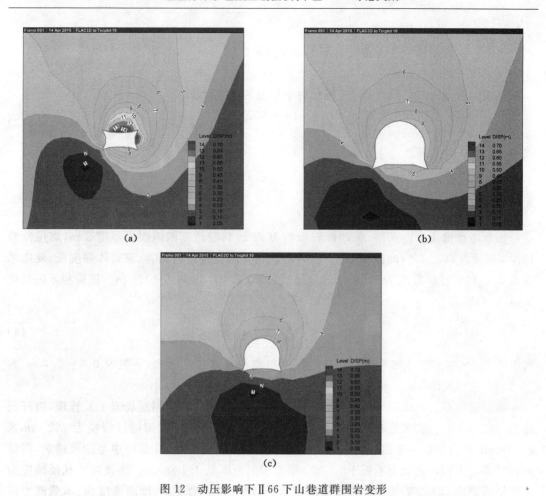

图 12 动压影响下Ⅱ66下山巷道群围岩变形

Fig. 12 Contour map of surrounding rock deformation of roadways after mining

(a) 回风下山变形；(b) 轨道下山变形；(c) 辅助下山变形

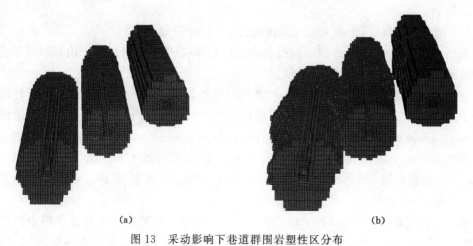

图 13 采动影响下巷道群围岩塑性区分布

Fig. 13 Plastic zone distribution of surrounding rock of roadway group after mining

(a) 4 煤回采结束；(b) 6 煤回采结束

f_c——煤层硬度系数;

h——煤层厚度或巷道轮廓范围内煤夹层厚度,m;

φ——煤体内摩擦角,(°)。

顶板岩层最大破坏深度 b[18],按相对于层理的法线计算:

$$b = \frac{(a+C)\cos\alpha}{k_y f_n} \tag{3}$$

式中 a——巷道跨距一半,m;

α——煤层倾角,(°);

k_y——锚固岩层稳定性系数;

f_n——锚固岩层硬度系数。

综合考虑地质生产条件,采动影响参数 B 取 2.4,巷道范围内煤层厚度 3 m,煤层倾角 10°,煤层硬度系数 2,内摩擦角 26°。同时考虑到 6 煤直接顶为泥岩,该岩体强度低,硬度系数取 2.2,稳定性系数取 1.5。带入式(2)、式(3)得 $C=1.33$ m,$b=1.9$ m。顶锚和帮锚长度通过式(4)确定[18]:

$$\left. \begin{array}{l} L_d = b + s \\ L_b = C + s \end{array} \right\} \tag{4}$$

式中 s——锚杆锚入破坏范围之外的深度与锚杆外露长度之和,一般取 0.5~0.7 m,本设计取 0.6 m。

将上述数据带入式(4)得:$L_d = 2.5$ m,$L_b = 1.93$ m。考虑到现场施工及管理,锚杆长度统一取 2.5 m。将原支护方案中的 ϕ20 mm×2 200 mm 螺纹钢锚杆替换为 ϕ22 mm×2 500 mm 中空组合注浆锚杆,锚索替换为 ϕ21.8 mm×6 300 mm 新型中空注浆锚索,间排距保持不变。锚杆预紧力不低于 55 kN,锚索预紧力不低于 120 kN。高强度小孔径预应力注浆锚索除结构本身能够满足高压注浆要求外,还具有主动支护、增阻速度快、承载能力高等特点。优化方案同时注重对底板及帮脚支护,在底板施作 3 根注浆锚索,帮脚补增 1 根注浆锚杆,与竖直方向夹角均为 15°(图 14)。水泥砂浆喷层厚度 50 mm,完全覆盖 M 型钢带、梯子梁和钢笆网。

(2)辅助下山、轨道下山支护方案优化

根据上述支护对策,结合现场调查和巷道群变形破坏分析,对辅助下山和轨道下山进行支护参数优化,改进方案如图 15 所示。

巷道刷扩后均采用 29U 型钢支架作为临时支护,排距 450 mm,两侧棚腿打锚杆配合 U 型棚卡缆进行锁固。铺设双层金属网,规格 1 600 mm×900 mm,直径 6 mm。锚杆选用 ϕ22 mm×2 500 mm 中空组合注浆锚杆,间、排距 800 mm×800 mm,预应力 50 kN。全断面喷射水泥砂浆,喷层厚度 250 mm 覆盖型钢支架,强度等级 C20。顶、帮选用 ϕ22 mm×9 200 mm,底板选用 ϕ22 mm×6 500 mm 高强度中空注浆锚索,排距 3 000 mm,预紧力不小于 150 kN。

辅助下山变形偏于右侧,基于巷道非对称破坏特征,断面非对称布置 9 根中空注浆锚索,如图 15(a)所示。底板施作 3 根,右侧肩窝处增设 1 根。将原支护方案中的 ϕ20 mm×2 200 mm 螺纹钢锚杆替换为 ϕ22 mm×2 500 mm 中空组合注浆锚杆,底板两侧各施作 1 根底脚注浆锚杆,右侧帮脚距底板 0.25 m 处按俯角 30°增补 1 根帮脚注浆锚杆。轨道下山掘

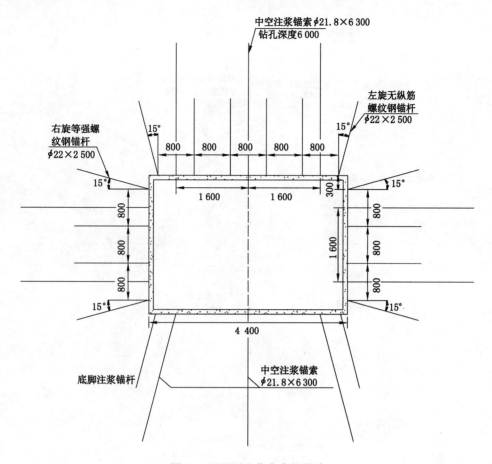

图 14 回风下山优化支护设计

Fig.14 Optimized support design of return airflow dip

进完成后表现为对称变形,受 6 煤采动影响后呈现非对称变形破坏,当停采线重新设计后,同时,Ⅱ66 西翼工作面停采线重新设计后改为距回风下山 120 m,减小支承压力对巷道群围岩稳定性影响。可不考虑回采影响。轨道下山两侧围岩均存在巷间应力叠加区,变形破坏程度较大,对左右两侧肩窝及帮脚均进行补强加固,如图 15(b)所示。通过锚索结合锚梁在帮部和肩窝位置对 U 型钢棚进行结构补强,增强支架稳定性,锚梁采用 20# 槽钢制作,内加 160 mm×160 mm×10 mm 钢板,相邻锚梁上下交错布置。

针对巷道群原生裂隙采取分区注浆加固措施。破碎区围岩通过中空组合注浆锚杆低压注浆胶结,注浆压力 1.5 MPa;塑性区围岩通过中空注浆锚索高压注浆胶结,注浆压力 4 MPa。浆液为高水速凝材料,待低压注浆 9 d,浅部围岩达到一定强度后再实施高压注浆。

4 工业试验

采用优化支护方案后,在下山巷道群布置测点,通过十字交叉法监测各巷道顶、底板及两帮移近量,并根据监测信息及时优化、修正支护方案,位移曲线如图 16 所示。

监测数据显示,各巷道顶、底板移近量均大于两帮。辅助下山测站顶、底板最大移近量为 413 mm,轨道下山测站顶、底板最大移近量为 448 mm,回风下山测站顶、底板最大移近

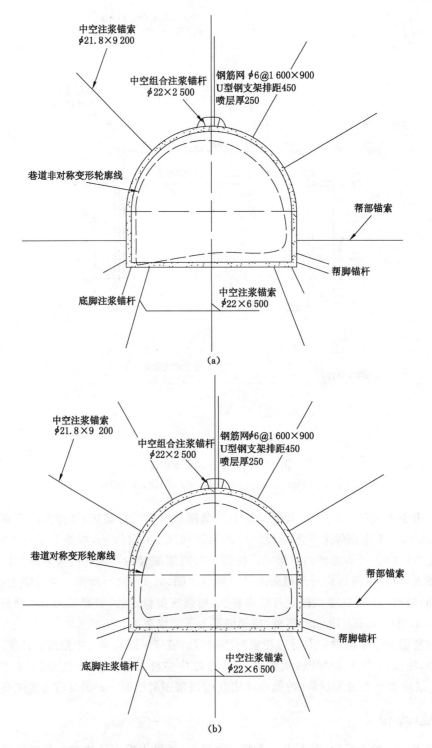

图 15 辅助下山、轨道下山优化支护方案
Fig.15 Optimized support designs of raided dip and track dip
(a) 辅助下山优化方案;(b) 轨道下山优化方案

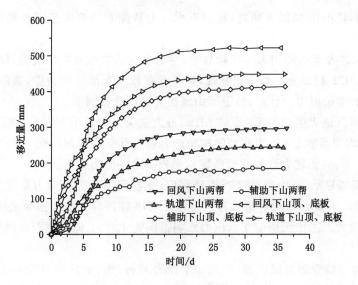

图 16 巷道群围岩位移监测曲线

Fig. 16 Monitoring curve of surface displacement of roadways

量为 522 mm,围岩变形得到有效控制。

在轨道下山肩窝处进行钻孔窥视(见图17),钻孔深度 2 m。与优化支护前难以成孔状态对比,围岩离层和破碎程度大大降低。

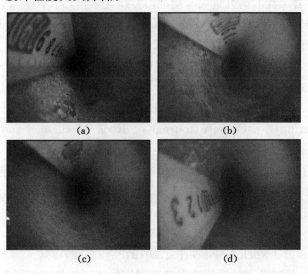

图 17 优化方案钻孔窥视结果

Fig. 17 Borehole imaging results of optimized support design

(a) 距孔底 1.9 m;(b) 距孔底 1.5 m;(c) 距离孔底 1 m;(d) 距孔底 0.5 m

5 结　论

(1)现场调查与实测分析发现,巷道群中后掘进巷道打破了先掘巷道围岩应力平衡,应力重新分布,在达到新的平衡过程中,围岩变形加剧。4煤回采期间,巷道群受保护煤柱底板集中应力影响较小;6煤回采期间,超前支承压力致使巷道群大范围失稳破坏,围岩非对称变形显著。

(2) 现场取芯及室内物理力学实验显示,巷道群围岩岩性软弱强度低,强度包络线均与煤样较为接近。CT 扫描及三维重构发现,煤样存在较多光滑板状裂隙,基质被裂隙网络分割和包围。X 射线衍射实验显示,轨道下山底板为弱膨胀性岩石。

(3) II 66 采区最大主应力为 31.87 MPa,最大主应力和最小主应力近似为水平应力,侧压系数 1.5,非均匀系数 1.61,为典型的构造应力场且具有明显方向性。巷道群与最大主应力方向夹角为 71°,夹角较大致使巷道断面承受较高水平应力作用。

(4) 数值模拟显示,下山巷道群开拓延深后,受巷间围岩二次应力场叠加影响,辅助下山呈非对称变形。4 煤回采对巷道群几乎没有影响,6 煤西翼工作面回采致使各巷道顶板和右帮位移急剧增大。工作面超前支承压力影响范围为 120 m,停采线距离回风下山仅 60 m,应重新设计。

(5) 基于巷道群变形失稳机理,提出中空组合锚杆(索)分区注浆全断面控制对策。依据冒落拱理论确定顶锚、帮锚长度不小于 2.5 m,1.93 m。增加了控底措施,并对应力集中及变形的始发部位加密支护,同时通过锚梁对 U 型钢棚进行结构补强。现场工程实践表明,优化后的支护方案使围岩变形得到了有效控制,围岩完整程度显著提高。

参 考 文 献

[1] 何满潮. 深部软岩工程的研究进展与挑战[J]. 煤炭学报,2014(39):1409-1417.
  HE Manchao. Progress and challenges of soft rock engineering in depth[J]. Journal of China Coal Society,2014(39):1409-1417.

[2] 何满潮,谢和平,彭苏萍,等. 深部开采岩体力学研究[J]. 岩石力学与工程学报,2005(16):2803-2813.
  HE Manchao, XIE Heping, PENG Suping, et al. Study on rock mechanics in deep mining engineering[J]. Chinese Journal of Rock Mechanics and Engineering, 2014(39):2803-2813.

[3] 黄兴,刘泉声,乔正. 朱集矿深井软岩巷道大变形机制及其控制研究[J]. 岩土力学,2012,33(3):827-834.
  HUANG Xing, LIU Quansheng, QIAO Zheng. Research on large deformation mechanism and control method of deep soft roadway in Zhuji coal mine[J]. Chinese Rock Mechanics and Rock Engineering,2012,33(3):827-834.

[4] 李学华,杨宏敏,刘汉喜,等. 动压软岩巷道锚注加固机理与应用研究[J]. 采矿与安全工程学报,2006(02):159-163.
  LI Xuehua, YANG Hongmin, LIU Hanxi, et al. Research on bolt-grouting reinforcement technology in dynamic pressure and soft rock roadway[J]. Journal of Mining & Safety Engineering,2006(02):159-163.

[5] 卢兴利,刘泉声,苏培芳,等. 潘二矿松软破碎巷道群大变形失稳机理及支护技术优化研究[J]. 岩土工程学报,2013(35):97-102.
  LU Xingli, LIU Quansheng, SU Peifang, et al. Instability mechanism and bracing

optimization for roadway groups with soft and fractured surrounding rock in Pan′er Coal Mine[J]. Chinese Journal of Geotechnical Engineering,2013(35):97-102.

[6] 李学华,姚强岭,张农.软岩巷道破裂特征与分阶段分区域控制研究[J].中国矿业大学学报,2009(38):618-623.
LI Xuehua, YAO Qiangling, ZHANG Nong. Fracture characteristics of a soft rock road way: staged and zoned control [J]. Journal of China University of Mining & Technology,2009(38):618-623.

[7] 王卫军,彭刚,黄俊.高应力极软破碎岩层巷道高强度耦合支护技术研究[J].煤炭学报,2011,36(2):223-228.
WANG Weijun, PENG Gang, HUANG Jun. Research on high-strength coupling support technology of high stress extremely soft rock roadway[J]. Journal of China Coal Society,2011,36(2):223-228.

[8] 柏建彪,王襄禹,贾明魁,等.深部软岩巷道支护原理及应用[J].岩土工程学报,2008(05):632-635.
BAI Jianbiao, WANG Xiangyu, JIA Mingkui, et al. Theory and application of supporting in deep soft roadways[J]. Chinese Journal of Geotechnical Engineering, 2008(05):632-635.

[9] 余伟健,高谦,朱川曲.深部软弱围岩叠加拱承载体强度理论及应用研究[J].岩石力学与工程学报,2010,29(10):2134-2142.
YU Weijian, GAO Qian, ZHU Chuanqu. Study of strength theory and application of overlap arch bearing body for deep soft surrounding rock[J]. Chinese Journal of Rock Mechanics and Engineering,2010,29(10):2134-2142.

[10] 鲁德丰,刘帅,王纪尧.刘桥一矿深井软岩下山巷道围岩控制技术[J].煤矿安全,2016,47(12):72-75.
LU Defeng, LIU Shuai, WANG Jiyao. Surrounding rock control technology of deep soft rock down-dip gateway in Liuqiao No. 1 coal mine[J]. Safety in Coal Mines, 2016,47(12):72-75.

[11] 张广超,何富连.大断面强采动综放煤巷顶板非对称破坏机制与控制对策[J].岩石力学与工程学报,2016,35(04):806-818.
ZHANG Guangchao, HE Fulian. Asymmetric failure and control measures of large cross-section entry roof with strong mining disturbance and fully-mechanized caving mining[J]. Chinese Journal of Rock Mechanics and Engineering, 2016, 35 (04): 806-818.

[12] 康红普,颜立新,张剑.汾西矿区地应力测试与分析[J].采矿与安全工程学报,2009,26(03):263-268.
KANG Hongpu, YAN Lixin, ZHANG Jian. Measurements and analysis of ground stresses in Fenxi coal mining area[J]. Journal of Mining & Safety Engineering,2009, 26(03):263-268.

[13] 陈菲,何川,邓建辉.高地应力定义及其定性定量判据[J].岩土力学,2015,36(04):

971-980.

CHEN Fei, HE Chuan, DENG Jian-hui. Concept of high geostress and its qualitative and quantitative Definitions[J]. Rock and Soil Mechanics, 2015, 36(04): 971-980.

[14] 孙玉福. 水平应力对巷道围岩稳定性的影响[J]. 煤炭学报, 2010, 35(06): 891-895.

SUN Yufu. Affects of in-situ horizontal stress on stability of surrounding rock roadway[J]. Journal of China Coal Society, 2010, 35(06): 891-895.

[15] 李迎富. 潘三矿深井动压回采巷道围岩稳定性分类及其支护设计[D]. 淮南: 安徽理工大学, 2006.

LI Yingfu. Classification on the stability of rock around deep gateway under dynamic pressure and its supporting design in Pansan mining area[D]. Huainan: AnHui University of Science and Technology, 2006.

[16] 臧龙. 煤柱底板巷道围岩稳定性分析及其控制技术研究[D]. 徐州: 中国矿业大学, 2014.

ZANG Long. Study on the surrounding rock stability analysis and its control technology of floor roadway under coal pillar[D]. Xuzhou: China University of Mining and Technology, 2014.

[17] 陈立伟, 彭建兵, 范文, 等. 基于统一强度理论的非均匀应力场圆形巷道围岩塑性区分析[J]. 煤炭学报, 2007(01): 20-23.

CHEN Liwei, PENG Jianbing, FAN Wen, et al. Analysis of surrounding rock mass plastic zone of round tunnel under non-uniform stress field based on the unified strength theory[J]. Journal of China Coal Society, 2007(01): 20-23.

[18] 康红普, 王金华, 林健, 等. 煤矿锚杆支护理论与成套技术[M]. 北京: 煤炭工业出版社, 2007: 73-75.

KANG Hongpu, WANG Jinhua, LIN Jian, et al. Rock bolting theory and complete technology for coal roadways[M]. Beijing: China Coal Industry Publishing House, 2007: 73-75.

大倾角煤层大采高工作面倾角对煤壁片帮的影响机制

王红伟[1,2]，伍永平[1,2]，罗生虎[3]，刘孔智[4]，解盘石[1,2]，刘茂福[2]

(1. 西安科技大学 西部矿井开采及灾害防治教育部重点实验室，陕西 西安，710054；
2. 西安科技大学 能源学院，陕西 西安，710054；3. 西安科技大学 理学院，陕西 西安，710054；
4. 重庆(贵州)煤电有限公司，贵州 毕节，551700)

摘 要：煤壁片帮是制约大倾角煤层大采高工作面安全高效开采的主要因素之一，结合艾维尔沟煤矿大倾角大采高工作面的实际，归纳了易造成煤壁片帮的主控参量，建立了倾向煤壁岩梁力学模型，分析了工作面倾角对煤壁片帮的影响机制，揭示了不同工作面倾角条件下煤壁应力、位移分布特征。研究表明：工作面倾角、采高、推进速度、支架支撑力、煤体内聚力及内摩擦角等是大倾角大采高工作面煤壁片帮的主控参量，受工作面倾角影响，煤壁支承压力沿倾向非对称分布，导致煤壁变形(挠度)呈现出非对称特性，在工作面中上部区域(约 0.66L 处)煤壁岩梁变形最大，随着工作面倾角增大，煤壁集中应力以较小幅度逐渐减小，变形量减小幅度明显，水平位移量不断增加，易形成片帮区域不断向倾斜上部转移。

关键词：大倾角煤层；大采高；煤壁片帮；倾角变化；水平位移

Mechanism of coal wall spalling induced by dipping angle of fully mechanized working face with great mining height in mining steeply inclined seam

WANG Hongwei[1,2], WU Yongping[1,2], LUO Shenghu[3],
LIU Kongzhi[4], XIE Panshi[1,2], LIU Maofu[2]

(1. Key Laboratory of Western Mine Exploitation and Hazard Prevention,
Ministry of Education, Xi'an 710054, China;
2. School of Mineral Engineering, Xi'an University of Science and Technology,
Xi'an 710054, China;
3. School of Science, Xi'an University of Science and Technology, Xi'an 710054, China;
4. Chongqing Energy (Guizhou) Coal Co. Ltd., Bijie 551700, China)

Abstract: Coal wall spalling is one of the main factors restricting safe and high efficient mining of working face with great mining height in steeply inclined seam. The main control parameters for coal wall spalling were generalized combined with the real conditions of steeply inclined working face with great mining height

---

基金项目：国家自然科学基金重点项目(51634007)；中国博士后科学基金资助项目(2018M633539)；陕西省自然科学基础研究计划项目(2016JQ5019)。

通信作者：王红伟(1983—)，男，湖北省随州市人，副教授，工学博士，从事难采煤层开采方面的研究。E-mail：646937403@qq.com。

in Ewirgol Mine, and the mechanical model of coal wall along tendency was established. The mechanism of coal wall spalling induced by dipping angle of working face was analyzed and the stress and displacement distribution characteristics of coal wall under different dipping angle were revealed. The results show that the main control parameters for coal wall spalling include dipping angle, mining height, advancing speed, support resistance and cohesion and internal friction angle of coal mass in mining steeply inclined seam with great mining height. Due to dipping angle of working face, the abutment pressure in coal wall is asymmetric along inclination, which leads to the asymmetry of deformation of coal wall, with the largest deformation at location of about 0.66L in upper area of working face. With the increase of dipping angle of working face, concentrated stress in coal wall reduced gradually with a small amplitude, deformation of coal wall reduced with an obvious amplitude, and the area with coal wall spalling continuously moved to upper area.

Keywords: steeply inclined seam; fully mechanized mining with great mining height; coal wall spalling; varied dipping angle; horizontal stress

煤壁片帮与端面冒顶是大倾角大采高工作面安全高效开采亟须解决的难题。受工作面倾角影响,煤壁片帮具有频次高、范围大、分区域、蔓延性与冒滑特点。针对煤壁片帮机理与控制方法,国内学者采用压杆理论、圆弧滑动面理论、拉裂理论、剪切理论、边坡随机性理论等进行了大量研究,文献[1]采用塑性滑移线理论分析了煤壁失稳的力学过程,确定了煤壁片帮的危险范围,认为端面距与砌体梁的回转变形压力是煤壁片帮的主要影响因素。文献[2,3]采用压杆理论分析了煤壁挠度特征、片帮形式,以及支架初撑力、工作阻力及移架工艺对片帮控制的影响。文献[4]采用圆弧滑移面理论分析了松软煤层开采煤壁片帮特征,认为在直接顶载荷、超前支承压力和基本顶失稳回转作用下,上部煤体破碎沿圆弧形滑动面发生剪切滑落。文献[5-7]采用拉裂理论分析了硬煤层、含夹矸煤层开采煤壁拉裂破坏形式,建立了煤壁拉裂力学模型,给出了煤壁稳定性主要影响因素。文献[8]认为煤壁片帮是煤体原生裂隙损伤积累到一定程度的宏观表达,建立了煤壁"楔形"滑动体力学模型,揭示了"楔形"滑动体稳定与煤体内聚力、内摩擦角、护帮板水平推力呈正比关系,与自重及顶板压力呈反比关系。文献[9]分析了大倾角软煤大采高工作面煤壁剪切滑移破坏机理,认为工作面倾斜中上部易发生非规则四棱锥体片帮。除了理论分析外,众多学者还通过数值计算、相似模拟、现场实测等对煤壁片帮特征、应力及位移特征等进行了系列研究[10-13]。

针对艾维尔沟煤矿 2130 煤矿大倾角大采高工作面煤壁片帮特征,采用数值计算、理论分析和现场监测方法,分析煤壁片帮主控参量及其作用机制,并以工作面倾角为主要研究指标,分析不同工作面倾角条件下煤壁应力、位移分布特征,揭示倾角变化对煤壁片帮诱导机制,为大倾角煤层大采高煤壁片帮控制与实践奠定理论基础。

1 煤壁片帮主控参量

煤壁破坏实质是煤壁所受的力超过自身能承受的极限强度而发生变形累积效应。因此,煤壁片帮主要受煤壁上部顶板支承压力及煤体自身重力、煤体强度及裂隙分布等直接因素影响,同时,受工作面倾角、推进速度、采高、回采工艺、煤体内摩擦角及内聚力等间接因素控制。

(1)工作面倾角影响煤壁上方支承压力沿倾向分布特征。在大倾角煤层走向长壁大采高开采过程中,煤壁所受支承压力沿倾向呈非对称分布(图1),表现为工作面中上部区域支承压力最大,上部区域次之,下部区域最小,随着倾角增大,支承压力逐渐减小,且分布曲率逐渐平缓。

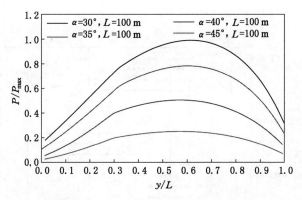

图 1 不同工作面倾角条件下煤壁支承压力分布规律

Fig. 1 Distribution of abutment pressure in coal wall under different dipping angle

同时,工作面倾角影响煤壁自重分量的分布。在特定地质开采条件下,工作面倾角 β(真倾斜布置时即为煤层倾角 α)为常数,煤壁受力 F 分解为平行岩层分力 F_1 和垂直岩层分力 F_2。随着 β 增大,F_2 随之减小,煤壁集中应力不断减小,F_1 不断增加,且煤壁垂高 H 增大,煤壁发生塑性破坏的概率增大,当 F_1 大于顶底板与煤体间摩擦力时,煤体容易在破碎后或整体向工作面倾斜下方滑移,形成煤壁滑帮。通过工作面伪斜布置,减小工作面倾角,减小煤壁垂高和平行岩层分力,减小煤壁滑帮概率。工作面仰伪斜开采时,覆岩作用力 F 在竖平面 ABD 内分解为平行岩层的分力 F_1 和垂直岩层的分力 F_2,F_2 在层面 ACD 内分解为平行工作面 AC 的分力 F_4 和垂直 AC 的分力 F_3,F_3 指向采空区,导致煤壁片帮可能性增大,且片帮深度增大(图 2)。

$$F_3 = F_1 \cdot \sin \gamma = F \cdot \sin \alpha \cdot \sin \gamma \quad (1)$$

$$F_4 = F_1 \cdot \cos \gamma = F \cdot \sin \alpha \cdot \cos \gamma \quad (2)$$

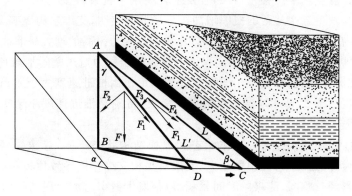

图 2 仰伪斜工作面煤壁受力状态

Fig. 2 Stress state of coal wall in oblique working face

(图中,AD 为真倾斜工作面,长度 L';AC 为仰伪斜工作面,长度 L;γ 为真倾斜工作面与仰伪斜工作间的夹角)

现场监测表明:艾维尔沟煤矿 25221 工作面煤壁片帮发生在工作面伪仰斜角度大于 3°的区域内,工作面伪斜角度越大,发生片帮的可能性也就越大,工作面煤壁片帮的影响范围也就越大、频次越高、强度越大(图 3)。

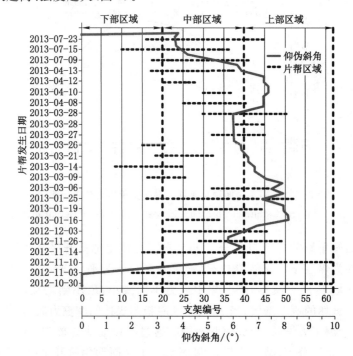

图 3　工作面煤壁片帮与仰伪斜角变化对照

Fig. 3　Contrast between angle of tilt is changed and coal wall spalling in working face

(2) 采高影响煤壁应力分布。随着采高增加,工作面前方煤体垂直应力不断向深部转移,应力峰值及其距煤壁距离增大,煤壁垂直应力值和垂直位移量不断减小,前方煤体垂直位移量增幅增大,煤壁水平位移量增加,且发生较大水平位移范围向深部延伸,同时煤壁变形自由空间增大,发生煤壁片帮的概率增加。现场监测表明,艾维尔沟煤矿 25221 工作面煤壁片帮基本发生在采高大于 3.5 m 的区域,片帮深度随着采高的增大呈非线性增加。

(3) 保持较快的推进速度能在一定程度上减缓矿山压力对煤壁附加载荷作用及煤壁变形量的累积。现场监测表明:艾维尔沟煤矿 25221 工作面推进速度大于 4 m/d 时不发生煤壁片帮,推进速度小于 2 m/d 时煤壁片帮概率增大,特别是推进速度小于 1 m/d 时煤壁片帮的范围和深度较大。

(4) 工作面回采过程中,支架工作状态和顶板结构状态不同,导致煤壁上的作用力不同。采煤机上行清煤过程中,控顶距最大,支架对顶板支撑作用导致顶板对煤壁的附加载荷较大,且煤壁变形经过割煤、清煤的时间累积,片帮发生概率增加。现场监测表明,艾维尔沟煤矿 25221 工作面在清煤过程中,煤壁片帮频繁,最频繁可达到 3~5 min 一次,但在割煤与移架过程中煤壁片帮次数较少。

(5) 煤体的单轴抗压强度 R_c 可由式(3)表示,当煤壁顶部所受载荷为均布载荷时,煤壁所能承受的极限载荷 q_{max} 可用式(4)表示,煤体的内聚力和内摩擦角越大,煤壁的极限承载能力越大,煤壁片帮的概率越小。

$$R_c = 2C\sqrt{\frac{1+\sin\varphi}{1-\sin\varphi}} \tag{3}$$

$$q_{\max} = \frac{2C\cos\varphi}{1-\sin\varphi} \tag{4}$$

式中 C——煤体内聚力,MPa;

φ——煤体内摩擦角,(°)。

25221 工作面开采 5 号煤层的硬度系数 $f=0.3\sim0.5$,属于软煤,较易发生煤壁片帮。

综上所述,大倾角煤层大采高工作面煤壁片帮的主要控制参量包括工作面倾角、采高、推进速度、支架支撑作用力、煤体内聚力与内摩擦角。

2 沿工作面倾向煤壁稳定性力学分析

受工作面倾角影响,大倾角煤层开采覆岩运动及力学演化特征呈现出近水平煤层开采时所不具有的分区和非对称变化特征[14],导致对煤壁稳定性影响更为复杂。根据煤壁受力特征,建立倾向岩梁力学模型(图4)。

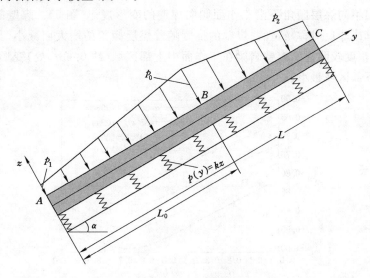

图 4 沿倾向煤壁岩梁力学模型

Fig. 4 Mechanical model of coal wall along inclination

由弯矩理论可得沿倾向煤壁岩梁 AB、BC 段的挠曲线近似微分方程为:

$$\frac{d^4 z_{AB}}{dy^4} + 4\beta^4 z_{AB} = \frac{p_1}{EI} + \frac{(p_0-p_1)}{EI}\frac{y}{L_0} \qquad 0 \leqslant y < L_0 \tag{5}$$

$$\frac{d^4 z_{BC}}{dy^4} + 4\beta^4 z_{BC} = \frac{p_2}{EI} + \frac{(p_0-p_2)}{EI}\frac{L-y}{L-L_0} \qquad L_0 \leqslant y < L \tag{6}$$

$$\beta = (k/4EI)^{1/4}$$

式中 L——工作面长度;

p_0、p_1、p_2——工作面煤壁 B、A、C 点上部顶板作用载荷;

k——弹性地基系数;

E——弹性模量;

I——惯性矩。

煤壁岩梁的边界条件为：
$$M_{AB}|_{y=0}=0, Q_{AB}|_{y=0}=0,$$
$$M_2|_{y=L}=0, Q_2|_{y=L}=0,$$
$$z_{AB}|_{y=L_0}=z_{BC}|_{y=L_0}, M_{AB}|_{y=L_0}=M_{BC}|_{y=L_0},$$
$$\theta_{AB}|_{y=L_0}=\theta_{BC}|_{y=L_0}, Q_{AB}|_{y=L_0}=Q_{BC}|_{y=L_0} \quad (7)$$

则 AB 段和 BC 段煤壁形成的岩梁挠曲线方程为：
$$z_{AB}(y)=e^{\beta y}(C_1\cos\beta x+C_2\sin\beta x)+e^{-\beta y}(C_3\cos\beta x+C_4\sin\beta x)+$$
$$\frac{p_1}{k}+\frac{(p_0-p_1)}{k}\frac{y}{L_0} \quad (8)$$

$$z_{BC}(y)=e^{\beta y}(D_1\cos\beta x+D_2\sin\beta x)+e^{-\beta y}(D_3\cos\beta x+D_4\sin\beta x)+$$
$$\frac{p_2}{k}+\frac{(p_0-p_2)}{k}\frac{L-y}{L-L_0} \quad (9)$$

式中，$C_1, C_2, C_3, C_4, D_1, D_2, D_3$ 和 D_4 为积分常数。结合 25221 工作面开采参数，由式(8)和式(9)可以得出不同煤层倾角下沿工作面倾向煤壁的变形规律(图5)。煤壁的变形(挠度)呈现出非对称(或非反对称)特性。煤壁的挠度随着煤层倾角的增大而减小，其变形量随倾角变化的减小幅度较明显。煤壁岩梁在工作面中上部区域(约 $0.66L$ 处)变形最大，上部区域变形次之，下部区域最小。

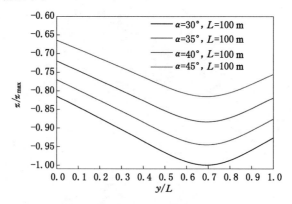

图 5 不同工作面倾角条件下煤壁沿倾向的挠度图

Fig. 5 Deflection diagram of coal wall along inclination of working face under different dipping angle

3 不同工作面倾角下煤壁应力分布特征

沿工作面倾向垂直应力重新分布，顶底板中形成非对称拱形应力释放区，回采巷道侧煤壁中形成应力集中区，集中应力 9.1~9.4 MPa。受工作面倾角影响，工作面中上部区域顶板垮落充分，顶板应力释放区范围大于下部区域，顶板应力分布拱形等值线的中心轴向上部区域偏移；下部区域底板易发生滑移破坏，底板应力释放区范围大于上部区域，底板应力分布拱形等值线的中心轴向下部区域偏移。随着工作面倾角增大，煤壁集中应力值及其影响范围逐渐减小，顶底板应力释放范围向工作面倾斜方向偏移并逐渐增大，如图6所示。

沿工作面倾向，上部区域煤体垂直应力小于下部区域，当工作面倾角为 30°、35°时，煤壁

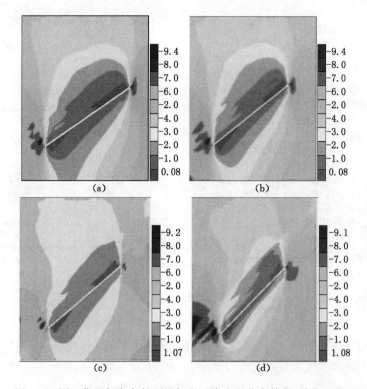

图 6 不同工作面倾角条件下沿倾向垂直应力分布特征(单位:MPa)

Fig. 6 Vertical stress in coal wall along inclination of working face under different dipping angle (Unit:MPa)

(a) 倾角 30°;(b) 倾角 35°;(c) 倾角 40°;(d) 倾角 45°

1.0 m 范围内垂直应力达到最大,最大值分别为 6.0 MPa、5.5 MPa,随着向工作面前方延深,煤体垂直应力不断减小;当倾角为 40°、45°时,煤壁处垂直应力较小,垂直应力向深部转移,在距煤壁 2.0~3.0 m 达到最大值 7.7 MPa、7.2 MPa。随着工作面倾角增大,垂直应力峰值向工作面前方转移,垂直应力值在倾角为 40°发生剧烈变化,如图 7 所示。

4 不同工作面倾角下煤壁位移分布特征

由图 8 可以看出,工作面倾角 30°时,顶板位移最大值 15.4 cm;倾角 35°时,顶板位移最大值 14.1 cm;倾角 40°时,顶板位移最大值为 10.1 cm;倾角 45°时,顶板位移最大值为 10.3 cm。随着工作面倾角增大,顶板最大位移不断减小,顶板位移等值线向水平方向倾斜,上端头围岩位移减小,底板位移等值线越向深部岩层移动。在顶板层位较低岩层中位移矢量密集、方向杂乱,在较高层位上位移矢量稀疏,在较高层位上形成一条近似的轮廓线。随着倾角增加,直接顶中位移矢量逐渐变少,并向工作面下端头方向缩减。高位顶板中位移矢量轮廓线则向更高层位移动,低位顶板岩层内部位移方向多变杂乱,挤压或拉伸变形破坏严重,岩层破碎度较大,易形成垮落带。

由图 9 可以看出,工作面上部区域,煤壁水平位移为负值(向工作面后方回采空间移动),易发生煤壁片帮,在工作面下部区域,煤壁水平位移为正值(向工作面前方煤体移动)。工作面倾角 30°时,上部区域煤壁 1.0 m 范围内最大水平位移-0.3 mm(由工作面运输平巷

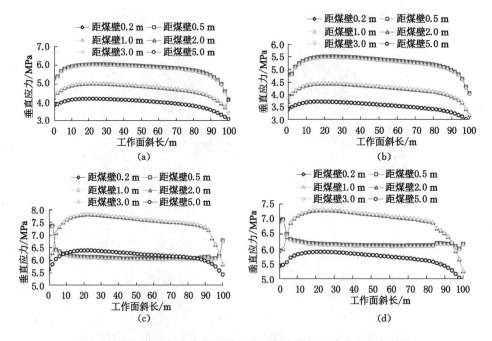

图 7 不同工作面倾角条件下沿工作面倾向煤壁垂直应力变化规律

Fig. 7 Variation law of vertical stress in coal wall along inclination of working face under different dipping angle

(a) 倾角 30°；(b) 倾角 35°；(c) 倾角 40°；(d) 倾角 45°

向上 70 m 处)，下部区域煤壁最大水平位移 0.4 mm；工作面倾角 35°时，上部区域煤壁 0.5 m 范围内最大水平位移 −0.3 mm(由工作面运输平巷向上 80 m 处)，下部区域煤壁最大水平位移 0.25 mm；工作面倾角 40°时，中、上部区域煤壁 2 m 范围内最大水平位移 −5.5 mm(由工作面运输平巷向上 85 m 处)，且向深部延伸水平位移值不断减小；工作面倾角 45°时，中、上部区域煤壁 2 m 范围内最大水平位移 −5.0 mm(由工作面运输平巷向上 90 m 处)。随着工作面倾角增大，工作面上部区域水平位移最大值不断增大，并向倾斜上部转移，易发生片帮区域面积增大，且在倾角 40°时，煤壁水平位移至发生明显变化。

由图 10 可以看出，工作面倾角 30°时，沿倾向上部区域煤壁 3 m 范围外垂直位移为负值，且 1 m 范围内垂直位移最大 2 mm，下部区域煤壁垂直位移为负值(向上移动)，且随着距煤壁距离的增加，垂直位移值不断增加；倾角 35°时，沿倾向上部区域煤壁 2 m 范围外垂直位移为负值，且 0.5 m 范围内垂直位移最大 0.7 mm，下部区域煤壁 0.5 m 范围内垂直位移为 −0.5 mm，且随着距煤壁距离的增加，垂直位移值不断增加；倾角 40°时，煤壁垂直位移 −27～−35 mm，煤壁前方煤体垂直位移随着距煤壁距离的增加不断增加；倾角 45°时，煤壁垂直位移 −28～−37 mm，煤壁前方煤体垂直位移随着距煤壁距离的增加不断增加。工作面倾角小于 40°时，上部区域煤壁发生垂直向下的位移，下部区域发生垂直向上的位移，且随着倾角增大，煤壁垂直位移量减小。工作面倾角 40°时，煤壁垂直位移至发生明显变化，随着工作面倾角增大，煤壁垂直位移增大。

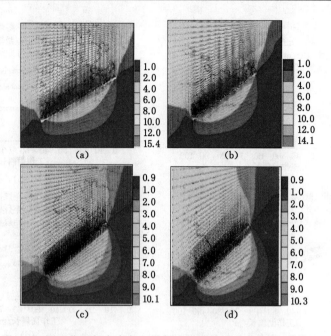

图 8 不同工作面倾角条件下沿倾向煤壁位移分布特征(单位：cm)

Fig. 8 Displacement in coal wall along inclination of working face under different dipping angle (Unit：cm)

(a) 倾角 30°；(b) 倾角 35°；(c) 倾角 40°；(d) 倾角 45°

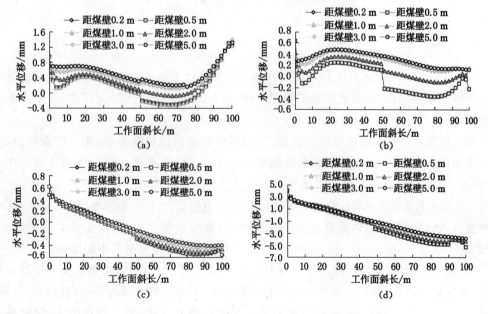

图 9 不同工作面倾角条件下沿倾向煤壁水平位移变化规律

Fig. 9 Variation law of horizontal displacement in coal wall along inclination of working face under different dipping angle

(a) 倾角 30°；(b) 倾角 35°；(c) 倾角 40°；(d) 倾角 45°

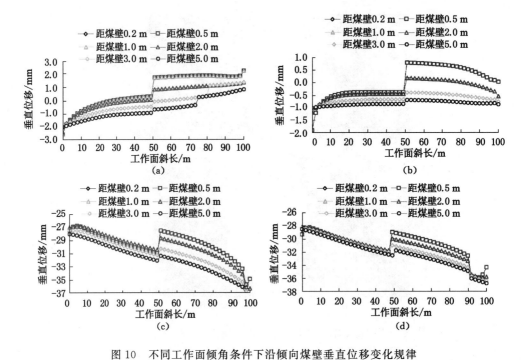

图 10 不同工作面倾角条件下沿倾向煤壁垂直位移变化规律

Fig. 10 Variation law of vertical displacement in coal wall along inclination of working face under different dipping angle

(a) 倾角 30°；(b) 倾角 35°；(c) 倾角 40°；(d) 倾角 45°

5 结　论

（1）大倾角煤层大采高工作面煤壁片帮主要受煤壁上压力、煤壁高度、煤体强度及裂隙分布等因素影响，其主控参量包括工作面倾角、采高、推进速度、支架支撑力、煤体内聚力及内摩擦角。

（2）建立了倾向煤壁岩梁力学模型，给出煤壁岩梁的挠曲线方程，受工作面倾角影响，煤壁支承压力沿倾向非对称分布，导致煤壁变形（挠度）呈现出非对称特性，在工作面中上部区域（约 $0.66L$ 处）煤壁岩梁变形最大。

（3）受工作面倾角影响，倾向上部区域煤体垂直应力小于下部区域，倾角小于 40°时煤壁垂直应力最大，前方煤体垂直应力不断减小，倾角大于等于 40°时煤壁垂直应力较小，距煤壁 2.0～3.0 m 煤体垂直应力达到峰值，且峰值位置随着倾角增大不断迁移。

（4）工作面上部区域煤壁水平位移为负值，易发生煤壁片帮，倾角小于 40°时，随着倾角增大，煤壁垂直位移量减小，煤壁水平位移不断增加，易发生片帮区域不断向倾斜上部转移。工作面倾角 40°时，煤壁垂直位移与水平位移至发生明显变化，随着倾角增大，煤壁垂直位移与水平位移增大。

参 考 文 献

［1］杨培举，刘长友，吴锋锋.厚煤层大采高采场煤壁的破坏规律与失稳机理［J］.中国矿业

大学学报,2012,41(3):371-377.

Yang Peiju, Liu Changyou, Wu Fengfeng. Breakage and falling of a high coal wall in a thick mined seam[J]. Journal of China University of Mining & Technology,2012,41(3):371-377.

[2] 尹希文,闫少宏,安宇.大采高综采面煤壁片帮特征分析与应用[J].采矿与安全工程学报,2008,25(2):222-225.

Yin Xiwen, Yan Shaohong, An Yu. Characters of the rib spalling in fully mechanized caving face with great mining height[J]. Journal of Mining & Safety Engineering,2008,25(2):222-225.

[3] 杨敬轩,刘长友,吴锋锋,等.煤层硬夹矸对大采高工作面煤壁稳定性影响机理研究[J].采矿与安全工程学报,2013,30(6):856-862.

YANG Jingxuan, LIU Changyou, WU Fengfeng, et al. The research on the coal wall stability mechanism in larger height coal seam with a stratum of gangue[J]. Journal of Mining & Safety Engineering,2013,30(6):856-862.

[4] 方新秋,何杰,李海朝.软煤综放面煤壁片帮机理及防治研究[J].中国矿业大学学报,2009,38(5):640-644.

Fang Xinqiu, He Jie, Li Haichao. A study of the rib fall mechanism in soft coal and its control at a fully mechanized top coal caving face[J]. Journal of China University of Mining & Technology,2009,38(5):640-644.

[5] 王家臣,杨印朝,孔德中,等.含夹矸厚煤层大采高仰采煤壁破坏机理与注浆加固技术[J].采矿与安全工程学报,2014,31(6):831-837.

Wang Jiachen, Yang Yinchao, Kong Dezhong, et al. Failure mechanism and grouting reinforcement technique of large mining height coal wall in thick coal seam with dirt band during topple mining[J]. Journal of Mining & Safety Engineering,2014,31(6):831-837.

[6] 王家臣,王兆会,孔德中.硬煤工作面煤壁破坏与防治机理[J].煤炭学报,2015,40(10):2243-2250.

Wang Jiachen, Wang Zhaohui, Kong Dezhong. Failure and prevention mechanism of coal wall in hard coal seam[J]. Journal of China Coal Society,2015,40(10):2243-2250.

[7] 杨胜利,孔德中.大采高煤壁片帮防治柔性加固机理与应用[J].煤炭学报,2015,40(6):1361-1367.

Yang Shengli, Kong Dezhong. Flexible reinforcement mechanism and its application in the control of spalling at large mining height coal face[J]. Journal of China Coal Society,2015,40(6):1361-1367.

[8] 袁永,屠世浩,马小涛,等."三软"大采高综采面煤壁稳定性及其控制研究[J].采矿与安全工程学报,2012,29(1):21-25.

Yuan Yong, Tu Shihao, Ma Xiaotao, et al. Coal wall stability of fully mechanized working face with great mining height in "three soft" coal seam and its control technology[J]. Journal of Mining & Safety Engineering,2012,29(1):21-25.

[9] 伍永平,郎丁,解盘石.大倾角软煤综放工作面煤壁片帮机理及致灾机制[J].煤炭学报,2016,41(8):1878-1884.

Wu Yongping, Lang Ding, Xie Panshi. Mechanism of disaster due to rib spalling at fully-mechanized top coal caving face in soft steeply dipping seam[J]. Journal of China Coal Society,2016,41(8):1878-1884.

[10] 常聚才,谢广祥,张学会.特厚煤层大采高综放工作面煤壁片帮机制分析[J].岩土力学,2015,36(3):803-808.

CHANG Jucai, XIE Guangxiang, ZHANG Xuehui. Analysis of rib spalling mechanism of fully-mechanized top-coal caving face with great mining height in extra-thick coal seam[J]. Rock and Soil Mechanics,2015,36(3):803-808.

[11] 黄庆享,刘建浩.浅埋大采高工作面煤壁片帮的柱条模型分析[J].采矿与安全工程学报,2015,32(2):187-191.

Huang Qingxiang, Liu Jianhao. Vertical slice model for coal wall spalling of large mining height longwall face in shallow seam[J]. Journal of Mining & Safety Engineering,2015,32(2):187-191.

[12] 殷帅峰,何富连,程根银.大采高综放面煤壁片帮判定准则及安全评价系统研究[J].中国矿业大学学报,2015,44(5):800-807.

Yin Shuaifeng, He Fulian, Cheng Genyin. Study of criterions and safety evaluation of rib spalling in fully mechanized top-coal caving face with large mining height[J]. Journal of China University of Mining & Technology,2015,44(5):800-807.

[13] 解盘石,伍永平.大倾角煤层长壁大采高开采煤壁片帮机理及防控技术[J].煤炭工程,2015,47(1):74-77.

Xie Panshi, Wu Yongping. Mechanism and control methods of rib spalling in steeply dipping thick seam in fully-mechanized longwall mining with large mining height[J]. Mining Engineering,2015,47(1):74-77.

[14] 王红伟,伍永平,解盘石,等.大倾角采场矸石充填量化特征及覆岩运动机制[J].中国矿业大学学报,2016,45(5):886-892+992.

Wang Hongwei, Wu Yongping, Xie Panshi, et al. The quantitative filling characteristics of the waste rock and roof movement mechanism in the steeply inclined working face[J]. Journal of China University of Mining & Technology,2016,45(5):886-892+992.

基于 VRP 的采矿过程矿石质量智能优化研究

李小帅[1]，郭连军[1]，徐振洋[1]，王雪松[1]，孙铭辰[2]

(1. 辽宁科技大学 矿业工程学院，辽宁 鞍山，114051；
2. 鞍钢集团鞍千矿业有限责任公司，辽宁 鞍山，114001)

摘 要：针对鞍千矿业有限责任公司采区数量多、分布面积广、各采区品位差异较大等问题，以系统方法为指导，采用 VRP 启发式优化方法，综合考虑入选品位、总成本、设备利用率等，建立多采区、多破碎站的采矿全过程协同配矿优化模型，并采用 C++ 语言开发矿石质量智能控制系统。实际应用表明：该方法实现了多采区协同优化配矿，解决了局部环节简单优化问题，降低了生产成本，达到了优化入选矿石质量的目的。

关键词：配矿优化；露天矿；入选品位；车辆路径；智能配矿软件

Study on intelligent optimization of ore quality in whole mining process based on virtual reality platform

LI Xiaoshuai[1], GUO Lianjun[1], XU Zhenyang[1],
WANG Xuesong[1], SUN Mingchen[2]

(1. School of Mining Engineering, University of Science and Technology
Liaoning, Anshan 114051, China;
2. Ansteel Anqian Mining Co. Ltd., Anshan 114001, China)

Abstract: Aiming at the problems in Anqian Mining Co. Ltd., such as the large number of mining areas, wide distribution area and the ore grade of each mining area is quite different, and so on. Based on the system method, the VRP heuristic optimization method is proposed. Considering the ore dressing grade, the total cost and the equipment utilization, the optimal model of mining coordinated distribution is established for multiple mining areas and multiple ore crushing plants. Development of ore quality intelligent control system with C++ language. Practical application shows that: this method realizes the synergistic optimization of ore blending in multiple mining areas, the problem of simple optimization of local links is solved, the production cost is reduced. Finally, the purpose of optimizing the quality of ore is achieved.

Keywords: ore-blending optimization; open-pit; beneficiation feed grade; vehicle routing; intelligent ore blending software

  矿石质量的优劣直接影响整个矿山乃至采选联合企业的生产效率和经济效益。长期

作者简介：李小帅(1994—)，男，硕士，主要从事爆破优化与设计研究工作。E-mail:516741249@qq.com。
通信作者：郭连军(1963—)，男，博士，教授，从事爆破工程和岩土工程研究与教学工作。E-mail:glj0412@126.com。

以来,矿山企业在制定采矿生产任务时,往往只简单考虑品位和生产量,忽视企业整体效益和效率。以致矿石质量难以得到有效控制,生产中经常出现入选矿石达不到最优品位、生产不稳定、生产成本高等问题。为此,必须着眼于矿石质量控制。对于存在多品位矿石供货来源的采选一体化企业来说,配矿优化[1-3]是实现精矿稳定生产、降低生产成本的重要内容。

近年来众多学者在配矿优化方面展开大量研究。王李管等[4]利用地质统计法对出矿点品位进行精细化预测,并建立配矿优化模型,通过C++语言编程实现多元素多卸矿点矿山快速配矿计算;杨驰等[5]通过二次开发3DMine软件,并结合线性规划单纯性大M法,实现配矿入选品位的稳定;徐铁军等[6]提出基于语言偏好和满意度的两步式模糊优化算法,灵活有效地解决了复杂配矿问题。鉴于前人相关研究,本文针对多采区大区域协调配矿问题,借助VRP(vehicle routing problem)求解中的两阶段构造启发式算法,借助C++开发多采区配矿软件,对多采区协同配矿进行控制,提供矿石质量控制的关键技术。

1 基于VRP的配矿优化模型构建

1.1 VRP算法研究现状

1959年Dantzig和Ramser研究了汽车配送汽油的路径优化问题,最早提出了车辆路径问题(VRP)以及线性规划模型求近似最优解的方法[7]。此后,大量学者针对不同的VRP问题开展相关研究[8-10],经过几十年的发展,VRP问题已经成为运筹学与最优化研究热点。由于VRP问题求解过程较复杂、处理时间较长,目前针对VRP求解方法往往采用启发式算法。

矿山配矿优化问题可以描述为:多个出矿点安排电铲、卡车向多个破碎站运送矿石,破碎站位置和处理量固定,卡车载重量一定,通过合理调配卡车使目标函数达到最优,即生产成本或总体做功最小,并满足所有约束条件。采矿环节配矿流程如图1所示。

1.2 目标函数建立

由于不同企业具体生产要求与条件各不相同,目标函数建立也略有差异,一般以经济利润、生产成本、能源消耗、生产质量、出矿量等因素建立目标函数居多。根据鞍千矿业有限责任公司(下称鞍千矿)生产实际,以生产成本最低为目标建立函数关系。鞍千矿现有3个采区,3个破碎站,共有待出矿点50个左右,设第i个出矿点参与配矿计算,第i个出矿点到第j个破碎站的距离为l_{ij},由第i个出矿点运送到第j个破碎站的矿石量为x_{ij},生产成本用V表示。目标函数表示为:

$$\min V = \sum_{i=1}^{n}\left[(a+bl_{ij})\cdot mx_{ij}+\frac{mx_{ij}}{d}\cdot c\cdot l_{ij}\right] \quad n=50, j=1,2,3 \tag{1}$$

式中,m为是否选择该出矿点参与配矿,$m=0$不选择,$m=1$选择;a为单位矿石铲装成本,元/t;b为单位矿石每公里运输成本,元/(t·km);d为每辆卡车的运输量,取80 t;c为每辆空车返回时每公里运输成本,元/km。

1.3 配矿约束条件分析

(1) 各出矿点出矿量约束

一般情况下矿山为了降低损失、贫化率,各出矿点矿石保有量一般可供电铲铲装2~

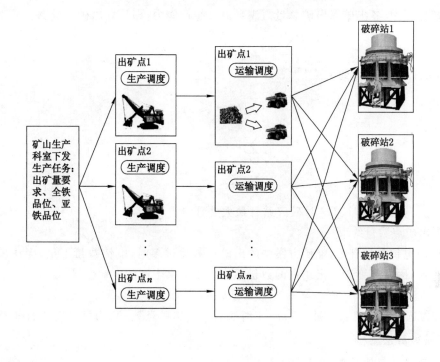

图 1 采矿环节配矿流程图
Fig. 1 Ore-blending flow in mines

3 d,在计算出矿量时不可超过该出矿点最大出矿量,也不可超过该电铲最大生产能力因此出矿点约束可表示为:

$$0 \leqslant m(x_{i1}+x_{i2}+x_{i3}) \leqslant \min(A_c, M_i) \quad i=1,2,3,\cdots,50 \quad (2)$$

式中,A_c 为电铲最大生产能力;M_i 为第 i 个出矿点最大出矿量。

电铲进行铲装工作时,频繁的移动会使工作效率会降低,通过约束出矿点最小出矿量来减少电铲频繁移动,使采装效率最大化。

$$m(x_{i1}+x_{i2}+x_{i3}) \geqslant C_i \quad i=1,2,\cdots,50 \quad (3)$$

式中,C_i 为出矿点最小出矿量。

(2) 目标品位约束

在鞍千矿实际生产中,3 个采区不同出矿点的矿石品位及元素种类差距较大,通过协调不同出矿点出矿量进行配矿工作,使总体全铁、亚铁品位满足规定目标品位。

$$\frac{\sum_{i}^{50} m x_{ij} g_i}{X} = G \quad j=1,2,3 \quad (4)$$

$$\frac{\sum_{i}^{50} m x_{ij} f_i}{X} = F \quad j=1,2,3 \quad (5)$$

式中,g_i 为第 i 个出矿点全铁品位;f_i 为第 i 个出矿点亚铁品位;G 为全铁目标品位;F 为亚铁目标品位;X 为计划出矿总量。

(3) 计划出矿总量约束

配矿方案中,各出矿点出矿总量需满足设计生产能力,即计划出矿总量 X。

$$\sum_{i}^{50} mx_{ij} = X \quad j=1,2,3 \tag{6}$$

(4) 破碎站处理量约束

$$\sum_{i}^{50} mx_{i1} \leqslant y_1 \tag{7}$$

$$\sum_{i}^{50} mx_{i2} \leqslant y_2 \tag{8}$$

$$\sum_{i}^{50} mx_{i3} \leqslant y_3 \tag{9}$$

式中,y_1、y_2、y_3 分别为 1、2、3 号破碎站口最大处理量。

(5) 出矿点数量约束

采区内电铲数量有限,配矿方案中的出矿点数目过多时,电铲数量不足,无法满足出矿点同时出矿,因此最终选取出矿点的数量不能大于可供生产的电铲数量。

$$N_k \leqslant N_c \tag{10}$$

式中,N_k 为选定的出矿点数量,即 $m=1$ 时所对应的出矿点数;N_c 为可供生产的电铲数量。

1.4 两阶段算法流程

两阶段算法具体步骤如图 2 所示。

(1) 确定现阶段矿山基本信息参数,如待出矿点全铁品位 g_i、亚铁品位 f_i、生产运输成本 a、b、c 等;

(2) 添加若干人工变量 $x_{i+1}, \cdots, x_{i+h}$,将原约束条件方程化为标准型,并构建只含人工变量的目标函数:$\min Z = \sum_{k=i+1}^{i+h} x_k$;

(3) 在原约束条件下求解,若人工变量均为 0 转步骤 4 继续进行第二阶段求解,否则无最优解;

(4) 以步骤(3)求得的基可行解为初始解,采用单纯形法对目标函数进行求解;

(5) 求出最优解 x_{ij},输出成本最低的配矿方案。

2 应用实例

2.1 矿山概况

鞍千矿业有限责任公司为大型铁矿石采选联合企业,拥有许东沟、哑巴岭、西大背三个采区,年设计采剥总量为 3 140 万 t,年出矿能力为 1 400 万 t,日设计矿石产量 41 320 t,境界内平均品位 28.58%,其中 TFe 平均含量 28.20%,SiFe 平均含量 2.03%,目前实际揭露品位为 24.5%;三个采区用于出矿的运输设备 50 余台,铲装设备 10 余台(均不含正在维修设备);该公司共有哑巴岭粗破、许东沟粗破、北破粗破三个破碎站,每个破碎站日最大处理量 20 000 t 左右;不同时期选矿厂要求全铁、亚铁入选品位会有所调整,目前全铁、亚铁目标品位分别为 26.8%,10.1%。

2.2 矿山三维模型建立

通过 C++独立开发三维建模软件,根据鞍千矿 3 个采区最新开采现状测量数据及品位

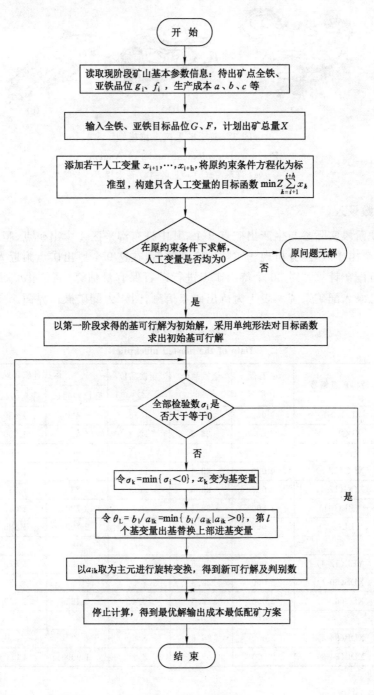

图 2 算法流程

Fig. 2 The arithmetic flow chart

信息建立可视化三维采区现状模型、出矿点品位模型等信息，为下一步的配矿计算提供一定的基础信息，并在这些所建立的模型基础上实现配矿方案信息可视化显示。鞍千矿采区现状三维模型如图 3 所示。

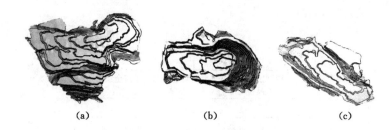

图 3 鞍千矿采区三维模型

Fig. 3 3D geological model of Anqian

(a) 许东沟;(b) 哑巴岭;(c) 西大背

2.3 配矿参数录入

鞍千矿现阶段实际参与生产出矿点 9 个,其中许东沟采区 4 个出矿点、哑巴岭 2 个出矿点、西大背 3 个出矿点,经分析现场实际生产需要后,在这 9 个原出矿点附近人工选择 20 个待出矿点进行配矿计算。对 20 个待出矿点进行矿石保有量估算、品位化验、距离测量后将待出矿点信息录入配矿系统。表 1 为待出矿点信息,图 4 为配矿操作界面。

表 1 待出矿点信息

Table 1 Data of the blasted muckpile

矿区	待出矿点编号	矿石保有量/t	全铁品位/%	亚铁品位/%	运往各破碎站距离/m		
					哑巴岭粗破	许东沟粗破	北破
许东沟	XDG(0-1)	7 000	28.6	13.5	4 430	1 630	1 529
	XDG(0-2)	8 000	28.5	12.8	4 400	1 500	1 450
	...	...	...	...	...	...	...
	XDG(12-3)	4 000	28.4	11.6	4 340	1 430	1 311
	XDG(12-4)	2 000	28.6	11.2	4 230	1 330	1 429
哑巴岭	YBL(0-1)	1 500	28.4	9.7	2 730	2 047	3 167
	YBL(0-2)	3 000	22.7	4.6	2 700	1 985	3 089
	...	...	...	...	...	...	...
	YBL(12-4)	2 000	23.5	3.9	2 600	1 879	3 107
西大背	XDB(60-1)	5 000	26.3	7.5	2 536	5 537	6 640
	XDB(60-2)	3 000	22.3	6.4	2 425	5 430	6 500
	...	...	...	...	...	...	...
	XDB(84-2)	5 000	25.6	8.6	2 000	5 000	6 100
	XDB(84-3)	2 500	27.1	8.9	1 900	4 850	5 990

2.4 配矿结果分析

设定全铁、亚铁目标品位分别为 26.8%,10.1%,设定出矿总量为 28 000 t。经济最优配矿方案如表 2 所示。配矿方案最终确定 9 个出矿点,其数量低于矿山可用电铲数量,符合生产要求;哑巴岭粗破、许东沟粗破、北破处理量分别为 6 000 t、7 000 t、15 000 t,均低于破碎站最大处理量;全铁最终品位为 26.808 525%,误差 0.085 25%;亚铁最终品位为 10.098 2%,误差 0.001 8%;总体误差率 0.974 902%。实际应用结果表明:此算法在满足

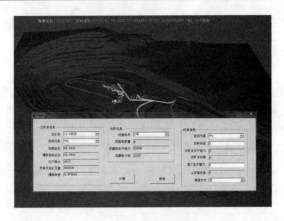

图 4 配矿界面图

Fig. 4 Ore blending operation interface

经济最优目标情况下,其整体误差率较低,基本满足矿山实际生产要求。

表 2 配矿方案

Table 2 Scheme of stope ore blending

矿区	出矿点编号	全铁品位/%	亚铁品位/%	运往各破碎站矿石量/t			成本
				哑巴岭粗破	许东沟粗破	北破	
许东沟	XDG(0-1)	28.6	13.5	0	0	4 000	6.116e+006
	XDG(0-2)	28.5	12.8	0	0	7 000	1.015e+007
	XDG(12-1)	28.6	11.2	0	2 000	0	2.66e+006
	XDG(12-3)	28.4	11.6	0	0	4 000	5.244e+006
哑巴岭	YBL(0-2)	22.7	4.6	1 500	0	0	3.6375e+006
	YBL(12-4)	23.5	3.9	3 000	0	0	6e+006
西大背	XDB(60-1)	26.3	7.5	1 500	0	0	2.85e+006
	XDB(84-2)	25.6	8.6	0	3 000	0	5.955e+006
	XDB(84-3)	27.1	8.9	0	2 000	0	3.758e+006
总计		26.808 524	10.098 2	6 000	7 000	15 000	4.637 05e+007

3 结 论

(1) 针对鞍千矿多采区、多破碎站、多元素的复杂配矿问题,在充分分析矿山实际生产约束条件后,以采矿过程生产成本最优为目标函数,构建了基于 VRP 两阶段算法的配矿数学模型,并结合自主开发的矿山三维建模软件,实现了出矿点自动选择并圈定、配矿信息显示以及线路规划,最终配矿方案符合矿山实际生产要求,为解决多采区协同配矿问题提供了一定方法和手段。

(2) 以采矿过程生产成本为目标函数,可以充分根据现阶段矿石价格及油价等经济因素,编制符合矿山经济利益的短期配矿方案,有利于降低矿山生产成本。

(3) 配矿数学模型构建过程中仅仅考虑了采矿过程并未将选矿过程融入其中,没有做到真正的系统化、一体化,并且忽略掉了一些次要的约束条件,今后的研究中可在这些方面加以改善,进一步提升配矿效果。

参 考 文 献

[1] Wilke F L, Reimer T H. Optimizing the short term production schedule for an open-pit iron ore mining operation[C]. Society of Mining Engineers, New York, 1979: 21-25.

[2] Gu Q H, Lu C W, Guo J P, et al. Dynamic management system of ore blending in an open pit mine based on GIS/GPS/GPRS[J]. Mining Science & Technology, 2010, 20(1):132-137.

[3] Souza M J F, Coelho I M, Ribas S, et al. A hybrid heuristic algorithm for the open-pit-mining operational planning problem [J]. European Journal of Operational Research, 2010, 207(2):1041-1051.

[4] 王李管,宋华强,毕林,等. 基于目标规划的露天矿多元素配矿优化[J]. 东北大学学报(自然科学版),2017,38(07):1031-1036.

[5] 杨驰,吴建胜,郭连军,等. 露天矿多采区协同开采资源配置优化[J]. 金属矿山,2017(06):18-23.

[6] 徐铁军,杨鹏. 基于模糊多目标优化算法的矿山配矿优化[J]. 北京科技大学学报,2009, 31(11):1363-1367.

[7] Dantzig G B, Ramser J H. The truck dispatching problem[J]. Management Science, 1959(6): 80-91.

[8] 石建力,张锦. 需求点位置随机的分批配送 VRP 优化[J]. 计算机应用研究,2018(11): 1-7.

[9] 梁凤婷,胡坚堃,黄有方. 低碳环境下电动汽车车辆路径问题[J]. 上海海事大学学报, 2018,39(02):34-40.

[10] 韩广,李雪杨,孙晓云,等. 铁路行车调度问题的改进粒子群优化研究[J]. 控制工程, 2017,24(09):1855-1859.

Assessing longwall shield-strata interaction from a basic understanding of shield characteristics and a physical modeling study

SONG Gaofeng, DING Kuo, SUN Shiguo

(School of Civil Engineering, North China University of Technology, Beijing 100144, China)

Abstract: The stability of longwall face and performance of longwall shields are the major ground control concerns in mining thick coal seams, and require a study of interactions between shield and strata. Physical modeling tests are frequently used for understanding the shield-strata interactions but the realistic hydraulic operations of longwall shields were widely ignored in the previous studies. This paper reproduces the behaviors of the longwall face and shield in response to roof loading through a physical modeling study with a consideration of the basic operating principles of the shield leg cylinder. A clear understanding of the extension and retraction of leg cylinder with relationship to mining height is provided in this research, as well as a detailed analysis of the important characteristics of longwall shields including shield capacity, setting force, vertical stiffness and yielding behavior. The results show that: (1) The vertical stiffness of longwall shields has a positive correlation with the shield capacity, and it decreases with the extension of hydraulic liquid inside the leg cylinder; (2) Higher capacity (stiffness) shields may develop more load on the shield in response to roof convergence, but remain some reserve capacity to prevent the shield from highly stressed or damage; (3) Yielding of a modern shield represents a 10% drop in the yielding pressure, and produces 5~10 mm of reduction in shield height and 100~250 tons of loss in support loading; and (4) Physical modeling study of shield-strata interaction with a consideration of leg hydraulic operation indicates a step function of roof deflection during shield yielding events. Face and roof displace in a slow rate during the initial shield loading stage but increase rapidly during the yielding events.

Keywords: longwall shield; leg cylinder operation; shield-strata interaction; face stability; physical modeling

1 Background and problem statement

Though the share of coal in China's energy mix has been decreasing in recent years due to the overcapacity-cutting and production-optimizing operations, coal still accounted for 62% of the national energy consumption in 2016 and remains the major source of energy and the dominant fuel in China[1]. Thick coal seams of 6~9 m are widely found in the north and west Chinese coal bases (Shendong, Shanbei, Huanglong and Xinjiang) with abundant proven reserves and preferable geological and mining conditions. Such seams should be mined most efficiently to achieve the sustainable development of Chinese coal industry. Due to the unfavorable cavability of the coal seams, the large-cutting-height

---

Supported by the National Natural Science Foundation of China under Grant(41772335,51574244).
Corresponding author: Tel: +86 010 88803280, E-mail: song.gaofeng@ncut.edu.cn.

longwall mining system for recovering thick coal seams in the coal bases has been recognized as one of the most important mining methods in terms of coal recovery, safety, productivity and efficiency[2-6]. Extra-large-cutting-height longwall face has further increased the full extraction height in a single pass to more than 6 m. The 6- and 7-m extra-large-cutting-height longwall mining applications have been extensively practiced in these mining areas in past few years with significantly improved economic benefits, production and efficiency. However, the goal is to mine the 9 m high full seam in a single pass for further improving the coal recovery. With the improvement of mining equipment and hazard management in the open face area, the largest mining height in China has increased to 8 m in recent years[7,8]. Currently, Shangwan coal mine in Shendong coal base is practicing the 8.8-m single-cut longwall mining method.

As compared with longwall faces with regular mining height, large-cutting-height longwalls may experience the most ground control problems and mine hazards in the open face area, which is observed in the field as face collapse on the coal wall, immediate roof cavity ahead of the shield, violent periodic roof weightings, tremendous high pressure level or even a massive impact loading on support units leading to an ironbound shield[9,10]. Fig. 1 gives a schematic showing the ground control problems in the face area and the

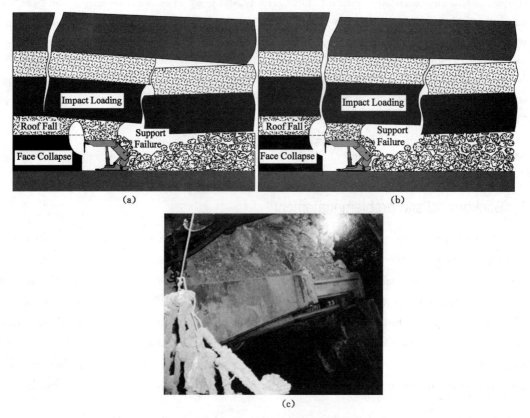

Fig. 1 Schematics of ground control problems and the ironbound shield in a large-cutting-height longwall face

ironbound state of a shield in a high-seam longwall face. Dislocation of the strong and massive roof may occur ahead of the faceline [Fig. 1(a)]; as the face advances to the dislocation position, yielding of the support may take place if the shield cannot provide adequate support to the roof [Fig. 1(b) and 1(c)]. These ground control problems have significantly negative impacts on the production and productivity of longwall faces, deteriorate the shied-strata interaction and reduce the life expectancy of longwall shields. This issue becomes extremely severe in the extra-large-cutting-height longwall face and causes significant loss of operating time and production. The tremendous potential of high seam longwall mining can be achieved only if a good ground control is maintained in the open face area.

2 Research hypothesis, goal and specific objectives

The stability and performance of the face and shields are the major concerns in mining thick coal seams, while a better understanding of the shield-strata interaction may assist improving the ground control in the face area. Longwall shield is the most important element in the face area not only because they occupy a massive capital investment but also function to maintain a safe working environment for miners and the shearer. A number of models have been developed to study how the longwall shields interact with the surrounding strata. The methods include the detached block model[11], equivalent system stiffness model[12], immediate roof-main roof coupling model[13], load cycle analysis[14,15], ground response curve[16,17], numerical modeling[18,19] and physical modeling studies[20,21]. Previous studies have largely improved our understandings on coupling of the shield with surrounding rocks in the face area, and the performance of face and shield in response to roof convergence. However, the hydraulic operating mechanism of the shield leg was not considered in these models. The authors believe that hydraulics mechanism of the leg cylinder in response to roof loading has an impact on the shield-strata interaction. Therefore, a basic understanding of the leg hydraulic operation principles as well as important characteristics of the shield should be considered when analyzing shield-strata interaction.

This paper attempts to study the shield-strata interaction from a physical modeling study considering the basic shield leg operations. This paper has three goals: (1) Provide a basic analysis of the extension and retraction of shield leg cylinder; (2) Analyze the important characteristic of the shields including the shield capacity, setting force, vertical stiffness and yielding behavior; and (3) Incorporate the shield leg operation in the physical modeling study for investigating the shield-strata interaction.

3 Pertinent literature review

Depending on shield-strata interaction and the degree of the powered shield supports may have on the ground behavior in the face area, the shield-strata interaction models can be characterized into 2 groups: force and displacement control[22]. The detached block model mentioned above fits in a force-controlled load environment where the shields need to provide enough support capacity to hold the dead weight of free block and maintain equilibrium of the rock mass. In this sense, the design of support is try to offset or overpower the ground forces with great support capacity. A number of studies have reported that a lack of adequate working resistance of the shield legs may be responsible for the poor performance or failure of face and shields[23-26]. But the shield capacity cannot increase unlimitedly and a high capacity shield does not necessarily guarantee a good ground control in the open face. On the other hand, it is realized in a displacement-control environment that the ground movement due to main roof convergence may not be completely controlled by the shield support, and the convergence may inevitably continue regardless of the shield capacity. As a matter of factor, the roof displacement is a combination of immediate roof deflection and main roof convergence, depending on the mechanical characteristics of the roof and coupling between different rock strata. The actual loading environment in the open face area may include both the displacement and force controlled loading aspects. The goal is to maintain the stability of face and shield in a specific time period (i.e. during the current mining cycle before shield advancement) by matching the support characteristics to the ground movement, which requires a basic understanding of the shield leg operation and a better understanding of shield-strata interaction.

Physical modeling of the roof caving behaviors has been frequently documented in the previous studies for understanding the progressive development of mining-induced fractures and mechanism of rock strata movement around underground openings. Song et al. (2015) studied the failure mechanism of roof and extension of caved and fractured zone heights with face advance[27]. Li et al. (2015) analyzed the characteristics of acoustic emission signals during the failure process of the hard roof in a physical model and found that the maximum AE energy occurred at roof breakage position[28]. Xu et al. (2017) reproduced the trapezoid-like configuration of the disturbed overburden strata and observed an overall decreasing trend of the fracture space from the model bottom to the top[29]. Yang et al. (2017) found in his physical model that the impact load on the support at roof weighting is about two times the support working resistance at normal advance of longwall face. Guo (2015) employed a 2-D physical model to evaluate the effects of shield pressure, joint spacing, and joint orientation with relation to mining direction on face stability during an excavation of a longwall. Kong (2017) developed a 3-D physical model to evaluate the face displacement and shield performance under different roof pressures and cutting heights. Shields have been incorporated in these physical modeling studies to reproduce the shield-strata interaction. However, the models did not consider the realistic operations of the shield as observed in the field, and therefore fail to record the convergence and loading of shield during roof loading and cannot simulate the shield-strata interaction in a proper way. Hence, it is necessary to understand the important characteristics of the shield and include the basic operations when performing the physical modeling study.

4 Important characteristics of longwall shield

4.1 Leg hydraulics operation

The moving and setting of a roof support include lowering the canopy, advancing the shield and setting the canopy to the roof again, which takes about 10 seconds for a single cycle, depending on the operation. The leg cylinder is the key element of a shield because it controls the capacity of the shield, and the raising and lowering of the canopy. More importantly, the hydraulic operation of the leg cylinder is also relevant to the coupling of the shield and roof strata. Therefore, the basic hydraulic cylinder operation is essential to understanding the shield-strata interaction, improving the shield design and maintaining a good ground control in the longwall face area.

(1) Extension and retraction of cylinders

The modern longwall shield typically utilizes the two-stage leg cylinders, which provides a wider range of operating heights and a greater maximum working height. Fig. 2 shows a two-stage, double-acting hydraulic cylinder typically utilized in the modern longwall shields. The main parts of the cylinder include the bottom stage, top stage, steel rod, raise port, lower port and several check valves. To extend the cylinders in order to raise the canopy, hydraulic fluid is first pumped into the cylinder cavity of the bottom stage through the pilot-operated check valve (POCV) at the raise port. The liquid flows inside the first cylinder and acts against bottom of the top stage to move it outward. After the first stage is fully extended, the check valve installed at the bottom of the top stage will open and allow the hydraulic liquid to flow from the first stage to the second, causing the steel rod to displace outward. The POCV functions to only allow the fluid to enter the cylinder but will prevent it from flowing back unless it is triggered upon lowering, therefore the second stage is held in position. Likewise, the bottom check valve will only release pressure at lowering. Thus the leg will stay extracted even after the fluid flow has stopped. After the canopy reaches the roof, the liquid trapped inside the cylinder will be pressurized and the hydraulic pressure starts to increase rapidly in direct proportion to roof loading. The yield valve at bottom of the leg will release the excessive pressure higher than the yield pressure through a loss of hydraulic liquid from bottom stage, which protects the leg cylinder and shield from damaging. If the roof support is forced down to a low position by roof fall, the yield valve will still close again and enable the leg cylinder to hold the load.

Since the leg cylinders in the shields are double acting, both the extension and retraction of the cylinder are hydraulically powered. To lower the canopy, the POCV is triggered and opens to allow the fluid inside the cylinder cavity that supports the mine roof to flow into the piston side of the cylinder. At the same time, pressure is applied to the lower port and fluid is pumped inside the retract annulus to lower the leg cylinder. It should be noted that, when retracting, the bottom stage needs to collapse fully to open the bottom valve. Only when the bottom valve is open can fluid be released from the top stage.

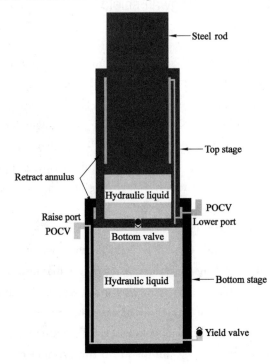

Fig. 2   A schematic of the hydraulic leg cylinder of longwall shields

(2) Operating principles of multistage cylinder

The operation of the multistage cylinder is critical to understanding the shield-roof interactions and how the shield develops support capacity. The extension and retraction of the cylinders can be best illustrated using Fig. 3, where the shield is assuming operating at different mining heights. The two-leg cylinder is initially at a fully collapsed position shown in Fig. 3(a). The bottom stage is extended first to its full stroke [Fig. 3(b)], followed by the partial stroke of second stage [Fig. 3(c)]. At this time, the shield is assumed to set against the roof and floor and working at the original mining height. When the shield is moved to the next cycle for a lower mining height, the bottom stage retracts through a loss of liquid while maintaining the same partial extension of top stage [Fig. 3 (d)]. Fig. 3(e) shows the shield returns to the original operating height by pumping the bottom stage to the full stroke again. For a higher mining height in the following mining cycle, the increase of the operating height is achieved by raising the second stage until the maximum operating height is reached at full strokes for both stages [Fig. 3(f)]. Figs. 3(g) and 3(h) show the support return to the original and lower operating height respectively by lowering the bottom stage. The staging performances, however, are totally different from the previously described positions even at the same operating heights [see Fig. 3(c) and 3(d)], since the bottom stage needs to collapse fully to lower the top stage. The cylinder is lowered to the fully collapsed position by retracting the bottom and top stage in sequence. This is plotted in Figs. 3(i) and 3(j).

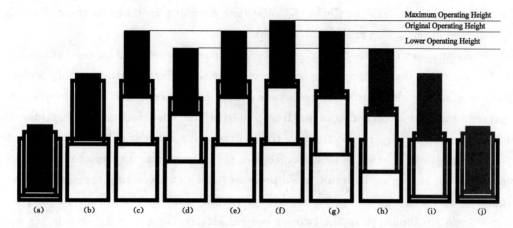

Fig. 3 A schematic of stage extension and retraction for different mining heights

Hence, the movement of a specific stage at a particular operating cycle depends on the history of the support operating heights. The extension and retraction of the bottom stage may represent the raising and lowering of the shield until a new higher mining height than the previous cycle is established. The top stage will be raised only after full stroke of the bottom stage, and be lowered only after the bottom stage is fully retracted. In any sense, the basic operating principle of the multi-staging functions is that the bottom stage extends and retracts first before the top stage. Therefore, the stages of the leg cylinder are extracting and retracting one after the other.

4.2 Shield capacity

The shield capacity has been continuously increasing since the start of longwall mining in China, which might be driven by the "bigger-the-better" design philosophy and the demand for extracting the full coal seam height of 6~9 m high at a single pass using the large-cutting-height longwall mining method. Table 1 gives the important parameters of

Table 1 Comparisons of important parameters for different shield roof supports

Shield Type	Setting Force/t (Pressure /MPa)	Capacity/t (Pressure /MPa)	Set-to-yield ratio	Operating range/m	Shield Width/mm	Shield Weight/t	Cylinder Diameter /mm
ZY10800/28/63D	791 (31.5)	1,080 (43.0)	73.2%	2.8~6.3	~1,750	45	400
ZY15000/33/72	1,237 (31.5)	1,500 (38.2)	82.5%	3.3~7.2	~2,050	~70	500
ZY18800/32.5/72D	1,237 (31.5)	1,880 (47.9)	65.8%	3.25~7.2	~2,050	69.5	500
ZY26000/40/88D	1,978 (35.0)	2,600 (46.0)	76.1%	4.0~8.8	~2,400	99	600

the 6-, 7- and 8.8-m high shields used in recent years in China. An increase in shield capacity and setting load is observed, as well as shield size (i.e. maximum operating height and width), weight, leg cylinder diameter and vertical stiffness that will be discussed later. The highest capacity shield in China is 2,600 tons currently installed in the 8.8-m high longwall face in Shangwan coal mine, Inner Mongolia.

The most majority of the shield vertical load capacity is provided by the hydraulic leg cylinder, with less than 5% coming from the caving shield-lemniscate assembly which is mainly used to provide a horizontal passive restraint to the face-to-gob roof movement and to prevent the caving rock fragments from entering the open face area. Therefore, the shield capacity can be increased through the hydraulic leg cylinder. Three approaches are typically used if to increase the shield capacity, i.e. (1) Increasing the number of legs in a shield; (2) Increasing the hydraulic yield pressure; and (3) Increasing the cross-sectional area of the leg cylinder. The four-leg shield corresponds to the first approach, and may have a higher capacity compared to the two-leg design. However, the two-leg shields are more preferred in the Chinese longwall coal mines because the inclined leg cylinders introduce an active horizontal friction force for arresting the slippage along weak planes in the roof strata by pushing the canopy towards the face. By contrast, the four-leg shields do not produce a horizontal force to the roof. The unbalanced load distribution between the front and rear hydraulic legs may also cause the rotation of canopy, leading to a loss of contact between roof and canopy and therefore a reduction in equivalent support capacity. The shield capacity can also be increased by increasing the yield pressure. Taking the shields ZY15000/33/72 and ZY18800/32.5/72D as an example (see Table 1), the 1,880-t shield sees a 380 tons increase in capacity from the 1,500-t shield. This is achieved by increasing the yield pressure to 47.9 MPa (1,880-t shield) from 38.2 MPa (1,500-t shield) while maintaining the same shield operating range, width, weight and cylinder diameter. Most typically, the approach used to increase the shield capacity is increasing the size of the shield leg cylinder while keeping the setting pressure at around 30~35 MPa and yield pressure at around 40~45 MPa. Thus, the resultant force in the leg and shield capacity is increased. This trend can also be found from Table 1 that the leg cylinder diameter is increased from 400 mm to 500 mm and the current 600 mm at maximum.

To accommodate the enlarged cylinder area, the modern shield has also found an increase in the overall size, mainly width, length and mass. The current largest shield (Shield ZY26000/40/88D) has increased the width and weight to 2.4 m and 100 tons from the previous 2-m and 70-ton shield design. A wider and heavier shield may not necessarily guarantee a ground control in the open face area. However, the potential benefits might be the improved shield stability in thick coal seam applications, and reduced cost and move time because the shield is wider and fewer are needed to be installed in a longwall face.

4.3　Setting forces

A setting force is provided to the roof and floor whenever a shield is set against the roof and floor. The magnitude of setting force depends on the setting pressure and the cross-sectional area of the leg cylinder. Setting force of a shield normally increases in direct proportion to shield capacity because the setting pressure is maintained at a constant level. According to the analysis of leg cylinder operation, the bottom stage might be at the pump pressure if it is not fully extended; while the top stage may represent the pump pressure if the bottom pressure is fully stroked. The setting force of the shied is a sum of the hydraulic leg cylinder forces, which is calculated as the setting pressure times the area of the cylinder that represents the pump pressure.

Different opinions on setting force have arisen since the start of longwall mining. This is because the setting pressure is manually chosen by the operator and a proper setting force is necessary to achieve a good ground control and make full use of the total shield capacity. The British coal mine preferred very low setting forces (typically 25% of the yield capacity), while German and American coal mines favored a high setting force (60%~80% of yield capacity). It is found from Table 1 that the setting pressure in China is generally 65%~80% of the shield yielding pressure. When a bad ground control in observed in the field, it is frequently recommended in the previous literatures that increasing the setting forces may help improving the ground control in the open face area. This is debatable because shield load development is independent with the setting pressure. Whenever a higher setting pressure is set for the shield, the iron bound state of shield may occur since the available pressure may be much lower. The optimum of the shield setting forces is actually a function of shield-strata interaction, depending on the physical and mechanical characteristics of the roof and floor.

4.4 Vertical stiffness

The increase of shield capacity typically results in a direct increase in vertical stiffness of the shield, which will be explained in detail below. A calculation of the shield vertical stiffness is provided in this work. The coefficient of compressibility (β) of the hydraulic oil inside the cylinder is given as:

$$\beta = -\frac{1}{V} \frac{\Delta V}{\Delta p} \tag{1}$$

Where V is the volume of hydraulic oil inside the cylinder; ΔV is the change in volume of the hydraulic liquid in response to the roof-to-floor convergence or manual hydraulic operations; Δp is the change of hydraulic pressure inside the cylinder due to an increase or loss of liquid. It is not unreasonable to assume the coefficient of compressibility to be 2,000 MPa. A simple reciprocal of the compressibility coefficient (β) yields the bulk modulus (K) of the hydraulic oil as:

$$K = \frac{1}{\beta} = -V \frac{\Delta p}{\Delta V} \tag{2}$$

Considering $\Delta p = \Delta F/A$ and $\Delta V = A * \Delta l$, where ΔF is the change of applied force, A is the cross-sectional area of the leg cylinder, and Δl is the change in height of the hydraulic liquid inside the cylinder, the bulk modulus can therefore be rewritten as:

$$K = -\frac{V}{A^2}\frac{\Delta F}{\Delta l} \quad (3)$$

The stiffness of the two-leg shield (k) is therefore given below:

$$k = -2 \times \frac{\Delta F}{\Delta l} = 2 \times \frac{A^2 K}{V} = 2 \times \frac{AK}{L} \quad (4)$$

It is readily found from the above equation that the stiffness of the shield increases in direct proportion to the cross-sectional area of the cylinder leg, which can be achieved by increasing the leg cylinder diameter. On the other hand, since increasing the size of cylinder leg is also the most used approach for increasing the shield capacity, it is therefore believed the vertical stiffness of the shield has a positive relationship with the shield capacity.

It is also noted that an increased shield operating height (typically in a thick coal-seam application) will result in reduced shield stiffness. However, when calculating the vertical stiffness of the shield using Eq. 4, care should be taken that L means the extension of hydraulic liquid inside the cylinder (L_2), not the operating height of the shield ($L_1 + L_2$). This can be easily understood from Fig. 4 but is widely overlooked in the previous studies where the full shield operating height (or the extraction height) is typically used for calculating the shield stiffness. The vertical stiffness of the shield is a combination of the steel piston rod and the hydraulic liquid inside the leg cylinder. The equivalent combined stiffness of the shield can be expressed as:

$$k_{shield} = \frac{1}{\frac{1}{k_{rod}} + \frac{1}{k_{liquid}}} \quad (5)$$

The piston rod stiffness is at least one to two orders of magnitude larger than the hydraulic liquid inside the cylinder. The hydraulic liquid stiffness can therefore represent the equivalent shield stiffness. This is understandable because the convergence of the shield is consumed in the displacement of the liquid but does not go into the piston rod. Based on Eq. 4, Fig. 5 plots the vertical stiffness of the shields as a function of the extension height of the hydraulic liquid inside the cylinder for the 1,080-ton, 1,500-ton, 1,880-ton and 2,600-ton shields, respectively. The maximum extension height of the hydraulic liquid is obtained by subtracting the minimum operating height of the shield from the full operating height. It is conformed again that a high capacity shield has a larger vertical stiffness.

The performance of the shied can be assessed from the setting load on initial setup of the canopy, the passive load developed in response to roof-to-floor convergence, and the reserve capacity before the shield load reaches the yield capacity. Eq. 6 gives an expression of the shield reserve capacity:

$$F_R = F_C - F_T = F_C - (F_A + F_P) = F_C - (F_A + k_{shield} \cdot L_1) \quad (6)$$

Where F_R is the reserve capacity; F_C is the shield yield capacity; F_T is the total shield load as a sum of the active load F_A and the passive load F_P; L_1 is the extension of the hydraulic liquid. Of those, the passive load F_P depends on the shield stiffness and roof-to-floor convergence. Since the

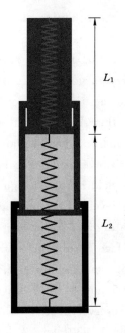

Fig. 4　The concept of equivalent stiffness of the shield

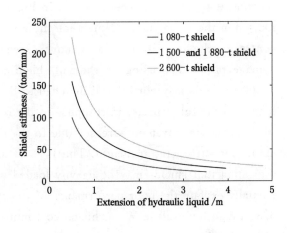

Fig. 5　Relationship between the shield stiffness and extension of hydraulic liquid

shield stiffness is not a deterministic value but varies as a function of the hydraulic liquid extension, the operating height of the shield should be determined if to calculate the shield stiffness. Table 2 lists the shield stiffness at a particular operating height, where the extension of the hydraulic liquid is computed as subtracting the minimum operating height from a particular operating height. It is interesting to note that, at a particular extraction height, the less-capacity 1,500-t shield with a smaller cylinder diameter of 400 mm may even have a larger stiffness than the 2,600-t shield with 600-mm leg cylinder, indicating that both the size of the leg cylinder and the extension of hydraulic liquid inside the cylinder have a major influence on the active shield stiffness.

Table 2　Vertical stiffness of the shield at a particular operating height

Shield Type	Min. operating height/m	Operating height/m	Extension of hydraulic liquid[1]/m	Shield stiffness /(ton/mm)	Vertical compression of coal upon onset of yield/mm	Developed load upon face yielding[2]/tons
ZY10800/28/63D	2.8	5.8	3.0	16.8	$5.8*10^3*0.5\%=29$	487.2
ZY15000/33/72	3.3	6.3	3.0	26.2	$6.3*10^3*0.5\%=31.5$	825.3
ZY18800/32.5/72D	3.25	6.8	3.55	22.2	$6.8*10^3*0.5\%=34$	754.8
ZY26000/40/88D	4.0	8.5	4.5	25.1	$8.5*10^3*0.5\%=42.5$	1 066.75

[1]: Extension of hydraulic liquid is computed as active operating height minus minimum operating height.

[2]: Developed load inside the leg cylinder upon face yielding is calculated as shield stiffness by vertical compression of coal upon onset of yield.

To evaluate the performance of the shield, a same load condition of 15 mm of roof convergence is assumed for the 4 shields mentioned in this paper. The developed load reacting at this convergence is shown in Fig. 6, with the setting load and reserve capacity plotted in the same figure. Note that the goal is to achieve a good or acceptable ground control in the open face area without extensively stressing the shield or causing any damage, rather than using up the full shield capacity. In this sense, both the 1,080- and 1,500-t shields may not be well matched to the ground condition, because both shields are loaded to the full capacity in response to the roof-to-floor convergence. The 1,500-t shield in particular, is extremely stressed due to the high set-to-yield ratio and high leg stiffness at the current operating height. The 1,500-t shield also develops the most passive load, indicating that shields may be heavily loaded when operating at a lower height. This may partially explain the poor performance of the shield and face during the extraction of the 6.3-m high longwall in Wangzhuang coal mine. The 1,880- and 2,600-t shields developed 333 and 377 tons of passive loads in response to the 15 mm of roof convergence, and remained 310 and 245 tons of additional capacity. The reserve capacity accommodates a further 14.0 and 9.8 mm of roof convergence for the 1,880- and 2,600-t shields, respectively. The reserve capacity may be helpful in the sense that it saves time for the shields to be moved forward and start the next mining cycle before the support is fully loaded or the roof and face deteriorates to an unacceptable level. This may somehow justify the modern "bigger-the-better" shield design philosophy.

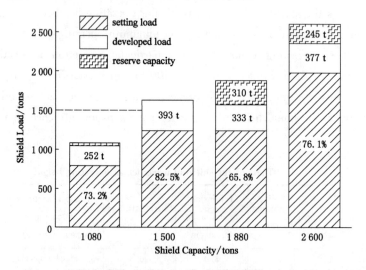

Fig. 6 Comparison of setting load, developed load and reserve capacity for different shields in response to 15 mm compression

According to the experimental studies on the mechanical behavior of coal, the onset of the yield in the unconfined coal may occur at about 0.5% of the axial strain, indicating that the onset of face fall occurs at about 5.8 m×0.5% strain=29 mm of vertical compression for a 5.8-m high longwall face, corresponding to 487 tons of passive load. The critical vertical compression and corresponding developed load for the other 3 shields are also listed in Table 2. It is obvious that the developed load for preventing the face yielding is unattainable for the modern shields. Therefore the yielding of longwall face is unavoidable. Face yielding typically further develops to face collapse in thick seam longwall mining, which is frequently observed in Chinese coal mines and remains one of the most difficult ground control problems. Considering the uncontrollable part of roof deflection, the shield may have to yield several times during a single mining cycle to accommodate this roof-to-floor convergence. Hence, it is necessary to understand the yield behavior of the shield support.

4.5 Yielding behavior

After the canopy has been actively set against the roof with hydraulic legs at the setting pressure, the shield starts to develop passive load in response to the roof-to-floor convergence by pressurizing the hydraulic oil inside the cylinder. When the overburden forces down, however, the weight of the overburden can easily raise the pressure inside the cylinder over the yield pressure, especially in an event of rapid loading of the roof support. To save the shield and cylinder from damage, the yield check valve at the bottom of the leg cylinder will be relieved to release the extra pressure higher than the yield capacity until the check valve reseats. During this yielding event, the pressure inside the cylinder will drop slightly by approximately 10% of the yield pressure due to the loss of hydraulic liquid. The yield valve will then close again at the reseating pressure (i. e. approximately 90% of the yield pressure), and enable the leg cylinder to pick up the load until the next yield cycle starts. In normal operations of each yield event, the canopy will only decline by a few millimeters, thus the cylinder is forced down in a controlled manner. The lowering of the canopy due to cylinder yielding can be understood through a simple calculation given below.

Eq. (2) can be rewritten in terms of the change of hydraulic oil volume during a yielding event as:

$$\Delta V = -V \frac{\Delta p}{K} \tag{7}$$

Where ΔV is the change in volume of hydraulic oil inside the leg cylinder upon yielding; Δp is the change in hydraulic pressure, or equivalently the yield pressure minus reseating pressure. The lowering in height of the canopy during a yielding event and the consequent reduction in support loading caused by yielding can be expressed as:

$$\Delta l = \frac{\Delta V}{A} = -\frac{\Delta p}{K}\frac{V}{A} = -\frac{\Delta p}{K}l \tag{8}$$

$$\Delta F = k_{shield} \cdot \Delta l \tag{9}$$

Where Δl is the change in height of the hydraulic liquid inside the cylinder during a yielding event, or equivalently the lowering height of the shield; A is the cross-sectional area of the leg cylinder; l is the extension height of the hydraulic liquid before yielding.

It can be seen from Eq. 8 that the size of the leg cylinder has no influence on decline of the canopy during the yielding event. The lowering of the canopy because of yielding is in direct proportion to pressure change inside the cylinder, and the extension height of the hydraulic liquid before yielding. Therefore, the reseating pressure and the operating height of the shield largely influence the shield lowering and support loading upon yielding. Table 3 gives the decline height and reduction in support loading during a yielding cycle of shield operation. It shows a 5~10 mm of reduction in shield height upon the open of yield valve, corresponding to approximately 100~250 tons of support loading loss. Note that the drop in support loading estimated by Eq. 9 is approximately 10% of the shield load capacity, corresponding to 10% drop of the yield pressure. On the other hand, it implies that the shields trend to compress in the order of 4~6 mm per 100 tons of applied roof loads, with low stiffness shields compressed less (the 1,500- and 2,600-t shields, for instance).

Table 3 Lowering of the shield and reduction of shield loading caused by yielding when support is operated at a particular height

Shield Type	Yield pressure /MPa	Min. operating height/m	Operating height/m	Extension of hydraulic liquid[1]/m	Compression of shield due to yielding /mm	Reduction in support loading caused by yielding/tons	Compression of shield per 100 tons of applied load/mm
ZY10800/28/63D	43.0	2.8	5.8	3.0	6.45	108.4	5.95
ZY15000/33/72	38.2	3.3	6.3	3.0	5.73	150.1	3.82
ZY18800/32.5/72D	47.9	3.25	6.8	3.55	8.50	188.7	4.50
ZY26000/40/88D	46.0	4.0	8.5	4.5	10.35	259.8	3.98

5 Physical modeling of shield-strata interaction

5.1 Physical modeling rig

Very few physical models have been attempted for simulating the proper operation of the shield, therefore failed to obtain the realistic performance of face and shield in response to roof convergence. The conventional 2-D physical model typically simulates a large area of strata profiles including several roofs and floors above and below the coal seam, therefore representing a small geometry similarity coefficient. As a result, the majority of the fractures and failures were observed on the roof instead of the face, and the face failure characteristics and interactions between face and shields were not well captured. A 3-D physical model designed by Kong (2016, 2017) increased the similarity coefficient by reducing the amount of the rock profiles involved in the model. The enlarged 3-D model enables simulating the problem at the particular area of interest, i. e. the open face area, and allows observing the development of face failure and performance of shield reacting at roof loading.

The physical modeling rig developed by Kong is adopted in this research. The test rig has an overall dimension of 800 mm × 1 500 mm × 1 300 mm and the physical model constructed in the test rig is 800 mm × 800 mm × 500 mm for simulating a 5-m high coal seam. Fig. 7 gives a schematic of the physical modeling rig and the finished physical model. The physical model is fixed along the left, right, rear and bottom sides. The front side of the model is free to allow face movement. Vertical loads are applied at the top of the model in Z direction using hydraulic rams via an individual steel platen. The platen is sitting on the top of the seam like a cantilever, thus roof pressures from the hydraulic rams can be transferred to both the coal seam and the shield. A total of 2 vertical rams are placed on the top of the face and 1 above the shield in this study, for simulating roof pressures on the coal and shield respectively. Roof shield is also modeled by means of hydraulic rams that can be controlled individually from the roof loading rams.

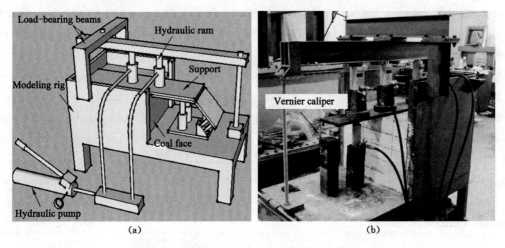

Fig. 7　A schematic of the physical modeling rig and the finished model

To accurately simulate the problem, appropriate geo-mechanical modeling materials that behave in a similar way to the real case should be carefully selected. The construction materials used in this research are a mixture of sand, lime, gypsum and water. The solid materials are fully mixed before the addition of water to ensure an overall homogeneity and isotropy. The physio-mechanical properties of the materials are determined by adjusting the proportion of the components through a trial and error process. Physical materials are first placed in the modeling steel frame and then compacted to the designed height. The physical model is constructed layer by layer to ensure the overall strength and height of the model.

5.2　Physical modeling procedure

The roof loading is simulated by pressuring the hydraulic rams on top of the physical model. During the test, the pressure of the hydraulic ram for simulating the shield may increase monotonously in response to roof convergence and show a linear load-deformation characteristic during the test. This however only represents the performance of a shield before yielding. The shield provides an active support to the roof after set-up and the load on the support unit increases before it reaches the shield load capacity. The pressure then drops to the reseating pressure for saving the shield from damage before it increases again and starts to pick up the roof loading. Yield of the shield may repeat before the shield advances to the next cycle. Therefore, proper operations should be conducted to incorporate the full cycle of the shield loading. To reproduce the support loading characteristic in a physically realistic way, the hydraulic ram for simulating the support experiences an initial loading stage and a cyclic loading stage in this study. After reaching the designed maximum load capacity (say, approximately 30 kN), the shield is manually controlled to reduce the working resistance to the reseating load at about 25 kN. The support repeats the loading and unloading process during the yield events until the face fall occurs.

The roof to floor convergence and deformation of the face are the major concerns for assessing the stability of the longwall face and shield. In this paper, a vernier caliper is used for monitoring the roof deflection [see Fig. 7(b)], which is measured at every count of the roof loading. The real-time horizontal displacement of the face is captured by using a wireless displacement monitoring system including the data acquisition equipment, data transition device and data receiving set (see Fig. 8). The data acquisition equipment comprises 3 displacement transducers which are installed at the designed height of the physical model during the model construction. The transducers are placed on the vertical middle plane of the physical model, 5 cm ahead of the coal wall, with Transducer A buried at 8 cm below the roof, Transducer B 16 cm below the roof and Transducer C 24 cm below the roof. Real-time face horizontal displacement is gathered by 3 of the transducers every 10 seconds and is then collected by the data receiving set through the data transition device.

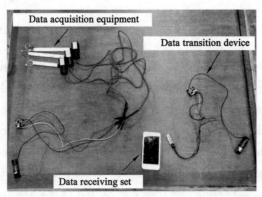

Fig. 8　Face horizontal displacement monitoring system

5.3　Results and analysis

Progressive development of the face failure is shown in Fig. 9. The cracks initiate on top of the face and extend downwards to the left edge of the physical model [Fig. 9(a)]. The apertures, trace length, as well as the number of the cracks gradually grow with the increase of the roof loads [Fig. 9(b)]. Fig. 9(c) provides a top view of the model with a near-vertical crack cutting through the entire face, indicating the horizontal movement of the face. As the roof loads continue to increase, roof bending is clearly reproduced and face fall occurs at the final loading stage [see Fig. 9(d)].

Fig. 9 Progressive growth of face failure

The support resistance over the counts of roof ram loading is plotted in Fig. 10(a). The set-up load of the support is about 7 kN, and the yield capacity and reseating load in the yielding events are 29 and 24 kN, respectively. The support experiences 8 unloading cycles and 81 counts of roof loading during the test. Massive face fall is observed at the final few counts of roof loading. Accordingly, roof deflection measured at every count of the roof loading is documented in Fig. 10(b). Roof vertical displacement increases slowly and gradually in the initial support loading stage, reaching 3.1 mm before the first support yielding operation. During the cyclic yielding events, however, roof deflection increases in a step function, with the increase increment standing at 5 mm or so. A total of 8 step function increases of roof deflection are found in the physical modeling test, corresponding to the 8 yielding events of the shield. The most roof vertical displacement is observed at the moment of support unloading or yielding, after which the support starts to pick up load again and the rate of roof vertical displacement as a function roof loading count slows down and returns to a level similar to that of the initial loading stage. The maximum roof deflection is found at 47 mm at the final roof loading stage when face fall occurs.

Fig. 11 plots the face horizontal movement recorded by using the displacement monitoring system. The horizontal movement of the face is unnoticeable (close to 0) during the starting 1500 s of the test. After that the difference between the face horizontal displacements recorded by the 3 transducers has become increasingly more significant. Keep in mind that 3 of the transducers were buried at different height of the face during

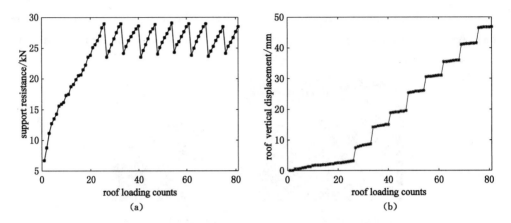

Fig. 10 (a) Relationship of the support resistance over roof ram loading counts;
(b) Roof vertical displacement measured from the vernier caliper during roof loading

the model construction, i. e. 8 cm, 16 cm and 24 cm below the roof. Coal face wall at the upper position of the face shows the largest displacement growth rate and therefore the largest horizontal displacement, followed by the middle position and the lower position of the face. At the final loading stage before face fall occurs, the maximum horizontal displacements monitored by Transducers A, B and C are 17.5, 14 and 11 mm, respectively. It should also be noted that the face movement starts to increase rapidly during the yielding events after 1 500 s of the test.

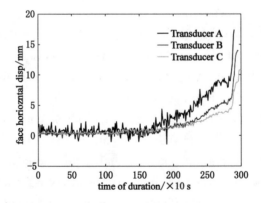

Fig. 11 Face horizontal displacement measured from the transducers at different face heights

6 Summary and discussion

A better understanding of the shield-strata interaction is the prerequisite for selecting a proper longwall shield and improving the ground control in the open face area. This paper assesses the coupling between the longwall shield and roof strata from a fundamental analysis of important characteristics of shields and a physical modeling study by incorporating the basic operation of shield hydraulic cylinders. The basic physical and mechanical behaviors of the shield are discussed in detail, including the leg hydraulics operations, shield capacity, setting force, vertical stiffness and yielding behavior, which are widely overlooked in the previous works. The physical model simulates the actual performance of the shield in the field by including both the initial and cyclic loading stages, thus the performance of face and shield and interaction between shield and strata are reproduced in a physically realistic way. Face horizontal displacement is recorded using a wireless displacement monitoring system, and is compared with the shield loading and roof movement. Important findings are listed below.

(1) The bottom stage of the shield leg cylinder extends first, followed by the top stage but only after the full stoke of the bottom. Upon retraction, the top stage needs to fully collapse before the bottom starts to retract. When the shield is advanced to the next mining cycle, the extension and retraction of bottom stage represent most of the raising and lowering operations of canopy, unless a new maximum mining height is established.

(2) Shield yielding capacity and setting force are improved by increasing the size of the leg cylinder in the industry, while maintaining the yielding and setting pressures at approximately the same level. Setting force can also be adjusted by manually changing the setting pressure. Increasing the setting force is typically recommended in the literature to improve the ground control in the open face area, but it remains debatable because the available/reserve shield load is much less at a high setting force, which may lead to the yielding of a shield. Selection of the setting pressure should consider the shield-strata interaction on a specific mine site.

(3) A longwall shield properly fitting to the geological mining environment should maintain at least an acceptable ground control in the open face area in the current mining cycle without extensively stressing the shield or causing any damage, rather than using up the full shield capacity.

(4) Vertical stiffness of longwall shield has a positive correlation with the shield capacity. The active shield stiffness increases linearly with the diameter of leg cylinder, but decreases exponentially with the extension of hydraulic liquid inside the leg cylinder. The length of the steel rod should be excluded when computing the shield stiffness. High capacity/stiffness shields develop more load on the shield in response to the same amount of roof convergence, but may remain more reserve capacity to prevent the shield from highly stressed or damage.

(5) Yielding of a modern shield represents a 10% drop in the yielding pressure (i.e. reseating pressure represents 90% of the yield pressure), and produces 5~10 mm of reduction in supporting height and 100~250 tons of loss in support loading. The compression of longwall shield per 100 tons of applied load is in the order of 4~6 mm.

(6) Physical modeling study of shield-strata interaction with a consideration of leg hydraulic operations shows that, during the shield initial loading stage, the face maintains good stability with few cracks but no major failure; roof deflection increases slowly and reaches only 3.1 mm at the end of initial loading stage; while face rarely shows any

horizontal movement.

(7) By comparison, most of the face failure and face falls are found during the shield cyclic loading stage. A step function of roof deflection is observed during shield yielding events, with the increase incremental at 5 mm. The total 8 step functions of roof deflection correspond to the 8 yielding events of the shield. Maximum roof deflection is about 47 mm when face fall occurs. The face horizontal displacement increases rapidly in this stage before face fall occurs. The maximum face movement is observed at face top position.

References

[1] BP. BP Statistical Review of World Energy 2017[R]. British Petroleum.
[2] Peng S S, Chiang H S. Longwall mining[M]. New York: John Willy & Sons, 1984.
[3] Wang J C. Theory and technology of thick seam mining[M]. Beijing: Metallurgical Industry Press, 2009.
[4] Wang J C, Zhong S H. The present status and the key issues to be resolved of thick seam mining technique in China[J]. Sciencepaper Online, 2008, 3(11): 829-834.
[5] Meng X R, Wu H T, Wang G B. Development and method selection of thick coal seam mining technology in China[J]. Coal Engineering, 2014, 46(10): 43-47.
[6] Song G F, Pan W D, Yang J H, et al. Mining methods selection in thick coal seam based on fuzzy analytic hierarchy process[J]. Journal of Mining and Safety Engineering, 2015, 32(1): 35-41.
[7] Wang G F, Li X Y, Zhang C C, et al. Research and development and application of set equipment of 8 m large mining height fully-mechanized face[J]. Coal Science and Technology, 2017, 45(11): 1-8.
[8] Yang J Z. Research on key mining technology of fully-mechanized working face with 8 m large mining height[J]. Coal Science and Technology, 2017, 45(11): 9-14.
[9] Song G F, Chugh Y, Wang J C. A numerical modeling study of longwall face stability in mining thick coal seams in China[J]. International Journal of Mining and Mineral Engineering, 2017, 8(1): 35-55.
[10] Song G F, Chugh Y. 3D analysis of longwall face stability in thick coal seams. Journal of the Southern African Institute of Mining and Metallurgy, 2018, 118(2): 131-142.
[11] Wilson A H. Support load requirements on longwall faces[M]. Mining Engineering, 1975: 479-488.
[12] Smart B G, Redfern A. The evaluation of powered support specifications from geological and mining practice information[J]. Ch. In Rock Mechanics: Key to Energy Production, Balkema, 1986(27): 367-377.
[13] Barczak T M, Oyler D C. A model of shiled-strata interaction and its implications for active shield setting requirements 1991[R]. Bureau of Mines, RI 9394.
[14] Park D W, Jiang Y M, Carr F, et al. Analysis of longwall shields and their interaction with surrounding strata in a deep coal mine[C]. In: Proceedings of the 11th

international conference on ground control in mining. Wollongong, Australia, 1992.

[15] Peng S S. What can a shield leg pressure tell us? [M]. Coal Age,1998: 54-57.

[16] Medhurst T P, Reed K. Ground response curves for longwall support assessment[J]. Trans. Inst. Min. Metall A Min. Tech. ,2005(114): 81-88.

[17] Barczak T M, Chen J, Bower J. Pumpable roof supports: developing design criteria by measurement of the ground reaction curve [C]. In: Proceedings of the 22nd international conference on ground control in mining. WV, USA, 2003.

[18] Gale W. Prediction and management of adverse caving about longwall faces[R]. Final report: ACARP Project C7020,2001.

[19] Klenowski G, Ward B, McNabb K E, et al. Prediction of longwall support loading at Southern Colliery, Queensland [C]. In: Proceedings of the 11th international conference on ground control in mining; Wollongong, Australia, 1992.

[20] Kong D. Development and application of a physical model for longwall coal face failure simulation [J]. International Journal of Mining and Mineral Engineering, 2007, 8(2): 131-143.

[21] Guo W. Stability of coal wall and interaction mechanism with support in fully mechanized working face with great mining height[D]. Thesis Submitted to China University of Mining and Technology,2015.

[22] Barczak T M, Tadolini S C. Longwall shield and standing gate designs-is bigger better? [C]. In: Proceedings of Longwall USA; PA, USA, 2007.

[23] Mondal D, Roy P N S, Behera P K. Use of correlation fractal dimension signatures for understanding the overlying strata dynamics in longwall coal mines[J]. International Journal of Rock Mechanics and Mining Science, 2017(91): 210-221.

[24] Verma A K, Deb D. Longwall face stability index for estimation of chock-shield pressure and face convergence[J]. Geotechnical and Geological Engineering,2010, 28(4): 431-445.

[25] Verma A K, Deb D. Analysis of chock shield pressure using finite element method and face stability index[J]. Mining Technology, Transactions of the Institute of Mining and Metallurgy: Section A,2007, 116(2): 67-78.

[26] Ghose A K. Why longwall in India has not succeeded as in other developing countries like China[J]. Ie. Journal-mn,2003(84): 1-4.

[27] Song G F, Yang S L. Investigation into strata behaviour and fractured zone height in a high-seam longwall coal mine[J]. Journal of the Southern African Institute of Mining and Metallurgy,2015, 115(8): 781-788.

[28] Li N, Wang E Y, Ge M C, et al. The fracture mechanism and acoustic emission analysis of hard roof: a physical modeling study[J]. Arabian Journal of Geoscience, 2015, 8(4): 1895-1902.

[29] Xu D J, Peng S P, Xiang S Y, et al. A novel caving model of overburden strata movement induced by coal mining[J]. Energies,2017, 10(4): 476.

Strata movement law and support capacity determination of a upward-inclined fully-mechanized top-coal caving panel

KONG Dezhong[1,2,3,4], LIU Yang[3], ZHENG Shangshang[3], HAN Chenghong[2]

(1. Key Laboratory of Safety and High-efficiency Coal Mining, Ministry of Education(Anhui University of Science and Technology), Huainan 232001, China;

2. Coal Mine Design and Research Institute, Guiyang 550025, China;

3. Mining College of Guizhou university, Guiyang 550025, China;

4. Faculty of Resources and Safety Engineering, China University of Mining and Technology (Beijing), Beijing 100083, China)

Abstract: Support capacity determination is the key point of stope surrounding rock control of a full-mechanized caving face during Topple Mining. In view of the difficulty of stope surrounding rock control, taking panel 8101 as background, The strata movement and roof broken law of panel 8101 are analyzed by using UDEC2D. What's more, the influencing factors of support capacity determination are analyzed and the support capacity is calculated. The result shows that: When working advancing is 20 m, top-coal is caving, while the immediate roof initial caving step distance is 25 m, the first weighting interval of main roof is 40 m and the periodical weighting interval 8~10 m; The abutment pressure influent range is larger and the stress concentration degree increases with the forward of abutment stress peak point in the upward-inclined mining; The roof is not easy to form structure, and the periodical weighting with more severe ground pressure behavior is more frequent, which is easy to cause coal face failure; The influencing factors of support capacity determination are the mechanical properties of top-coal, topple mining angle and the dirt band, the support capacity of 5 858 kN is determined by solving the ultimate bearing capacity of the top-coal coal body.

Keywords: fully-mechanized top-coal caving face; topple mining; strata movement; support capacity determination

In recent years, with the advantages of high yield, high efficiency and excavation of the whole thickness of coal seam, the fully-mechanized top-coal caving technology has become one of the main mining methods for thick coal seams in China[1-4]. But if it encounters the complex geological conditions, the roof is not easy to form stable structure and the broken main roof will directly act on the support when the working face advances in uphill, the stability of coal face will become lower as well[5-8]. Therefore, it is needed to study the movement law of the overlying strata and the roof structure of the inclined top-coal caving

---

Supported by the Scientific Research Foundation of Guizhou Provincial Department of Science and Technology and Guizhou University (QianKehe LH[2017]7280), the Annual Academic Training and Special Innovation Program of Guizhou University in 2017 (Guizhou Kehe[2017]5788) and the Fund of Key Laboratory of Safety and High-efficiency Coal Mining, Ministry of Education(JYBSYS2017101).

Corresponding author: Tel:18511072876, E-mail: 1361316170@qq.com.

face to obtain the reasonable support capacity, so as to ensure the safe, efficient and rapid advancing of working face[9-11]. At present, some researchers have been carried out the studies on the upward-inclined fully mechanized caving mining technology and coal face stability control in China. A systematic study on the disaster mechanism and control technology of coal face spalling by professor Wang is made[12-13], which pointed out that there are shear failure and tensile failure forms of coal face. Yang Shengli[14] established the coal face failure mechanical model of the upward-inclined mining face and achieved a good control effect after strengthening coal face using the "manila + grouting" technology. The mechanical model and the condition of the coal face and end face roof stability are established by Guo Weibin[15] and the relationship between the initial support force of the support, the inclined angle in uphill and the rake angle of the column is discussed. The factors affecting the roof stability of the end face is obtained and the corresponding face roof control technology is put forward through theoretical analysis by Pan Weidong[16]. In view of this serious problem caused by coal face spalling and caving face, Suo Yonglu[17] has analyzed the surrounding rock stress characteristics at the six kinds of upward mining angle using FLAC3D.

At present, the researches on the inclined fully mechanized caving mining have found that the stability of the coal face can be guaranteed by studying the end face roof stability of working face. However, it is not enough to confirm the movement law of overburden rock movement and the support work resistance of the upward-inclined caving face. Therefore, taking the panel 8101 as the engineering background, the movement law of the overlying strata of the upward-inclined fully mechanized caving face at different advancing distance is simulated using UDEC2D so as to determine the roof broken characteristics, and then determine the working resistance of the support.

1 Engineering background

The coal seams mined in upward-inclined fully-mechanized top-coal caving panel 8101 are coal seam 7+8#. With the mining depth of 250~312 m, the average thickness of coal seams 7+8# is 9.1 m, and the hardness coefficient of working face is less than 1.2. The length of the working face is 160 m, and the advancing distance of working face is 510 m. The geological and mining conditions of the working area are more complex. With the average angle of 14°, the maximum angle of working face is 23°, and the upward inclined angle of working face is 16°. With a smaller variation in thickness, the immediate roof is limestone and the thickness is 14 m, besides, there are joints and cracks in the immediate roof. The mining method of panel 8101 is the longwall mining of upward inclined fully mechanized top-coal caving method and the ratio of the cutting and caving height is 1:2. There are a dense joint and fracture distribution of the working face. The coal face failure degree is serious when the working face is inclined forwarded. Therefore, it is needed to study the movement law of overlying strata and support working resistance to ensure the stability of the stope surrounding rock. The coal seam occurrence characteristics are shown in Fig. 1.

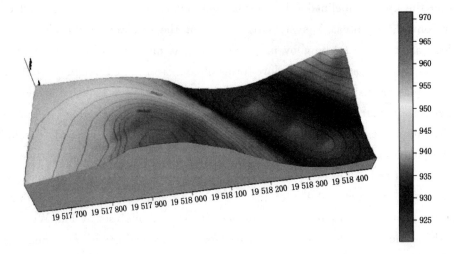

Fig. 1　The coal seam occurrence characteristics of panel 8101

2　The movement law of overlying strata simulation using UDEC

2.1　Model building

Taking the geological and mining conditions of working face 8 101 as background, considering the boundary effect, the model adopts the plane strain model. The calculation length of the model is 160 m and the height is 100 m. The model consists of 3 354 blocks, 5 304 units and 11 833 interfaces, and the shape of unit is quadrangular. The selected calculation memory of computer is 512 MB. There is four boundaries in the model, of which the upper boundary is stress boundary with the vertical stress of 5 MPa on it. The left, right and lower boundary of model is the displacement boundary. With the gravitational acceleration of 9.8 m/s², the calculated model is Mohr-Coulomb constitutive model. The numerical model is shown in Fig. 2 and the physical and mechanical parameters of coal and rock mass are shown in Table 1.

Table 1　　　　Physical and mechanical parameters of coal and rock mass

Rock	density/(kg/m³)	bulk/GPa	shear/GPa	cohesion/MPa	tension/MPa	friction/(°)
siltstone	2 712	8.1	4.1	1.45	3.15	40
mudstone	2 240	5.2	2.40	2.53	1.78	35
coal	1 440	2.3	1.38	0.78	0.64	25
limestone	2 630	8.3	4.12	3.43	2.14	37

2.2　Simulation result analysis

　　(1) Movement law of overlying strata

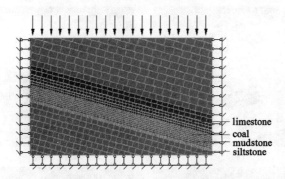

Fig. 2　Numerical model diagram

Combined with the actual top-coal caving mining condition, the cutting of the front coal mining machine and the rear top coal caving proceeded simultaneous, and the spacing is 5 m. With the interval of 5 m between mining and caving, the cutting of coal seam without caving has carried out in the first 20 m of advancing distance in numerical simulation. The overburden strata failure characteristics of working face at different advancing distance is shown in Fig. 3.

It is concluded from Fig. 3:

① When the advancing distance of working face is 10 m, the cracks of top coal have begun to develop but the top coal is not collapse. When the advancing distance is 20 m, top coal caving appears and the immediate develops upward with bed-separation space occurs. With the advancing of working face, the caving zone is gradually rising, and the caving area is also increased with the caving angle of 75°or so.

② When the working face advances to 25 m, the interface separation between roof limestone and coal seam occurred with the appearance of immediate roof caving. When the working face advancing is 30 m, the overall separation and caving of immediate roof occur with the caving height of 4.5 m. The caving height and area of the overlying strata have increased with the sustained increase of advancing distance. When the advancing distance is 40 m, the immediate roof has been all caving to the bottom and the main roof has broken, forming the first weighting. While the advancing distance of working face is 50 m, the first periodic fracture occurs at the top of the roof, which is the first periodic weighting of the working face. After that, the periodic fracture of main roof will occur once every 8～10 m of working face is advanced, that is, the periodic weighting length of main roof is 8～10 m.

(2) Abutment pressure distribution

Fig. 4 is the curve of the abutment stress peak and the peak position at different advancing distances of working face.

As shown in Fig. 4:

① With the advancing of working face, the roof hanging area is gradually increasing and forming a stable structure when it reaches a certain value, at the same time, the

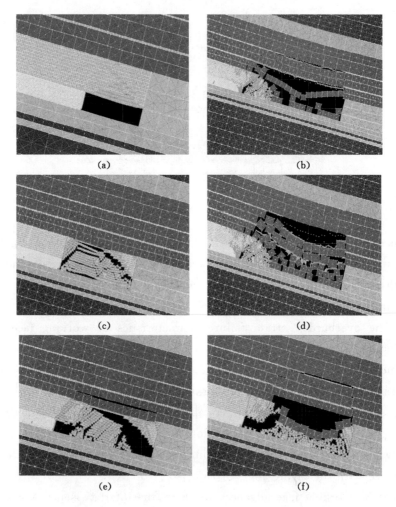

Fig. 3 Movement law of overlying strata under different advancing distance
(a) Advancing distance: 10 m; (b) Advancing distance: 20 m;
(c) Advancing distance: 25 m; (d) Advancing distance: 30 m;
(e) Advancing distance: 40 m; (f) Advancing distance: 50 m

abutment stress peak value has little change; The variation of stresses in the bottom coal are very large during the first advancing distance of 20 m, but the variation in the main roof is little. It shows that the top coal seam has entered into the caving zone due to the influence of mining action, and the caving zone height is increasing and developing to the upper immediate roof, and some strata of main roof has entered into the caving zone simultaneously. Meanwhile, the fractured zone and bend subsidence zone have developed to the upper strata; The abutment stress peak decreases gradually from the bottom coal to the top coal, immediate roof, main roof and the bottom coal zone is the most stress concentrated area.

② The higher is the peak value of abutment stress, the distance from the coal face is greater from the bottom coal up to the top layer. On the whole, the peak of the abutment

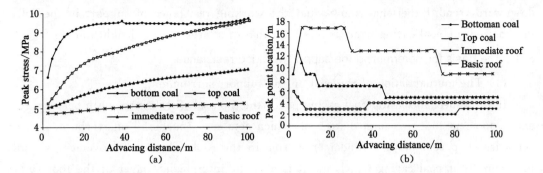

Fig. 4 The variation curve of peak value of abutment pressure and position of different strata
(a) The curve of stress peak value and advancing distance;
(b) The curve of the peak point location and advancing distance

stress in the coal seam is far away from the coal face with the advancing of working face, and the peak of the abutment stress in immediate and main roof are gradually close to the coal face. This may be because that the vertical displacement of coal and rock trending to the goaf is greater than the horizontal displacement, and the limit span is large due to the great main roof stiffness, however, the immediate roof and coal seam is relatively soft, and the peak value of abutment stress of main roof is far from the coal face due to the intense mining disturbance when transporting to the goaf; The peak value location of abutment pressure in the coal seam is closest to the coal face, and the location of abutment stress peak in the bottom coal is 2～3 m far away from the coal face. The peak value location of abutment stress in the top coal seam keeps 3～4 m far away from the coal face during the normal advancing of working face. The abutment stress peak value location of the immediate roof and main roof increase at first then decrease with the advancing of working face, and the peak value location of abutment stress in the immediate roof and main roof are 5～13 m and 9～17 m far away from the coal face.

Under the conditions of the thickness, inclination of main roof lithology and the same mining technology, the periodic weighting interval of the main roof is smaller than that of the horizontal mining. During inclined mining, the fracture rock has the trend to movement to the goaf and the fracture crack is enlarged, which is not easy to be articulated with the unbroken rock in front. The roof subsidence and pressure will be further increased, causing coal face spalling and roof caving, deteriorating roof condition.

3 Calculation of support working resistance

Through the above analysis, the roof is not easy to form a stable structure after breaking off during upward inclined mining, due to the influence of inclination angle. The roof pressure of the overlying strata can be divided into two components, one component is perpendicular to the support, and the support resistance is determined based on the component. Another component paralleling to the support beam, will make support have a downward trend. Experience shows that the working resistance of support in the fully-mechanized top-coal caving panel is less than that of a large-cutting-height mining face.

3.1 The basis for determining the support working resistance

(1) The mutual action of top-coal on the support-surrounding rock

In fully-mechanized top-coal caving panel, the top coal with low strength between roof and support is added as immediate roof, which acts as a "cushion" of overburden strata activities. Top coal plays a fundamental role in the roof subsidence movement of fully mechanized top coal caving panel, and it is also the intermediate layer of the roof rotary sinking on the support, the mechanical properties of which play a key role in the relationship of the support and surrounding rock. After the test and research on the ground pressure of fully-mechanized top-coal caving panel, the top coal with the low strength and the multi structure face has obvious influence on the support-surrounding rock. In the study of the interaction, we should fully understand and study the physical and mechanical properties of the top coal.

(2) The influence of the inclined angle

If the working face has an inclined angle, it is upward inclined or downward inclined mining, the pressure on the support by top coal is divided into two parts. One part is perpendicular to the support force, and the support working resistance is not less than the part pressure, or the danger of hydraulic support pushed down would occur. The other part is a force parallel to the support beam, which can be offset by the working face anti slip device. Therefore, it can be determined that in a certain depth of buried and roof-floor combination of the case, the support working resistance in the upward or downward inclined mining face is less than of the support working resistance of the horizontal working face.

(3) Dirt Band effect

If the dirt band is very thick and the occurrence is stable, which is like a natural anchor occurred in the coal seam, playing a role enhancing coal strength. With the advancing of working face, the damage degree of top coal is greatly weakened during the migration and destruction process of top coal, and the ability to transfer vertical stress is stronger due to the insufficient of top coal failure. However, in the actual production process, if the coal gangue thickness is generally small and the strength is not high, the effect on working resistance by dirt band is small.

3.2 Determination of support working resistance

A large number of experimental results show that the support working resistance of fully mechanized caving face depends on the deformation pressure of the immediate roof and top coal transferring to the support without dynamic load. So the influence of static load on the support working resistance is only considered in the calculation. As shown in Fig. 5.

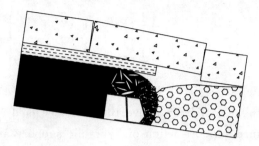

Fig. 5 The roof mechanical model of a upward inclined top coal caving mining face

After mining, the top coal body is caving under the roof pressure and the sustained support action of the hydraulic support and become loose solid due to the loss of mechanical contact with the front coal body. So it can be considered that the horizontal stress of the top coal in the back of the hydraulic is equal to zero. And the horizontal stress of coal face is approximately equal to the initial stress. It is assumed that the horizontal stress from the upper part to the back of is reduced by exponential form. The initial horizontal stress is $\sigma_h = \mu \gamma H$, taking the coal face as the original point, the horizontal stress near the goaf side is $\sigma_3 = \sigma_h e^{ax}$.

The relationship between the three-axial compressive strength and the confining pressure and uniaxial compressive strength of the top coal can be obtained formula (1):

$$R = R_c + \sigma_3 \frac{1+\sin\varphi}{1-\sin\varphi} \tag{1}$$

The coal body of top coal presents different failure degree under the action of roof pressure. It is assumed that the broken coal body of top-coal behind the support is completely destroyed, remaining residual stress, and the top coal body above the coal face is in a false three direction pressure with slight strength. In order to simplify the calculation, according to the previous studies, the variation stress of the top coal above the top of the control area is in accordance with the negative exponential curve. Thus, the maximum principal stress from coal face to the roof caving area is $\sigma_1 = R e^{bx}$.

According to the elastic mechanics, the stress which is perpendicular to the support beam can be expressed in formula (2):

$$\sigma_n = \frac{1}{2}(\sigma_1 + \sigma_3) + \frac{1}{2}(\sigma_1 - \sigma_3)\cos 2\alpha \tag{2}$$

Putting σ_1, σ_3 into the following formula (2), the top-coal stress of a fully mechanized caving face with the inclination angle of θ which is perpendicular to the support beam can be obtained, as shown in formula (3):

$$\sigma_n = \frac{1}{2}(R e^{bx} - \sigma_h e^{ax}) + \frac{1}{2}(R e^{bx} + \sigma_h e^{ax})\cos 2\theta \tag{3}$$

The support working resistance of a upward inclined caving face can be obtained by the integration of the support control caving area, as shown in formula (4):

$$P = \int_0^{L_k} \sigma_n dx = \int_0^{L_k} \left[\begin{array}{l} \frac{1}{2}(Re^{bx} - \sigma_{hi}e^{ax}) + \\ \frac{1}{2}(Re^{bx} + \sigma_{hi}e^{ax})\cos 2\theta \end{array} \right] dx \qquad (4)$$

The initial horizontal stress is 1 MPa, and the compressive strength of coal body is 13.8 MPa. The strength of coal body above the coal face is decreased under the mine pressure and the sustained support action of hydraulic support, and the strength of coal face is 2.8 MPa, which is about 1/5 of the intact coal strength. The top coal residual strength is 0.05 MPa and the accused of zenith distance is 5.08 m with the support width of 1.5 m, and the inclined angle of working face is 15 degrees.

Putting the above parameters into the formula (4), the support working resistance can be obtained: $P = 5858$ kN. Thus, it can be seen that the selected support working resistance for 7 000 kN can meet the production requirements.

4 Conclusions

(1) UDEC2D simulation was used to study the movement law of overlying strata and the roof caving characteristics of the upward-inclined fully-mechanized top-coal caving panel under different advancing distances, the research results show that: When the advancing distance is 20 m, the top coal is caving. The first caving length of the immediate roof is 25 m and the first weighting length is 40 m. And the periodic weighting length is 8~10 m; Compared with the horizontal full-mechanized top-coal caving panel, the roof is not easy to form structure, and the periodic weighting with more severe ground pressure is more frequent.

(2) Compared with the horizontal full-mechanized top-coal caving panel, the abutment stress influent range is larger and the stress concentration degree increases with the forward of abutment stress peak point in the upward-inclined mining. The reason is that loosening range of coal face enlarges due to the large horizontal stress caused by syncline structure. The coal face spalling prevention of a upward-inclined mining face is the key point of the stope surrounding rock control.

(3) Physical and mechanical properties, the upward-inclined mining angle, the dirt band and the other factors affecting the determination of support working resistance are analyzed through theoretical analysis. The static model of the force on the support in the upward-inclined fully-mechanized caving mining is established to determine the reasonable working resistance of the support by calculating the ultimate bearing capacity of top-coal. The engineering practice shows that the selected support working resistance can meet the requirements.

References

[1] WANG Jia-chen. The mining theory and technique of thick coal seam[M]. Beijing:

metallurgical industry press,2009.

[2] Wang Jiachen,Wang Lei,Guo Yao. Determining the support capacity based on roof and coal face control[J]. Journal of China Coal Society,2014,39(8):1619-1624.

[3] YAN Shao-hong. Research on Side and Roof Falling Mechanism and Control Approaches in Full-mechanized Caving Mining with Large Mining Height[J]. Coal Mining Technology,2008,13(04):10-13.

[4] ZHANG Jin-wang,SUN Shao-long. The caving mechanism of loose top coal in upward-inclined fully-mechanized caving[J]. Coal technology,2015,34(3):14-16.

[5] Kong Dezhong,Yang Shengli,Gao Lin,et al. Determination of support capacity based on coal face stability control[J]. Journal of China Coal Society,2017,42(3):590-596.

[6] CHANG Jucai,XIE Guangxiang,ZHANG Xuehui. Analysis of rib spalling mechanism of fully-mechanized top-coal caving face with great mining height in the extra-thick coal seam[J]. Rock and Soil Mechanics,2015,36(3):803-808.

[7] PANG Yihui,WANG Guofa. Hydraulic support protecting board analysis based on rib spalling "tensile cracking-sliding" mechanical model[J]. Journal of China Coal Society,2017,42(8):1941-1950.

[8] CHEN Hao,KANG Liyun,CHANG Baisheng,et al. The prevention and control measures and the effect of upward-inclined mining on coal wall spalling complex geological conditions[J]. Coal mine safety,2014,45(1):118-121.

[9] LI Jianguo,TIAN Quzhen,YANG Shuansuo. The Effecting Principle of Dip and Rise Mining on Rib Fall in Full-Mechanized Caving Face[J]. Journal of Taiyuan University of Technology,2004,35(4):407-409.

[10] LI Shugang,LI Haotian,DING Yang,et al. Physical Simulation Experiment on Laws of Stress Distribution and Crack Evolutionary of Overburden Strata in Up-dip Mining [J]. Safety in Coal Mines,2015,46(2):25-27,31.

[11] RAN Yujiang, ZHAO Lilong, XU Tao, et al. The study and application of surrounding rock control technology and large slope inclined mining of full-mechanized top-coal caving panel[J]. Mining Safety & Environmental Protection,2012,39(1):64-66.

[12] WANG Jia-chen. Mechanism of the rib spalling and the controlling in the very soft coal seam[J]. Journal of China Coal Society,2007,32(8):785-788.

[13] Wang Jiachen, Yang shengli, Kong Dezhong. Failure mechanism and control technology of Longwall coal face in large-cutting-height mining method [J]. International Journal of Mining Science and Technology,2016,26(1):111-118.

[14] YANG Shengli, KONG Dezhong, YANG Jinghu, et al. Coal wall stability and grouting reinforcement technique in fully mechanized caving face during topple mining[J]. Journal of Mining & Safety Engineering,2015,32(5):827-833+839.

[15] GUO Weibin,LU Yan,HUANG Fuchang,et al. Stability of surrounding rock in head face of upward fully-mechanized caving face and its control technology[J]. Journal of

Mining & Safety Engineering,2014,31(3):406-412.

[16] PAN Weidong,KONG Dezhong,WANG Zhaohui,et al. Stability Control of Upward Fully Mechanized Top Coal Caving Face Ends in Soft and Thick Coal Seam[J]. Coal Engineering,2014,46(3):65-67+71.

[17] SUO Yonglu, LIU Mingliang, YANG Zhanguo, et al. Characteristics of Stress Distribution in Surrounding Rock at Uphill Fully Mechanized Caving Face[J]. Coal Technology,2011,30(11):69-71.

厚硬岩层直覆大倾角综采顶板失稳机理分析

魏祯,杨科,池小楼,付强,刘文杰

(安徽理工大学 煤矿安全高效开采省部共建教育部重点实验室,安徽 淮南,232001)

摘 要:为了研究大倾角厚煤层综采过程中直覆厚硬顶板的稳定性,运用损伤力学的基本理论对基本顶失稳规律进行了研究分析。本文以潘北矿11513工作面为工程背景,建立了基本顶损伤参量的计算模型,阐述了综采直覆厚硬顶板在上层覆岩压力作用下损伤演化与稳定性之间的关系,并利用控制变量法讨论了开采深度、煤层倾角、岩石强度、原始裂隙对基本顶稳定性的影响。

关键词:厚硬顶板;损伤力学;覆岩压力;稳定性;控制变量法

Analysis of instability mechanism of fully-mechanized top-coal caving mining face in thick and hard rock strata

WEI Zhen, YANG Ke, CHI Xiaolou, FU Qiang, LIU Wenjie

(Key Laboratory of Ministry of Education for Safe and Efficient Mining of Coal Mines, Anhui University of Science and Technology, Huainan 232001, China)

Abstract: In order to study the stability of straight overburden thick roof in fully mechanized coal seam with large dip Angle, the basic theory of damage mechanics is used to study and analyze the basic roof instability rule. Based on the engineering background of 11513 working face of Panbei Mine, established the basic roof damage parameter calculation model, this paper expounds the fully mechanized direct overlying thick and hard roof in the upper strata pressure under the action of damage evolution and the relationship between the stability, and USES the method of control variables discussed the mining depth, dip Angle of coal seam, rock strength, the effect on the stability of the original crack of roof.

Keywords: Thick hard roof; damage mechanics; overburden pressure; stability; control variable method

随着采用长壁工作面综采技术以来,采场围岩控制一直是煤矿工程技术人员关注的核心问题,而采场顶板岩层的控制是矿压理论研究的主要目的之一。针对倾斜煤层的开采,众多专家学者对其顶板破断规律开展了大量研究工作。钱鸣高院士等[1]提出了关键层理论,建立了不同支撑条件下的弹性板模型,得到了基本顶的"O-X"破断规律。尹光志等[2]建立了(急)倾斜煤层上覆岩层变形的力学模型,得到了采空区顶板变形特征;张益东等[3]采用力

基金项目:国家自然科学基金重点项目(51634007);国家重点研发计划重点专项项目(2016YFC0801400);安徽省重点研究与开发项目(1704a0802129)。

作者简介:魏祯(1992—),男,现为硕士研究生。E-mail:sdweizhen0302@163.com。

通信作者:杨科(1979—),男,博士,教授,主要从事矿山压力与岩层控制、煤与瓦斯共采方面的教学与研究工作。E-mail:yksp2003@163.com。

学分析与数值模拟相结合的方法,研究了大倾角俯(仰)斜开采时顶板的应力状态,并提出了采场顶板破断准则。

上述利用弹性薄板理论研究直覆厚硬顶板的变形失稳,将存在不能满足薄板理论厚宽比的局限性[4]。鉴于基本顶的稳定性与损伤参量之间的关系,本文以潘北矿 11513 工作面为工程背景,运用损伤力学基本理论,以损伤参量为基本顶的失稳指标进行研究。理论分析结果与现场实际相吻合。

1 基本顶损伤参量与稳定性关系

随着近二十多年来损伤力学的兴起与发展,学者们应用损伤力学研究了上覆岩层的变形和破坏过程,进而研究支承压力分布和矿压显现规律[5-6]。

如图 1 所示,顶板裂隙随着工作面推进距离的增加而增大。如果顶板在上覆岩层压力作用下岩层内部裂隙越发育,则冒落矸石的块度越均匀,基本顶稳定性越差[7]。

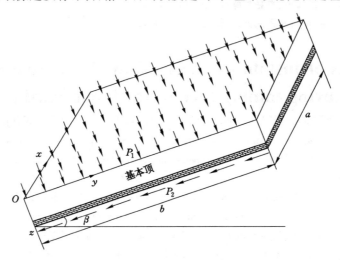

图 1　基本顶简化受力模型

Fig. 1　Basic top simplified force model

1.1　损伤参量一般分析

通常我们将材料内部裂隙的发育过程称为损伤,材料裂隙的变形发育,使其失去承载能力可以通过损伤参量 D 表示。当损伤参量 $D=0$ 时,认为材料未破坏;当损伤参量 $D=1$ 时,认为材料达到损伤破坏极限。

1.2　节理岩体损伤特点

大量实验研究表明:节理岩体的损伤破坏通常都是由原有的裂隙开始发育,不断产生新的裂纹继而扩展,最后发展成贯穿性裂纹导致岩体失稳破坏[8]。当岩石体内节理裂隙越多,其弹性模量和承载力越小,越易发生失稳破坏。

2　损伤参量模型建立

定义初始损伤为岩体中分布的原始节理裂隙。随着岩石尺度的逐渐减小,非线性力特征逐渐降低,当减小到一定程度时,这种尺度效应消失,表现出线性的弹脆性[9]。

为了便于损伤参量计算模型的建立,取微元岩体,假定微元岩体满足:

(1) 微元岩石体的应力与应变关系服从广义胡克定律。

(2) 微元岩石体服从 Mises 屈服破坏准则。

引入统计学理论,岩石强度服从 Weibull 分布[10],即:

$$P(\varepsilon) = \frac{m}{\varepsilon_0}\left(\frac{\varepsilon}{\varepsilon_0}\right)^{m-1}\exp\left[-\left(\frac{\varepsilon}{\varepsilon_0}\right)^m\right] \quad (1)$$

式中,m,ε_0 分别是 Weibull 分布的标度和形态参数;ε 是岩石单元体的应变;$P(\varepsilon)$ 为岩石应变为 ε 时的破坏概率。

假设微元岩石体服从广义胡克定律[11]:

$$\left.\begin{array}{l}\sigma_x = \lambda(\varepsilon_x+\varepsilon_y+\varepsilon_z)+\dfrac{E}{1+\mu}\varepsilon_x \\ \sigma_y = \lambda(\varepsilon_x+\varepsilon_y+\varepsilon_z)+\dfrac{E}{1+\mu}\varepsilon_y \\ \sigma_z = \lambda(\varepsilon_x+\varepsilon_y+\varepsilon_z)+\dfrac{E}{1+\mu}\varepsilon_z\end{array}\right\} \quad (2)$$

在大倾角工作面直覆厚硬顶板上取一微元,进行力学分析,如图 2 所示,在等围压作用下,即 $\sigma_x = \sigma_z$。

式(2)可以变形成以应力表达应变的广义胡克定律得:

$$\varepsilon_y = \frac{1}{E}[\sigma_y - \mu(\sigma_z+\sigma_x)] \quad (3)$$

微元岩石体服从 Mises 屈服破坏准则,即:

$$(\sigma_x-\sigma_y)^2+(\sigma_y-\sigma_z)^2+(\sigma_x-\sigma_z)^2 = 2\sigma^2 \quad (4)$$

式中,σ 是基本顶岩石体的单轴抗压强度,在围压相等情况下,上式可变为

$$|\sigma_x-\sigma_y| = \sigma \quad (5)$$

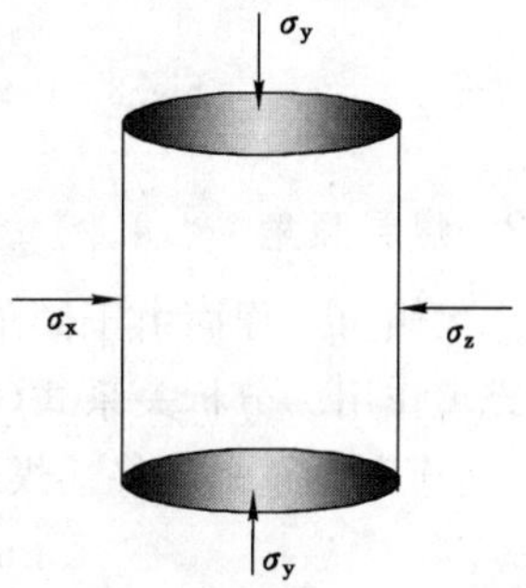

当基本顶微元在外力作用下,应变增大到 Mises 破坏准则时,岩石体发生破裂损伤,此时岩石的受损伤的面积为:

$$S = S_0\int_0^\varepsilon P(x)\mathrm{d}x \quad (6)$$

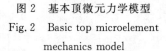

图 2 基本顶微元力学模型

Fig. 2 Basic top microelement mechanics model

式中,S_0 是岩石未受损伤时的面积,根据损伤力学基本理论,岩石的损伤参量可以用材料损伤面积 S 和未损伤时面积 S_0 之比表示,即:

$$D = \frac{S}{S_0} = \int_0^\varepsilon P(x)\mathrm{d}x \quad (7)$$

为了计算接近现场工程背景取基本顶岩体泊松比 $\mu=0.2$,联立式(1)、(5)、(7)得:

$$D = 1-\exp\left[-\left(\frac{\varepsilon_y-\dfrac{3\sigma_x}{5E}}{\varepsilon_0}\right)^m\right] \quad (8)$$

3 基本顶稳定性分析

采用控制变量法逐个分析其对基本顶稳定性的影响规律。假设支承压力峰值处基本顶

原始损伤为 D_0,则基本顶变形失稳过程中总的损伤可表示为:

$$D_{\text{总}}=D_0+D \tag{9}$$

3.1 开采深度

当开采深度增加,基本顶垂直应力 σ_y 也增大,故基本顶损伤参量随着开采深度增大而增大,稳定性也越来越差(图3)。基本顶稳定性随着采深加大而越来越差,这与现场实际相符。

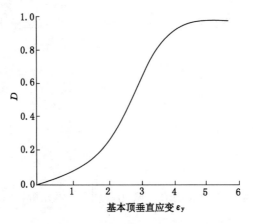

图3 开采深度与基本顶稳定性关系图

Fig. 3 Relationship between mining depth and basic roof stability

3.2 煤层倾角

大倾角工作面由于倾角的存在,使得上风巷和下机巷的埋深不同,不考虑部分冒落矸石的充填作用。分析关系式(8)知,随着岩层水平应力 σ_x 的增大,基本顶的损伤参量 D 逐渐减小,其稳定性越来越好。煤层倾角与损伤参量关系如图4所示。

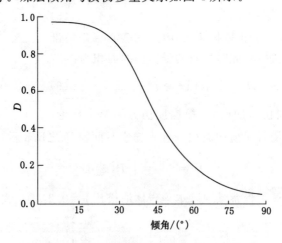

图4 煤层倾角与基本顶稳定性关系图

Fig. 4 Relationship between coal dip and basic roof stability

3.3 岩石强度

岩石体之间存在差异,著名的 Weibull 分布中 ε_0 是微单元岩石强度分布的形态参数,也可反映出岩石的强度性能[13]。如图所示,横坐标用参量 ε_0 代表岩石强度。基本顶稳定性与参量 ε_0 呈负指数关系。如图 5 所示。

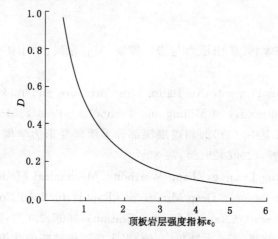

图 5 岩石强度与基本顶稳定性关系图

Fig. 5 Relationship between rock strength and basic roof stability

3.4 原始裂隙

在外力作用下,岩石的破坏往往通过这些原始裂隙弱面发育形成。Weibull 分布中参量 m 表示岩石微单元强度的集中程度,参量 m 越大,岩石材料匀质程度越好[14]。基本顶稳定性与原始裂隙关系如图 6 所示。

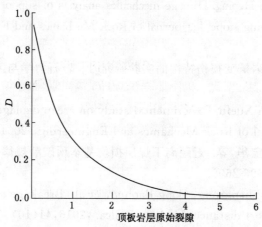

图 6 原始裂隙与基本顶稳定性关系图

Fig. 6 Relationship between original crack and basic roof stability

4 结 论

(1) 研究了开采深度、煤层倾角、岩石强度、原始裂隙等对基本顶稳定性的影响。

(2) 运用 Weibull 分布中的参量 ε_0 反映岩石体强度,得出基本顶稳定性随着岩石强度的增大而增加。

(3) 由于倾角的存在使得工作面上、下巷埋深不同,工作面下部顶板覆岩压力大于上部,下部岩石体的损伤程度大于上部。

参 考 文 献

[1] 钱鸣高,石平五,许家林.矿山压力与岩层控制[M].徐州:中国矿业大学出版社,2010:180-188.
Qian Minggao, Shi Pingwu, Xu Jialin. Mine pressure and rock strata control[M]. Xuzhou: China University of Mining and Technology Press,2010:180-188.

[2] 尹光志,王登科,张卫中.(急)倾斜煤层深部开采覆岩变形力学模型及应用[J].重庆大学学报(自然科学版),2006,29(2):79-82.
Yin Guangzhi, Wang Dengke, Zhang Weizhong. Mechanical Model and Application of Overburden Deformation in Deep Mining of (Rapid) Inclined Coal Seams[J]. Journal of Chongqing University(Natural Science Edition),2006,29(2):79-82.

[3] 张益东,程敬义,王晓溪,等.大倾角仰(俯)采采场顶板破断的薄板模型分析[J].采矿与安全工程学报,2010(04):487-493.
Zhang Yidong, Cheng Jingyi, Wang Xiaoxi, et al. Analysis of Thin Plate Models with Broken Roof in Large Angle Inclined (Dipping) Mining Stope[J]. Journal of Mining & Safety Engineering, 2010(04): 487-493.

[4] 陈忠辉,谢和平.综放采场支承压力分析的损伤力学分析[J].岩石力学与工程学报,2000,19(4):436-439.
Chen Zhonghui, Xie Heping. Damage mechanics analysis of supporting pressure analysis in fully mechanized caving stope[J]. Journal of Rock Mechanics and Engineering,2000,19(4):436-439.

[5] 曹树刚,鲜学福.煤岩蠕变损伤特性的实验研究[J].岩石力学与工程学报,2001,20(6):817-821.
Cao Shugang, Xian Xuefu. Experimental study on creep damage characteristics of coal and rock[J]. Journal of Rock Mechanics and Engineering, 2001, 20(6): 817-821.

[6] 何尚森,谢生荣,宋宝华,等.近距离下煤层损伤基本顶破断规律及稳定性分[J].煤炭学报,2016,41(10):2596-2605.
He Shangsen, Xie Shengrong, Song Baohua, et al. Breakage law and stability of coal seam damage in short distance[J]. Acta. Sinica. ,2016,41(10): 2596-2605.

[7] 刘长友,杨敬轩,于斌,等.多采空区下坚硬厚层破断顶板群结构的失稳规律[J].煤炭学报,2014,39(3):395-403.
Liu Changyou, Yang Jingxuan, Yu Bin, et al. Instability Law of broken Roof Group structure of hard thick layer under multiple Goaf[J]. Acta. Sinica. ,2014 (3): 395-403.

[8] 李兆霞.损伤力学及其应用[M].北京:科学出版社,2002.
Li Zhaoxia. Damage mechanics and its application[M]. Beijing: Science Press, 2002.

[9] 徐芝纶. 弹性力学(第4版)[M]. 北京:高等教育出版社,2006
Xu Zhiguan. Elasticity (Fourth Edition)[M]. Beijing: Higher Education Press,2006.

[10] 陈忠辉,唐春安,傅宇方. 岩石破裂过程中声发射的围压效应[J]. 岩石力学与工程学报,1997,16(1):65-70.
Chen Zhonghui, Tang Chunan, Fu Yufang. Confining pressure effect of acoustic emission during rock failure[J]. Chinese Journal of Rock Mechanics and Engineering, 1997,16(1): 65-70.

[11] 曹文贵,赵明华,刘成学. 基于统计损伤理论的莫尔-库仑岩石强度判据修正方法之研究[J]. 岩石力学与工程学报,2005,24(14):2403-2408.
Cao Wengui, Zhao Minghua, Liu Chengxue. Study on Mohr Coulomb Rock strength Criterion Correction method based on Statistical damage Theory[J]. Journal of Rock Mechanics and Engineering, 2005, 24(14): 2403-2408.

长壁充填开采充填步距对覆岩位移的影响分析

贾林刚[1,2]，高庆丰[3]

(1. 煤炭科学技术研究院有限公司安全分院，北京，100013；
2. 煤炭资源高效开采与洁净利用国家重点实验室(煤炭科学研究总院)，北京，100013；
3. 神木市能源局，陕西 神木，719300)

摘　要：长壁膏体充填的充填步距是充填开采设计的关键参数，与沉降控制程度和充填生产效率等密切相关，本文采用弹性地基梁理论，建立了顶板的位移方程和支撑力方程，根据充填体随充填时间加长强度逐渐增强的力学性能，结合顶板位移方程，推导得出了顶板的最大跨距，以及顶板最大跨距与控顶距、充填步距的数学关系式，进而得出了最大充填步距，为充填步距的合理确定提供了依据。通过应用实例分析，得出了实例条件下充填步距的理论值为9.6 m，增大充填步距，可提高开采效率。通过数值模拟计算得出了充填步距与顶板最大下沉值之间线性相关，不同的充填步距可影响到地表沉陷位移，空顶期及充填体未承压前的顶板位移较大。因此，充填步距的合理设置对于充填开采控制开采沉陷及提高开采效率至关重要。

关键词：膏体充填；充填步距；岩梁跨度；充填体强度；顶板下沉

Influence analysis of filling interval on overburden displacement in longwall filling mining

Abstract: The filling interval of the long wall paste filling is a key parameter in the design of filling mining. It is closely related to the degree of settlement control and the efficiency of the filling production. In this paper, the displacement equation and the support force equation of the roof are established by using the elastic foundation beam theory, and the strength of filling body increases gradually with the filling time prolonged. Strong mechanical properties, combined with the displacement equation of the roof, derived the maximum span of the roof, as well as the mathematical relation between the maximum span of the roof and the distance of the roof and the filling interval, and then the maximum filling interval is obtained, which provides a basis for reasonable determination of filling interval. Through practical example analysis, it is concluded that the theoretical value of filling interval distance is 9.6 m under actual conditions, Increasing the filling distance can improve the mining efficiency. Through numerical simulation, the linear correlation between the filling interval and the maximum subsidence of roof is obtained. Different filling interval can affect the surface subsidence displacement, and the roof displacement is larger in the empty roof period and before the filling body being not pressurized. Therefore, the reasonable setting of filling interval is very important for controlling mining subsidence and improving mining efficiency.

作者简介：贾林刚(1978—)，男，内蒙古呼和浩特人，副研究员，主要从事"三下"采煤、开采沉陷损害防治等研究工作。Tel：13811087627，E-mail：jlg010@163.com。

Keywords: paste filling mining; filling interval; rock beam span filling body strength; roof subsidence

充填开采是一种绿色开采工艺,可以通过充填材料置换出煤炭资源,同时控制覆岩下沉、减少地表移动,从而保护地表建筑物和矿区生态环境[1-3]。长壁充填开采基本不改变原有开采方式,可在原有基础上增加充填系统,减少了基础建设及投资。近年来我国专家学者对充填体支撑顶板理论及覆岩破坏机理等进行了研究,文献[4]利用差分法薄板理论建立了长壁工作面充填开采顶板活动力学模型,分析研究了工作面长度对关键层变形及破坏的影响。文献[5]基于长壁充填采煤岩层移动特征,推导出了采场顶板下沉方程与煤柱承受载荷计算公式,并结合极限强度理论,提出了煤柱失稳判据。文献[6]提出了充填开采上覆岩层连续曲形梁理论模型,阐述了连续曲形梁的几何特征和力学特征,指出了顶板形成连续曲形梁的条件,分析得到连续曲形梁与关键层的量化关系。文献[7]针对"三下"煤炭充填开采引起的岩层移动与控制问题,考虑充填体、煤柱和承重岩层的共同协调作用,提出"充填体+煤柱+承重岩层"协作支撑系统概念。文献[8]推导出了充填体与煤柱耦合作用下顶板岩层的挠曲方程,陈绍杰[9]研究了条带煤柱膏体充填开采覆岩结构模型及运动规律。

这些研究成果从不同角度深入分研究了充填开采岩层移动控制的力学特性和原理,推动了膏体充填开采的发展和应用。然而,检索文献未发现对长壁充填开采充填步距未有深入研究,长壁膏体充填开采的充填步距是充填开采工艺中的主要参数,它不仅涉及充填的效率,也涉及充填开采的沉降与控制效果,它不仅与顶板的力学性质相关,也与充填材料的强化过程相关。目前,充填步距的确定主要依据充填生产工艺的要求,不仅如此,充填步距的确定还应考虑充填材料强度渐变的时间特性,以及充填材料与顶板岩体的耦合作用。因此,研究分析合理的充填步距以及充填步距对覆岩移动的影响,对于长壁膏体充填开采具有重要的理论和工程实际意义。

1 长壁膏体充填开采最大充填步距理论推导

1.1 顶板受力分析

长壁充填开采推进过程中,充填膏体按照充填步距充入采空区,充填膏体从初凝到终凝过程中,充填体强度逐步增强。在这过程中顶板下沉,覆岩对充填体产生压力,使充填体产生压缩变形,同时充填体给顶板反作用力,支撑顶板并限制顶板继续下沉移动,在充填体的连续线支撑作用下,顶板不发生贯穿性断裂,处于连续变形状态(图1)。

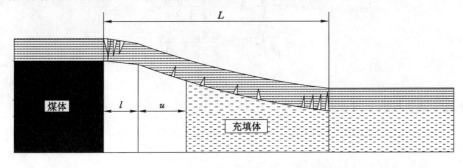

图1 充填开采顶板位移示意图

Fig.1 Schematic diagram of roof displacement in filling mining

当充填体充满采空区时,在顶板向下位移过程中,充填体与顶板紧密接触,即充填体的压缩变形量值与顶板岩梁的挠度处处相等,此时顶板覆岩与充填体形成弹性地基梁力学模型,充填体支撑反力服从 E.Winker 假设,即:

$$F = kz \tag{1}$$

式中　F——为单位面积上的压力强度,MPa;
　　　k——充填体地基弹性系数,MPa/m;
　　　z——充填体的压缩量,m。

充填体的压缩量与充填体的支撑反力 F 呈正比,由于梁的各点都支承在弹性地基上,因而可使梁的变形减少,此时,顶板的下沉量 Z(即充填体的压缩量)与充填体的反力 F 之间近似呈线性关系。因顶板岩梁与充填体地基是共同变形的,根据 Winker 假设,顶板下沉挠度与荷载、地基支撑力之间的相关方程为:

$$EI\frac{\mathrm{d}z^4}{\mathrm{d}x^4} + kz = q \tag{2}$$

解微分方程可得:

$$z = \mathrm{e}^{-\alpha x}(A_1\cos\alpha x + A_2\sin\alpha x) + \mathrm{e}^{\alpha x}(A_3\cos\alpha x + A_4\sin\alpha x) + \frac{q}{k} \tag{3}$$

式中,$\alpha = \sqrt[4]{\dfrac{k}{4EI}}$,其中 A_1、A_2、A_3、A_4 为待定系数。

$x\to\infty$ 时,即充填稳定区顶板沉降值即顶板地基梁的挠度为 $z = \dfrac{q}{k}$,要满足该条件,公式(3)中必有 $A_3 = A_4 = 0$,考虑充填前顶板的下沉量 z_0,此时顶板下沉方程为:

$$z = \mathrm{e}^{-\alpha x}(A_1\cos\alpha x + A_2\sin\alpha x) + \frac{q}{k} + z_0 \tag{4}$$

同理可得支架控顶区顶板位移方程为:

$$z_\mathrm{c} = \mathrm{e}^{\beta x}(B_1\cos\beta x + B_2\sin\beta x) + \frac{q}{k_\mathrm{c}} \tag{5}$$

式中,$\beta = \sqrt[4]{\dfrac{k_\mathrm{c}}{4EI}}$,$k_\mathrm{c}$ 为支架的弹性系数。

考虑边界条件和顶板位移连续特性,在充填区与支架控顶区交界处,顶板下沉位移值、挠度弯曲、弯矩、剪切应力相等原则,对公式(4)和(5)求一阶、二阶、三阶微分,可解得:

$$A_1 = -\frac{q}{k}, A_2 = \frac{(\alpha - \beta)q}{(\alpha + \beta)k},$$

$$B_1 = \frac{\alpha^2 q}{\beta^2 k}, B_2 = \frac{\alpha^2(\beta - \alpha)q}{\beta^2(\alpha + \beta)k}$$

$$z_0 = \frac{q}{k_\mathrm{c}} + \frac{\alpha^2 q}{\beta^2 k}$$

充填区域顶板地基梁位移方程为:

$$z = \frac{q}{k}\left[\mathrm{e}^{-\alpha x}\left(\cos\alpha x + \frac{\alpha - \beta}{\alpha + \beta}\sin\alpha x\right) + \frac{\alpha^2 + \beta^2}{\beta^2}\right] + \frac{q}{k_\mathrm{c}} \tag{6}$$

支架控顶区顶板地基梁的位移方程为:

$$z_\mathrm{c} = \frac{\alpha^2 q}{\beta^2 k}\mathrm{e}^{\beta x}\left(\cos\beta x + \frac{\beta - \alpha}{\alpha + \beta}\sin\beta x\right) + \frac{q}{k_\mathrm{c}} \tag{7}$$

充填体的支撑力为：

$$F = q[1 - \mathrm{e}^{-\alpha x}(\frac{\beta-\alpha}{\alpha+\beta}\sin \alpha x + \cos \alpha x)] \tag{8}$$

充填体的弹性地基系数 k 可按照以下公式进行计算：

$$k = \frac{E_0}{(1-\mu^2)h} \tag{9}$$

式中　E_0——充填体压缩模量，MPa；

　　　μ——充填材料泊松比；

　　　h——充填体高度，m。

支架的弹性系数 k_c 表示压缩支架 1 mm 的工作阻力变化量[10]，其计算公式为：

$$k_c = \frac{F_k - F_0}{1\,000(l_0 - l_k)S} \tag{10}$$

式中　F_k——安全阀开启前压力，kN；

　　　F_0——初始加载力，kN；

　　　l_0,l_k——分别为初始加载时和安全阀开启前的顶板与底座距离，mm；

　　　S——支架控顶面积，m²。

1.2　长壁充填开采最大充填步距研究

顶板在覆岩压力和自重作用下下沉移动，随着充填材料从初凝到终凝的过度，承压能力逐渐加强，顶板位移下沉幅度逐步减小并趋于稳定，在这过程中，顶板弹性地基梁可简化成一边由工作面前方煤壁支撑，一边为可移动的悬臂梁，此时，由于充填体下沉位移呈线性连续增大，对顶板支撑力基本为三角形形态增加，在充填体支撑力及顶板均布载荷上下合力作用下，顶板岩梁荷载可等效为覆岩与充填体综合作用下的三角形分布荷载 $q_{综}$（图2），顶板弯矩最大值为：

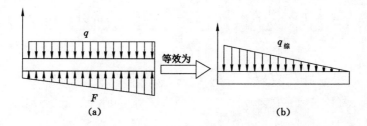

图 2　应力等效图

Fig. 2　Stress equivalent diagram

$$M_{\max} = \frac{1}{6}q_{综} L^2 \tag{11}$$

L——顶板岩梁最大跨距，m。

在顶板岩梁中的应力由顶板弯矩产生，即 $\sigma = \frac{M}{W_0}$，其中单位宽度矩形截面梁截面系数 $W_0 = \frac{h^2}{6}$，当顶板岩梁的正应力达到抗拉强度极限 $\sigma_{\max} = R_T$ 时，最大弯矩为：

$$M_{\max} = \frac{1}{6} q_{综} L^2 = \frac{1}{6} R_T h^2 \tag{12}$$

顶板岩梁挠度方程为：

$$\omega = \frac{q_{综} x^2}{120 EIL} \times (10L^3 - 10L^2 x + 5Lx^2 - x^3) \tag{13}$$

最大挠度值：

$$\omega_{\max} = \frac{q_{综} L^4}{30 EI} \tag{14}$$

式中 L——顶板岩梁最大跨距。

顶板覆岩达到平衡稳定后，顶板最大位移由式(6)可得：

$$z_{\max} = \frac{q}{k} \cdot \frac{\alpha^2 + \beta^2}{\beta^2} + \frac{q}{k_c} \tag{15}$$

顶板位移即为挠度值，则：

$$\omega_{\max} = z_{\max} = \frac{q_{综} L^4}{30 EI} = \frac{q}{k} \cdot \frac{\alpha^2 + \beta^2}{\beta^2} + \frac{q}{k_c} \tag{16}$$

惯性矩 $I = \frac{h^3}{12}$，由式(12)和式(16)可推导得出顶板岩梁最大跨距 L 为：

$$L = \sqrt{\frac{5Ehq[(k+k_c)\beta^2 + k_c \alpha^2]}{2 R_T \beta^2 k k_c}} \tag{17}$$

在支架和充填体协同支撑作用下顶板的弯矩方程为：

$$M = \frac{1}{2} q L^2 - \frac{1}{2} F_c l^2 - \int_f^L F(x) x \mathrm{d}x \tag{18}$$

式中，$f = u + l$，u 为充填步距，l 为支架控顶距。

当顶板弯矩值达到最大值时，将最大弯矩式(12)、充填体支撑力式(8)和位移方程式(15)带入式(18)可得：

$$\frac{1}{6} R_T h^2 = \frac{1}{2} q L^2 - \frac{1}{2} F_c l^2 - q \int_f^L [1 - \mathrm{e}^{-\alpha x}(\frac{\beta - \alpha}{\beta + \alpha}\sin \alpha x + \cos \alpha x)] x \mathrm{d}x \tag{19}$$

解方程得：

$$f(L) - f(f) = (\frac{1}{2} q L^2 - \frac{1}{6} R_T h^2 - \frac{1}{2} F_c l^2)/q \tag{20}$$

其中：

$$f(L) = \frac{1}{2} L^2 - \frac{L}{\alpha(\alpha + \beta)} \mathrm{e}^{-\alpha L}(\alpha \sin \alpha L - \beta \cos \alpha L) + [\frac{\beta - \alpha}{2\alpha^2(\alpha + \beta)}\cos \alpha L - \frac{1}{2\alpha^2}\sin \alpha L] \mathrm{e}^{-\alpha L} f(f)$$

$$= \frac{1}{2} f^2 - \frac{f}{\alpha(\alpha + \beta)} \mathrm{e}^{-\alpha f}(\alpha \sin \alpha f - \beta \cos \alpha f) + [\frac{\beta - \alpha}{2\alpha^2(\alpha + \beta)}\cos \alpha f - \frac{1}{2\alpha^2}\sin \alpha f] \mathrm{e}^{-\alpha f}$$

按照公式(17)可求出顶板在充填体的支撑作用下的跨距，根据式(20)充填工作面顶板跨距与充填步距之间的关系式，可求得 f 值，支架控顶距长度 l 为定值，进而求得最大充填步距 u 值。

2 实例应用分析

2.1 工程背景

陕西某煤矿 2307 工作面位于该矿井的西南部，主采 3 号煤层位于侏罗系中侏罗统延安组,平均埋深 160 m,采厚 3.7 m,煤层顶板岩性为砂岩、粉砂岩、泥质砂岩和泥岩,直接顶板厚 3 m,煤层倾角为 1°,该区域地质构造简单,地面为风积沙,属于半沙漠丘陵地貌,生态系统比较脆弱。为了控制地表沉陷,保持含水层不被破坏,采用综合机械化长壁膏体充填开采工艺进行回采,工作面长 1 150 m,宽 160 m,充填材料为风积沙、粉煤灰、水泥和辅料制成的膏体,在地面制作成膏体料浆,通过管路泵送至井下采空区,充填材料全部充填占据采空区空间。

2.2 最大充填步距的计算

根据对充填体力学参数的测定,初凝时间为 12 h,初凝强度为 0.3 MPa,终凝时间约为 28 d,终凝强度为 7.5 MPa,内摩擦角为 30.9°,黏聚力 $c=0.79$ MPa,充填体压缩模量 E_0 通过压实实验来求取,从初凝到终凝不同龄期的压缩模量为 $5.85\sim17.72$ MPa,泊松比为 $0.18\sim0.22$,充填体平均弹性系数 $k=3.6$ MPa/m;工作面顶板覆岩平均重度 25 kN/m³,采深 $H=160$ m,顶板荷载为 4 MPa,顶板弹性模量 $E=18$ GPa,顶板岩梁最大拉应力 $R_T=15$ MPa;支架支撑力为 1.2 MPa,支护长度为 $l=4$ m,支架弹性系数为 20.3 MPa/m。带入以上推导所得公式,经计算可得,该条件下长壁充填开采顶板最大跨距为 141.47 m,最大充填步距为 9.6 m。

2.3 充填步距对开采效率的影响分析

长壁充填开采是采煤和充填交替进行,充填步距的长短影响到采煤机的割煤时间和开采效率,采煤机每刀 0.8 m 的截深,若充填步距设为 4.8 m,工作面推进 9.6 m 时,采煤机需要开机工作约 12 h,2 个步距的充填体初凝共需要 24 h;而如果充填步距从 4.8 m 提高到 9.6 m,采煤机工作时间为 12 h,凝固时间为 12 h,总时间为 24 h,开采充填共节省 1/3 左右的时间,考虑检修设备和充填时间,可总体提高充填开采效率约 20%。因此,提高充填步距有利于提高开采效率。

3 充填步距对顶板位移影响的数值模拟计算分析

不同的充填步距,充填前后顶板覆岩的位移和采场空间的压缩量不同。以上述工程实例为背景,采用数值模拟计算方法分析不同充填步距条件下,顶板的位移值以及对地面沉陷的影响。数值模拟计算分析是岩土工程和采矿工程研究的有效手段,通过数值模拟模型的建立,可直观清晰的研究分析不同充填步距开采充填过程中,顶板覆岩的位移变化规律和力学演化特性。

3.1 数值分析模型建立

计算分析的模型尺寸为 360 m×10 m×190 m,模拟煤层埋深为 160 m,煤层厚度 3.7 m,模型底部留设 30 m 作为底部边界,模型的两侧各留 60 m 的煤柱,以降低边界区域的影响,其中 60~300 m 范围为开采充填区域,煤层开采厚度为 3.7 m,数值分析模型如图 3 所示。模型共 2 880 个单元,6 050 个节点。

模型中各煤岩层及充填体的力学参数以实际岩块实验室测定结果为基础,考虑结合矿井地质构造及物理力学参数进行赋值计算选用的物理力学参数见表 1。

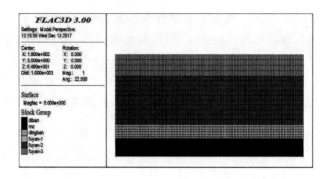

图 3　计算模型

Fig. 3　Calculation model

表 1　　　　　　　　　　岩体力学参数

Table 1　　　　　　　　Rock mechanics parameters

序　号	密度/(kg/m³)	弹性模量/GPa	泊松比	抗拉强度/MPa	黏结力/MPa	内摩擦角/(°)
松散层	1 970	1.26	0.35	0.03	0.08	28
砂质黏土	2 000	0.3	0.23	0.20	0.25	22
细砂岩	2 300	4.3	0.22	5.2	7.0	29
砂质泥岩	2 150	3.5	0.28	1.6	2.0	27
煤	1 400	1.5	0.3	0.65	0.8	22
砂岩	2 200	2.8	0.24	1.45	1.58	30
初凝充填体	1 550	1.3	0.32	0.45	0.6	19
过渡充填体	1 600	1.45	0.30	0.80	0.94	22
终凝充填体	1 700	1.60	0.29	1.35	1.67	25

计算模型边界条件确定如下：

① 模型前后和左右边界施加水平约束,采用等效应力来添加约束。

② 模型底部边界固定,即底部边界水平、垂直位移均为零。

③ 模型顶部为自由边界,全煤岩层建模,没有等效载荷。

3.2　模拟分析方案

模拟计算方案的充填步距分别设置为 4.0 m、6 m、8 m、9.6 m。首先按照设定充填步距进行开挖,模型开挖后计算运行的时间和时步与充填步距设置的长短密切相关,4.0 m 步距计算 300 时步,6 m 步距计算 450 时步,8 m 步距计算 600 时步,9.6 m 步距计算 750 时步。然后通过改变材料属性,对采空区进行充填,再计算 300 步,视为充填体凝固时间,按照设置的充填步距,逐步向前推进开挖,开采充填交替进行。由于充填体随着凝固时间的不同而强度不同,在变换充填体材料时,要随着工作面的向前推进,分别按照不同时间采用不同的材料参数替代煤层参数,逐步加强充填体强度,最终模型计算运行至平衡。在模型中的煤层顶板和地表设置两条岩层移动测线,开采推进过程中,通过 FLAC 中的 FISH 语言程序分别求取记录控顶区最大垂直位移以及充填区最大位移。

3.3 模拟结果分析

从模拟结果中分别取各充填步距的垂直位移图进行分析(图4)。

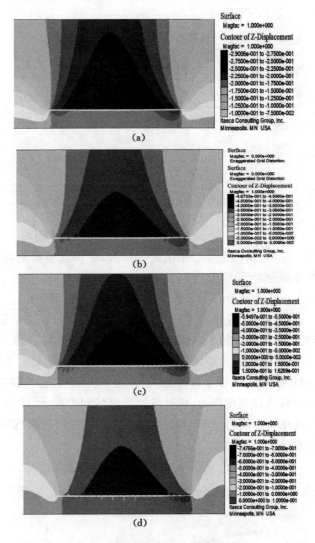

图 4 不同充填步距垂直位移图

Fig.4 Vertical displacement of different filling steps

(a) 4.0 m 充填步距的垂直位移图；(b) 6 m 充填步距的垂直位移图；
(c) 8 m 充填步距的垂直位移图；(d) 9.6 m 充填步距的垂直位移图

垂直位移图显示,随着充填体凝固后发挥支撑作用,顶板下沉受到抑制,顶板沉降位移幅度逐渐减小,在模型的边界区域受原煤岩体的影响作用,位移相对较小,充填区域中间垂直位移最大。垂直位移云图显示,充填步距从 4.0 m 到 9.6 m 变化过程中,顶板的最大下沉值分别为 289 mm、467 mm、590 mm 和 747 mm,地面最大沉降值分别为 244 mm、371 mm、451 mm 和 558 mm。不同充填步距控顶区的最大垂直位移随步距变化的曲线如图 5 所示。

充填步距从 4.0~9.6 m 的变化过程中,充填前及充填体未凝固承压时,采空区顶板最

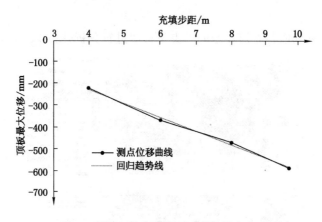

图 5 不同充填步距控顶区顶板最大位移

Fig. 5 Maximum displacement of roof at different filling steps

大位移分别为 219 mm、363 mm、466 mm 和 578 mm，9.6 m 的步距与 4.0 m 步距相比，顶板最大下沉值增加了 2.64 倍，采用最小二乘法进行拟合，得到充填步距与控顶区最大位移之间的回归曲线图，充填步距与下沉值之间的回归方程为：

$$y = -60.674x + 21.255 \quad (R^2 = 0.99) \tag{21}$$

回归方程显示，下沉位移与充填步距长度基本呈线性相关，相关系数 R^2 为 0.999，顶板最大下沉位移值与充填步距线性回归曲线相关性良好。

从计算结果的垂直位移云图中取顶板和地表沉降位移值，绘制出不同充填步距的顶板位移曲线和地表沉陷位移曲线进行分析（如图 6 和图 7 所示）。各位移曲线显示，在工作面开采初期，顶板垂直位移增加幅度较大，在开采边界区域斜率较大，中间区域由于充填体的作用，在工作面开采充填交替进行过程中，充填体的强度逐渐增强，位移过渡较平缓，达到终凝强度后，顶板位移最终趋于相对稳定。

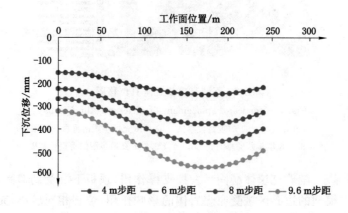

图 6 不同充填步距地表测线位移

Fig. 6 Surface displacement of different filling step distance

不同充填步距的地表位移曲线形态相似，地表沉陷整体呈现较平缓，不同步距的下沉量值存在较大差异，随着充填步距的加大，地面下沉位移逐渐加大，因此，充填步距的不同，可

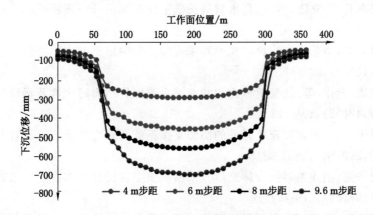

图 7 不同充填步距顶板测线位移

Fig. 7 Roof displacement of different filling steps

以反映到地面的位移不同,在该地质采矿及参数模拟方案条件下,顶板的最终下沉位移值中,充填前空顶期及充填体未承压时的顶板位移在顶板最终沉降值的占比较大,说明增大充填步距影响到覆岩及地表沉陷位移。因此,合理设置充填步距对于控制充填开采沉陷效果及提高开采效率至关重要。

4 结 论

(1)根据长壁充填开采工作面的支架和充填体协同支撑体系特征,建立了长壁充填开采顶板弹性地基梁力学模型,推导得出了顶板不同区域的位移方程和支撑力方程。

(2)以长壁充填开采顶板弹性地基梁力学模型的基础,考虑充填体强度线性增强的力学性能和覆岩均布荷载,把顶板岩梁力学模型进行了等效转化,推导得出了顶板的最大跨距,进而求取得出了顶板最大跨距与控顶距、充填步距的数学关系式,推导得出了最大充填步距计算公式。

(3)通过长壁充填开采工程实例分析,应用分析研究推导公式,计算得到了该条件下的最大充填步距为 9.6 m,同时分析了不同充填步距可影响到开采效率。

(4)通过数值模拟计算分析了不同充填步距顶板及地表的位移特征,回归分析得到充填步距与顶板最大下沉值之间线性相关,不同的充填步距可影响到地表沉陷位移,空顶期及充填体未承压前的顶板位移较大。因此,充填步距的合理设置对于充填开采控制开采沉陷及提高开采效率至关重要。

参考文献

[1] 闫少宏,张华兴. 我国目前煤矿充填开采技术现状[J]. 煤矿开采,2008,82(03):1-3+10.

[2] 缪协兴,钱鸣高. 中国煤炭资源绿色开采研究现状与展望[J]. 采矿与安全工程学报,2009,26(1):1-14.

[3] 胡炳南. 我国煤矿充填开采技术及其发展趋势[J]. 煤炭科学技术,2012,456(11):1-5+18.

[4] 许猛堂,张东升,马立强,等.超高水材料长壁工作面充填开采顶板控制技术[J].煤炭学报,2014,39(03):410-416.
[5] 刘建功,赵家巍,李蒙蒙,等.煤矿充填开采连续曲形梁形成与岩层控制理论[J].煤炭学报,2016,41(02):383-391.
[6] 孙强,张吉雄,殷伟,等.长壁机械化掘巷充填采煤围岩结构稳定性及运移规律[J].煤炭学报,2017,42(02):404-412.
[7] 余伟健,冯涛,王卫军,等.充填开采的协作支撑系统及其力学特征[J].岩石力学与工程学报,2012,262(S1):2803-2813.
[8] 常庆粮,周华强,柏建彪,等.膏体充填开采覆岩稳定性研究与实践[J].采矿与安全工程学报,2011,97(02):279-282.
[9] 陈绍杰,郭惟嘉,周辉,等.条带煤柱膏体充填开采覆岩结构模型及运动规律[J].煤炭学报,2011,202(07):1081-1086.
[10] 徐刚.采场支架刚度实验室测试及与顶板下沉量的关系[J].煤炭学报,2015(07):1485-1490.

近水平厚煤层沿空巷道位置选择及支护技术研究

王志强,苏越,苏泽华

(中国矿业大学(北京)资源与安全工程学院,北京,100083)

摘　要:针对近水平厚煤层区段间留设煤柱尺寸较大造成的资源浪费和回采巷道沿底板布置造成的支护困难等问题,结合察哈素煤矿工程背景和实践经验,运用极限平衡理论计算了下区段进风巷顶煤的弹塑性变形区,为沿空掘巷的位置提供了选择范围。并采用注水测漏法对下区段进风巷顶煤破裂区范围进行验证,最终确定下区段进风巷的合理位置。结合实际情况,设计了区段间相邻巷道联合支护方案。采用FLAC3D5.01数值模拟软件对沿空掘巷位置和联合支护方案效果进行模拟。最终结果表明,相较于综放开采沿底板布置回采巷道,采用错层位外错式布置下区段回采巷道并采用区段间相邻巷道联合支护方案进行巷道支护可以有效减小塑性区的范围,保证了沿空掘巷的稳定性。其研究方法和结论可以为类似工程背景提供一定的参考。

关键词:沿空掘巷;错层位;联合支护;数值模拟

Study on location choice and support technology of roadway along gob in near horizontal thick coal seam

WANG Zhiqiang, SU Yue, SU Zehua

(College of Resources and Safety Engineering, China University of
Mining and Technology(Beijing), Beijing 100083, China)

Abstract: In order to solve the problem of resource waste because of the large size of the coal pillar between the sections of the horizontal thick coal seam and the difficulty of supporting of the roadway setting along the bottom, this paper calculates the elastic-plastic deformation zone of the upper coal by using the limit equilibrium theory Combining with the background and practical experience of Chahasu coal mine to provide selection range for the location of the air intake of next section. In this paper, water injection method is used to verify the range of the upper coal fracture zone in the lower section, and finally the reasonable position of the wind tunnel is determined. According to the actual engineering situation, a new support scheme is designed. And the FLAC3D 5.01 numerical simulation software is used to simulate the roadway layout scheme and the combined support scheme. The final result shows that compared with the traditional roadway layout method, there is an obvious improvement in the split-level mining method with the adoption of joint support schemes for adjacent roadways between sections. It effectively reduces the area of the plastic zone

基金项目:国家自然科学基金面上项目(51774289);中央高校基本科研业务费专项资金项目(2011QZ06)。
作者简介:王志强(1980—),男,内蒙古呼伦贝尔人,副教授。Tel:13810796225,E-mail:wzhiqianglhm@126.com。
通信作者:苏越(1994—),男,河北邯郸人,现就读于中国矿业大学(北京)。Tel:18811798598。E-mail:1163648570@qq.com。

and ensures the stability of roadway driving along gob. The research methods and conclusions can provide some references for similar engineering background.

Keywords: gob-side entry driving; split-level mining; combined support; numerical simulation

我国煤炭储量中厚煤层(即煤厚≥3.5 m)的储量约占全部煤炭储量的44%,且其产量占煤炭总产量的45%左右。且目前我国的千万吨级矿井均以厚煤层开采为主[1-2]。

而我国厚煤层开采的主要方法有分层开采,放顶煤开采和大采高开采[3-4]。虽然综放开采具有高产高效、生产集中、成本低、对地质条件的适应性强等显著优势,但是同时也存在着许多技术难题。其中放顶煤采煤法的回采率,瓦斯,发火和粉尘等问题一直难以得到有效解决[5]。同时也存在着区段间留设煤柱尺寸过大而造成矿井煤炭回采率低的问题。随着我国对煤炭开采的强度增加,我国煤炭资源可采储量日益衰减,如何有效地提高煤炭的采出率从而延长矿井服务年限,早已成为煤炭行业科技工作者共同研究的重要课题。

综采放顶煤工作面一般沿煤层底板掘进回采巷道,以煤层为顶板,顶煤一般强度比较低、松软破碎、稳定性差,故其沿底板巷道的支护受到地质与回采技术条件的限制。且在开采过程中,受采动应力和原岩应力的双重影响,留设区段煤柱内会出现应力集中,使得煤柱煤体发生屈服软化,进而可能导致回采巷道垮塌,无形之中增加了巷道维护的难度和成本。因此,工作面回采巷道的支护在综放开采时也显得极为重要。

针对察哈素煤矿某工作面现场地质条件,拟采用错层位外错式布置形式对该采区综放工作面巷道布置位置进行优化并采用区段间相邻巷道联合支护技术对回采巷道进行支护方案优化,以期在提高回采率的同时,降低其支护成本,为厚煤层回采巷道优化设计及支护提供一定的参考。

1 错层位巷道布置法[6]

为了解决综放开采存在的回采率低等问题,厚煤层错层位巷道布置采全厚采煤法于20世纪90年代被提出。该方法[7]保留了放顶煤开采的高产高效优势,同时借鉴了分层开采的经验,可以实现无煤柱开采,降低了巷道掘进、维护的难度与成本,提高了采区采出率;在安全生产方面,减少了端头丢煤,降低了瓦斯积聚和发火危险[8]。

在空间形式上,错层位巷道布置改变了传统厚煤层采煤方法两巷布置在煤层同一层位的现状,其两工作面之间的相邻巷道具有"一高一低"立体化空间关系,如图1所示。

2 沿空掘巷合理位置设计

在研究沿空掘巷技术时,留设合理宽度的护巷窄煤柱不仅关系到综放沿空巷道围岩结构的整体稳定性,而且适当的煤柱宽度可以使综放沿空巷道处于较低的围岩应力环境下,有效避免冲击矿压等动力灾害。因此,掌握采场采空侧煤体内侧向支承压力的分布规律是合理设计护巷窄煤柱宽度及选择沿空掘巷位置的基础。已有很多学者对沿空掘巷煤柱留设做出了研究[9-14]。

在错层位外错式巷道布置情况下,煤层顶煤的应力分布状态如图2所示,可以分为破碎区Ⅰ,塑性区Ⅱ,弹性区应力升高部分Ⅲ以及原岩应力区Ⅳ。外错式布置方式也属于沿空掘巷的一种,与现有技术形式上的区别在于煤柱两侧的回采巷道位于煤层中的不同层位,也即

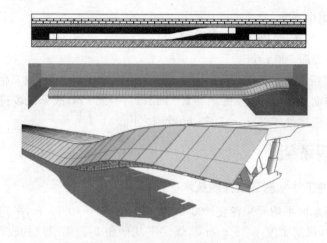

图 1 错层位外错式巷道布置及三维示意图

Fig. 1 External-misaligned stagger arrangement of roadway layout and three-dimensional schematic diagram

两巷之间有一定的高差,通过现有公式估算 x_1 和 x_2 确定破碎区和塑形区的位置,拟将下区段巷道位置控制在 x_1 和 x_2 之间,此时下区段进风巷顶板的锚杆锚索可以固定在具有一定强度的煤体中,并且下区段进风巷的顶煤也处于一个应力较小的位置,有利于下区段进风巷的围岩控制。

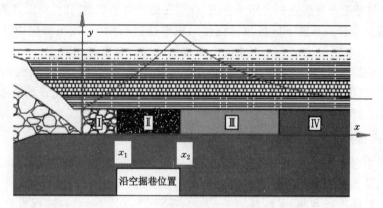

图 2 顶煤应力分区示意图

Fig. 2 Schematic diagram of stress partition

经典的沿空掘巷塑性区和破裂区计算公式如下[15]:

$$x_1 = \frac{mA}{2\tan \varphi_0} \ln\left(\frac{\gamma H + C_0/\tan \varphi_0}{C_0/\tan \varphi_0 + P_x/A}\right) \quad (1)$$

$$x_2 = \frac{mA}{2\tan \varphi_0} \ln\left(\frac{k\gamma H + C_0/\tan \varphi_0}{C_0/\tan \varphi_0 + P_x/A}\right) \quad (2)$$

式中 m——煤层采厚,取 3.5 m;

A——侧压系数,$A = \mu/(1-\mu)$,μ 为泊松比,取 $A = 0.3$;

φ_0——煤层界面的内摩擦角,取值为 29°;

C_0——煤层界面的黏聚力,1.35 MPa;

k——应力集中系数,取 2.5;

γ——岩层平均重度,25 kN/m³;

H——巷道埋深 800 m;

P_x——对煤帮的支护阻力,因上区段巷道采用锚杆锚索支护,取其值为 0.25 MPa。

结合综放工作面数据,计算得到 $x_1=2.34$ m,$x_2=7.58$ m,故下区段进风巷与上区段回风巷水平外错距离应该处于 2.34~7.58 m 范围之内。

3 工程背景与现场实测

3.1 工程背景及井下注水测漏法探测技术

察哈素煤矿某采区平均开采深度 800 m,煤层厚 8.50~10.03 m,平均厚度为 9.03 m,普氏系数 $f=1.59$,煤层密度 1.36 g/cm³,煤层平均倾角 2°。煤层结构较复杂,煤层底部含 0~2 层厚 0.15~0.25 m 的夹矸,夹矸岩性为砂质泥岩、泥岩。煤层厚度总体变化较小,局部有波状起伏。煤层直接顶为粉砂岩,厚 19.87 m,单轴抗压强度 32.13 MPa;基本顶为细砂岩,厚 4.2~4.5 m,单轴抗压强度 144.85 MPa,整体性强;伪底为泥岩,厚 1.45 m;直接底为粉砂岩,裂隙发育,单轴抗压强度 70.91~117.49 MPa;基本底为细砂岩,厚 3.35 m。

该采区左翼布置的 301 工作面长 2 800 m,宽 220 m,采用传统方式沿底布置回采巷道,割煤高度 3.5 m,放煤高度 5.53 m,由于实际生产中,巷道变形量大且难以维护,受采动影响,锚杆破断失效情况较普遍,且原计划留设 20 m 煤柱会造成严重的资源浪费。从采区 302 工作面开始采用错层位巷道布置形式布置回采巷道,以期在提高回采率的同时,降低其支护成本。

为了对错层位巷道布置下区段顶部煤体的完整性进行划分,找出低应力区内煤体较完整区;采用仰孔注水测漏法对 301 工作面实体煤侧顶煤破裂区进行测定。

井下注水测漏法实质是在井下工作面的回采巷道或硐室中向需要探测的地方打设钻孔,钻孔穿过预计的破碎区范围,以确定确切的破碎区宽度。采用钻孔双端堵水器对钻孔进行逐段封隔注水,测定各孔段漏失量变化情况,以此探测确定实体煤侧破坏规律。

3.2 观测工程布置方案及测量结论

观测工程布置在本工作面回风巷道实体煤侧帮。根据观测设计方案,现场进行了从准备、钻窝施工、钻孔施工与钻孔注水观测的全部工作。

钻孔成孔后即进行观测,采前观测 1~2 次,采后 1~2 次观测,且进行检验,对部分有疑问的观测结果,可进行部分或全孔段的重复观测。按上行方式观测,观测段 0.5~1.0 m,从孔口处直到孔底,不允许空段或"跳跃式"观测。工作自 2017 年 2 月 10 日开始,到 2017 年 4 月 20 日结束,历经 69 d,圆满完成了综放工作面实体煤侧破裂区的观测研究任务,获得了必要的基础资料。实际钻孔情况如表 1 所示。

表 1 实际钻孔情况

Table 1　Actual drilling conditions

钻窝号	孔号	孔性	孔径/mm	孔深/m	观测时间
1	1	采前孔	ϕ89	30	2017-03-25
1	2	采后孔	ϕ89	30	2017-04-19

由图3知,通过对钻孔1,2注水漏失量进行对比,综放工作面实体煤侧0~2.4 m的范围内注水漏失量较大,分析认为该区域为实体煤侧破碎较严重区域(破碎区);2.4~7.8 m 范围内注水漏失量较小,认为该区域为实体煤侧的塑性区,该区域内煤体较为完整且存在承载较小的区域;7.8~30 m范围内注水漏失量较前两个区域相比明显减小,认为该区域为煤体完整区,注水漏失仍然存在主要由于煤体内本身存在的裂隙所致。

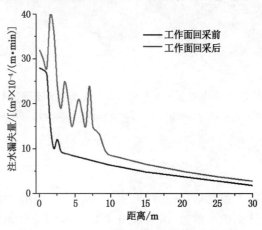

图3 注水漏失曲线图
Fig.3 Water leakage curve diagram

下区段进风巷的顶煤应处于低应力区内煤体较完整区,结合理论计算与现场实测的结果,拟确定错层位外错式巷道布置形式布置下区段进风巷。下区段进风巷(304工作面进风巷)沿煤层底板布置,其与上区段回风巷(302工作面回风巷)水平外错距离为2.5 m。

4 支护方法拟定

4.1 错层位立体化相邻巷道联合支护技术机理分析

在特厚煤层中采用错层位外错式沿空掘巷技术在巷道围岩稳定特征、承载方面均出现了显著的变化,且沿空掘巷的顶部为上一工作面回风巷实体煤侧的破碎区或塑性区;靠采空区一侧的帮部为处于上区段遗留下的三角煤体。利用错层位外错式巷道布置时,上下两巷具有"一高一低"的立体化空间关系,可以实现上区段工作面回风巷与接续工作面沿空掘巷二者之间的"共同支护"效果,有利于下区段沿空巷道的稳定性维护。

如图4所示,在采用错层位外错式沿空掘巷技术时,上区段回风巷道沿煤层顶板布置,锚杆、索可以打入顶部稳定的岩层内,更易发挥悬吊作用;同时,通过向回风巷道实体煤侧巷帮打入锚杆、索,甚至注浆,对联合锚固区进行加固,控制采动影响下的实体煤内破碎区与塑性区的扩展,提高二者的残余强度,为沿空掘巷顶板支护创造条件。下区段进风巷通过向顶部打锚杆、索,可以深入煤层顶部稳定岩层内进行锚固,其锚杆、索可与上区段回风巷道实体煤侧巷帮支护体之间形成一个联合锚固区,可充分发挥巷道顶部锚杆、索的支护作用。

错层位外错式沿空掘巷技术的应用,理论上具有巷道围岩稳定、承载小、利于巷道支护等优点且实现了煤柱两侧相邻巷道的联合支护,可能解决现有沿空掘巷技术存在的围岩稳定状态与承载大小二者之间的矛盾、沿底巷道顶部支护难以找到有效锚固点与仅考虑单条

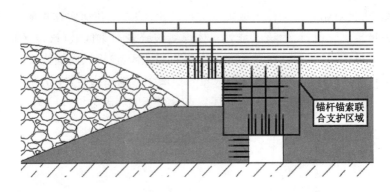

图 4 错层位外错式相邻巷道立体化的联合支护技术

Fig. 4 Combined support technology for adjacent roadways between sections

巷道支护设计几个方面的问题。

4.2 支护方案

结合 301 工作面实际工程背景,其两回采巷道断面均为矩形,沿煤层底板布置,巷道尺寸为 4.5 m×3.2 m,设计支护参数为:顶板采用 ϕ22 mm×2 500 mm 左旋螺纹钢高强锚杆 5 根,间、排距 850×800 mm,锚索采用 ϕ18.9 mm×6 300 mm 左旋钢绞线,间、排距 2 400 mm×800 mm;帮部安设锚杆 4 根,间、排距 750 mm×800 mm,巷帮打设 ϕ18.9 mm×4 300 mm 高预应力锚索,间距 2 400 mm。

鉴于在实际生产中,巷道变形量大且难以维护,受采动影响,锚杆破断失效情况较普遍等问题,对 302 工作面巷道支护方案与参数进行优化,即将巷道顶部与帮部锚索更换为 ϕ18.9 mm×8 000 mm,安设下回风巷道顶板锚索 3 根,其与锚杆在巷道前进方向上错开 400 mm 进行布置。具体布置情况如图 5 和图 6 所示。

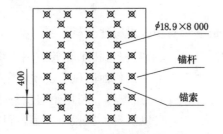

图 5 相邻巷道联合支护示意图

Fig. 5 Adjacent roadway joint support diagram

在研究巷道围岩支护时原岩应力场、采动应力场以及支护应力场相互作用,由于原岩应力场和采动应力场在数量级上比支护单元体形成的支护应力场大很多,会将支护应力场"覆盖",不利于观察支护单元体形成的支护应力场。因此为了验证优化后支护方案的支护效果,首先在零原岩应力模拟优化后支护方案的支护效果,如图 7 所示。

图 7 为区段间相邻巷道锚杆锚索联合支护方案对区段间煤柱的支护作用效果图。观察图 7(a)中的垂直应力分布云图,可以明显发现,区段间各巷道的锚杆,上区段回风巷的侧帮锚索,下区段进风巷的顶板锚索三者的支护压应力场相互叠加。首先在上区段回风巷顶板

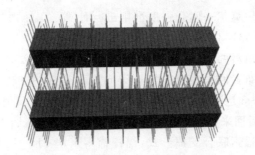

图 6 相邻巷道联合支护图
Fig. 6 Adjacent roadway joint support diagram

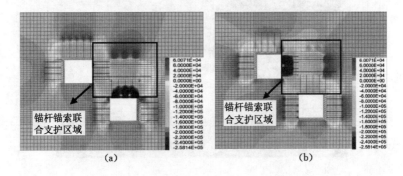

图 7 零原岩应力场联合支护效果图
Fig 7 Zero-rock stress field combined support effect
(a) 垂直应力云图；(b) 水平应力云图

附近形成了较为明显的压应力场(100 kPa)。其次，由于上区段回风巷侧帮锚杆和下区段进风巷顶板锚索的支护作用，从上区段回风巷实体煤侧帮至下区段进风巷上部顶煤的区域(联合锚固区)被明显控制，在其周围形成了 120 kPa 左右的支护压应力场。

观察图 7(b)，水平应力云图同样在锚杆锚索联合支护区域内出现了较大的应力集中现象，形成了支护压应力场。通过以上分析发现，认为采用本小节中的支护方案可以有效地控制区段间相邻巷道的围岩，尤其是对从上区段回风巷实体煤一侧到下区段进风巷上部的实体煤部分(锚杆锚索联合支护区域)起到良好的控制作用。

5 巷道布置方案分析

5.1 模型建立

为了避免人为设定网格生成对数值模拟效果的影响，拟借助韩国开发的优秀有限元软件 MIDAS/GTS 强大的前处理功能，进行几何建模和网格划分，尝试将节点文件以及单元文件通过数据接口转换成 FLAC 3D 可以识别的节点单元文件，从而完成模型从 MIDAS/GTS 到 FLAC 3D 的转换，提高数值模拟的效果。

模型尺寸为 500 m(长)×300 m(宽)×100 m(高)。在模型顶部均施加 800×0.025 MN/m³＝20 MPa 的应力用来模拟未建模的上覆岩层重力。在模拟底部约束纵向和横向位移。由于模型埋深较大，需要考虑水平应力的影响，在模型的前后左右边界施加水平应力并限制横向位移。模型采用莫尔—库仑强度准则进行计算。工作面长 220 m,两回采巷道断面均为矩形,尺寸为 4.5 m×3.5 m。

为了更加明显有效地对两种方案进行对比,采用控制变量法进行建模。在模型(a)中,按 301 工作面实际情况布置回采巷道并留设 20 m 区段煤柱,并按实际 301 工作面回采巷道支护方案对回采巷道 1~3 进行支护;在模型(b)中,按该矿 302 工作面巷道布置形式布置回采巷道 1~3,并按优化后回采巷道支护方案对回采巷道 1~3 进行支护。

巷道支护采用锚杆锚索联合支护形式,锚杆锚索采用 FLAC 3D 内置单元体锚杆锚索(cable)进行模拟,煤岩与支护体参数如表 2 所示,模型如图 8 所示。

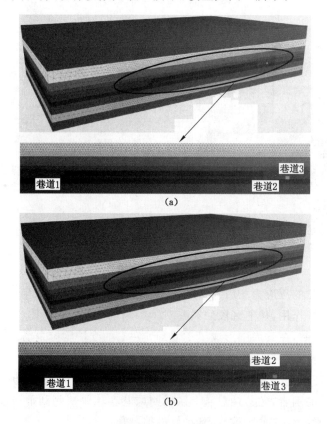

图 8　数值模拟模型

Fig. 8　Numerical simulation model

(a) 传统沿煤层底板布置回采巷道;(b) 错层位外错式布置回采巷道

表 2　各岩层力学参数

Table 2　Rock mechanical parameters

序号	岩层	体积模量 /GPa	剪切模量 /GPa	密度 /(kg/m³)	内摩擦角 φ /(°)	内聚力 /MPa	抗拉强度 /GPa
1	细砂岩	15.80	11.30	2 700	42	5.50	2.40
2	黏土岩	12.40	8.10	2 600	35	3.50	1.80
3	细砂岩	15.80	11.30	2 700	42	5.50	2.40

续表 2

序号	岩层	体积模量/GPa	剪切模量/GPa	密度/(kg/m³)	内摩擦角 φ/(°)	内聚力/MPa	抗拉强度/GPa
4	粉砂岩	8.40	5.90	2 350	39	2.60	1.20
5	煤	11.90	7.10	1 400	29	1.35	0.68
6	粉砂岩	8.40	5.90	2 350	39	2.60	1.20
7	细砂岩	15.80	11.30	2 700	42	5.50	2.40
8	黏土岩	12.40	8.10	2 600	35	3.50	1.80
9	细砂岩	15.80	11.30	2 700	42	5.50	2.40

5.2 不同方案下巷道布置应力分布分析

图 9 所示为两种巷道布置形式垂直应力分布对比云图。

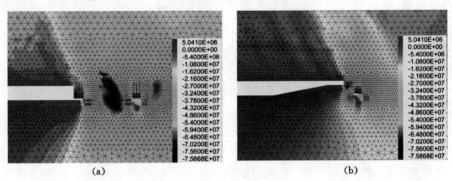

图 9 沿底巷道与错层位外错式开采应力分布对比云图

Fig. 9 Comparison of stress distribution of roadway driven along bottom and external-misaligned stagger arrangement of roadway driven along goaf

(a) 沿底布置回采巷道；(b) 错层位外错式沿空掘巷

从图 9(a)中发现,传统沿底布置下区段回采巷道时,在下区段进风巷两帮形成了不同程度的应力集中。虽然区段间留设 20 m 区段煤柱,但在区段煤柱侧出现了较大的应力集中,区段煤柱中垂直应力达到了 70 MPa 左右,应力集中系数 k 接近 3.5。

从图 9(b)中发现,因为下区段进风巷沿煤层底板布置,避开了由于上区段工作面回采所形成的应力集中区域,并且下区段进风巷距离上区段工作面较近,对上区段回采形成的侧向支承应力起到了卸压的作用。所以采用错层位外错式布置下区段进风巷时,仅在上区段回风巷与下区段进风巷之间出现了小范围应力集中现象,且应力集中系数较小。

对比回采巷道沿底布置和错层位外错式沿空掘巷开采应力分布云图,发现将回风平巷沿煤层顶板布置,有效降低了高支承应力峰值影响区域,待掘的底板沿空巷道整体位置均处于低应力值范围内,从承载方面证明了错层位外错式沿空掘巷的优势。

5.3 不同方案下巷道围岩稳定性分析

图 10 所示为两种方案下塑性破坏对比云图。

从图 10(a)中发现,传统沿底布置下区段回采巷道时,与图 9(a)中的垂直应力分布云图

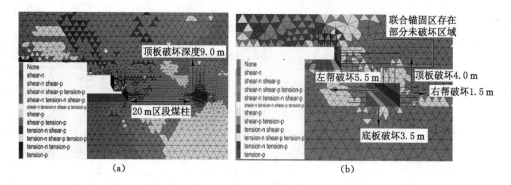

图 10 实际工程背景支护效果对比图

Fig.10 Comparison of actual engineering background support effect

(a) 沿底布置回采巷道；(b) 错层位外错式沿空掘巷

相对应,留设的 20 m 区段煤柱全部发生了剪切和拉伸破坏。下区段进风巷顶板破坏深度达到 9 m,导致顶板安设的 6.3 m 锚索失效,不能对巷道顶部围岩起到良好的控制作用。

从图 10(b) 中发现,采用错层位外错式布置下区段进风巷时,由于避开了应力集中区域,巷道围岩发生破坏的范围不大。巷道左帮由于处于应力较大位置,破坏深度为 5.5 m;巷道右帮破坏深度为 1.5 m;顶板破坏深度 4.0 m;底板破坏深度为 3.5 m。在图 9(b) 中的联合锚固区存在未破坏区域。

对比图 10 沿底布置和错层位外错式沿空掘巷塑性破坏云图。发现采用沿底巷道进行开采,接续工作面相邻巷道围岩几乎全部处于塑形破坏状态,在现有支护体规格的前提下,几乎无法控制破坏围岩破坏范围;而采用错层位外错式沿空掘巷及立体化巷道的联合支护技术,整个塑性区尺寸大幅度降低,且巷道顶部几乎保持稳定状态,而两帮仅仅出现较小范围的破坏,经过对比发现错层位外错式沿空掘巷相邻立体化巷道的联合支护技术在控制围岩稳定性方面具有显著的优势。

6 结 论

(1) 根据对厚煤层顶煤进行极限平衡计算及现场实测,得出区段间相邻巷道合理的水平错距为 2.5 m。

(2) 利用错层位外错式巷道布置空间关系的特点,采用区段间相邻巷道联合支护技术对下区段进风巷的支护方案进行设计。在零原岩应力场条件下进行数值模拟,体现了外错式布置条件下联合支护方案的支护效果。

(3) 通过模拟计算分析,发现采用错层位外错式沿空掘巷和联合支护技术显著降低了下区段进风巷的围岩破坏区。通过对新提出的巷道布置及支护方案和传统沿底板布置回采巷道的方案进行对比,肯定了错层位外错式巷道布置及联合支护方案的效果,为近水平厚煤层沿空掘巷提供了一定的参考。

参考文献

[1] 中华人民共和国国家统计局. 中华人民共和国 2016 年国民经济和社会发展统计公报[M]. 北京:统计出版社,2017.

[2] 张海洋.我国煤炭工业现状及可持续发展战略[J].煤炭科学技术,2014,42(S0):281-284.

ZHANG Haiyang. China Coal Industry Status and Sustainable Development Strategy[J]. Coal Science and Technology,2014,42(S0):281-284.

[3] 孟宪锐,吴昊天,王国斌.我国厚煤层采煤技术的发展及采煤方法的选择[J].煤炭工程,2014,46(10):43-47.

MENG Xianrui,WU Haotian,WANG Guobin. Development and Method Selection of Thick Coal Seam Mining Technology in China[J]. Coal Engineering,2014,46(10):43-47.

[4] 王家臣,仲淑姮.我国厚煤层开采技术现状及需要解决的关键问题[J].中国科技论文在线,2008,3(11):829-834.

WANG Jiachen, ZHONG Shuheng. The present status and the key issues to be resolved of thick seam mining technique in China[J]. science-paper online,2008,3(11):829-834.

[5] 赵景礼.错层位立体化巷道布置技术发展与展望[J].煤炭工程,2014,46(1):1-3.

ZHAO Jingli. Development and Outlook of 3-D Stagger Arrangement Roadway Layout Technology[J]. Coal Engineering,2014,46(1):1-3.

[6] 赵景礼,吴健.厚煤层错层位巷道布置采全厚采煤法[P].中国专利:ZL98100544.6,2002-01-23.

[7] 赵景礼.厚煤层错层位巷道布置采全厚采煤法的研究[J].煤炭学报,2014,29(2):142-145.

ZHAO Jingli. Research on full-seam mining adopted roadway layout of stagger arrangement in thick coal seam[J]. Journal of China Coal Society,2014,29(2):142-145.

[8] 李磊,柏建彪,王襄禹.综放沿空掘巷合理位置及控制技术[J].煤炭学报,2012,37(9):1564-1569.

LI Lei, BAI Jianbiao, WANG Xiangyu. Rational position and control technique of roadway driving along next goaf in fully mechanized top coal caving face[J]. Journal of China Coal Society,2012,37(9):1564-1569.

[9] 张科学,姜耀东,张正斌.大煤柱内沿空掘巷窄煤柱合理宽度的确定[J].采矿与安全工程学报,2014,31(2):255-263.

ZHANG Kexue, JIANG Yaodong, ZHANG Zhengbin. Determining the reasonable width of narrow pillar of roadway in gob entry driving in the large pillar[J]. Journal of Mining & Safety Engineering,2014,31(2):255-263.

[10] 祁方坤,周跃进,曹正正,等.综放沿空掘巷护巷窄煤柱留设宽度优化设计研究[J].采矿与安全工程学报,2016,33(3):475-480.

QI Fangkun, ZHOU Yuejin, CAO Zhengzheng, et al. Width optimization of narrow coal pillar of roadway driving along goaf in fully mechanized top coal caving face[J]. Journal of Mining & Safety Engineering,2016,33(3):475-480.

[11] 冯吉成,马念杰,赵志强,等.深井大采高工作面沿空掘巷窄煤柱宽度研究[J].采矿与安全工程学报,2014,31(4):580-586.
FENG Jicheng,MA Nianjie,ZHAO Zhiqiang,et al. Width of narrow coal pillar of roadway driving along goaf at large height mining face in deep mine[J]. Journal of Mining & Safety Engineering,2014,31(4):580-586.

[12] 张科学,张永杰,马振乾,等.沿空掘巷窄煤柱宽度确定[J].采矿与安全工程学报,2015,32(3):446-452.
ZHANG Kexue,ZHANG Yongjie,MA Zhenqian,et al. Determination of the narrow pillar width of gob-side entry driving[J]. Journal of Mining & Safety Engineering,2015,32(3):446-452.

[13] 张广超,何富连.大断面综放沿空巷道煤柱合理宽度与围岩控制[J].岩土力学,2014,37(6):1721-1729.
ZHANG Guangchao,HE Fulian. Pillar width determination and surrounding rocks control of gob-side entry with large crosssection and fully-mechanized mining[J]. Rock and Soil Mechanics,2014,37(6):1721-1729.

[14] 张广超,何富连,来永辉,等.高强度开采综放工作面区段煤柱合理宽度与控制技术[J].煤炭学报,2016,41(9):2188-2194.
ZHANG Guangchao,HE Fulian,LAI Yonghui,et al. Reasonable width and control technique of segment coal pillar with highintensity fully-mechanized caving mining[J]. Journal of China Coal Society,2016,41(9):2188-2194.

[15] 康红普,姜鹏飞,蔡嘉芳.锚杆支护应力场测试与分析[J].煤炭学报,2014,39(08):1521-1529.
KANG Hongpu,JIANG Pengfei,CAI Jiafang. Test and analysis on stress fields caused by rock bolting[J]. Journal of China Coal Society,2014,39(08):1521-1529.

[16] 刘金海,姜福兴,王乃国,等.深井特厚煤层综放工作面区段煤柱合理宽度研究[J].岩石力学与工程学报,2012,31(5):921-927.
LIU Jinhai,JIANG Fuxing,WANG Naiguo,et al. Research on reasonable width of segment pillar of fully mechanized caving face in extra-thick coal seam of deep shaft[J]. Chinese Journal of Rock Mechanics and Engineering,2012,31(5):921-927.

[17] 孔德中,王兆会,李小萌,等.大采高综放面区段煤柱合理留设研究[J].岩土力学,2014,35(增刊2):460-466.
KONG Dezhong,WANG Zhaohui,LI Xiaomeng,et al. Study of reasonable width of full-mechanized top-coal caving with large mining height[J]. Rock and Soil Mechanics,2014,35(Sup 2):460-466.

[18] 秦永洋,许少东,杨张杰.深井沿空掘巷煤柱合理宽度确定及支护参数优化[J].煤炭科学技术,2010,38(2):15-18.
QIN Yongyang,XU Shaodong,YANG Zhangjie. Rational coal pillar width determination and support parameter optimization for gateway driving along goaf in deep mine[J]. Coal Science and Technology,2010,38(2):15-18.

[19] 李学华,张农,侯朝炯. 综采放顶煤面沿空巷道合理位置确定[J]. 中国矿业大学学报,2000,29(2):186-189.
LI Xuehua, ZHANG Nong, Hou Chaojiong. Rational position determination of roadway driving along next goaf for fully-mechanized top coal caving mining[J]. Journal of China University of Mining & Technology, 2000,29(2):186-189.

沿空留巷可缩性墩柱破坏形态及加固分析

郭东明,凡龙飞,王晓烨,韩笑,刘杰

(中国矿业大学(北京)力学与建筑工程学院,北京,100083)

摘　要:基于梧桐庄矿墩柱式沿空留巷,为研究墩柱在竖向荷载作用下的力学性能,利用有限元软件 ANSYS 对墩柱在竖向荷载作用下的力学性能进行数值模拟,并将数值模拟结果与试验结果进行对比,验证了数值模拟中建模、相关参数和单元选取的正确性。根据此数值模型对不同直径、高度、搭接长度墩柱的力学性能进行模拟,绘制不同各因素作用下墩柱的荷载曲线。并针对其中部发生弯曲破坏、下端发生鼓曲破坏,提出了加劲肋的加固措施,并通过 ANSYS 对加固后的墩柱进行数值模拟。数值模拟结果表明:墩柱承载力随直径增加呈现线性关系增加,随搭接长度增大呈二次抛物线增大,随墩柱高度增大呈二次抛物线减小。通过加劲墩柱中部变形减少 75%,墩柱底部基本不发生变形。加劲肋能明显减少墩柱变形量,提高墩柱承载力。

关键词:可缩性墩柱;数值模拟;加劲肋;承载力

Analysis of failure forms and reinforcement of retractable pier columns in gob-side entry retaining

GUO Dongming, FAN Longfei, WANG Xiaoye, HAN Xiao, LIU Jie

(School of Mechanic & Civil Engineering, China University of Mining and Technology(Beijing), Beijing 100083, China)

Abstract: Based on the pier column gob-side entry retaining technology in Wutongzhuang mine, to study the mechanical properties of piers under vertical loads, the finite element software ANSYS was used to numerically simulate the mechanical properties of the piers under vertical loads, the simulation results are compared with the experimental results to verify the correctness of modeling, related parameters and unit selection in numerical simulation. According to the numerical model, the mechanical properties of the pier columns with different diameters, heights and joints are simulated, and the load curves of the piers under different factors are drawn. In this paper, the strengthening measures of stiffened rib are put forward, and a numerical simulation is carried out by ANSYS for the reinforcement. The results of numerical simulation show that the bearing capacity of the pier increases with the increase of the diameter, and the increase of the length of the pier increases with the increase of the second parabola, and the second parabola decreases with the height of the pier. The deformation of the central pier was reduced by 75%, and the bottom of the pier was not deformed. The strength can obviously reduce the deformation of the pier column and improve the bearing capacity of the pier column.

Keywords: retractable piers column; numerical simulation; stiffening rib; bearing capacity

基金项目:国家重大专项资助项目(2017YFC0804204-04)。
通信作者:郭东明(1974—),男,江西新余人,教授,博士生导师。Tel:010-62339225,E-mail:dmguocumtb@126.com。

我国的煤炭含量很丰富，每年生产量位居世界之首，大约占世界产量的三分之一。绝大多数采用地下井工作业进行煤炭开采[1]。为了避免水，地压等因素破坏回采巷道，基于开采安全的前提下一般会留煤柱。煤柱的留设降低了煤炭的开采率，这部分损失量一般占40%左右，另外宽煤柱的设立既增加煤炭的损失量，也加大维护煤柱的难度。同时宽煤柱内部空间会产生集中应力，使得底板维护困难也影响下一煤层开采时的维护，严重时会导致岩爆和瓦斯突出等灾害，同时宽煤柱内部空间会产生集中应力使得底板维护困难也影响下一煤层开采时的维护，严重时会导致岩爆和瓦斯突出等灾害，影响正常开采相对于传统煤矿开采[2]。无煤柱开采可以有效地减少采煤时所留设的煤柱损失，近些年来逐步得到了运用推广。在无煤柱开采技术中，沿空留巷技术占有重要地位，一条巷道得到两次利用，无须留设煤柱，减小了煤炭的损失量，提高煤炭采出率、缓解采掘接替紧张、降低瓦斯危害等[3-5]。墩柱式沿空留巷技术在巷道支护中的广泛应用，大大减少煤柱的留设，提高了煤炭采出率，缓解了采掘接替紧张，降低了巷道的维护难度以及瓦斯危害[10]。目前，针对墩柱研究主要集中在理论计算、现场试验。有关墩柱承载特性的数值分析研究较少，本文利用ANSYS有限元软件对墩柱的承载特性在不同因素的影响下的变化进行了数值分析[13]。考虑的影响因素有：直径、高度、搭接长度。找出了墩柱承载特性的一般影响规律。

1 墩柱数值计算分析

1.1 墩柱的结构

可缩式墩柱的整体结构如图1所示，为了适应不同巷道高度，墩柱外部为两节嵌套的无缝钢管。上下两节高度为2 000 mm，2节钢管之间的搭接长度不得低于500 mm。上管的内径稍大于下管的外径。墩柱内部充满砂石灰，填沙子：石子：生石灰比例为1.0：1.6：0.5，且上出料口主要用于充填，下出料口用于充填材料的释放。便于操作，分别在墩柱不同的地方焊接了8个吊环，为了保护墩柱在养护期间能够正常直立，在距离上部的钢管顶部40 cm处焊接了2个单体支撑座（8号槽钢）。

1.2 模型建立及网格划分

本工程选用的钢管为上下两节高度为2 000 mm，2节钢管之间的搭接长度500 mm。故墩柱体的总高度为3 500 mm。选用ANSYS中的SOLID45实体单元来模拟钢管。由于钢管材质本身具有弹塑性，可以应用多线随动强化法则。采用vo.n Mises屈服准则和相关流动法则。屈服强度为325 MPa，抗拉强度为520 MPa，弹性模量为$2.06×10^5$ MPa，泊松比取0.3。对核心的充填砂石材料我们采用ANSYS中的SOLID65三维实体结构单元来进行模拟。对于模型钢管与充填材料—砂石桩的界面进行解决，需要采取ANSYS中的接触单元CONTAC173及目标单元TARGE170。经过多次计算对比，选取接触刚度的比例因子为0.1，充填的核心砂石充材料与钢材表面的摩擦系数取0.3，把材料的最大摩擦力取值为砂石充填材料的抗拉应力2.57 MPa。

模拟时用到的SOLID65单元是一种弥散裂缝模型，它在模拟过程中是以最大拉应力开裂作为破坏依据的。所以，在模拟时要对模型的单元网格进行合理划分，以防止材料因应力集中而发生破坏。以大量的计算为依据，得出的结果，在该模型的网格单元尺寸定位大于2 cm时，可以有效地避免出现应力集中的现象。在对模拟计算的可靠性和运算速度加以思

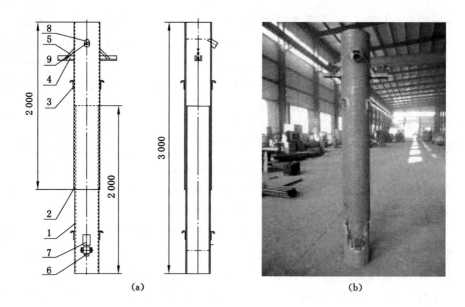

图 1 墩柱结构图

Fig. 1 Pillar structure

(a) 几何尺寸;(b) 实体

1——下节钢筒;2——上节钢筒;3——吊钩;4——单体支撑座;
5——加强筋;6——垫板;7——卸压口;8——上档口槽;9——上档口板

考之后,将本模型的单元尺寸确定为 2 cm。墩柱模型及网格划分如图 2 所示。

图 2 墩柱模型及网格划分图

Fig. 2 Pier model and meshing diagram

1.3 结果分析

图 3 是数值模拟计算结果,图 4 是分别在 25 MPa 的加载下墩柱试件变形情况,由数值和实验结果知,其模拟结果和墩柱室内实验结果基本类似。均是整体发生弯曲破坏,搭接处弯曲严重,在墩柱最下端发生鼓曲破坏。

从竖向变形的角度进行分析,由于墩柱自身的可缩性,模型整体变形量较小,墩柱上部区域的每一阶段变形的范围较大,下部区域变形范围较小且密集,且在墩柱底部达到变形最大值。结合墩柱横向变形数据分析,墩柱底部一定范围内的纵向变形达到 5 cm,横向变形达到 5 cm,对钢管的破坏较大。由表 1 和图 5 得,模拟值和实验值在极限荷载和极限位移

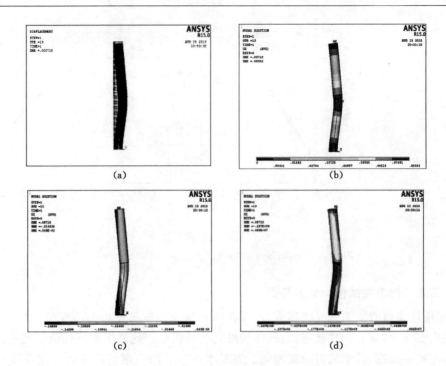

图 3 墩柱应力应变云图

Fig. 3 Pier stress and strain contours

(a) 变形图;(b) X 向位移云图;(c) Z 向位移云图;(d) Z 向应力云图

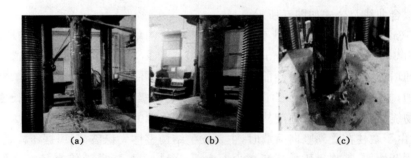

图 4 墩柱静载实验

Fig. 4 Static load test on pier column

(a) 整体破坏现象;(b) 搭接处压缩弯曲;(c) 底部鼓曲

上误差均较小,这就说明墩柱类钢混的力学性能问题用此计算模型模拟是可行的。

表 1 试验结果与有限元结果比较

Table 1 Comparison of test results with finite element results

试件编号	极限荷载/kN			极限位移/mm		
	试验值	模拟值	相对误差	试验值	模拟值	相对误差
C1-3	2 700	2 500	7%	19	21	10%
C2-3	2 300	2 200	4%	33	36	9%

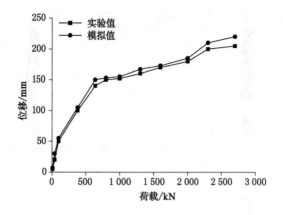

图 5 试件荷载-滑移曲线比较

Fig. 5 Comparison of load-slip curves of specimens

1.4 同因素下的墩柱试件承载力公式

由于墩柱是由两节钢管嵌套而成,所以在中间搭接处容易发生偏心受压引起的弯曲破坏,墩柱高度越大,偏心荷载就越大,破坏程度就会越明显。此外根据压杆稳定理论,墩柱高度越大,越容易失稳,其承载力也就越低。如果增加上下两节钢管的搭接长度及墩柱直径,可以有效提高中间的抗弯能力,提高其承载力。为了探究墩柱承载力与各影响因素的关系,用数值模拟的方法进行多种的模拟。搭接长度梯度为 5 cm,由于墩柱的搭接长度不得低于 50 cm,所以就从 50 cm 到 75 cm。单管高度梯度为 0.3 m。钢管直径梯度为 10 mm。根据模拟结构拟合曲线及公式如图 6 所示。

由图 6(a)可得,墩柱的极限荷载随着墩柱搭接量的增加呈现二次抛物线增加,当墩柱搭接长度由 500 mm 增加到 750 mm 时,荷载由 3 580 kN 增大到 4 000 kN,这是由于随着搭接长度的增加,墩柱中部抗弯性能提高,承载力提高;并随着搭接长度的增加增大,提高系数也在增加,幅度在 7%～17%;由图 6(b)可得,墩柱的极限承载力随着单管长度的增加呈现二次抛物线关系减少,单管高度从 500 mm 增加到 3 400 mm,承载力由 3 500 kN 减少到 1 500 kN;由于墩柱在受到竖向荷载时,主要发生结构整体失稳破坏,根据压杆理论,随着墩柱高度的逐渐增加,其临界压力逐渐减少。由图 6(c)可得,随着钢管直径的增加呈现线性增加,管径从 320 mm 增加到 400 mm,其极限承载力由 2 500 kN 增加到 4 000 kN,这是由于随着钢管直径的增大,抗弯刚度逐渐增大,抗弯性能提高,承载力提高。

2 墩柱加固分析

由室内试验及模拟结果可得,墩柱在竖向荷载作用下,当达到极限荷载时,墩柱中部搭接处发生弯曲破坏,弯曲变形量最大点位于下节钢管最上端紧靠上节钢管处,这是由于墩柱本身由两节钢管嵌套而成,上节钢管在向下移动时,容易出现偏向移动,以及内部材料充填不均匀,使得墩柱受到偏心压力,弯曲量增大;鼓曲变形最大位置位于最下端,钢管最下端收到的上部压力最大,故最容易发生鼓曲破坏。

薄壁圆筒应力计算:

$$\sigma_r = \frac{a^2}{b^2-a^2}\left(1-\frac{b^2}{r^2}\right)p_i - \frac{b^2}{b^2-a^2}\left(1-\frac{a^2}{r^2}\right)p_0 \tag{1}$$

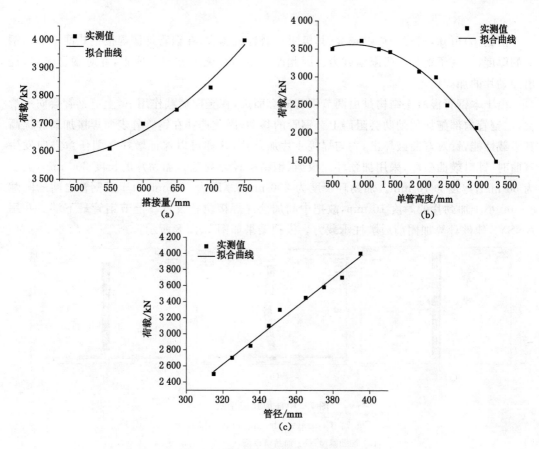

图 6 不同因素下的试件承载力曲线

Fig. 6 The bearing capacity curve of specimens under different factors

(a) 搭接长度-承载力曲线;(b) 单管高度-承载力曲线;(c) 直径-承载力曲线

$$\sigma_\theta = \frac{a^2}{b^2-a^2}(1+\frac{b^2}{r^2})p_i - \frac{b^2}{b^2-a^2}(1+\frac{a^2}{r^2})p_0 \tag{2}$$

$$\sigma_{\theta\max} = \sigma_{\theta r=a} = \frac{b^2+a^2}{b^2-a^2}p_i \tag{3}$$

其中,$p_i = k \cdot q = (1-\sin\theta) \cdot q$;$k$ 为侧压系数;$k=1-\sin\theta$。

$$\sigma_{\theta\max} = \frac{b^2+a^2}{b^2-a^2}(1-\sin\theta)q = f_y \tag{4}$$

$$q = \frac{f_y(b^2-a^2)}{b^2+a^2} \cdot \frac{1}{1-\sin\theta} = \frac{f_y(b-a)(b+a)}{b^2+a^2} \cdot \frac{1}{1-\sin\theta}$$

$$= \frac{f_y(b+a)t}{b^2+a^2} \cdot \frac{1}{1-\sin\theta} \tag{5}$$

其中 a——钢管内径,m;

b——钢管外径,m;

r——任意直径,m,$a<r<b$;

p_i——钢管内压力;

p_0——钢管外压力,$p_0=0$;

t——钢管厚度,m。

由式(5)可知,钢管的壁厚越大,其极限承载也就越大,在钢管外侧焊接钢片即增加了钢管的厚度,提高了钢管的极限承载力。因此课题组结合施工方便及成本,针对钢管破坏,提出加劲片的加固措施[14-16]。

由于本钢管混凝土结构是由两节钢管嵌套而成,在竖向荷载作用下,上节钢管会向下滑移,在钢管内部施加加劲肋会阻碍上节钢管的移动,因此选择在钢管外表面焊接加劲片提高其承载性能,针对弯曲破坏出,均匀焊接4片加劲片,这样可以保证墩柱在朝任意方向发生弯曲时,针对鼓曲破坏,采用加劲环。根据《钢结构设计规范》,加劲片的长度200 mm、宽度为60 mm,厚度为10 mm。加劲环长度为200 mm、厚度为10 mm,由于两钢管之间的间隙8 mm,小于加劲片的厚度10 mm,故把中间加劲片焊接位置设置在上节钢管最下端。并用ANSYS软件计算加固前后墩柱承载力。模拟结果如图7、图8所示。

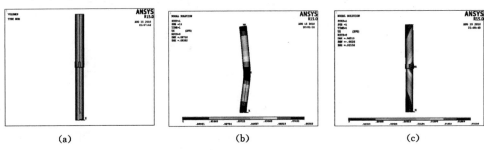

图 7 中部加劲模型图

Fig. 7 Central stiffening model diagram

(a) 加劲模型;(b) 加劲前模型;(c) 加劲后模型

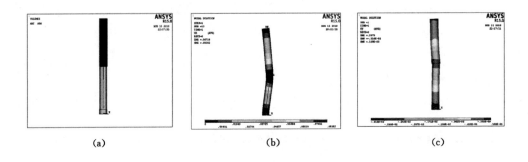

图 8 下端加劲模型图

Fig. 8 Bottom end stiffening model diagram

图7、图8均是在上部荷载为25 MPa,约束为在墩柱下端面为x,y,z约束,上端面在x,y方向约束,由图7、图8可知,在加劲前,墩柱最大的弯曲挠度为8 cm,且最大挠度发生在下节钢管距底端1.9 m处。加筋后墩柱的最大挠度为2 cm,通过加筋使横向变形减小75%。所以在墩柱中间加劲能有效调高墩柱的抗弯性能。由图7、图8可知,在墩柱的下端焊接加劲环后,使钢管壁厚由原来的8 mm增加到18 mm,使其径向位移由原来的2 cm减少到0.031 6 cm,基本可以使其保持不发生变形。加劲肋有效地减少了钢管的变形量,提高墩柱承载力。

3 结 论

(1) 通过和实验对比,确定了数值模拟的可行性,并进一步进行多种搭接长度、单管高度、钢管直径的模拟,得出搭接长度在一定程度上能提到墩柱承载力,另外墩柱的极限承载力随着搭接长度的增加呈现二次抛物线关系增加,随着单管长度的增加呈现二次抛物线关系减少,随着钢管直径的增加呈现线性增加。

(2) 通过模拟,验证了在墩柱破坏处进行加劲肋的加固措施,使弯曲变形减小75%,下端几乎不发生鼓曲变形,加劲肋能有效控制墩柱的变形,提高墩柱承载力。

参 考 文 献

[1] 王宏岩,王猛.深部矿井开采问题与发展前景研究[J].煤炭技术,2008(01):3-5.
[2] 朱红青,张民波,王宁,等.Y型通风高位钻孔抽采被保护层卸压瓦斯研究[J].煤炭科学技术,2013,41(2):56-59.
[3] 陈勇,柏建彪,王襄禹,等.沿空留巷巷内支护技术研究与应用[J].煤炭学报,2012,37(6):903-910.
[4] 费旭敏.我国沿空留巷支护技术现状及存在的问题探讨[J].中国科技信息,2008(7):48-49.
[5] 张农,韩昌良,阚甲广,等.沿空留巷围岩控制理论与实践[J].煤炭学报,2014,39(8):1635-1641.
[6] 何满潮,陈上元,郭志飚,等.切顶卸压沿空留巷围岩结构控制及其工程应用[J].中国矿业大学学报,2017(5):959-969.
[7] 贾民,柏建彪,田涛,等.墩柱式沿空留巷技术研究[J].煤炭科学技术,2014,42(1):18-22.
[8] 陈勇.沿空留巷围岩结构运动稳定机理与控制研究[D].徐州:中国矿业大学,2012.
[9] 李胜,李军文,韩永亮,等.综放工作面沿空留巷支护阻力研究[J].安全与环境学报,2015,15(1):133-136.
[10] 王化亮,贾德福.沿空留巷巷旁充填支护技术研究[J].黑龙江科技信息,2015(5):133-136.
[11] 王振,张宁.钢管混凝土的界面黏结-滑移性能数值分析研究[J].中外公路,2018(1):192-199.
[12] 郑亮,张大鹏,郭宏,等.高应力下圆钢管混凝土柱界面黏结性能研究[J].郑州大学学报(工学版),2018(1):18-23.
[13] 徐顾.加劲肋-钢管混凝土短柱力学性能及参数分析[D].哈尔滨:哈尔滨工程大学,2014.
[14] 木标.加劲钢管混凝土组合柱动力力学性能分析[D].哈尔滨:哈尔滨工程大学,2016.
[15] 张俊光,陈建华.设加劲肋钢管混凝土轴压短柱试验研究及有限元分析[J].公路交通科技(应用技术版),2015(12):199-203.
[16] 郭兰慧,张素梅,徐政,等.带有加劲肋的大长宽比薄壁矩形钢管混凝土试验研究与理论分析[J].土木工程学报,2011(1):42-49.

优化露天矿山爆堆单元划分方法与出矿品位确定

王雪松,徐振洋,李小帅,张金铭

(辽宁科技大学 矿业工程学院,辽宁 鞍山,114051)

摘 要:为解决露天矿山配矿出矿点位置爆堆品位分布不均匀的问题,按照出矿位置对应生产能力与铲装设备,确定最小配矿单元大小,将爆堆离散化形成独立的配矿单元,使离散化的配矿单元作用于铲运与配矿工作,通过距离幂反比法估值对离散单元中矿石各组分的品位进行估值,采用PSO算法确定距离反比法中的最优幂次,将不同计算方法下的品位值与实测结果进行对比,优化距离幂反比法估值爆堆品位方法的准确性,减少配矿单元中品位估算的误差对配矿结果的影响。

Optimization of open pit mine blasting reactor unit division method and determination of ore output grade

WANG Xuesong, XU zhenyang, LI Xiaoshuai, ZHANG Jingming

(School of Mining Engineering, University of Science and Technology Liaoning, Anshan 114051, China)

Abstract: In order to solve the problem of uneven distribution of blasting heap grade at the ore distribution point of open pit mine, the minimum size of ore distribution unit is determined according to the corresponding production capacity and shovel loading equipment at the ore distribution location, and the blasting heap is discretized to form an independent ore distribution unit, so that the discrete ore distribution unit acts on the shovel and ore distribution work and is inversely proportional to the power of distance. The PSO algorithm is used to determine the optimal power of the inverse distance ratio method. The grade values under different calculation methods are compared with the measured results. The accuracy of the inverse distance ratio method is optimized to reduce the error pair of grade estimation in the ore blending unit. Influence of ore matching results.

1 引 言

目前矿业形势下,粗放式管理显然无法满足矿山工作要求,由粗放式向精细化管理的转变成为矿业发展的新趋势[1],在露天铁矿山的日常生产工作中,采场矿石质量控制作为一项重要工作深受矿山企业的重视,而实施配矿工作离不开爆堆的矿石质量情况等基础信息[2]。

作者简介:王雪松(1995—),男,硕士研究生,从事露天开采与爆破技术的研究。E-mail:974360483@qq.com。
通信作者:徐振洋(1982—),男,博士,副教授,从事爆破工程与岩石破碎的研究。E-mail:xuzhenyang10@foxmail.com。

露天采场因出矿点数量众多、品位分布不均匀,这就为日常配矿工作造成了很大影响,从采场出矿点源头产生的问题直接导致了配矿效率低下、配矿方案可行性差等问题。

因此,从采场出矿点的源头处,研究优化露天矿山爆堆配矿单元的划分与其出矿品位的确定方法应当趋于精细化,配矿单元的划分工作主要分为确定配矿单元大小、推进方式及出矿品位确定几个方面。其中距离幂次反比法广泛应用于出矿品位的确定中,但此方法中幂次的确定受到了广泛的关注,基于此,通过PSO算法对距离幂次反比法中的幂次进行计算,PSO算法的简单易行、收敛速度快、优化效率高等优势,处理非线性问题中得到很好的效果[3],在确定幂次的过程中,设定目标函数,给出算法中所需的各项参数,使结果收敛于最优解,将此方法所得的幂次作用于距离幂次反比法之中,将得到的品位值与实测值及平均系数法计算值进行比较。

2 配矿单元的划分

采用平均品位表示整个爆区的品位情况,显然无法真实地反映供矿品位,因此,需要将整个爆区划分为若干个配矿单元,配矿单元的划分中涉及配矿单元矿量、相邻单元步距、爆堆推进方式几个方面,露天铁矿山采区具有多个出矿位置,每个出矿位置含有供电铲工作至少2~3 d的矿量[4]。

划分配矿单元的时,国内的很多研究按照炮孔数量划分,以每个单元内存在3~4个或7~8个炮孔为规则,区域内含有多个品位控制点[5]。区域划分时对铲装设备、推进方式、生产能力等方面考虑不足。因此,确定配矿单元大小、计算出矿点品位、划分爆堆的推进方式需要按照一定的规则进行。

2.1 确定配矿单元大小

露天矿山日开采计划各出矿点需要满足品位稳定、矿石质量均衡等条件,因此,在编制日开采计划时需要将各出矿点爆堆划分为合适大小的配矿单元,避免矿石暴露时间过长导致的矿石质量降低,增加铲装效率,减少不必要的损失。

配矿步距确定以铲装位置爆堆为起始点,采用固定大小的矩形划分配矿单元,相邻配矿单元之间的步距为电铲工作半径(图1)。

$$L \leqslant r_g$$

式中 L——配矿单元步距;

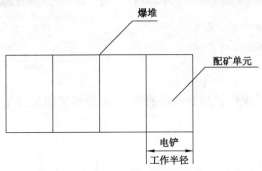

图1 配矿单元示意图

Fig.1 Schematic diagram of ore blending unit

r_g——电铲工作半径。

随着矿山机械化的发展,矿山设备趋于大型化,电铲在工作过程中一般移动较少,以此减少不必要的花费,每个出矿点只有一个电铲进行工作,为生产效率,电铲日工作量一般为其生产能力的80%以上,考虑到出矿点各处地形等原因,以电铲电铲生产能力确定配矿单元大小的范围,出矿位置的划分方式根据出矿台阶所谓对应电铲生产能力的60%,且不大于电铲的最大日生产能力,即:

$$0.6A_r < W_{min} < A_r$$

式中　A_r——对应电铲的最大生产能力;

W_{min}——最小配矿单元所含矿量。

配矿单元划分大小确定后,将可出矿的爆堆按照大小规则划分,这样的划分方式便于统计出矿点矿量,此处电铲生产能力所编制计划时间的长短有所不同,这样的方法有利于生产计划的编制,便于矿山企业在生产中进行管理。如图2所示。

2.2 配矿单元开采顺序

露天矿推进方式指根据台阶主推进方向及设备能作业方向可能的采掘方式,采场提供的台阶主推进方向以及自由面,根据适用电铲效率的采掘宽度和电铲日生产能力,产生出多种可能的推进方式[6]。按照采场推进方式将划分后的出矿点排序,从台阶自由面的最右侧起,按照次序排列,优先开采起始的下一排配矿矿单元,即采用电铲移动距离最小的顺序方式排序配矿单元,如图3所示。

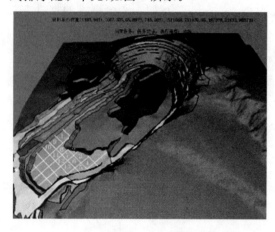
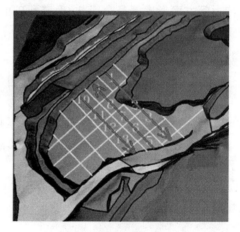

图2　划分配矿单元示意图　　　　图3　开采顺序示意图

Fig.2　Schematic diagram of dividing ore distribution unit　　　　Fig.3　Schematic diagram of mining sequence

此种开采方式可减少电铲频繁移动的次数,也使得矿山进行生产计划编制时进行配矿单元的选择。

2.3 出矿品位的确定

矿石出矿品位是配矿工作的基础参数,国内大部分矿山采用平均系数法完成出矿点品位的预测,此方法便于操作计算量小,但具有局限性,有时无法准确地估算出矿石的品位。在地质学领域中,许多具有空间相关性的变量一般采用距离幂次反比估值法进行[7]。相比平均系数法,距离幂次反比法引入预估点品位的影响与距离相关,距离越近影响越大,反之影响较小[8]。针对距离幂次反比法,从不同的层次和角度提出相应的解决途径,主要在于确定最优参估样品数据是此方法在实际应用中必须要考虑的问题[9]。

两种品位估值方法广泛应用,具有较好的品位预测能力,相比于露天矿山出矿点的实际测量值,两种方法均存在误差,对比不同方法下计算结果误差,介绍不同计算方法的区别,对比不同方法下计算结果的误差。

距离幂次反比方法的基本流程如下:

以观测点的中心为原点,最小配矿单元的范围大小相同的区域作为估值区域,连接每个探测孔与观测点中心确定距离,采用距离幂反比公式计算估值区域品位。

$$Z = (\sum_{i=1}^{n} \frac{1}{d_i^m} g_i) / (\sum_{i=1}^{n} \frac{1}{d_i^m})$$

其中,Z 为配矿单元中心点的品位;d_i 为配矿单元中第 i 个炮孔距中心点的距离;g_i 为第 i 个炮孔的矿石品位;m 为幂指数。

图 4 为距离幂次反比法示意图。

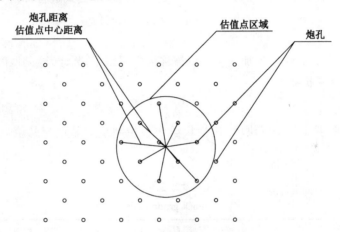

图 4　距离幂次反比法示意图

Fig. 4　Schematic diagram of distance power inverse ratio method

3　基于 PSO 算法确定最优幂次

粒子群优化算法的基本思想是通过群体中个体之间的协作和信息共享来寻找最优解,PSO 的优势在于简单容易实现并且没有很多的参数调节,在距离幂次反比法中的幂次确定中,各控制点距离中心的位置对中心点的品位值影响并不相同,随着距离的增大呈现近似正相关,在幂次计算时制约计算结果的除各控制点的品位值以外,还有所处位置的距离,故利用 PSO 算法进行求解非常适合最优幂次的求解中。

3.1　PSO 算法

粒子群算法[10],也称作粒子群优化算法或鸟群觅食算法,是基于鸟群觅食行为而提出的,开发的一种新的进化算法,其特点是从随机解出发,通过迭代寻找最优解,达到运算目的。PSO 算法具有参数少、实现容易、收敛速度快、便于理解等优点,算法高效适合运用于寻找品位计算中最优幂次的计算中[11]。

PSO 的运算原理是初始化一群随机粒子,从初始解开始寻找最优解,运算中,通过两种方法更新粒子,第一种方法是由粒子本身所找到的最优解,此解成为个体极值,另一个方法是整个种群目前找到的最优解,也就是全局极值 9 999。

假设存在一个 D 维的目标搜索空间,与一个由 N 个粒子组成的群落,其中第 i 个粒子表示为一个 D 维的向量,即:

$$X_i = (x_{i1}, x_{i2}, \cdots, x_{iD}) \quad i = 1, 2, \cdots, N$$

粒子的速度也为一个 D 维的向量,表示位置的改变,即:

$$V_i = (v_{i1}, v_{i2}, \cdots, v_{iD}) \quad i = 1, 2, \cdots, N$$

粒子个体的最优极值记为 p:

$$p_{\text{beat}} = (p_{i1}, p_{i2}, \cdots, p_{iD}) \quad i = 1, 2, \cdots, N$$

整个粒子群搜索到最优位置称为全局极值记为 g,即:

$$g_{\text{best}} = (p_{g1}, p_{g2}, \cdots, p_{gD})$$

两个最优位置找到后,粒子的速度和位置更新通过以下公式进行,即:

$$v_{id} = wv_{id} + c_1 r_1 (p_{id} - x_{id}) + c_2 r_2 (p_{gd} - x_{id})$$

$$x_{id} = x_{id} + v_{id}$$

式中,w 为惯性系数;c_1 与 c_2 为加速系数也被称为学习因子;r_1 与 r_2 是区间为[0,1]服从均匀分布的两个独立随机数。

3.2 应用实例

本试验共选取了 8 个取值块段,包括块段中各控制点的品位与其距离中心点的距离。如表 1 所示。

表 1　　1 号块段品位表
Table 1　　1 block grade table

序号	矿石品位	距离中心点距离
1	24.18	10.20
	25.45	12.40
	22.80	5.60
	22.23	7.50

针对距离幂次反比中的幂次确定问题,采用 PSO 算法,并进行对比,实验共进行 1 000 次迭代计算。其参数设置如表 2 所示。

表 2　　算法参数设置表
Table 2　　Algorithm parameter setting table

c_1	c_2	W	种群数	迭代次数
2	2	0.6	100	1 000

在本问题中,粒子的适应值函数可定义为预估点与实测点的差值平方,满足该函数值最

小时,即为函数的最优解,进而得到最优幂次。即:

$$f(x) = \sum_{i=1}^{n}(g_r - g_i)^2$$

其中,g_r 为点位实测品位;g_i 为控制点 i 的品位估计值。

3.3 结果分析

根据 3 种结果的对比结果如表 3 所示,对比表中的数据可发现由 PSO 方法计算幂次反比误差方法明显优于平均系数法,并且结果趋向于控制点的取样品位。

表 3 计算结果表
Table 3 Calculation results table

序号	控制点取样品位	平均系数法估计品位值	PSO法距离幂次反比法估计品位值	平均系数法估计误差	PSO法距离幂次反比法估计误差
1	21.80	23.67	22.32	1.87	0.52
2	21.96	22.47	21.36	0.51	−0.60
3	24.44	24.15	24.22	−0.29	−0.22
4	22.56	24.99	23.04	2.43	0.48
5	23.74	26.32	25.60	2.58	1.86
6	22.35	23.31	22.25	0.96	−0.10
7	19.24	21.85	21.03	2.61	1.79
8	20.34	22.48	21.86	2.14	1.52

通过误差图像进行分析发现,相比平均系数法估计品位值,8 组数据中 PSO 法所计算的距离幂次法均更接近控制点取样的品位值实际数据,但个别点位的误差仍然较大。总的来说,PSO 算法计算得出的幂次具有良好的预测性,此方法适用于该数值的确认,提升出矿品位值的计算精度。如图 5 所示。

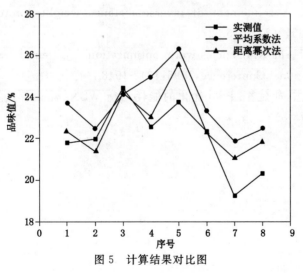

图 5 计算结果对比图
Fig. 5 Comparison of calculated results

4 结 论

(1) 进行配矿单元划分,以电铲工作半径确定配矿单元步距,确定配矿单元的最小矿量为电铲生产能力的60%,得出以"蛇形"方式推进的开采顺序。

(2) 根据出矿点品位计算的特征,使用距离幂次反比法确定品位值,使用PSO算法设置各项参数,得到了以偏差最小的目标函数。

(3) 以PSO算法得出的品位值相较于平均系数法得出的品位值更接近块段实测品位值。

参 考 文 献

[1] 王李管,宋华强,毕林等.基于目标规划的露天矿多元素配矿优化[J].东北大学学报(自然科学版),2017,38(7):1031-1036.

[2] 井石滚,卢才武,顾清华.基于GIS的露天矿生产配矿数字化管理系统[C].国际遥感大会.杭州,2010:426-429.

[3] 陈前宇,陈维荣,戴朝华,等.基于改进PSO算法的电力系统无功优化[J].电力系统及其自动化学报,2014,26(2):8-13.

[4] 李章林,王平,张夏林.距离幂次反比法的改进与应用[J].金属矿山,2008,38(4):88-92.

[5] 陈彦亭,巩瑞杰,南世卿,等.露天矿智能配矿系统研发与应用[J].现代矿业,2016(4):206-210.

[6] 崔方宁,邵明国,赵然磊,等.基于三维可视化建模技术的矿量计算方法[J].现代矿业,2014(8):61-63.

[7] 王仁铎,胡光道.线性地质统计学[M].北京:地质出版社,1988.

[8] Shepard D. A two dimensional interpolation function for irregularly spaced data[C]. Proceedings of ACM 23rd National Conference,1968:517.

[9] Watson D F. A refinement of inverse distance weighted interpolation[J]. Geo-Processing,1985,2(2):315-327.

[10] Kennedy J,Eberha R. Particles warm optimization[C]. Neural Networks,Proceedings IEEE International Conference,1995:1942-1948.

[11] 朱永红,丁恩杰,胡延军.PSO优化的能耗均衡WSNs路由算法[J].仪器仪表学报,2015,36(1):78-86.

动载作用下岩石破坏模式与强度特性分析

王 军

（山东理工大学 建筑工程学院，山东 淄博，255049）

摘 要：岩石在高应变率下的破坏类型及动态强度理论是工程爆破中的一个重要基本问题。以往研究岩石的破坏类型只停留在静载作用下，而强度理论也多采用静态强度理论。文章通过分析SHPB装置对常见的四种岩石进行的冲击试验，发现岩石在冲击荷载作用下破坏分为四种类型：压剪破坏、拉应力破坏、拉应变破坏和卸载破坏，而且破坏强度随冲击速度的增加而提高。

关键词：岩石；高应变率；破坏模式；强度理论

Research on the failure pattern and strength properties of rock under dynamic loading

WANG Jun

(*School of Architecture Engineering, Shandong University of Technology, Zibo* 255049, *China*)

Abstract：The failure patterns and dynamic strength theory of rock is an important problem in engineering blasting. The failure of rock was formerly only studied on the bases of static loading and the static strength theory was adopted. A large amount of impact test was carried out on four kinds of common rock by using SHPB. Results shown that the rock failure under impact loading are classified into four patterns, i. e. compress-shearing damage, tensile stress damage, extensible strain damage and unloading damage, and the rock damage strength increases with increasing impact velocity. At last, the strength theory of rock are discussed.

Keywords：rock；high strain rate；failure pattern；strength theory

1 引 言

高应变率下岩石的本构特性和强度特性是研究控制爆破机理、岩石破碎机理以及爆炸应力波传播的重要问题，而强度及破坏理论一直是岩土力学与工程界的重要课题。自Attewell用冲击试验研究岩石的破裂以来的几十年中，国内外很多学者对高应变率下的岩石动态特性进行了大量研究。对岩石的破坏模式，以往都停留于静载作用的情况，对在动载作用下特别是爆炸冲击载荷作用下岩石破坏模式研究甚少；这主要是由于试验设备和试验技术的限制。对于岩石动态强度特性的研究，W. A. Osson[1]进行的实验结果表明，应变率

作者简介：王军（1971—），男，博士，副教授。Tel：13589557069，E-mail：lgdwangjun@126.com。

小于某一临界值时,强度随应变率的增加而缓慢增加;当应变率大于此值时,强度迅速增加,用公式表示为:

$$\sigma_f \propto \begin{cases} \varepsilon^{0.007} & \varepsilon < \varepsilon* \\ \varepsilon^{0.35} & \varepsilon > \varepsilon* \end{cases} \quad (1)$$

式中,ε*为动态与静态发生变化的临界应变率。

R. D. Perkin、S. T. Green、Lankford 等学者也都得出过类似的结果[2]。在国内,陆岳屏、寇绍全等在20世纪80年代初用SHPB装置对砂岩、石灰岩等进行了动态破碎应力和杨氏模量的测试[3];于亚伦、李夕兵等也先后利用SPHB装置对不同类型岩石进行了动态强度试验。在其试验范围内,得到了岩石强度近似与ε*呈正比结论[4]。

现有的岩石强度理论,主要还是采用古典静态强度理论,这显然与岩石在爆炸与冲击作用时的情形有很大差异,考虑到岩石在爆炸与冲击作用下其强度理论的复杂性[5],以及实验设备与实验技术的限制,本文提出了一种简单易行的处理方法,就是仍然采用静态或准静态的强度理论但对有关常数考虑应变率应做些综合修正,计算结果表明,这种简易处理方法是可行的。

2 岩石在动载作用下的破裂类型

研究岩石在动载作用下的破裂类型目前比较困难,这是因为动荷载特别是冲击荷载一般比较高,试件在冲击荷载作用下裂纹的扩展速度远远低于载荷的传递速度,使得岩石试件在完全破坏之前大量的裂纹得以产生并扩展,导致试件在其作用下的破裂程度远远高于静载作用下的破坏,因此难以收集并拼接破坏后的试件。

岩石在静载作用下的破裂类型有很多研究报道。J. C. Jaeger、N. G. W. Cook[6]、陈庆寿[7]、郭志[8]等在此方面做了很多工作。

文献[2]利用SHPB装置对花岗岩和大理岩在冲击荷载作用下的破坏模式进行了大量试验研究,发现以下四种破坏模式:压剪破坏,拉应力破坏,张应变破坏和卸载破坏。但是,在实验中经常发现上述四种破坏模式很难单独出现,往往都是两种或两种以上的形势同时出现。对于大理岩,压剪破坏和拉应力破坏常常同时发生,但是压剪破坏起主导作用;对花岗岩,同时发生的往往是拉应力破坏、拉应变破坏和卸载破坏三种模式。

岩石在动载作用下特别是冲击载荷作用下会产生如此复杂的破坏模式,原因有三个方面:

(1) 由于岩石本身的物理性质决定的,岩石试块内部不可避免地存在各种微结构,为结构发育的地方往往能够控制破裂的发展。

(2) 冲击载荷与静载不同,它能使岩块在极短的时间内得到很高的能量,而且加载速度高于破裂发展速度,这就使冲击载荷能促使岩石中各个方向各个层次的裂隙发展。

(3) 岩石材料的力学性质对岩石试件的冲击破坏模式起着重要作用。强度高、脆性大的岩石一般不会出现压剪破裂;强度低、柔性大的岩石很少出现张应变破裂和卸载破裂。

3 岩石动态强度试验结果和分析

利用SHPB装置对煤矿常见的四种岩石进行了大量的单轴冲击抗压强度试验。实验装置及由试验的得到的每种岩石的动态应力-应变曲线见文献[9]。本文主要研究岩石的力学特性。

由于每次试验中应变率变化很快,难以用确定的数据来衡量一次试验的应变率大小,因此,本文岩样的强度与应变率的关系用它与冲击速度的关系来代替。

由试验结果可以看出,不论何种岩石,随着冲击速度(或应变率)的提高,其抗压强度基本上相应增大。但不同的岩石,其增长速度不同。说明岩石的破碎程度主要取决于应变率(或者加载速率)和岩石类型(岩石的物理性质)。很多学者通过大量的实验结果均已证明了上述结果的正确性,并且有的学者还给出了破碎强度与应变率的简单关系式。

通过冲击试验可以看出,不同岩石类型对应变率的敏感率不同,文献[7]给出了岩石对冲击速度的敏感性系数 K_1:

$$K_1 = (\sigma_d - \sigma_{do})/(V - V_o) \tag{2}$$

式中 σ_{do}, σ_d——分别表示岩石产生可见裂隙和强烈破碎时的强度,MPa;

V_o, V——分别表示与 σ_{do} 和 σ_d 相对应的冲击速度,m/s。

K_1 值越大的岩石对冲击速度的变化越敏感,表示该类岩石采用提高冲击速度来提高其破碎程度是有效的;反之,对于软岩 K_1 值较小,表明提高冲击速度的办法来达到改善破碎效果是比较困难的,这一结论符合炸药与岩石阻抗匹配的原则。

由冲击试验还可以看出,即使静压强度相似的不同类岩石,其动态破碎程度亦存在较大差异。因此,在爆炸与冲击领域,应采用相应应变率下的岩石破碎程度等动力参数作为衡量岩石动力破碎过程难易程度的综合指标,而不宜采用岩石的静态参数。

4 岩石冲击破坏强度准则的讨论

地下结构工程的岩石,一般处于复杂应力状态下,研究其强度理论具有重要的意义。自1773年提出库仑理论以来,至今已提出了许多强度理论。对于岩石在动载作用下破坏的强度理论,目前广泛应用的仍然是古典静态强度准则。比如,人们在分析岩体中装药爆破的内部作用和外部作用,以及冲击凿岩中压头的破岩机理时总是运用压剪强度理论和最大拉应力理论。

近几年有少数人将在金属材料中流行的累积损伤准则引入岩石材料中来分析岩石的冲击破碎,其表达形式:

$$\int_0^t (\sigma - \sigma_o)^a dt = K \tag{3}$$

式中 a, σ_o, K——均为材料常数;

σ_o——材料破坏时所需的下限应力。

在上式中,当 $a=1$ 时相当于冲量准则,$a=2$ 时相当于 Stewarding-Lehnigk 推导的能量准则。

愈茂宏[10]在总结自己30多年研究的基础上,提出了岩土类材料的统一强度理论——双剪统一强度理论。

正如前言所述,许多学者给出了岩石动态强度与应变率的简单指数函数或幂函数表达式,这些关系式本身就是岩石动态强度准则的一种表达,包括作者所做的实验结果,都表明一定的应变率决定一定的强度,并随着应变率的增加而增加。但是由于在实际冲击破碎过程中往往同时包括几个不同数量级的应变率(加载率)的变化,使得强度与应变率的一一对应关系很难得到应用。

只有当一个强度理论被简化到如式(3)或库仑准则等那样简洁并具有明确物理意义时才会真正被多数人接受。笔者认为,不妨采用静态或准静态准则,对有关常数考虑应变率效应的综合修正,以此来研究岩石在动载作用下的强度理论。对于本文中的大理岩,其破坏形式以压剪方式为主,兼有拉应力破坏,因此库仑准则比较适用。通过计算知,大理岩在静载作用下的内摩擦角 φ_s、内摩擦系数 $\tan\varphi_s$ 和黏聚力 C_s 分别为 35°、0.7、10.7 MPa;而在冲击载荷作用下三者的数值为:50°、1.2、19.4 MPa。从上述计算结果可以看出,冲击载荷作用下大理岩的内摩擦角、内摩擦系数和黏聚力都比静载下有明显提高。正因为内摩擦系数和黏聚力的提高,才使岩石的单轴抗压强度、抗剪强度得到提高,这与所有学者的实验结果是一致的。

对于花岗岩,其冲击破坏模式一般以拉应力、张应变和卸载破坏为主,一般不会出现剪压破坏,因此不能应用库仑准则进行分析。从破坏模式来看用拉应力或拉应变准则比较合适,但要考虑卸载破坏,现有一切强度理论似乎都不适用。因此对于花岗岩的冲击破坏强度准则有待进一步研究。

5 结　论

(1) 利用 SHPB 装置对花岗岩等四种常见岩石的冲击破坏试验可以看出,岩石的破坏模式主要有以下四种:压剪破坏,拉应力破坏,张应力破坏和卸载破坏。而且在每次试验中上述四种破坏很难单独出现,往往是两种或两种以上同时出现。对于大理岩,压剪破坏和拉应力破坏常常同时发生,但剪压破坏起主导作用;对于花岗岩,往往后三种破坏模式同时发生。

(2) 解释了岩石在冲击载荷作用下破坏模式比较复杂的原因,主要有三点:岩石的物理性质,力学性质,以及在冲击载荷作用下能够使岩石中的各个方向各个层次的裂隙都得到发展。

(3) 由冲击试验可以看出,不论何种岩石,随着冲击速度(应变率)的提高,其抗压强度也相应增大,但不同岩石增长幅度不同,说明岩石的破碎强度主要取决于应变率和岩石类型。

(4) 静压强度相似的不同岩石,其动态破碎强度可能存在较大差异,因此,在爆炸与冲击领域内,应采用相应应变率下的岩石的静态参数,这样更符合实际。

参 考 文 献

[1] Olsson W. A The Compresssive Strength of Tuff as a Function of Strain Rate from 10^{-6} to 10 sec[J]. Int. J. Rock Merth. Min. Sci. & Geomech. Abstr., 2001,28(1):115-118.

[2] 李夕兵,古德生. 岩石冲击动力学[M]. 长沙:中南工业大学出版社,2002.

[3] 谢和平. 岩石分形学[M]. 北京:科学出版社,2001.

[4] 马晓青. 冲击动力学[M]. 北京:北京理工大学出版社,1988.

[5] 于亚伦. 岩石动力学[M]. 北京:冶金工业出版社,1990.

[6] 张奇. 岩石爆破的粉碎区及空腔膨胀[J]. 爆炸与冲击. 2007,17(1):68-72.

[7] Chen W, Lu F. A technique for dynamic proportional multiaxial compression on soft materials[J]. Experimental Mechanics, 2004, 40(2): 226-230.

[8] 席道英, 郑永来. 大理岩和砂岩动态本构关系的试验研究[J]. 爆炸与冲击, 2005, 15(3): 259-263.

[9] 宗奇. 岩石内爆炸应力波破裂区半径的计算[J]. 爆破, 1993(5): 15-17.